The Design of *Iowa*-Class Battleships Vol. 2

Armor of the *Iowa*-Class Battleships

John M. Miano

Table of Contents

Preface

This book examines the armor of the Iowa-class battleships. This is the second book in a series dedicated to the structural aspects of the *Iowa*-class battleships. This book assumes that the reader is familiar with the overall layout of the Iowa-class battleships as described in volume 1, *A Visual Tour of Battleship USS New Jersey*. The deck plans in volume 1 will be useful for understanding the locations specified here in volume 2 for armor components.

The scope of this book is limited to armor ordered from mills. The designers liberally incorporated additional armor into the structure of the ship, both during the original construction and during the 1980's reactivations. There is also a network of armored trunks feeding gun mounts from their magazines and trunks connecting directors to the fire control system. Due to space limitations, these additional features of the *Iowa*-class protection scheme will have to wait for a future volume for coverage.

While the history of the *Iowa*-class has been well documented in books, their construction details have largely been ignored. Published descriptions of the armor frequently contain errors or are misleading. Armor is usually addressed in a tabular form that does not convey how a ship is protected. Such tabular descriptions of armor are often inconsistent in what is actually being measured. In reality, the armor configuration of the *Iowa*-class (or any other class of battleships) is too complex to be represented in tables. A two-foot change in impact location could result in a projectile striking an entirely different armor configuration.

Sadly, it appears that the Navy destroyed all the original plans for the Iowa-class battleships. At the time I am writing this, the last remaining full set of copies, along with all the spare parts, is sitting on USS *Charleston* which is awaiting scrapping. Some of the plans have survived in blueprint copies and in a microfilm copy of plans for USS *New Jersey*. Little care was taken by those who created the microfilm. Most of the images are difficult to read and, in many cases, are totally unreadable. Some digits are difficult to tell apart, especially in the microfilm. Three often looks like five and seven often looks like nine.

There were several ways to determine the correct values. Sometimes the values could be resolved using multiple plans. The same feature was often replicated across several plans. Just one plan was needed to have a readable value. If an unreadable measurement was at the edge of a feature, the same measurement could often be found on the matching edge of the plan for an adjoining feature. The correct value could often be found using pixel measurements of digitized plans. Scanned printed plans almost always had enough resolution to determine digits this way. Microfilm measurements could usually be resolved in the same manner. In the worst case, three-dimensional computer models were used to resolve measurements. Obscured denominators in fractions were fairly easy to interpret because they are almost always powers of two.

The contents of this book are analogous to that of a dig report in archaeology. The goal here was to present enough detail that one could create computer models of the armor to analyze its behavior across a wide range of attacks or even build their own *Iowa*-class armor. My hope is that the effectiveness of the *Iowa*-class's protection eventually will be subjected to serious computer analysis.

Consequently, the level of detail here is much greater than what one would normally find in a book on the structural aspects of a ship. It is impracticable to include every detail shown in the original plans. However, I have tried to show all the important details and have endeavored to err on the side of including when in doubt. The major area of contraction here is that I have eliminated most of the repetition in the original plans. The Navy created a plan for each armor plate that completely describes the plate that would be used to order the plate from the mill. Common features among plates are reproduced on each plan where I have tried to show such features only once. In the many cases where plates are mirrors, I have only included plans for one side of the ship.

I have used reproductions from original plans as illustrations wherever possible. In nearly all cases this was not possible because the quality of the originals is so poor. In a handful of cases, the quality of the original plan was good but it was not possible to reproduce it within the format of the book.

The level of originality among the book's plans varies. Plans in this book can be:

- Reproductions of original plans
- Recreations of original plans to make them readable
- Mergers of features from multiple plans.

- Divisions of original plans into multiple plans for clarity.
- Analogs of original plans that have been enhanced by using three-dimensional models.
- Fresh drawings based on measurements in the original plans to make features clearer.
- Plans made by inference that fill in details not shown in available plans.

Regarding that latter category, four *Iowa*-class battleships were completed (USS *Iowa*, USS *New Jersey*, USS *Missouri*, and USS *Wisconsin*) and two more were left unfinished (USS *Illinois* and USS *Kentucky*). Most of the surviving plans are for USS *New Jersey*. The plans for USS *New Jersey* and USS *Iowa* were made at the same time and the differences for other ships were indicated. The USS *New Jersey* plans were updated with subsequent changes for the later ships. Plan sheets entirely devoted to the other ships have not appeared. In most cases involving armor, the differences for USS *Iowa*, USS *Missouri*, and USS *Wisconsin* can be inferred from changes indicated on the USS *New Jersey* plans. Drawings relying on inferences, rather than original plans, are indicated as such.

Let me illustrate how such inferences may occur in the drawings. There are complete plans for the conning tower tubes for USS *New Jersey*. Among the sources are plans showing how the ends of the tubes are different on USS *Missouri*. At the time of writing, no actual plans for the tubes for USS *Missouri* have turned up. Using the known deck heights and the USS *New Jersey* plans, one can calculate what the tube di-

mensions would need to be for USS *Missouri* for them to fit on the ship. However, if there were some allowances for shimming in the original plans, then these inferred measurements would be slightly off. Because the measurements do not come directly from plans, they are marked as being inferred. In any event, there are no places where I have guessed at a measurement.

The USS *New Jersey* plans also contain some indications of changes for USS *Illinois* and USS *Kentucky*. Due to the extremely limited amount of information available about the construction of these two ships, the reader should not assume the information given in this book applies to those ships unless specifically indicated.

The primary source for the book is the plans. The shipyards did not follow the plans exactly. Where inspection has turned up discrepancies between the plans, I have indicated those in the book. It is a certainty that I have not identified all the plans–construction discrepancies.

The figure captions include their scale in parentheses where applicable. Plate numbers are indicated by surrounding them with a box Original plans include the New York Navy Yard plan number in parentheses.

Readers will likely notice that, in many cases, this book contains data that contradicts that shown in other published sources. That is because the original sources frequently contradict what has been published in books. Once an error gets into print, that error gets copied repeatedly. Hopefully, this book will bring to an end some of those errors.

A project this size always requires the involvement of many people. I was fortunate to find that so many people were willing to help,

particularly from the staff of the *Iowa*-class battleships. On USS *New Jersey*, Ryan Szimanski, Libby, Jones, and Alaina Noland were always ready to help and explore the ship. On USS *Iowa*, David Way, Jim Kurrasch, and Mike Getscher provided photographs and technical information. Keith Nitka led the way for information on USS *Wisconsin*. Franklin Clay did the same for USS *Missouri*. The Henry Ford Library provided obscure information I was seeking. Crewmember Henry M. Strub III was a constant source of information. Drs. Ralph and Margaret Miano reviewed all aspects of the book.

Two more volumes on the Iowa-class battleships are underway in parallel and more are in the planning stage.

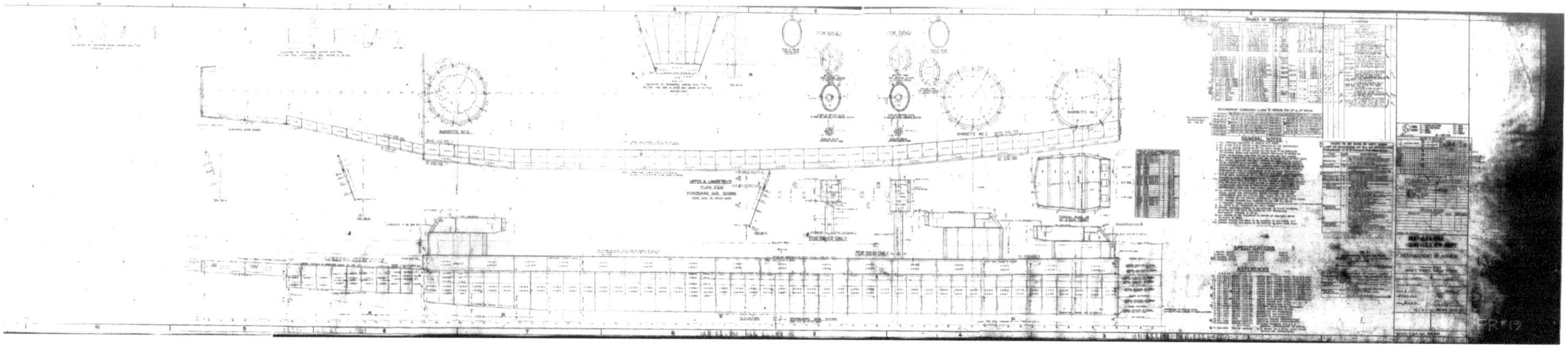
GENERAL NOTES
SPECIFICATIONS

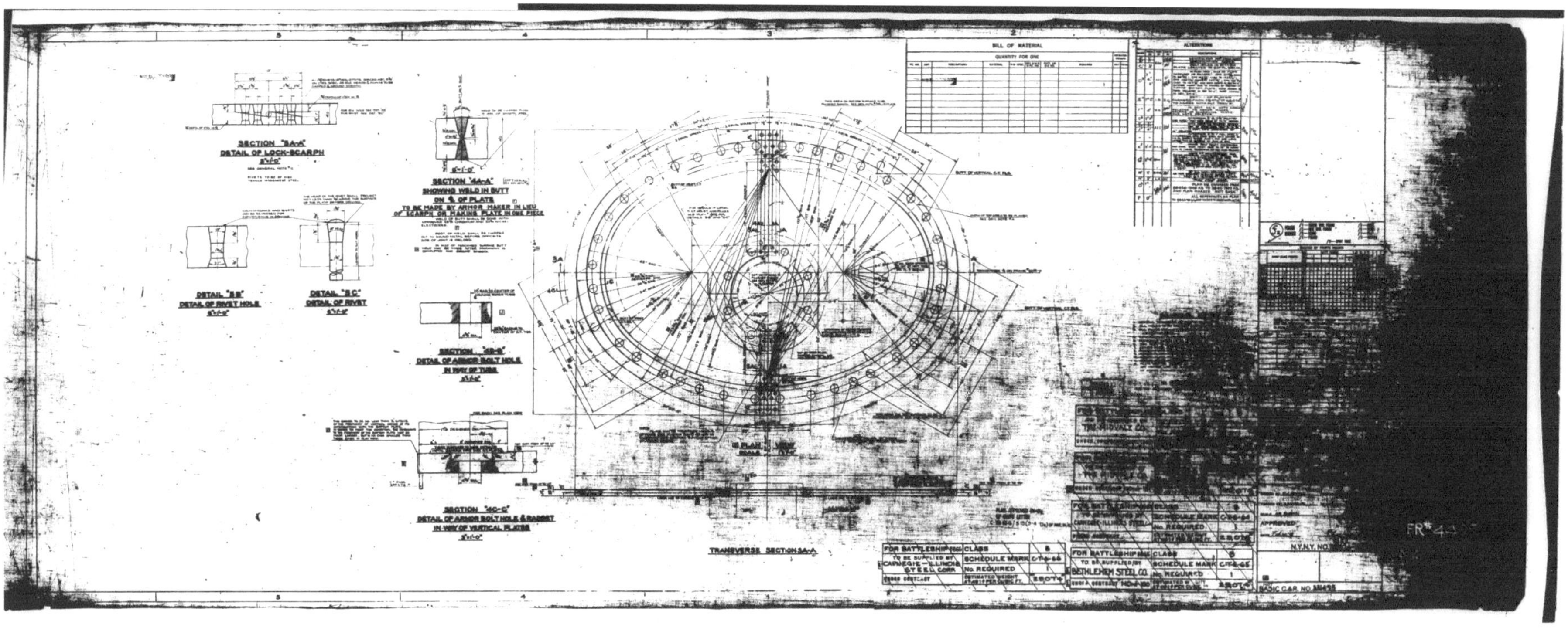
BILL OF MATERIAL
ALTERATIONS
TRANSVERSE SECTION 5A-A
FOR BATTLESHIP
BETHLEHEM STEEL CO.

Chapter 1: Armor Basics

Robust armor protection is the feature that distinguished battleships from all other types of ships. The design of a battleship's armor scheme always involves compromise. It would have been impossible to build a battleship that had sufficient armor to protect against all possible enemy weapons and float at the same time. The designers of the *Iowa*-class also had the hard limitation of the size of the Panama Canal locks and the soft limitation of the London and Washington Naval Treaties. Fortunately, it will never be known how successful the designers of the *Iowa*-class were with their design because they never faced the kind of opponent they were designed to face and none of the ships was ever subjected to more than pinpricks from the enemy.

Producing the armor for a battleship was a major undertaking that taxed the resources of the nation's steel makers. Armor for the *Iowa*-class was divided among Carnegie-Illinois, the Midvale Company, and Bethlehem Steel. Counter-intuitively, the navy did not have the mills specialize in specific armor plates. Among the various ships, the same plate could have come from any mill. There were long lead times for delivering armor. Plans for the armor sections for USS *New Jersey* were drawn up in the summer of 1940 with shipyard delivery of parts scheduled between May 1941 and September 1942.

It will become apparent that the *Iowa*-class battleships represent the antithesis of what America was known for producing during

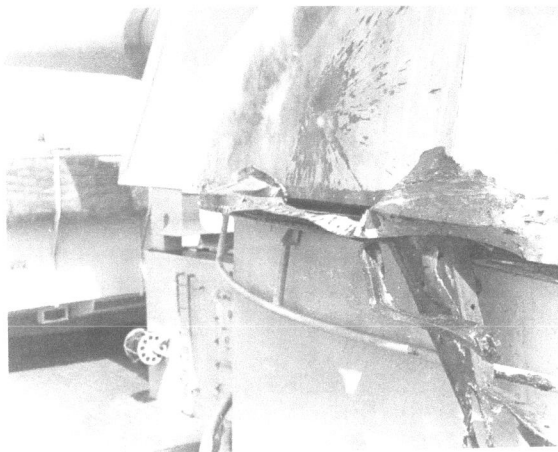

World War II. These were not like the Sherman tank or *Essex*-class carriers that were designed for mass production. The *Iowa*-class battleships are hand-crafted machines where no expense was spared in their construction. If their designers could have found a place for one, they would probably have had an elaborately carved figurehead at the bow.

1.1. Armor Basics

1.1.1. Sloped Armor

Armor tends to be more effective against projectiles that strike at an angle than against projectiles that strike directly. Increasing the angle between the path of an impacting projectile and the face of the armor by sloping the armor provides two benefits. First, the slope increases the tendency for impacting projectiles to skip off the armor's surface rather than penetrate. The angle between the path of the projectile and the armor face decreases the component of force acting to penetrate the armor and increases the

Photograph 1.1: The effect of a Japanese 6-inch gun to Turret No. 2 of the USS Iowa. This occurred during the bombardment of Mili Atoll in 1944. The armor has been scratched and the seal between the turret and the barbette has been damaged. (Nat'l Archives)

component of force acting to push the projectile along the armor's face. While this is an oversimplification of the factors at work, if a projectile impacts the face of armor at a 45-degree angle, the force acting to deflect it along the face is the same as the force acting to penetrate the armor. Second, the slope increases the distance that the projectile must pass through the armor in order to penetrate. A projectile striking a 17-inch armor plate at a forty-five degree angle must pass through over 24 inches to penetrate. The drawback of sloped armor is that it tends to reduce the usable volume within the ship. Armor thicknesses for the *Iowa*-class battleships were always specified measured normal from the inside face of the plate to the outside.

1.1.2. Immunity Zone

The thickness of armor a projectile can penetrate depends upon the range of the target and the position of the armor that it impacts. A ship's armor is roughly vertical along the sides and roughly horizontal over the decks. Because of the angle of impact and the effects of gravity and drag, it turns out that it is generally easier for a projectile to penetrate side armor at closer ranges and easier to penetrate deck armor at longer ranges.

As soon as a shell leaves the muzzle of a gun the forces of drag from air resistance and gravity start to alter its velocity and direction. These forces cause a shell to follow a curved path. As a projectile travels longer, gravity and drag have more time to act so the curvature of the flight path becomes greater. The force of drag from air resistance is roughly proportional to velocity so it has the greatest effect at the start of flight while the force of gravity is constant throughout. The combined effect of drag and gravity cause a projectile's path curve to be closer to a straight line at the beginning of its trajectory and most curved at the end.

In theory, a battleship could adjust the range of a shell by altering the firing angle of the guns or by changing propellant charge. In practice, the propellant charge was kept the same and range was adjusted solely by adjusting the gun angle. Because of the constant propellant and because battleship guns were not able to fire like mortars at angles over forty-five degrees, there was one general flight path for each target range. Increasing the angle of the gun when it is fired increases the distance the projectile will travel before it reaches the surface. Consequently, the angle between the flight path and the Earth's

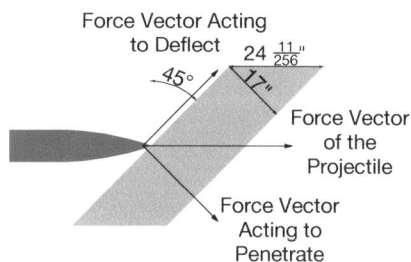

Figure 1.1: Sloped armor increases the distance a projectile must penetrate and acts to deflect a projectile.

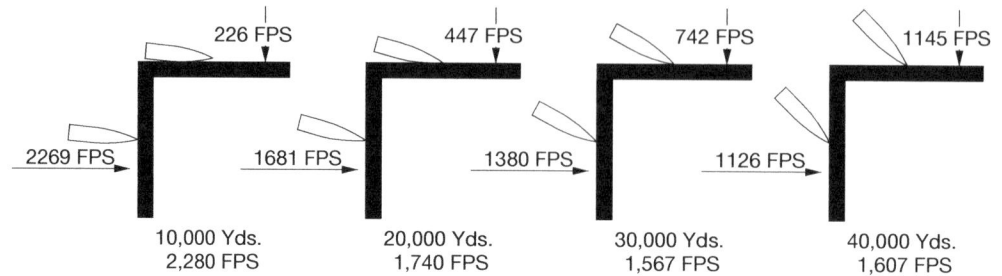

Figure 1.2: As the range increases a projectile's vertical velocity increases due to gravity accelerating longer and horizontal velocity decreases due to air drag acting longer. A projectile is more likely to penetrate side armor at short ranges and horizontal armor at longer ranges. Figures from 16-Inch Range Tables.

(ocean's) surface when a projectile reaches the target is roughly constant for any given range.

At the closest practicable range, drag has less time to affect the projectile so it strikes the target's armor with the highest velocity. At the same time, the projectile's direction at impact will be closest to perpendicular with the armor face. As the range increases, air resistance drag acts for a longer time so the projectile's velocity at impact decreases along with its penetrating power. At the same time, increasing the range increases the angle at which the projectile impacts the side armor, further reducing its penetrating power.

The effect of range is completely different when a projectile strikes deck armor. At close ranges the projectile's path is close to horizontal so deck armor is at its most effective protection. A projectile is more likely to skip across deck armor than penetrate it. As the range increases, the projectile's flight path becomes more curved and the angle at impact starts to become closer

to vertical. Also, as the range increases, the angle at which the gun is fired increases so the projectile travels to a higher altitude. The projectile's vertical velocity increases as the range increases because gravity has a longer time to accelerate the projectile downward. The combination of striking closer to vertical and increased velocity makes projectiles more likely to penetrate deck armor at longer ranges than close ranges.

When a typical battleship gun is used against a typical battleship's armor there is usually a range gap where the gun's projectiles cannot penetrate the target's armor at all. Within this gap, projectiles do not have enough horizontal velocity to penetrate the side armor. At the same time they do not have enough vertical velocity and do not arrive at a steep enough angle to penetrate the deck armor. This gap is called the *immunity zone*. If the target ship can keep itself within the immunity zone, in theory, its opponent cannot penetrate its hull armor. Obviously,

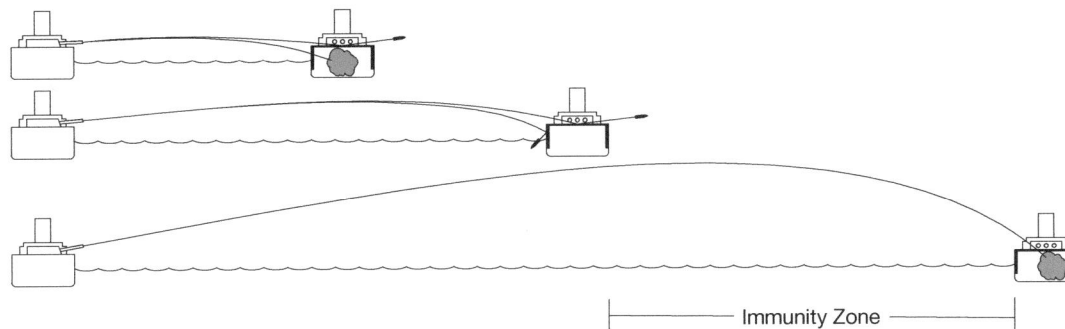

Figure 1.3: (Top) At short battleship engagement ranges projectiles impact the side armor directly and have enough energy to penetrate. However, projectiles strike the deck at an angle so they skip off. (Middle) As the range increases there will be a point where the armor cannot be penetrated. The projectiles do not have enough energy to penetrate the side armor and the angle of impact is too great to penetrate deck armor. This is the start of the immunity zone. (Bottom) As the range increases further the angle of impact with the deck decreases and gravitational acceleration causes the vertical speed of the projectile to increase to the point where it has enough energy to penetrate deck armor. This the end of the immunity zone.

the inner and outer ranges of the immunity zone depend upon type of gun on the opposing ship.

One practice for evaluating the effectiveness of a ship's armor scheme in the first half of the twentieth century was to determine its immunity zone relative to the ship's own guns. If a ship's armor scheme provided reasonable protection against the ship's own guns it was considered to have a *balanced design*.

Unfortunately, unexpected things happen during the design process. When the design of the *North Carolina*-class battleships began, their main armament was limited to fourteen-inch guns under the Second London Naval Treaty. Consequently, the armor of these ships was designed to protect against fourteen-inch guns. The treaty contained a provision to increase the gun limit to sixteen-inches if any signatories to the earlier Washington Naval Treaty naval trea-

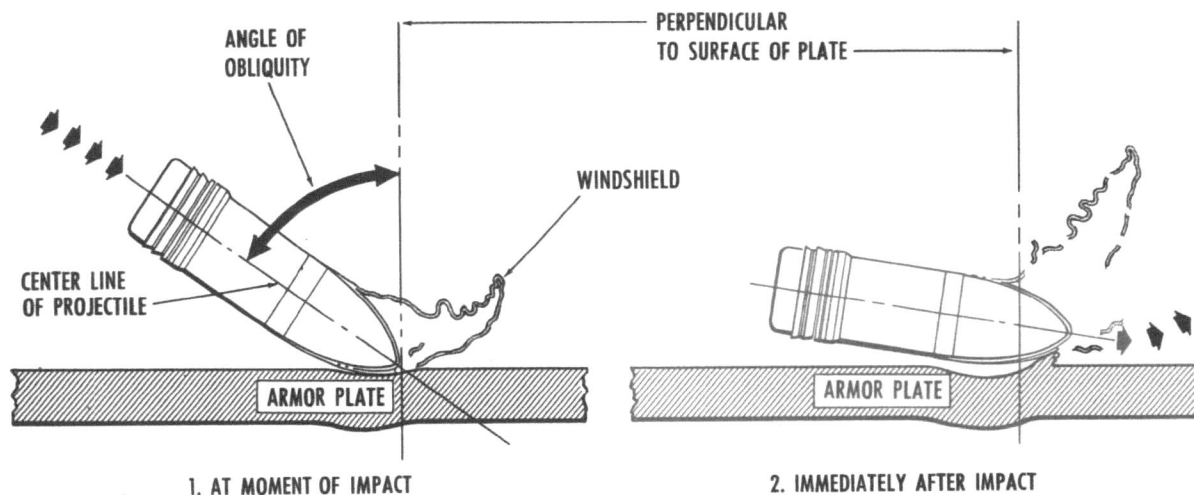

Figure 1.4: Projectiles impacting armor at an angle tend to rotate, reducing their penetration ability. (U.S. Navy)

ties did not sign this later treaty by 1937. When Italy and Japan declined to sign, this provision went into force. At that time, construction on the ships was about to begin. The U.S. Navy was faced with the choices of retaining the fourteen-inch guns as planned and risk having he ship encounter better armed enemies; redesigning the entire ship and delay the start of construction; or keeping the same design but upgrading to sixteen-inch guns. The navy chose the last option with the result that the *North Carolina*-class had an unbalanced design.

A similar unexpected event occurred to the *Iowa*-class. At the time the *Iowa*-class was designed, the Navy was using the Mk. 5 2,240 lb. armor-piercing projectile in sixteen-inch guns. The *Iowa*-class's armor was designed to protect against this projectile and had an immunity zone 21,700 to 32,100 yards. Before the *Iowa*-class entered service, the Navy introduced the Mk. 8 2,700 lb. armor-piercing round that had much greater armor penetration. Against the heavier Mk. 8 projectile, the *Iowa*-class's immunity zone shrank to between 23,600 and 27,400 yards. With such a tiny immunity zone, the Iowa-class no longer had a balanced design—yet the only change was the projectile against which their performance was measured.

1.2. The All or Nothing Principle

Battleships were originally designed with armor that extended the entire length of the ship. Designers would vary the thickness of the armor depending upon the perceived need for protection. HMS *Dreadnought*'s belt (or side) armor extended the length of the ship. The armor was eleven inches thick amidships around the ma-

chinery space but was only six inches at the bow and four inches at the stern.

American ship designers quickly realized there were areas of a battleship that were not worthy of protection. A projectile penetrating the paint locker will create a spectacular mess. If a projectile makes it to the laundry, the crew might have to wear dirty underwear for a while. However, these types of damage would have only a minor impact on the ship's fighting ability. The U.S. Navy developed an approach to ship armor that became known as the *all or nothing* principle. Under this system areas of the ship were either completely protected in an area called a *citadel* or left unarmored. The *Nevada*-class battleships of 1912 were the first to implement this system. All battleships built after the First World War adopted it to some degree as well. Taken to the extreme, under the all or nothing principle the entire citadel would have the same level of protection. In reality, most battleships, including the *Iowa*-class, did not have uniform protection within the citadel.

The *Iowa*-class's citadel forms an inverted box of armor that protects the areas the ships need to keep functioning in order to remain in a fight. The straight sides of the citadel have knuckles so that the citadel's plan roughly follows the curve of the hull. One of the unusual aspects of the *Iowa*-class's armor protection is that (aside from the turrets) it is almost entirely enclosed within the hull. The tops of the barbettes and the top of the conning tower are the only parts visible.

Except for the barbette tops and the top of the conning tower, the citadel is concealed by the decks and plating. The sides of the citadel consist of linear segments that roughly follow the shape of the hull. The key areas in the *Iowa*-class battleships that needed protection by the citadel were the propulsion machinery, command and control spaces, magazines, and rudder machinery. All of those spaces, except the rudder, are located between Turret No.1 and Turret No. 3. However, the hull form forced the rudders and their machinery to be located over a hundred

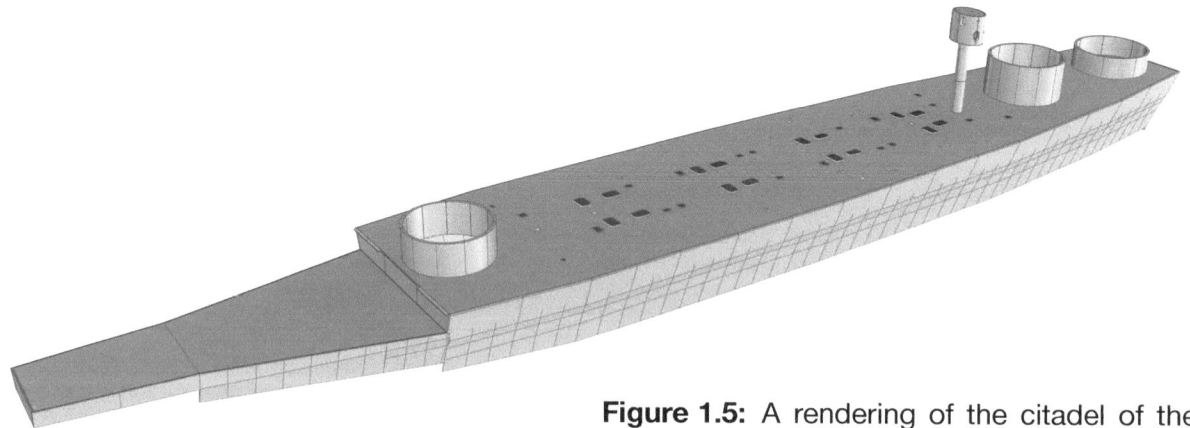

Figure 1.5: A rendering of the citadel of the Iowa-class.

Plan 1.1: (1302B) Iowa-Class Armor Plates.
This plan shows the different plates ordered
for USS Iowa and USS New Jersey.

6

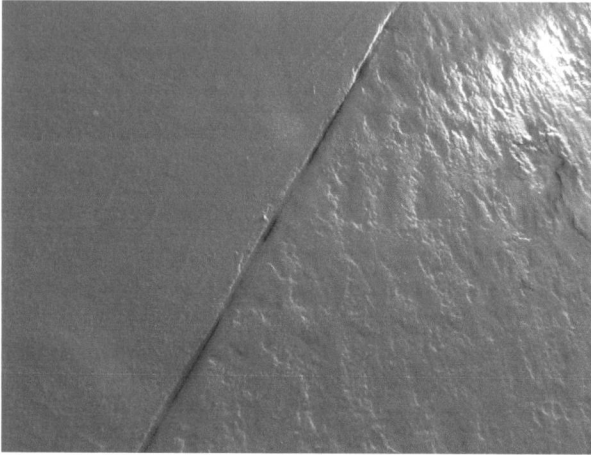

Photograph 1.2: Turret no. 1 on USS New Jersey. The Class A side armor is to the right and the Class B front armor is to the left. The Class B armor is smooth while the Class A armor is pitted. This difference make it easy to differentiate between these two types of armor.

feet aft of Turret No. 3. If the designers had followed the all-or-nothing principle slavishly, there would have been uniform protection from Turret No. 1 to the rudder machinery. However, such an implementation would have been wasteful because much armor would have gone to protect non-vital spaces between Turret No. 3 and the rudders. Instead, the designers broke the citadel into three sections with different protection schemes. The main section extends from just forward of Turret No. 1 at FR50 to just aft of Turret No. 3 at FR166. The horizontal protection in this region is at the Second Deck. The vertical protection extends down to just above the Hold. Another section protects the rudder machinery. It extends from FR189 to FR203. In this area the citadel only extends to the Third Deck. Because there are only non-vital storage areas between the rudder machinery and Turret No. 2, the designers could have created a separate armored space for the rudders (as done on some other battleships). However, for structure, buoyancy, and shaft protection, the *Iowa*-class battleships have a third citadel section from FR166 to FR189 that links the other two. This section extends to the Third Deck so that the citadel has a step down from the Second Deck at FR166. Transverse armored bulkheads at FR 50, FR166, and FR 203 protect the ends of the citadel and the step down from the Second Deck to the Third Deck.

1.3. Steel

There is no single measurement that will evaluate the performance of material used in armor. Instead, several measurements have to be considered. Three such measurements are tensile strength (resistance to breaking) and yield strength (resistance to deformation), and hardness. Tensile strength and yield strength are usually measured by devices that apply an increasing force to pull a sample apart. Tensile strength is the pressure required to break the sample. Yield strength is the pressure required to strain the material to the point it will not return to its original shape. The tensile strength is always greater than or equal to the yield strength. Armor hardness is usually measured on the Brinell scale in which a hard ball is dropped on a sample and the resulting indentation is measured. There tend to be tradeoffs among these measurements. Increasing hardness tends to make a material more brittle, which tends to result in a reduced yield strength.

Steel is an alloy consisting of the metal iron and carbon. Steel is vastly superior to iron in structural applications because it is much stronger and resists rust better. Steel had been in use since ancient times but, until the Bessemer process was invented in the mid-nineteenth century, there had been no means of mass-producing steel. The Bessemer process came out roughly the same time as warships started incorporating metal for armor, so the age of iron warships was very short. In most cases, such warships were ironclad over wood. By the end of the 19th century, steel had replaced iron in ship construction.

The amount of carbon in steel is relatively small (0.05% to 2% by weight) but it has dramatic effects on the performance of the material, making it much stronger and harder than pure iron. Depending upon the composition, the tensile strength of steel can be several times that of pure iron. Once the carbon content of an iron alloy exceeds about 2% it is called *cast iron* and iron with a very low carbon content is *wrought iron*. Other elements, such as chromium and nickel, are routinely incorporated in steel to adjust the material properties, such as tensile and yield strength and rust resistance. Steel may also contain contaminants, such as sulfur, that tend to degrade the material quality. For the purpose of understanding the structure of warship armor, the carbon content is the only aspect of metallurgy that will be discussed further here. For now, one can just say the properties of steel can be adjusted by varying the amount of constituent elements and by the process used to form the crystalline structure of the material. The rest of the discussion on armor only states constituent elements (other than carbon) for identification purposes.

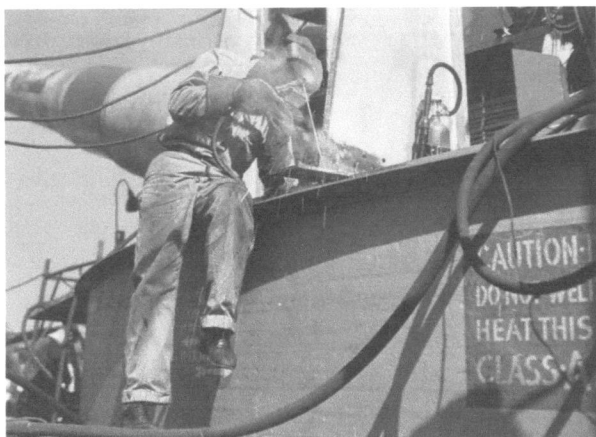

Photograph 1.3: A welder at work on USS Iowa in October 1942. The barbette has a sign warning not to weld or heat the Class A armor.

Steel gets harder and more brittle as its carbon content increases. Harder armor is desirable because it tends to break up incoming projectiles. Brittle armor is undesirable because impacts can cause shattering. Based upon carbon content, structural steels are classified as mild steel (0.06%–0.25% carbon), medium steel (0.25%–0.55% carbon), and high carbon (0.55%–1.00% carbon). Extremely hard steels with a carbon content close to 2% are frequently used for cutting tools.

1.3.1. Face-Hardened Armor

As steel armor became standard in warships, techniques were developed to create steel plates where the outer face was much harder than the inner face. This kind of material is called *face-hardened armor*. Armor whose hardness is the same throughout is known as *homogenous armor*. The theory behind face-hardened armor is that the hard outer face would shatter projectiles while the ductile backing would keep the plate from shattering.

The U.S. Navy tested the first face-hardened armor in 1890 using a process developed by Hayward Harvey. The Harvey process started with nickel-steel plates. The outside face was covered with charcoal and the plate was heated to about 2000° F. for several weeks. This treatment caused the steel to absorb carbon through one face. At the end of the process the carbon content at the outer face increased by about 1% (to around 1.1%–1.2%). The additional carbon made the outer face hard while the inner face remained ductile. The additional hardness could add the equivalent of a couple inches or more of protection with no increase in weight. Harvey called the process of adding carbon *cementing* and plates face-hardened in this manner or similar were called *cemented armor*.

The Harvey process was soon supplanted by one developed by Krupp in Germany. The main improvement was that Krupp replaced Harvey's nickel-steel with chromium-steel which was more effective in absorbing carbon. Another change was that Krupp used a hydrocarbon gas (*e.g.*, acetylene) rather than charcoal as the carbon source. However, U.S. manufacturers produced Krupp Cemented Armor using solid carbon rather than gaseous hydrocarbon. Thus, American cemented armor tended to use the Krupp formula but followed the Harvey process. While cementing was the most common method of creating face-hardened armor for ships, other physical processes were developed to face-hard armor. The Japanese used a heat treatment process to create some face-hardened armor plates. Likewise, the Germans used heat treatment to face-harden tank armor during World War II.

One issue with cemented armor is the hardened face had the potential to crack. Welding and machining could cause such cracking. The areas of an outside face that required machining were left soft. This could be accomplished by placing a material, such asbestos, which would block the absorption of carbon over the areas of a plate that were not to be hardened.

The U.S. Navy called face-hardened armor *Class A* regardless of whether it was cemented or non-cemented. The Navy called the alternative—homogenous armor—*Class B*. These terms suggest that one type was better than the other. In reality they describe which was more expensive. Class A armor can be easily identified because it has a pockmarked surface while Class B armor is smooth.

Two significant weaknesses were found with Class A armor. First, by 1896 the Navy was testing projectiles that incorporated caps over the nose that could defeat face-hardened armor. The cap spreads the force of the impact over a wider area than that of an uncapped projectile so the projectile has less of a tendency to shatter. The cap also tends to dig into the plate, which causes the projectile to orient more towards normal to the face of the armor and reduce the deflection effect resulting from a strike at an angle. As a result capped projectiles could nullify the advantage of face hardening. Initially caps were made of a soft metal but the later practice for the U.S. Navy was to make the caps as hard as possible—to the point that the caps tended to be harder than the armor they had to penetrate. Second, oblique projectile impacts can cause enormous damage to face-hardened armor. A projectile skipping

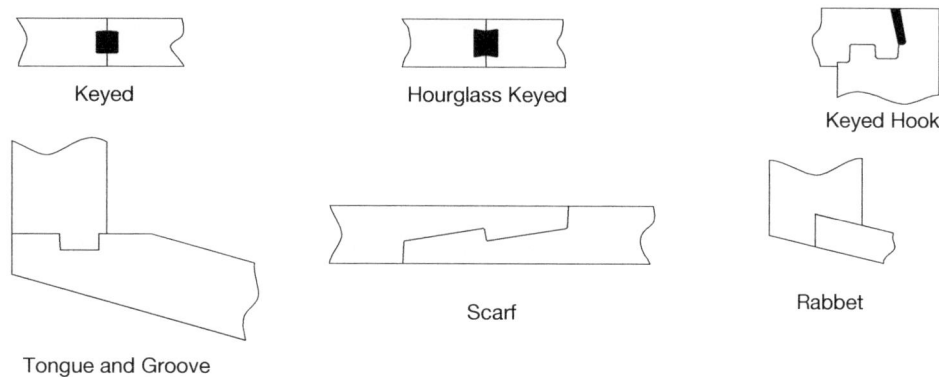

Figure 1.6: These are examples of the types of joints used at the edges of armor plates. The joints in the top row use keys. The most numerous type of joint is the Simple Keyed Joint shown at the upper left. A rectangular key was place in keyways machined into adjoining plates. The keys are often tapered when they are aligned vertically. This type of joint maintains plate alignment. At the upper center is an Hourglass Joint. Once the key was hammered in from above, its hourglass shape and friction prevented any movement. When a tapered key was used, this is the only type of joint that was not welded or bolted. Joints with a straight key were welds. A Keyed Hook Joint is shown at the upper right. This type of joint was used to attach horizontal plates to side plates. In the lower row (from left to right) are a Tongue and Groove Joint, Scarf Joint, and Rabbet.

along the surface might not penetrate, but it can cause extensive fracturing. Class B armor is more ductile than Class A armor so it has the tendency to bend as a projectile moves along its surface rather than break. Consequently, Class B armor tends to be more effective than Class A armor in resisting oblique strikes and strikes by capped projectiles.

Putting aside differences in protection, Class B armor had many practicable advantages over Class A armor. First, the added step of the face-hardening process was lengthy and had to be done with great care, making Class A armor much more expensive than Class B armor. Sec-ond, Class B armor was much easier to work with. The hardened face of Class A armor could crack from the heat of welding so it generally required complex machining to join the plates into larger assemblies. Third, because of its rigidity and difficulty in assembly, Class A armor was generally dead weight while Class B armor easily could be used for both structural and protective purposes. Some sources, incorrectly state that Class A armor could not be used structurally. However, there are places in the *Iowa*-class battleships where Class A armor is used structurally and where Class B armor is dead weight. The general, though not universal, rule is that the

designers of the *Iowa*-class battleships specified Class A armor where projectiles were likely to strike normal to the surface and Class B where projectiles were likely to strike at an angle.

At this point it is worth noting that most research into the behavior of projectiles striking armor involves tanks or body armor. It was much cheaper for the Army to fire three-inch projectiles into captured German tanks than it was for the Navy to fire sixteen-inch projectiles into mockups of ship armor. Furthermore, there is continued financial incentive to conduct research into these smaller scale applications while battleship armor is now entirely an area of academic, historical research. While the tank and body armor research indicates certain behaviors of projectiles striking armor, it is not currently known how these behaviors would scale up to the size of battleship armor.

One such behavior is decapping. A defense against capped projectiles is to add a thinner layer of steel outside the actual armor. While a projectile of the size used on a battleship can easily pass through an inch or two of steel, the process of penetration can cause the cap to disintegrate or separate making it more difficult for the projectile to penetrate the face-hardened armor behind. A projectile impacting the armored belt of an *Iowa*-class battleship might have to pass through three protective layers before reaching the armor. The effectiveness of this system in decapping projectiles is a subject of debate.

Another behavior that is not well understood at battleship scales is the affect of layered armor. When firing a projectile normal to an armor face, a single plate tends to resist penetration better than two or more laminated plates giving the

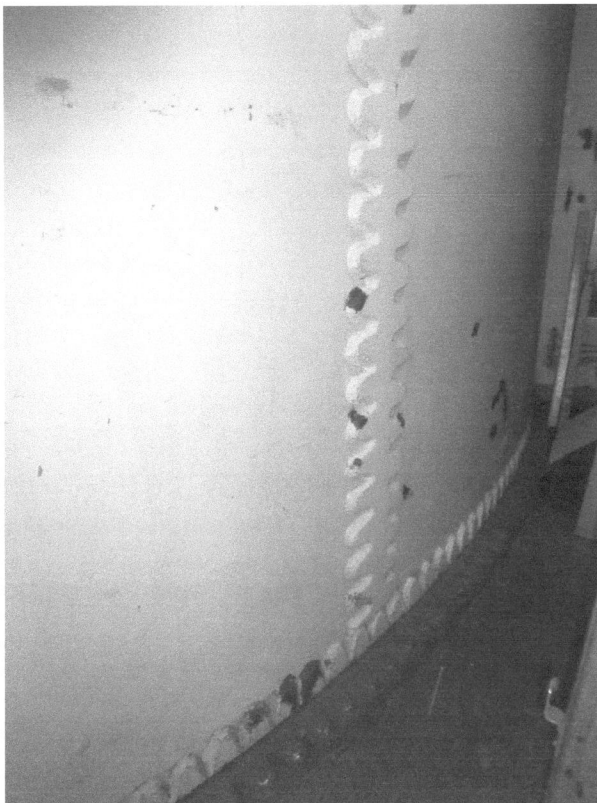

Photograph 1.4: The barbette support for Turret 1 of USS New Jersey at the Third Deck. This shows two examples of scalloped plates. The vertical scallop secures a butt joint between plates. The turret support is welded to the deck along a scalloped angle. The scallops on the plates are staggered across the sides and those on the angles are staggered across the flanges. The scallops create additional welding surface. Riveted cover plates on the Iowa-class are often scalloped as well to save weight. This extra step was normal in US warships of the era.

same thickness. Laminated plates resist penetration better than spaced plates. There are various rules of thumb for determining the relative protection of those three different configurations.

However, armor behavior gets complicated as many parts come together. A thinner layer of armor outside the main armor plate can cause a projectile to twist such that it strikes the main armor on the side rather than the nose. A projectile striking a laminated plate at an angle may simply tear the upper layer away where it would rip open a thicker single layer. Current tanks frequently use laminated (often using different materials) and spaced armor. However, there is little research on their behavior at battleship scales.

1.3.2. Steel Plates

Six types of steel plates were specified for the construction of the Iowa-class. The government set standards for each steel type. First, most of the ships' structure specified medium steel. Medium steel is defined by its carbon content. Second, high tensile steel was specified for some structural elements. High tensile steel is defined by its tensile strength rather than its carbon content. High tensile steels can be either mild steel or medium steel. They tend to incorporate nickel, chromium, and molybdenum for strength. Third, Special Treatment Steel (STS) was an alloy specified by formulation that was developed for armor protection. STS is a medium steel with nickel and chromium. The Iowa-class battleships used STS liberally for dual structural and protective purposes. Because of the extensive use of STS, the ships contain much additional armor that does not show up in tabular armor comparisons. Fourth, HY-80 was used as armor added during the 1980's reactivations. HY-80 is specific formulation of mild steel developed after World War II for naval purposes that incorporates manganese, silicon, nickel, chromium, and molybdenum. Fifth, Class A armor used in the Iowa-class was cemented, face hardened armor of Krupp chromium-steel. Finally, Class B armor used in the Iowa-class was used interchangeably with STS. The navy tended to call homogenous armor STS when it was fabricated at the yard and Class B when it was fabricated at the mill and supplied by the Bureau of Ordnance. For example, the lower belt plates on USS Iowa and USS New Jersey were specified as STS while the same plates used in the other ships were specified as Class B. The only real difference was where the plates where shaped.

1.3.3. Forming Steel

Three methods were used to form the steel used in the Iowa-class battleships. Most of the plates were rolled. A steel ingot was heated until the metal was soft and was forced through a pair of rollers spaced to create the desired thickness of plate. Rolling could only produce plates a few inches thick. The navy only considered plates at least 4-inches thick to be armor. This reflected the limits of rolling to create plates. Such thicker plates were forged. These plates started out as ingots of steel. A typical ingot weighed about a hundred tons. However, the ingots used for large plates weighed about three hundred tons. Heating the ingot to about 1900°F softened it enough for forging in a hydraulic press. This forging process of heating and pressing was repeated until the plate had the correct shape and thickness. The inside faces of the armor plates on the Iowa-class are flat, follow a circular arc, or are a combination of those profiles. After forging, the plates were heat-treated (tempered, annealed, quenched) as needed. Class A plates would then be cemented and heat treated again. Casting was used to shape armor used in the turret hoods. For these, the steel was melted, poured into a mold, and allowed to cool. Casting allowed creating complex shapes but the end product was not as resistant to penetration as forged or rolled plates and there were practicable limits to the thickness of steel castings. Every armor plate or casting required final machining. Such machining included creating joints at the edges, creating openings, and drilling and tapping bolt holes.

1.4. Assemblies

The armor plates had to be joined into assemblies. This meant either joining the plates to each other or to the rest of the ship. The armor plates were laboriously machined to incorporate joints that would be more familiar to woodworkers than today's metal workers. There is no place where two armor plates meet without some kind of machined joint. With the exception of Hourglass Joints with tapered keys, the machining of the joint alone would not hold the plates in place. Structural armor plates had to be joined to their neighbors by welding, bolting or both. If a plate was not structural (dead weight), it had to be secured to the structure of the ship using bolts or pins.

The *Iowa*-class uses two methods to attach nonstructural armor plates. The first method was used for Class B deck plates that rested directly on top of the structural deck. In this situation quilting pins, spaced at regular intervals, hold the armor plates to the structural deck. The pins were driven upwards from below through holes drilled in the deck and the armor plates. The pins were welded at the top of the armor plate and a head at the bottom of the pin prevents the plate from pulling up. Pins function like rivets with the Navy's distinction being that the end of a rivet was forged while the end of a pin was welded. *Quilting* referred to the use of the pins away from the edge of the plate. There are quilting rivets in the hull.

The second method was used where it was necessary to compensate for thickness variations in thicker armor plates. Such thickness variations would prevent the plate from sitting flush against its structural supporting bulkhead. If the plate does not sit flush, the joints machined into its edge would not line up with those in the adjoining plate. To allow for such variations in armor the plates were offset from the bulkhead by 1" to $1\frac{1}{4}$". The gap compensated any irregularities so that a plate would fit in its designed location. Holes were drilled and tapped at regular intervals along the face of the armor plate and corresponding holes were drilled through the bulkhead. Bolts screwed through the bulkhead side secured the armor. Caps were welded over the bolts and the space between the bulkhead and armor was filled with concrete.

The dimensions for much of the armor were specified with great precision so that the plate could be assembled like puzzle pieces. However, this was not the case with the deck and lower belt armor. In these areas the plate dimensions approximated the shape that was required and the shipyard had to do extensive machining to create the final shapes for the armor plates.

1.5. Welding

Welding is the process of joining two base pieces of metal by heating them to a high temperature and introducing a filler material (weld metal) to span the gap between them. In the welding process the bases and the filler cool to fuse into a single piece of metal. The welded joint becomes as strong as the rest of the base material. While welding is the normal method for jointing structural metals now, at the time the *Iowa*-class battleships were constructed, it was a relatively new process for ship construction. The *Iowa*-class battleships were largely assembled using rivets with welding being reserved for place where riveting was impracticable. Over the construction period for the *Iowa*-class, welds gradually replaced many of the riveted joints.

Class A armor created special problems for welding. The heat from the welding process could cause the hardened face to crack. Structural Class A plates were welded. However, the welding was only done on their inner face, with one exception. Where the outer face had to be watertight against a deck, a caulk weld was run along the edge. No documentation has appeared that suggests special precautions, such as cooling the outer face, were taken when welding Class A plates.

While welding was used extensively, the ships' structure was largely riveted. A riveted joint required overlapping plates. If the plates did not naturally overlap, cover plates or angles had to be placed to the joint to create an overlap. Holes were drilled through the overlap along the joint and white-hot steel rivets were driven through those holes. A die was use to expand the rivet on the other side so it would not pull out.

The joints had to be caulked to make them watertight.

Rivets had two advantages over welding. First, they were easy to inspect. Just strike the rivet with a hammer and one could tell from the sound if it were correctly installed. Second, riveting creates a barrier against defect propagation. If a plate cracked, the crack could easily propagate across a welded joint but would stop at a riveted joint. Riveted construction tends to create a less rigid structure than welding, which was percieved as an additional advantage during this era.

The disadvantages of riveting greatly outweigh the few advantages over welding. Riveted joints are not nearly as strong as welds. Riveting is much more labor intensive and dangerous than welding. The cover plates and angles add significant weight without adding any strength. Consequently, riveting has largely vanished from, ship construction.

Plans give the appearance of having a distrust of welded joints. Welding frequently followed the riveting model where a cover plate or angle was placed over the joint between two pieces of metal to be joined. Cover plates were generally called *seamstraps* except that at plate butts they were usually called *butt straps*. The two pieces were welded to the angle or strap, rather than directly to each other. The edges of the angle or cover plate were scalloped to increase the welding surface. The plans specified the sizes of the scallops with great specificity. However, it was assumed that the these specification meant *about this size*. Scalloped straps were usually cut from a single sheet such that each cut formed the edge of two cover plates. The cutting process would create lossage making the cover plates smaller than specified. The actual ship construction frequently deviated from the plans in cases like this where the difference did not matter. In many places, two layers of scallops were used to weld angled joints. The outer layer would be a flat kick plate that was scalloped on the outer edge only. The kick plate would be welded to one plate along the scallop and both plates along the flat edge. The use of straps over welded joints contributed little for strength but added considerable dead weight to the ships.

1.6. Lifting Pads

Nearly all of the armor plates are slabs with no openings. Such plates inherently have no place to link to a a crane to move them. Temporary crane attachments were part of the plate designs. Plates were drilled and tapped so that lifting pads or lifting eyes could be screwed into them. The *Iowa*-class reused pads that had been created for the *North Carolina*-class battleships and were previously reused on the *South Dakota*-class. Cables could be run through the lifting eyes or shackles on the pads so a crane could hook on to the plate. After the pad and eyes were removed, plugs were screwed into their holes.

1.7. Measurement

The *Iowa*-class battleship used some forms of measurement common in shipbuilding but may be unfamiliar to readers. Armor plate thicknesses were measured in decimal inches. Other plates were measured in pounds per square foot. A one-inch thick plate weighs about 40 pounds. The thickness 25# indicates the plate is $\frac{5}{8}$"(25/40) thick.

Longitudinal locations are measured in *frames*. Frame locations are spaced four feet on the *Iowa*-class. The term *frame* also describes the transverse structural elements for the hull. The numbered frame locations generally correspond to structure frames. However, many structural frames are located at fractional fame locations ($\frac{1}{4}$, $\frac{1}{2}$, and $\frac{3}{4}$) and some frame locations lack a structural frame. In this book, *frame* is only used as a measurement unit. The *Forward Perpendicular* (FP) is the location where the bow intersects the waterline and the *Aft Perpendicular* (AP) is the location where the stern intersects the waterline. There are 215 frames between these locations. Fractions (halves and quarters) are used to indicate foot offsets. FR100 $\frac{1}{2}$ indicates a plane 402 feet aft of the FP. Positive or negative offsets were used to indicate other locations (*e.g.*, FR50 + 11"). Elevations are measured in feet/inches from the Baseline (BL). The baseline runs lengthwise above the lowest point of the keel plating. The Centerline (CL) runs lengthways through the center of the ship. Horizontal locations are measured in feet and inches from the CL. Such measurements are also called *halfbreadths*.

The structure of the ship is defined using two-dimensional *molded lines*. The molded lines establish reference points for components of the ship without accounting for plate thickness. The location of the top of a deck's plating was specified as an offset above the molded line. One side of a straight bulkhead will be aligned to a molded line. The relationship between the molded line and the actual plating can get complicated in some cases.

Area	Type	Thickness
Upper Belt	Class A	12.1"
Lower Belt FR50–FR166	STS (Class B *USS Missouri* and USS *Wisconsin*)	12.1" at the top tapering at knuckles to 1" at the bottom
Lower Belt FR166–FR172	STS (Class B USS *Missouri* and USS *Wisconsin*)	13" at the top tapering at knuckles towards 5"
Lower Belt FR172–FR189	STS (Class B USS *Missouri* and USS *Wisconsin*)	13.5" at the top tapering at knuckles towards 5"
Steering Gear FR189–FR203	Class A	13.5"
Transverse FR50	Class A	11.3" at the top tapering from a knuckle to 8.5" (14.5" tapering to 11.7" USS *Missouri* and USS *Wisconsin*)
Transverse FR166 and FR203	Class A	11.3" (14.5" USS *Missouri* and USS *Wisconsin*)
Deck FR50–FR166	Class B	4.75"
Deck FR166–FR189	Class B	5.6"
Deck FR166–FR203	Class B	6.2"
Conning Tower Sides	Class B	17.3"
Conning Tower Roof	Class B	7.25"
Conning Tower Floor	Class B	4"
Conning Tower Tube	Class B	16"
Barbettes	Class A	17.3" at sides tapering to 11.6" at front and back
Turret Front	Class B	17"
Turret Sides	Class A	9.5"
Turret Rear	Class A	11"
Turret Roof	Class B	7.25"

Table 1.1: Iowa Class Armor Summary

Chapter 2: Belt Armor

The belt armor protects the side of the citadel from FR50 to FR203. The belt is arranged in eight facets that roughly conform to the curve of the hull plan over this region. The belt armor is inclined outboard with its top 19 degrees (more precisely 19° 0' 25") away from the centerline to increase protection. This chapter also covers the vertical transverse armor at FR166 and FR203 because they are integrated with the belt.

Along its length the belt has four different configurations:

1. FR50–FR166: The main belt section extends from the bottom of the ship to second deck and consists of a Class A upper belt, a Class B or STS lower belt, and a 40# STS lower strake.
2. FR166–FR172: The belt is a single tier that extends from second platform to third deck.
3. FR172–FR189: Nearly identical to that of FR166–FR172 except that the armor is thicker.
4. FR189–FR203: Around the steering gear the belt is a single tier of Class A armor between first platform and third deck.

This chapter also covers the transverse armor at FR166 between second and third deck because it is integrated with the upper belt. It also covers the transverse armor at FR203 because it is integrated with the steering gear armor.

All of the belt plates (with the exception of the Class A upper belt) are welded and mechanically linked together to form a solid unit. All of the structural members of the ship are intercostal

to the belt plates.

The structure of the belt armor is closely aligned with the ships' underwater protection. Four torpedo bulkheads are a major part of that protection. These are numbered from outside in as bulkhead 1, bulkhead 2, bulkhead 3, and bulkhead 4. Bulkhead 4 is also called the *holding bulkhead*. An attacker would have to penetrate five layers (including the hull shell) of steel and four spaces between them to reach vital components through the side of the ship. The torpedo bulkheads vary in height and in some places are structural but not watertight. The descriptions of the torpedo bulkheads here are of their general layout and omit their configuration changes over their entire lengths.

Looking at the protective layers from outside in, the hull shell is a patchwork of style types and thicknesses. Near the waterline along the citadel the shell plating is 60# STS. Fuel storage is behind the shell. Bulkhead 1 is 25# high tensile steel outside the citadel. The bulkhead extends to third deck shelf and runs from FR50 to FR189. Fuel storage is located behind that bulkhead as well. Bulkhead 2 extends to second deck shelf and is 10# or 12# medium steel. This bulkhead runs from FR20 to FR189. There is a void space between bulkhead 2 and bulkhead 3 so that the force of a blast will not press non-compressible fuel or water against bulkhead 3. Bulkhead 3 extends from the bottom of the ship to main deck until FR166 and up to third deck from FR166 to FR203.

The belt armor and bulkhead 3 are intertwined. TThe upper belt is attached to the torpedo bulkhead. Behind the upper belt the bulkhead serves the dual purpose of underwater

protection and as the backing bulkhead that supports the upper belt. This area of the bulkhead is either 35# or 40# STS. In other locations, the lower belt is also the torpedo bulkhead. There is another void space between bulkheads 3 and 4. Bulkhead 4 is 25# STS. This bulkhead runs from FR50 to FR181 and extends to second deck.

Between FR62 and FR166 the transverse supports for the belt armor are staggered between the ships' structural frames while the frames support the holding bulkhead 4. This prevents a force against the belt armor from being directly transmitted to the holding bulkhead. Between FR50 and FR62 the hull is too narrow for that kind of arrangement. In this area the belt and holding bulkhead are both attached to frames, giving less protection.

The belt plates are grouped in eight facets that roughly follow the curve of the hull. The plates are aligned along their inside faces. At facet joints, the change in angle longitudinally causes the other faces in adjoining plates to be misaligned. Changes in plate thickness at FR166, FR172, and FR189 cause additional misalignment. In order to place scallops for welding or cover plates for tap riveting over such joint, the plates had to be planed. In most cases a corner just had to be grinded off from one plate but where there were changes in plate thickness, the amount of planing could be extensive. Readers should note that the tabular values the plans give for plate locations are heights/halfbreadths rounded to the nearest $\frac{1}{16}$". As the facet directions change relative to the centerline, rounding errors cause the tabular data to produce slightly inconsistent, but noticeable, differences in plate angles from the centerline.

Photograph 2.1: An outboard support for the armor belt. This support is at FR109 starboard on USS New Jersey. Torpedo Bulkhead 2 is to the left and the lower belt armor is to the right.

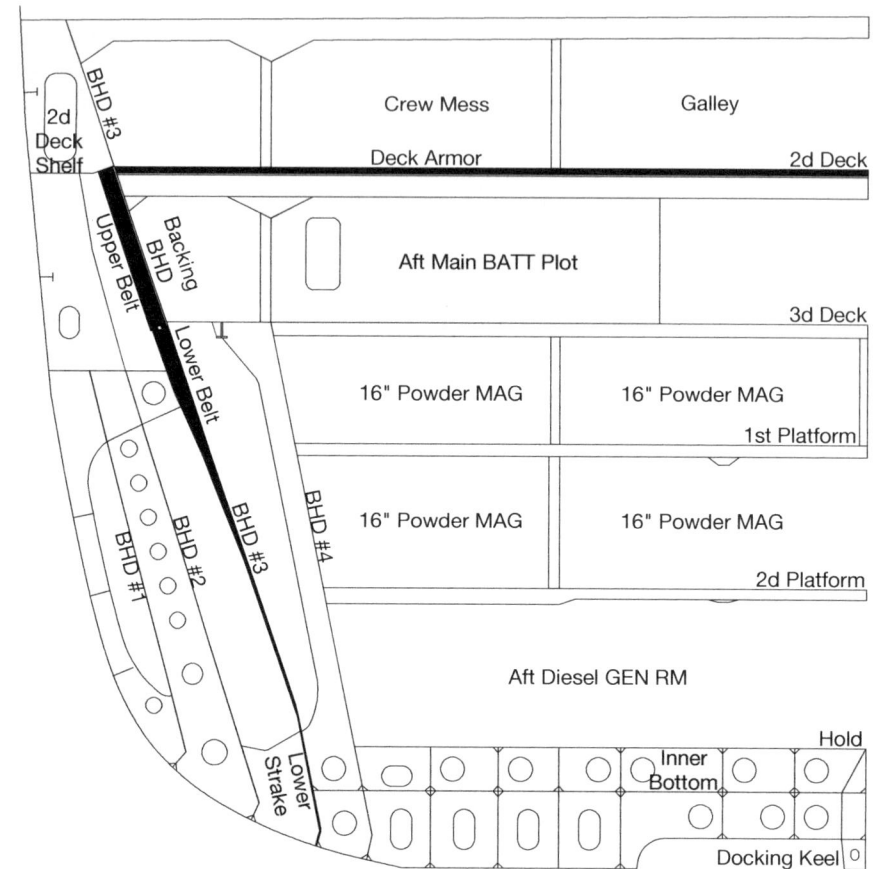

Figure 2.1: A cross section at FR154 $\frac{3}{4}$ looking aft. This location is forward of turret 3. Armor is shown in black. The belt armor has this configuration from FR50 to FR166. The four torpedo bulkheads are labeled. In this region BHD #3 extends from main deck to the bottom of the ship. The upper belt is attached to the bulkhead while the lower belt forms part of the bulkhead. The inboard and outboard supports for the lower belt alternate between the inside and outside. At this frame the support is located outside. The inside supports do not rest on BHD #4 so that the force of a blast against BHD #3 would not be directly transferred to BHD #4. The deck armor is 4.75 inches and rest on top of the 50# second deck. The upper belt is 12.1 inches. The lower belt is 12.1 inches at the top but tapers at horizontal knuckles to 1 inch at the bottom.

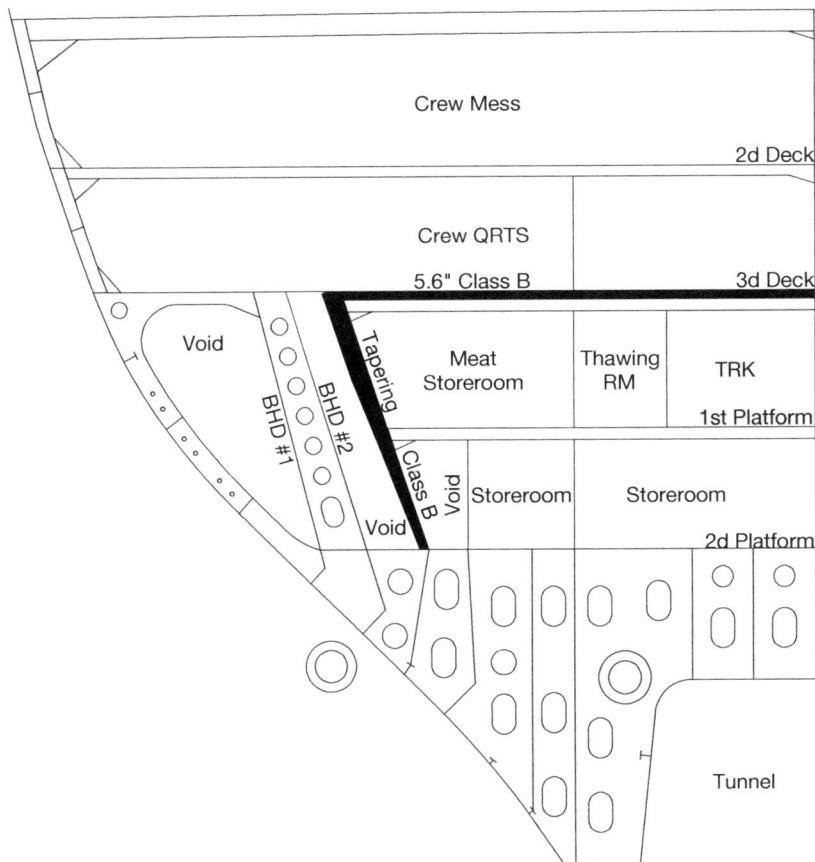

Figure 2.2: FR178 looking aft. The armor has this configuration from FR166 to FR189. At FR166 the citadel drops down from second deck to third deck and steps up from the bottom of the ship to second platform. The armor tapers from top to bottom at diagonal knuckles. Between FR166 and FR172 the belt is 13 inches at the top. From FR172 to FR189 it is 13.5 inches at the top. Otherwise those two ranges are structurally the same. The deck armor is 5.6 inches and rest on top of the 30# third deck.

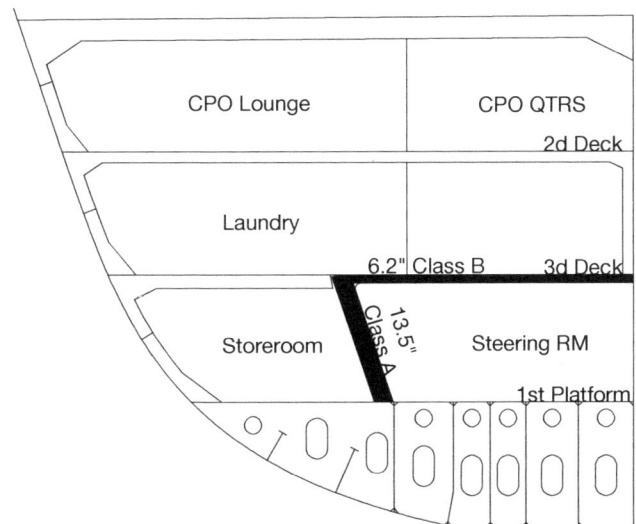

Figure 2.3: FR198 looking aft. This is the armor configuration around the steering gear from FR189 to FR203. Second platform ends at FR189 so the bottom of the belt armor in this area takes a step up to first platform. The deck armor is 6.2 inches. Both the belt and the deck armor are structural in this region. This configuration contradicts the assertion that Class A armor could not be used structurally.

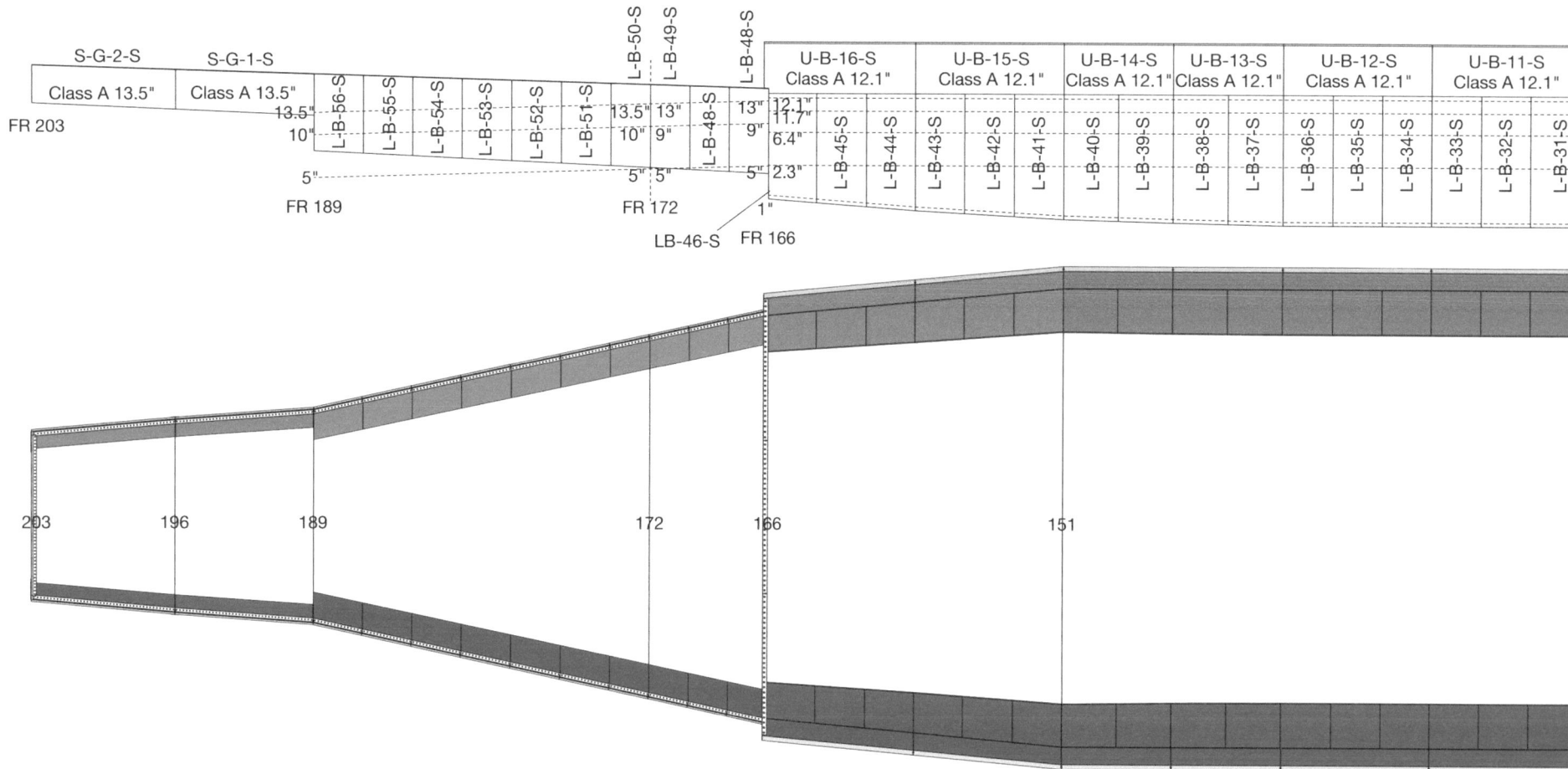

Figure 2.4: A rendering of the belt, FR189 and FR203 transverse plates. The diagram above identifies the plates that make up the belt. The locations of the four belt configurations are apparent in the upper diagram In the forward configuration from FR50 to FR166, the upper edge of belt slopes upwards at the bow to follow the deck shear. The 12.1" segment of the forward lower belt plates is longer than that of the plates further aft and the lower edge of the belt slope upward aft as the hull rises to accommodate the propellers and rudders. The plates taper at horizontal knuckles. Note that the lower belt plates between FR50 and FR166 are grouped with an upper belt plate. Aft of FR166 the belt has only one tier. Between FR166 and FR172 and between FR172 and FR189 the belt tapers at angled knuckles. Between FR189 and FR203 the belt has a constant thickness of 13.5 of Class A armor.

U-B-10-S Class A 12.1"			UB-9-S Class A 12.1"			U-B-8-S Class A 12.1"			U-B-7-S Class A 12.1"			U-B-6-S Class A 12.1"			U-B-5-S Class A 12.1"			U-B-4-S Class A 12.1"			U-B-3-S Class A 12.1"			U-B-2-S Class A 12.1"			U-B-1-S Class A 12.1"		
LB-30-S	L-B-29-S	L-B-28-S	L-B-27-S	L-B-26-S	L-B-25-S	L-B-24-S	L-B-23-S	L-B-22-S	L-B-21-S	L-B-20-S	L-B-19-S	L-B-18-S	L-B-17-S	L-B-16-S	L-B-15-S	L-B-14-S	L-B-13-S	L-B-12-S	L-B-11-S	L-B-10-S	L-B-9-S	L-B-8-S	L-B-7-S	L-B-6-S	L-B-5-S	L-B-4-S	L-B-3-S	L-B-2-S	L-B-1-S

12.1"
11.7"
6.4"
2.3"

1"
FR 50

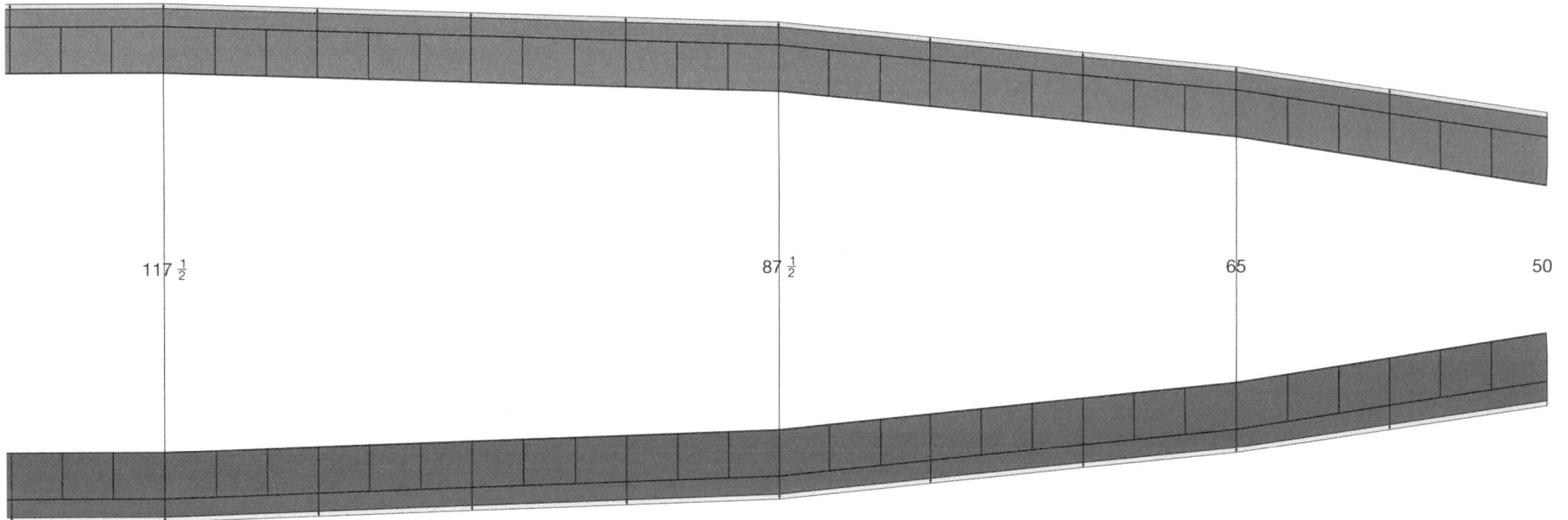

$117\frac{1}{2}$ $87\frac{1}{2}$ 65   50

Cement Fill Between
Class A Armor and
Backing Bulkhead

Connecting
Bolts

Class A
Upper Belt

Backing Bulkhead

Key

12.1"

11.7"

Class B/STS
Lower Belt

6.4"

Outer Face   Inner Face

2.3"

1"

Hull Shell   Lower Strake #40 STS

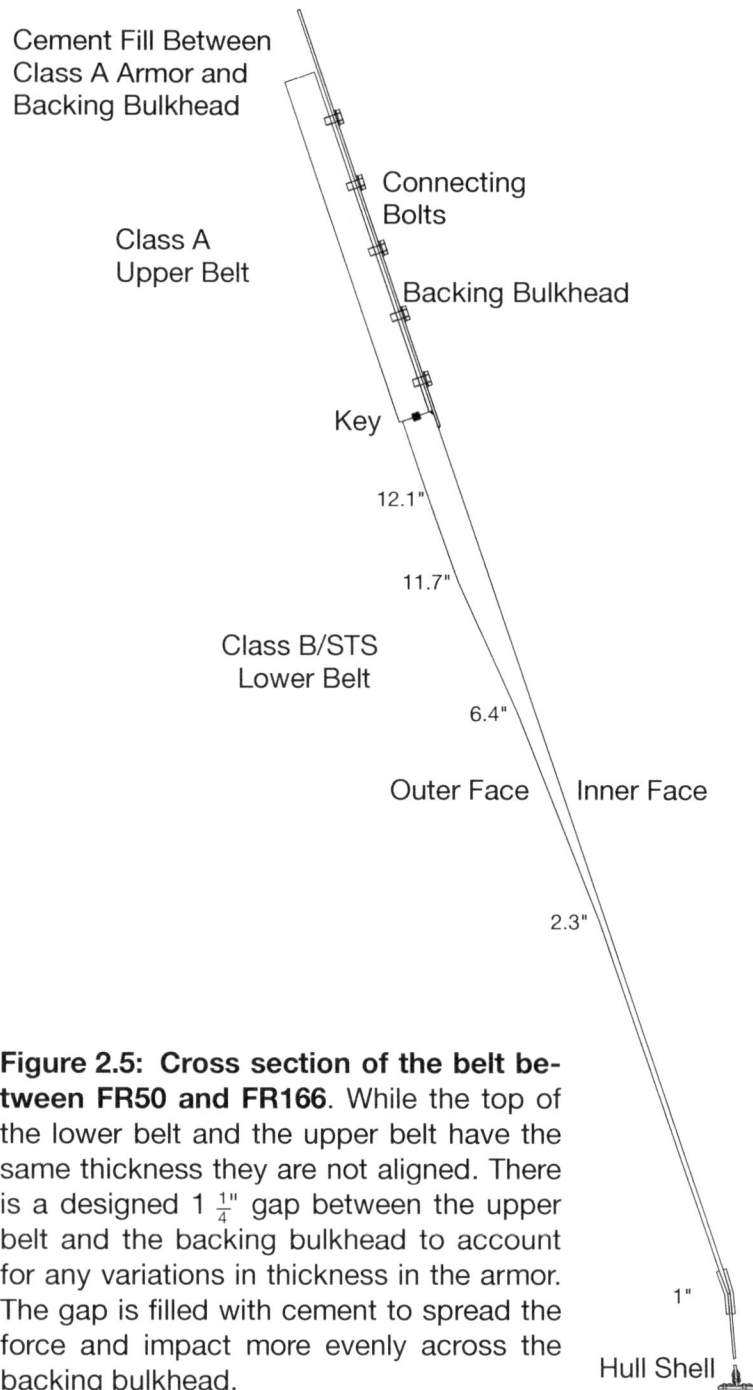

Figure 2.5: Cross section of the belt between FR50 and FR166. While the top of the lower belt and the upper belt have the same thickness they are not aligned. There is a designed $1\frac{1}{4}$" gap between the upper belt and the backing bulkhead to account for any variations in thickness in the armor. The gap is filled with cement to spread the force and impact more evenly across the backing bulkhead.

Figure 2.6: A rendering of a section of belt armor. Covers are welded over the bolts securing the upper belt to the backing plate to prevent them from becoming projectiles after an impact. The sections of the backing plate are riveted together using cover plates on the inside face. The scallops in these plates are for weight saving. The lower belt plates are welded together with scalloped cover plates. These scallops there increase the welding surface. The lower belt is riveted to the bottom strake of torpedo bulkhead 3.

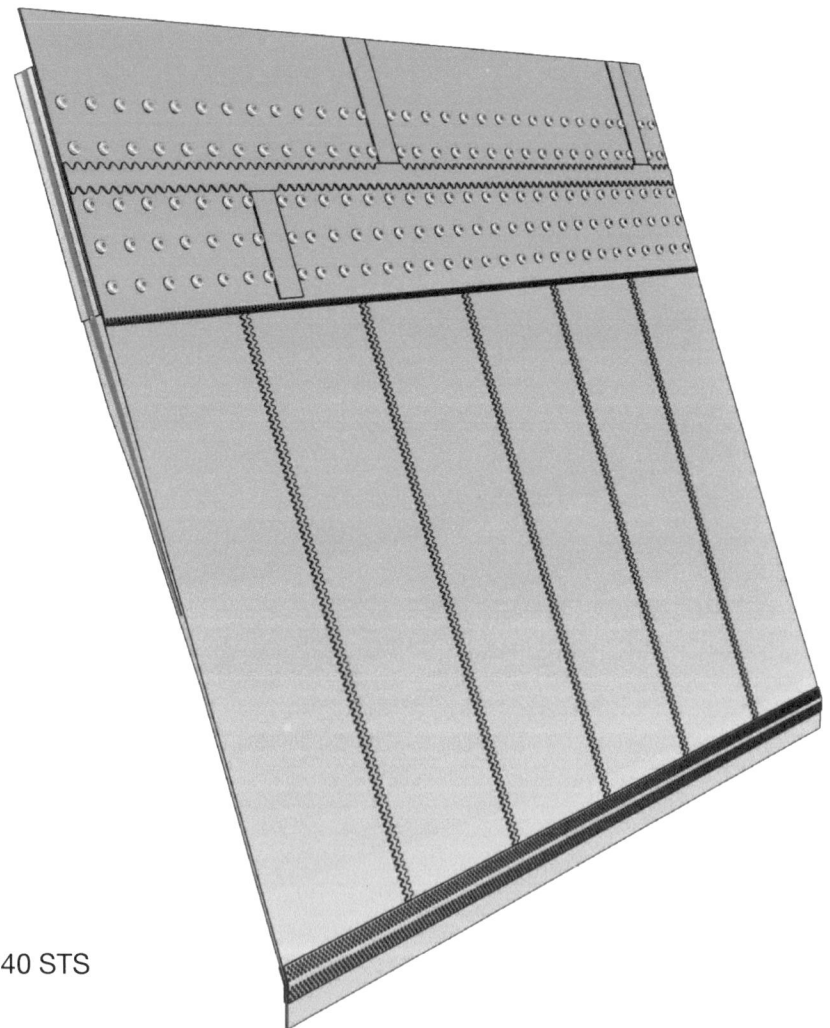

Plan 2.3: (1303BC) The inboard supports for the belt above third deck.

Plan 2.2: (1302BS) A typical inboard support for the belt from inner bottom to third deck. Figure 2.1 shows what the outboard supports look like.

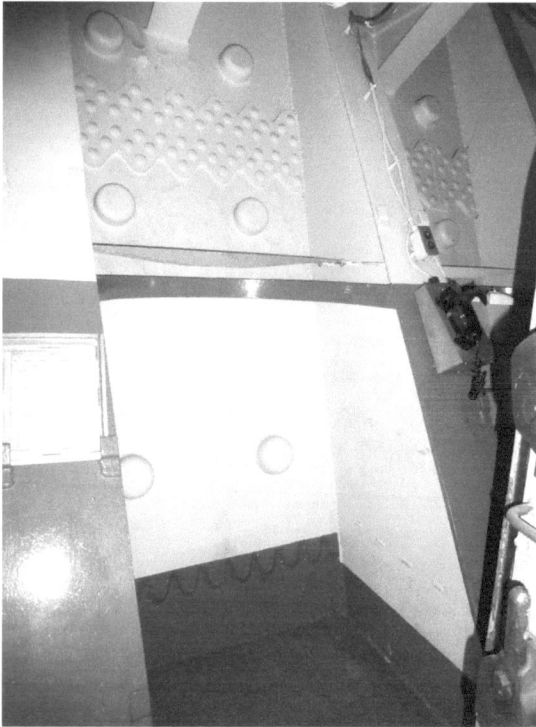

Photograph 2.4: The third deck boiler room shop at FR107 starboard on USS New Jersey looking outboard at torpedo bulkhead 3. Most of what is visible is where the torpedo bulkhead also serves as the backing bulkhead for the upper belt. The rows of rivets running horizontally attach a cover plate over the joint between two strakes of the torpedo bulkhead. The cover plate is scalloped to save weight. The circular caps were welded over the bolts connecting the upper belt to the bulkhead. The caps were intended to prevent the bolts from becoming projectiles if the armor were impacted. The triangular structures are the inboard supports for the bulkhead (see Plan 2.3). A small part of the lower belt is visible beneath scalloped lower edge of the backing bulkhead.

Photograph 2.5: Third deck shelf looking at FR103 starboard on USS New Jersey. The belt armor is to the right and torpedo bulkhead 4 is to the left. The divide between the upper and lower belts runs through the center of the photograph. A seamstrap linking lower belt plates is at the bottom right.

2.7.1. Lower Belt

There are 46 lower belt plates on each side of the citadel between FR50 and FR166. The plates are aligned in groups of two or three with an upper belt plate above. The lower belt plates on USS *Iowa* and USS *New Jersey* were milled at the shipyards from STS. Those for USS *Missouri* and USS *Wisconsin* were Class B supplied by the Bureau of Ordinance. The plates came from the mill as quadrilaterals with no features. The plates were ordered oversized by $\frac{3}{8}$" in width and $\frac{3}{8}$" in height added at the bottom. The outer faces of the plates tapered at knuckles from 12.1" at the top to a theoretical knuckle thickness of 1.62" at the bottom. The knuckles were located normal to theoretical horizontal lines measured along the inside (flat) face of the plates. The plates were grinded by hand to the desired thickness so the thicknesses and knuckles are not mathematically precise. The hull rises at the stern to accommodate the propellers and rudders so the aftmost

Offset From the theoretical Bottom Inside Edge	Plate Thickness
0"	1.62"
$147\frac{7}{8}$"	2.3"
$231\frac{7}{8}$"	6.4"
$285\frac{7}{8}$"	11.7"
$318\frac{7}{8}$"	12.1"

Table 2.2: Distance of Lower Belt Knuckles from the Theoretical Bottom of the Plates.

nine plates on each side do not extend all the way down to the theoretical 1.62" knuckle line.

2.7.1.1. Machining at the Shipyard

Like the deck plates, the lower belts were not ordered to fit together like a puzzle. The lower belt plates required extensive machining at the shipyard, which was directed by the mold loft. The excess material at sides and bottom had to be trimmed off at the yard. On USS *Iowa*, the sides were grinded so that the butts were normal to the centerline (as with the upper belt). On the other ships this was simplified by grinding the butts normal to their inside faces, except at facet knuckles where the butts were grinded normal to the centerline. The designed shape of the plates was such that, except for the run of plates parallel to the centerline between FR117$\frac{1}{2}$" and FR151, all the plates would have to be trimmed on both sides in order to fit. As delivered, the knuckles on outer faces of adjoining plates were aligned. After the plates were trimmed at the side, the knuckles became slightly misaligned. The yard also had to address both excess length and varying thicknesses at the bottoms of the lower belt plates. After the plates were trimmed to the correct length, the bottom edges varied in thickness. The bottom nine inches of the inside face were tapered to a uniform one-inch thickness so they could be attached to the lower strake of the torpedo bulkhead. It is likely that plates towards the aft end had to be grinded on both sides to fit the lower strake.

The constructions plans for the four completed ships instructed that the rest of the machining be done in groups of two or three plates that were aligned with single upper belt plate.

A tapered keyway was grinded into the sides to align each plate with its neighbor. The side keyways extend from the top of the plate to the point where the plate is four inches thick. After side keyways for a group were created, the group was assembled with small, temporary plates welded over the joints to hold them in place. Then the upper keyway was machined for the entire group. A welding groove was cut along the upper inside edge for eventual welding to the torpedo bulkhead. After edge machining was finished, two of these 2–3 plate groups were joined to form a 4–6 plate group where they were lifted into place. Bolt holes for lifting pads were drilled and tapped into the top edge as needed. When the pads were removed, plugs were screwed into the bolt holes and grinded flat.

For USS *Illinois* and USS *Kentucky*, the upper keyway was omitted. In its place, brackets were welded to the top of the lower belt that fit into the keyways in the upper belt.

2.7.1.2. Plate Installation

The plates at each side of the lower belt were welded together to form a solid unit. The first step in joining two plates was to insert a key into the keyways at the side to align the plates. Both sides of the butts were welded using 30# scalloped seamstraps. The outer seamstrap is STS and the inner is medium steel. Because the facets in the belt cause the inside faces to have different angles from vertical, the horizontal knuckles in the plates do not align across facets. At these locations, the outer surface of the plates had to be grinded so that the seamstrap could sit flat.

Plate groups were installed with connections to the rest of torpedo bulkhead 3 and plate groups already in place. A key was inserted into the vertical keyway between plate groups to align them. The bottom edge was riveted to the 40# STS lower strake of the torpedo bulkhead using 32.5# seamstraps on both sides. The plate group was welded to the upper backing plate of the torpedo bulkhead at the outside along the welding group in the plates at the inside along the scalloped lower edge of the backing plate. Adjoining plate groups were welded together using the same scalloped cover plates used to join plates with a group.

The forward edge of the lower belt was trimmed to match the tapered profile of the transverse armor at FR50. The connection to the transverse armor is described in that section. The connection to the belt armor aft of FR166 is described in that section as well. The part of the lower belt that extends below second platform was riveted to the bulkhead at FR166 using angles on both sides.

Third deck and third deck shelf were welded directly to the lower belt plate or the backing bulkhead. The location depended upon the deck shear. The decks were cut away from the seamstraps holding the plats together.

Frame	Lower		Upper	
	HT	HB	HT	HB
50	7' 6"	14' 3$\frac{5}{16}$"	35' 0$\frac{1}{8}$"	23' 8$\frac{15}{16}$"
57$\frac{1}{2}$	7' 6"	18' 11$\frac{9}{16}$"	34' 2$\frac{3}{4}$"	28' 2"
65	7' 6"	23' 7$\frac{13}{16}$"	34' 0$\frac{1}{2}$"	32' 9$\frac{1}{2}$"
72$\frac{1}{2}$	7' 6"	26' 7$\frac{5}{8}$"	33' 10$\frac{1}{4}$"	35' 8$\frac{9}{16}$"
87$\frac{1}{2}$	7' 6"	32' 7$\frac{1}{4}$"	33' 10$\frac{1}{4}$"	41' 8$\frac{3}{16}$"
117$\frac{1}{2}$	7' 6"	36' 7$\frac{1}{4}$"	33' 10$\frac{1}{4}$"	45' 8$\frac{3}{16}$"
140	7' 6"	36' 7$\frac{1}{4}$"	33' 10$\frac{1}{4}$"	45' 8$\frac{3}{16}$"
145$\frac{1}{2}$	7' 9$\frac{3}{4}$"	36' 8$\frac{9}{16}$"	33' 10$\frac{1}{4}$"	45' 8$\frac{3}{16}$"
151	8' 7"	36' 11$\frac{3}{4}$"	33' 10$\frac{1}{4}$"	45' 8$\frac{3}{16}$"
158$\frac{1}{2}$	10' 2$\frac{3}{4}$"	34' 9$\frac{9}{16}$"	33' 10$\frac{1}{4}$"	42' 11$\frac{3}{16}$"
166	12' 6"	32' 9$\frac{15}{16}$"	33' 10$\frac{1}{4}$"	40' 2$\frac{3}{16}$"

Table 2.3: Locations of knuckles in the molded lines inside face of the lower belt. The plates at FR50 extend forward to align with the face of the transverse plates. The positions here are reflect the position before the bottom edge of the plate was tapered.

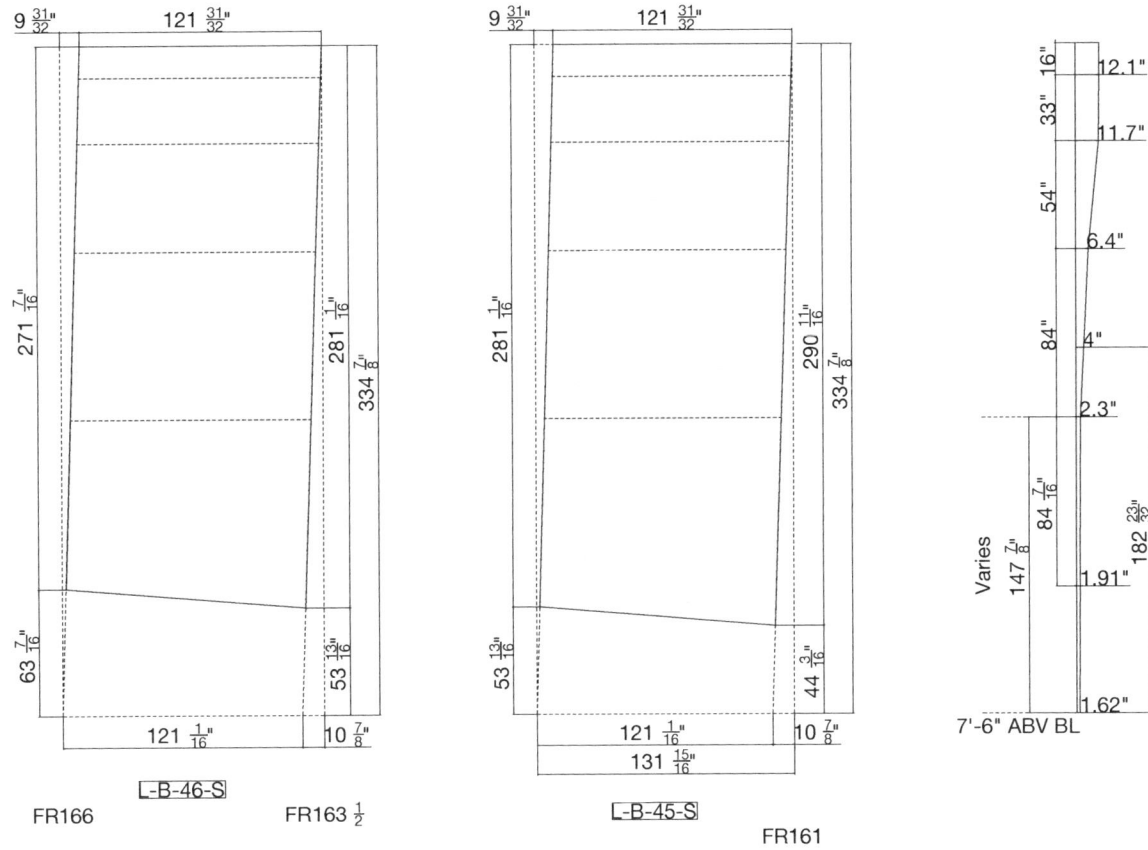

Figure 2.7: Mill Sizes for Lower Belt Plates FR161–FR166 starboard (1:96). Port side plates (-P) are mirrors. The yard had to trim the plates at both sides and the bottom as directed by the mold loft. The horizontal knuckles line up from plate to plate as milled. However, they became misaligned after the plates were trimmed.

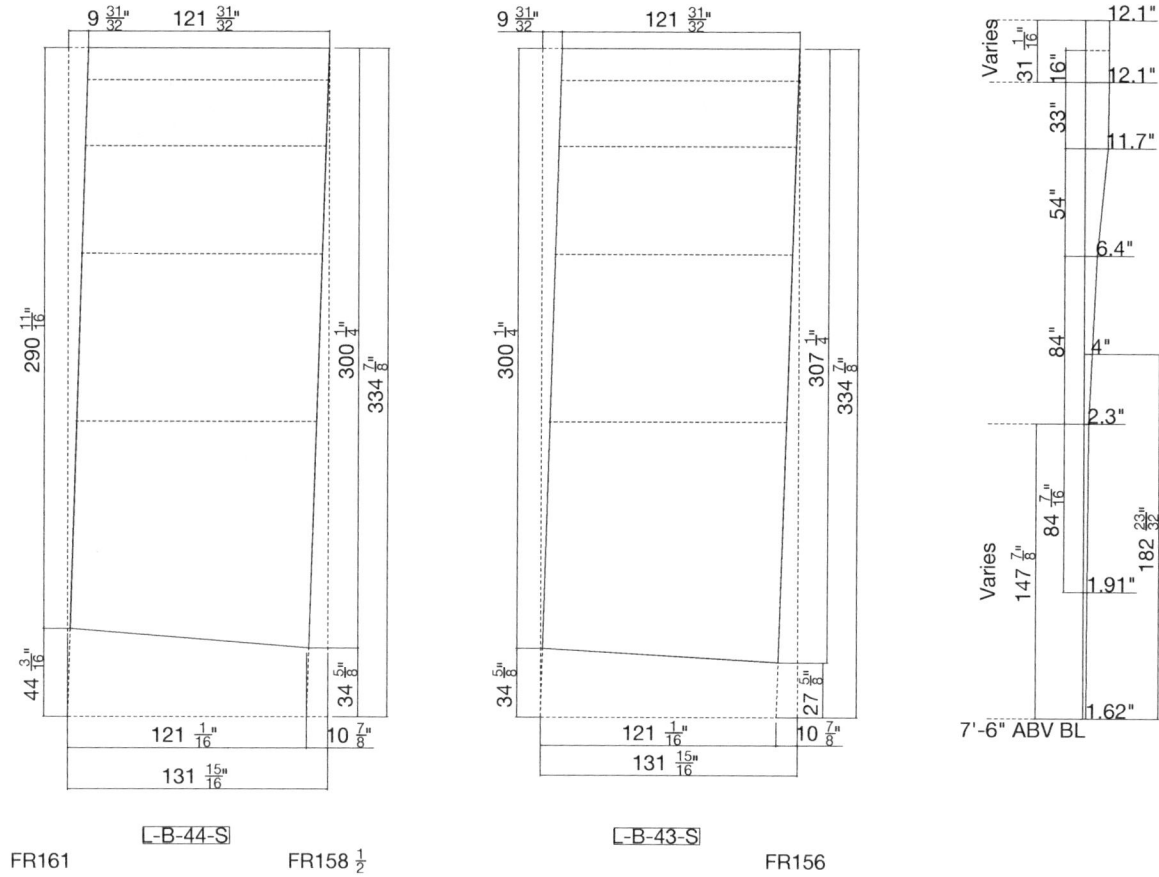

Figure 2.8: Mill Sizes for Lower Belt Plates FR156–FR161 starboard (1:96). Port side plates (-P) are mirrors.

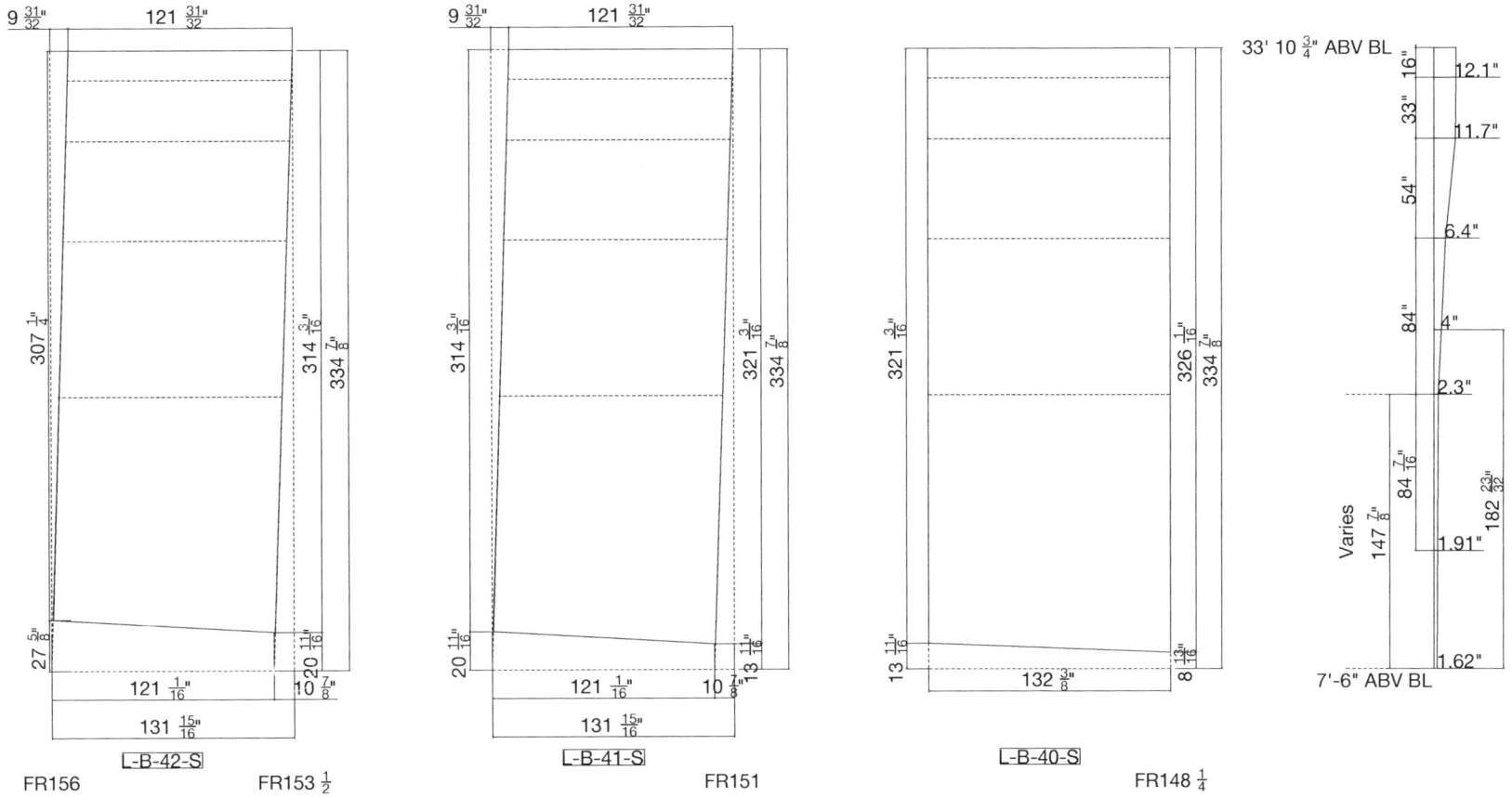

Figure 2.9: Mill Sizes for Lower Belt Plates FR156–FR148 $\frac{1}{4}$ starboard (1:96). Port side plates (-P) are mirrors.

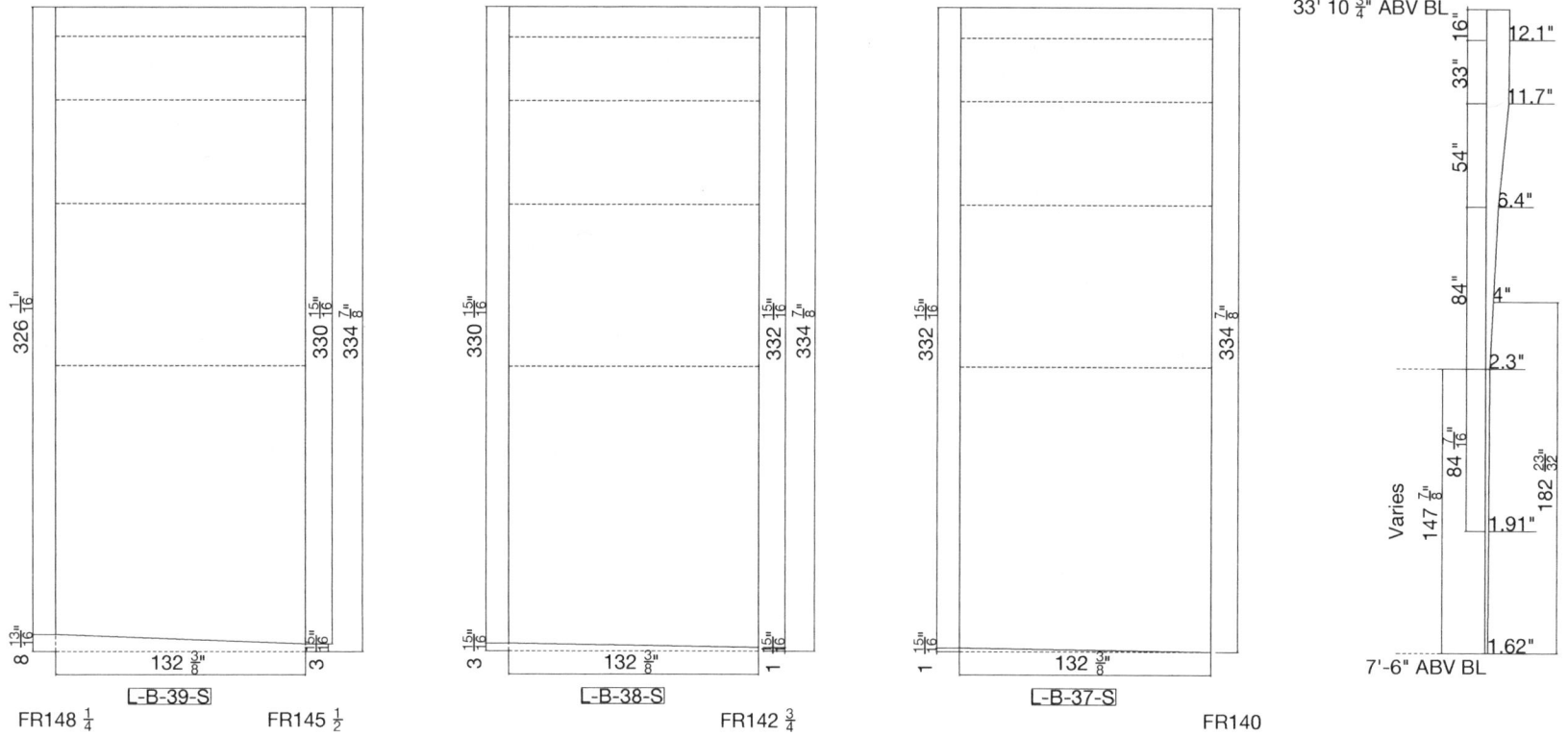

Figure 2.10: Mill Sizes for Lower Belt Plates FR148 $\frac{1}{4}$ –FR140 starboard (1:96). Port side plates (-P) are mirrors.

120 $\frac{13}{16}$"   3 $\frac{5}{8}$"

122 $\frac{5}{32}$"   10 $\frac{25}{32}$"

33' 10 $\frac{3}{4}$" ABV BL

16"   12.1"

33"   11.7"

54"   6.4"

84"   4"

2.3"

334 $\frac{7}{8}$"   334 $\frac{7}{8}$"   334 $\frac{7}{8}$"

147 $\frac{7}{8}$"   182 $\frac{23}{32}$"

1.62"

7'-6" ABV BL

120 $\frac{3}{8}$"

3 $\frac{15}{16}$"   120 $\frac{1}{2}$"

124 $\frac{7}{16}$"

11 $\frac{13}{16}$"   121 $\frac{1}{8}$"

132 $\frac{15}{16}$"

FR140   FR117 $\frac{1}{2}$

FR87 $\frac{1}{2}$

FR72 $\frac{1}{2}$

L-B-28-S
L-B-29-S
L-B-30-S
L-B-31-S
L-B-32-S
L-B-33-S
L-B-34-S
L-B-35-S
L-B-36-S

L-B-16-S
L-B-17-S
L-B-18-S
L-B-19-S
L-B-20-S
L-B-21-S
L-B-22-S
L-B-23-S
L-B-24-S
L-B-25-S
L-B-26-S
L-B-27-S

L-B-10-S
L-B-11-S
L-B-12-S
L-B-13-S
L-B-14-S
L-B-15-S

Plate L-B-10-S was made
3" short on USS New
Jersey. Plates L-B-11-S
and L-B-12-S were made
wide to compensate.

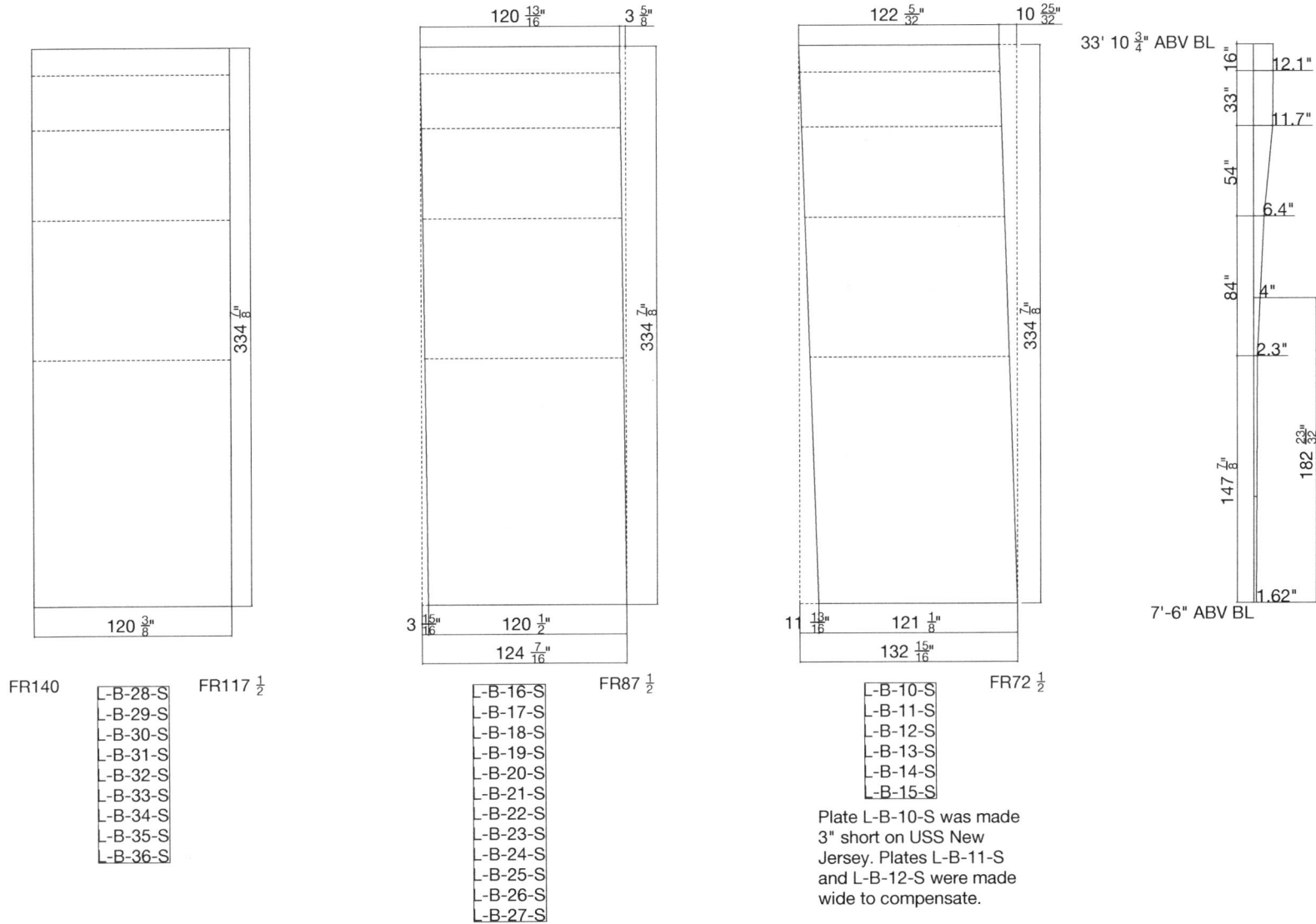

**Figure 2.11: Mill Sizes for Lower Belt Plates
FR87 $\frac{1}{2}$–FR117 $\frac{1}{2}$ starboard (1:96).** Port side
plates (-P) are mirrors.

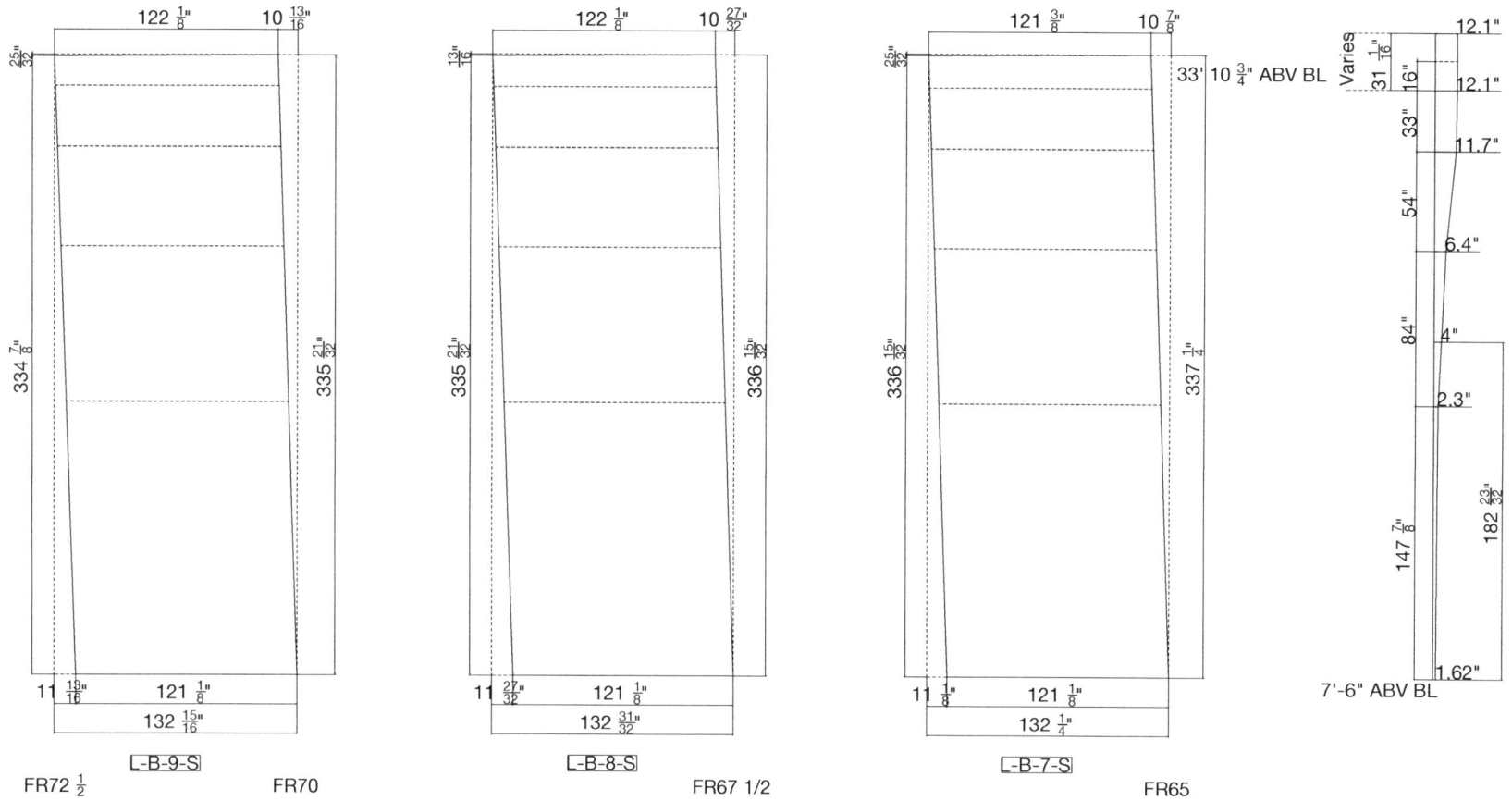

Figure 2.12: Mill Sizes for Lower Belt Plates FR65–FR72 $\frac{1}{2}$ starboard (1:96). Port side plates (-P) are mirrors.

$123 \frac{21}{32}"$ $17"$

$\frac{25}{32}"$

$337 \frac{1}{4}"$ $338 \frac{1}{32}"$

$18 \frac{9}{16}"$ $122 \frac{3}{32}"$

$140 \frac{21}{32}"$

L-B-6-S

FR65   FR62 $\frac{1}{2}$

$123 \frac{23}{32}"$ $17 \frac{1}{32}"$

$\frac{13}{16}"$

$338 \frac{1}{32}"$ $338 \frac{27}{32}"$

$18 \frac{21}{32}"$ $122 \frac{3}{32}"$

$140 \frac{3}{4}"$

L-B-5-S

FR60

$123 \frac{23}{32}"$ $17 \frac{1}{16}"$

$33' 10 \frac{3}{4}"$ ABV BL

$\frac{25}{32}"$

$338 \frac{27}{32}"$ $339 \frac{5}{8}"$

$18 \frac{11}{16}"$ $122 \frac{3}{32}"$

$140 \frac{25}{32}"$

L-B-4-S

FR57 $\frac{1}{2}$

Varies $31 \frac{1}{16}"$ $16"$ $33"$ $54"$ $84"$ $147 \frac{7}{8}"$ $182 \frac{23}{32}"$

12.1"
12.1"
11.7"
6.4"
4"
2.3"
1.62"

7'-6" ABV BL

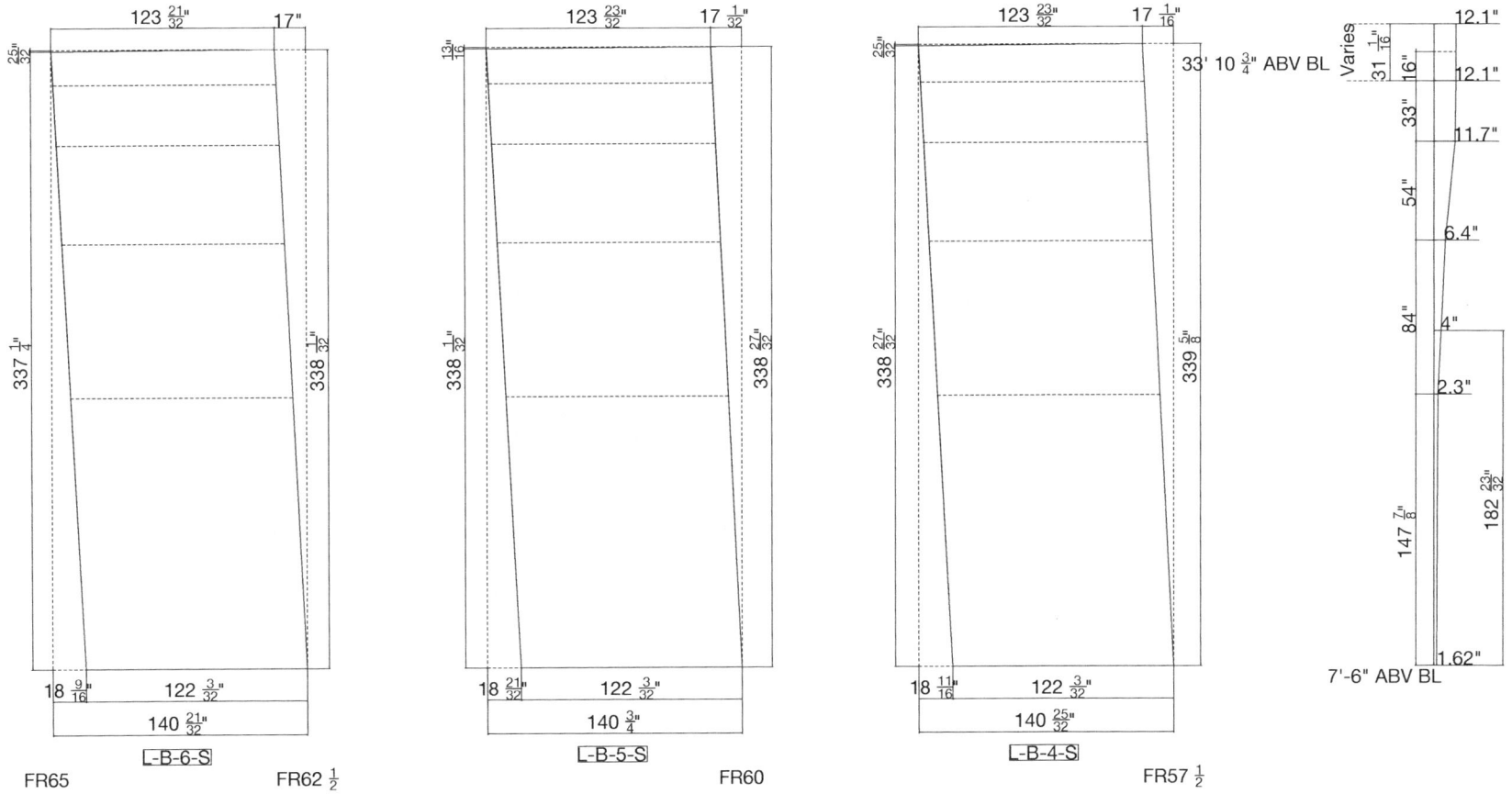

**Figure 2.13: Mill Sizes for Lower Belt Plates
FR57$\frac{1}{2}$–FR65 starboard (1:96).** Port side
plates (-P) are mirrors.

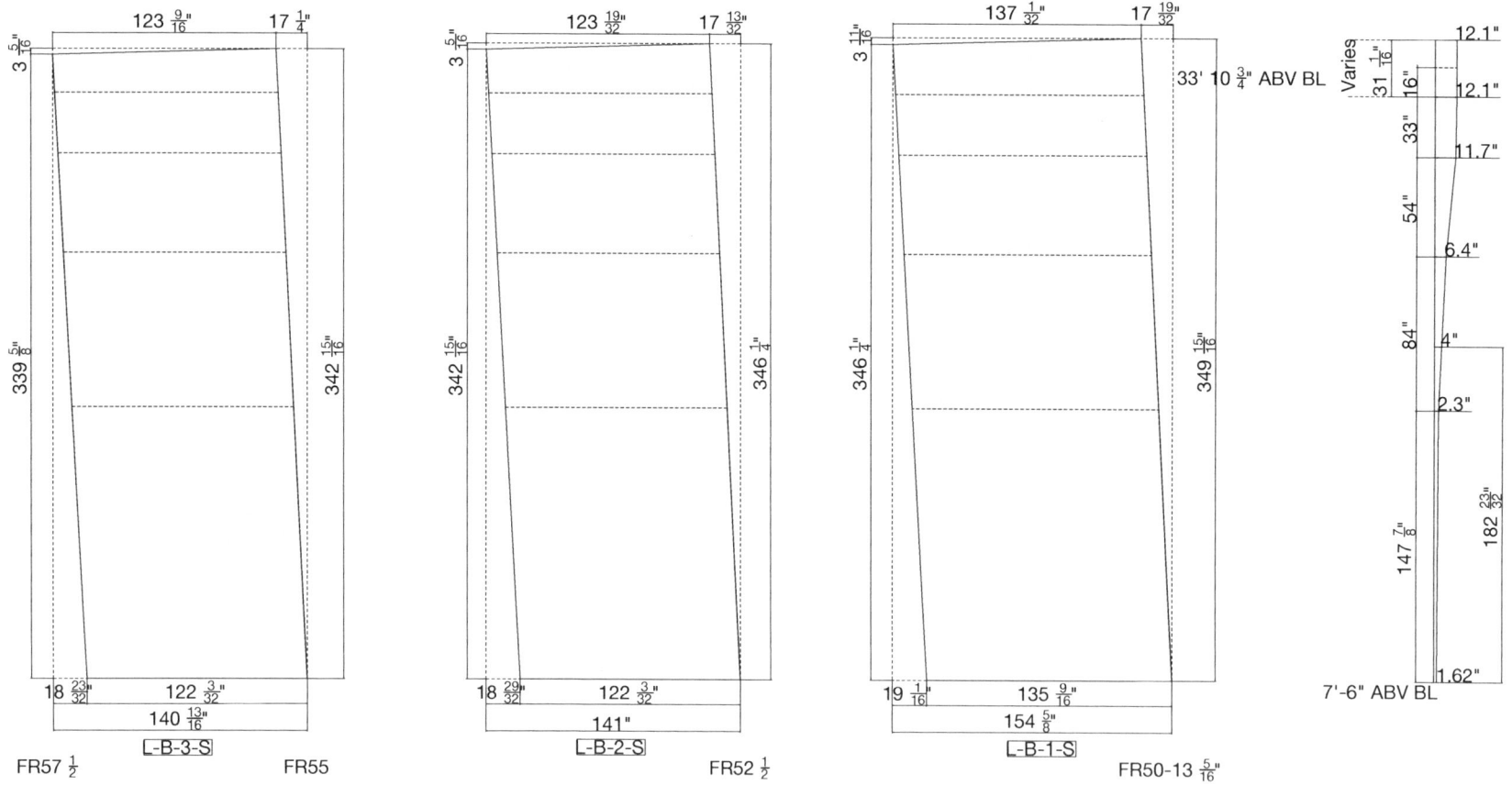

Figure 2.14: Mill Sizes for Lower Belt Plates FR50–FR57 $\frac{1}{2}$ starboard (1:96). Port side plates (-P) are mirrors.

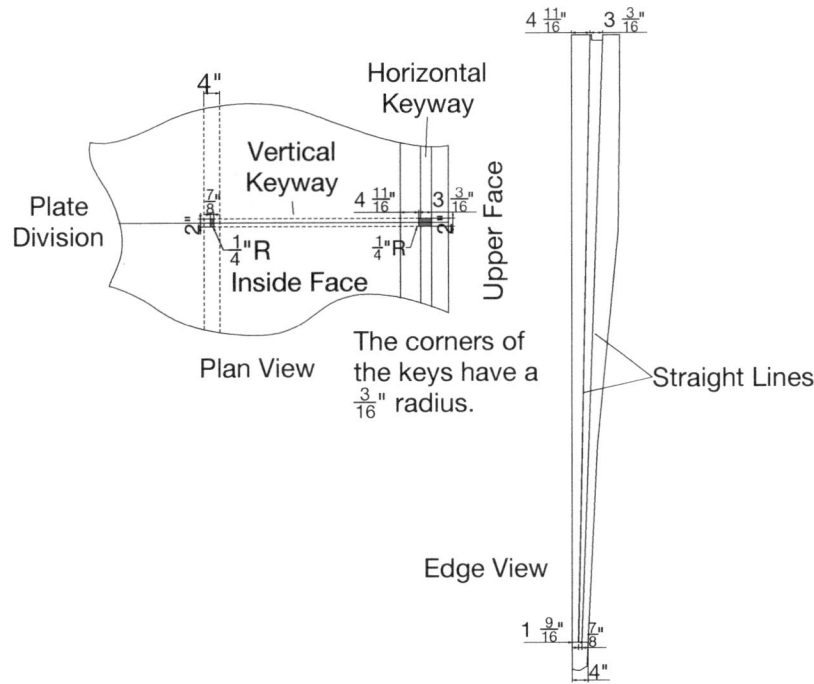

Figure 2.15: Side Keyway (1:48). The yard machined tapered keyways into the side of lower belt plates to maintain alignment. While the keys were tapered, they were not hammered into the keyway from the top. The taper in the key was necessary because of the taper in plate thickness. The keys extend to the point where the plates are 4" thick.

Frame	Length
$52\frac{1}{2}$	14'-2"
55	13'-11"
$57\frac{1}{2}$	13'-7"
60	13'-6"
$62\frac{1}{2}$	13'-6"
65	13'-5"
$67\frac{1}{2}$	13'-4"
70–$163\frac{1}{2}$	13'-2"
166	12'-11"
168	18'-11"
170	18'-9"
172	18'-7"
174	18'-5"
$176\frac{1}{2}$	18'-3"
179	18'-1"
$181\frac{1}{2}$	17'-11"
184	17'-9"
$186\frac{1}{2}$	17'-7"
189	9'-8"
196	9'-5"

Table 2.4: Lengths of Vertical Keys in Lower Belt Plates. These lengths suggest that the horizontal keys were intercostal to the vertical keys

DET. OF UPPER TEMPORARY STRAP FOR BUTTS
SCALE: 3" = 1'-0"

DET. OF LOWER TEMPORARY STRAP FOR BUTTS
SCALE: 3" = 1'-0"

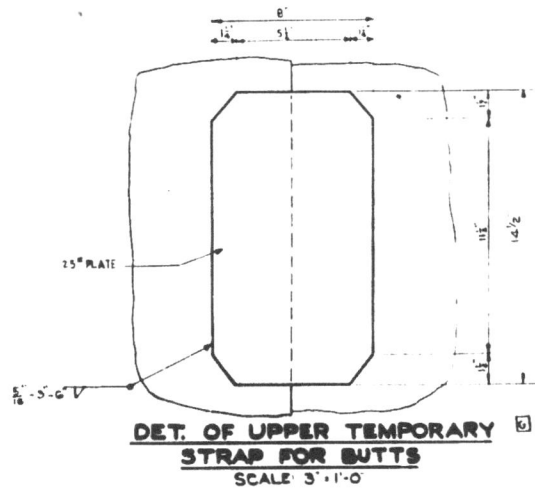

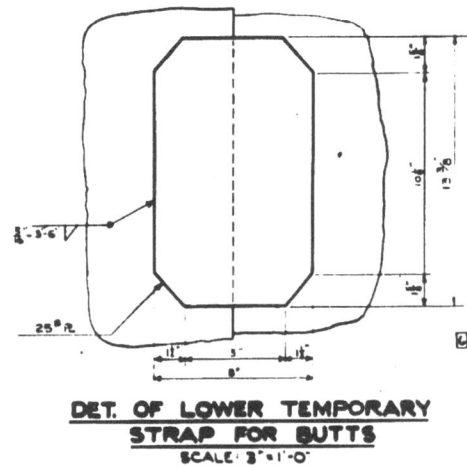

Plan 2.6: (1302CI) Shows how groups of two or three lower belt plates were temporarily welded together before the yard machined the upper edge.

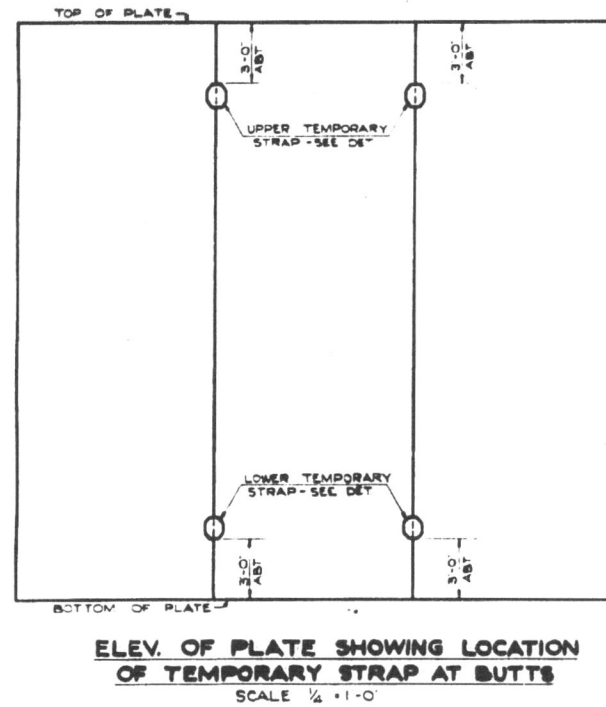

ELEV. OF PLATE SHOWING LOCATION OF TEMPORARY STRAP AT BUTTS
SCALE: 1/4" = 1'-0"

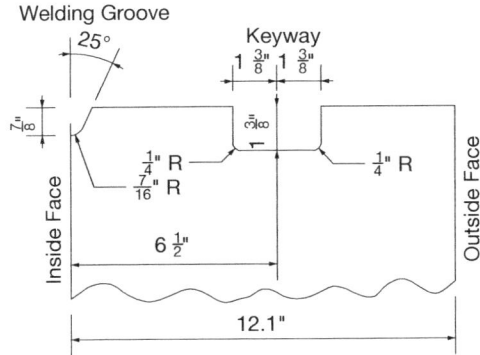

Figure 2.16: Upper Edge of Lower Belt (1:6). The yard machined the top edge of the lower belt plates in groups of two or three plates temporarily welded together.

Figure 2.17: Connection of Lower Belt to Lower Strake of Torpedo Bulkhead 3 (1:6). The thicknesses of the bottom edges varied with the aft plates being thicker than those forward. The lower nine inches of all the plates were tapered to a uniform one inch thickness. This drawing shows how the plates were attached to the lower strake of the 40# torpedo bulkhead 3 that extends to the bottom of the ship.

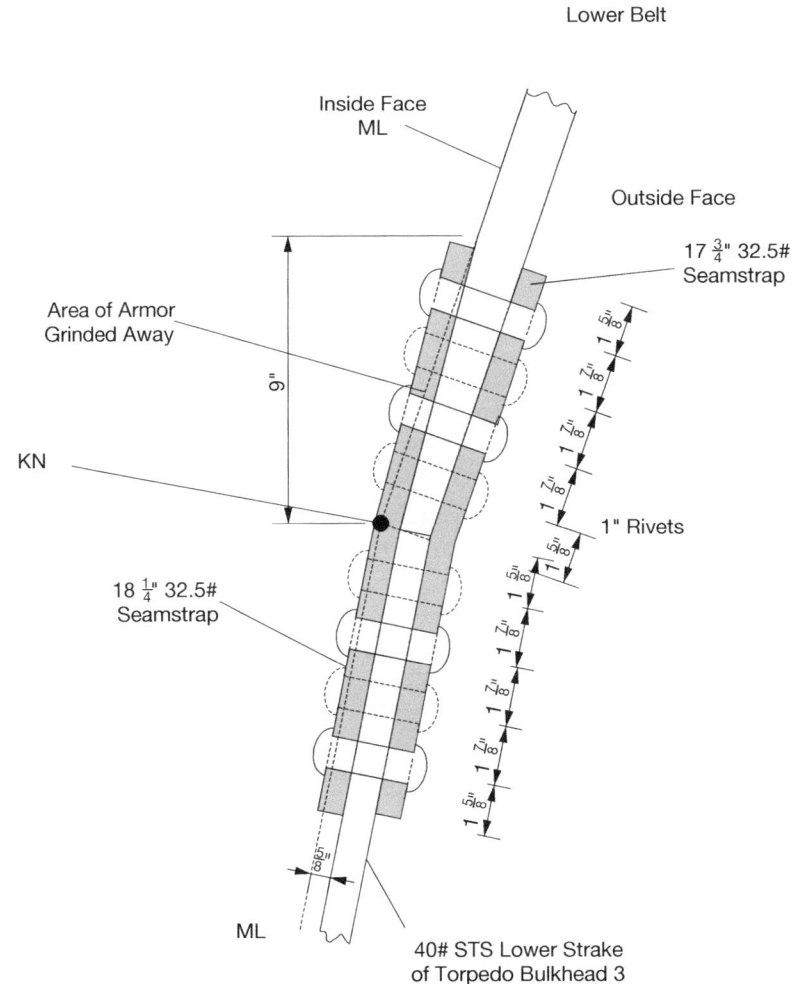

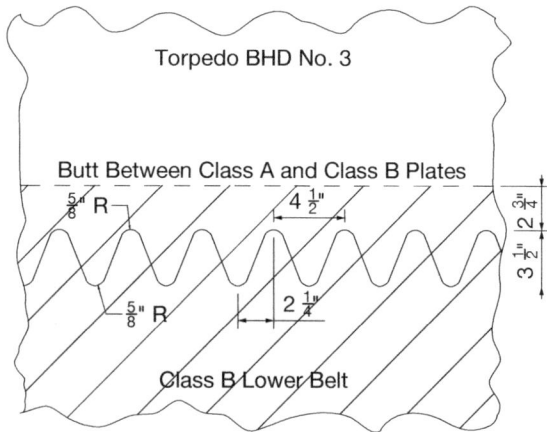

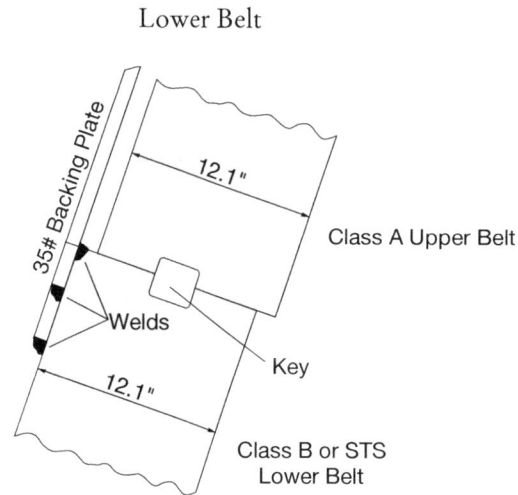

Figure 2.18: Attachment of Lower Belt to Backing Bulkhead (1:12). The upper edge of the lower belt was welded to the backing plate of the torpedo bulkhead along the welding groove on the outside and along the scalloped lower edge of the bulkhead on the inside.

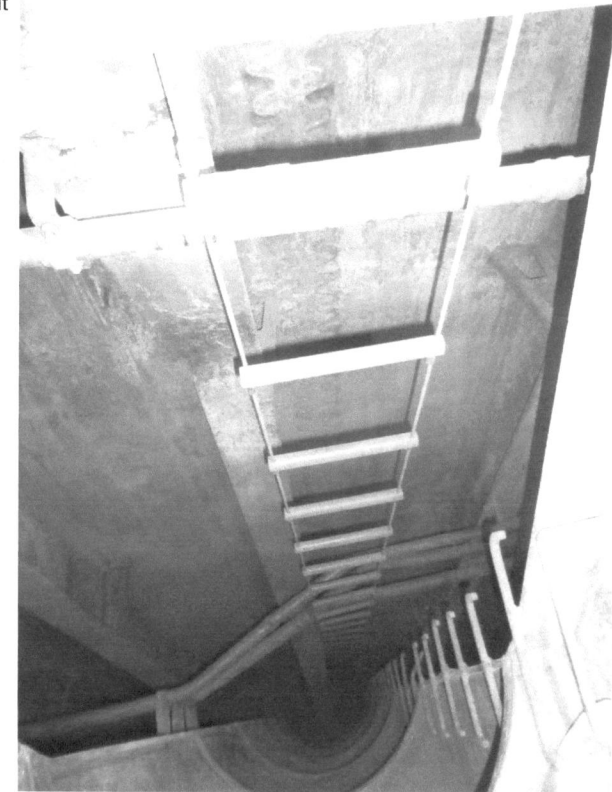

Figure 2.19: Seamstrap (1:6). Adjoining lower belt plates were welded together using scalloped seamstraps on both sides of the joint. This is the scallop pattern used for seamstraps. Structural members, such as framing and decks, were constructed around the seamstraps.

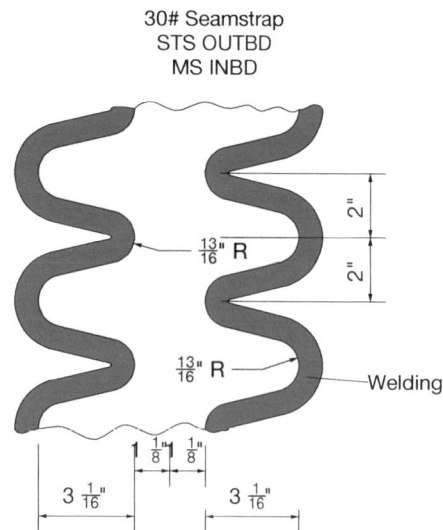

Photograph 2.7: Lower belt of USS New Jersey FR149 port side. A scalloped strap is visible behind the wire rope ladder. Bulkhead 4 is at the bottom of the photograph.

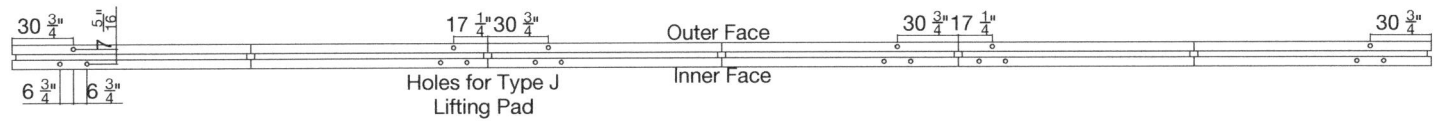

30 $\frac{3}{4}$" $\frac{5}{16}$" 7" 17 $\frac{1}{4}$"30 $\frac{3}{4}$" Outer Face 30 $\frac{3}{4}$"17 $\frac{1}{4}$" 30 $\frac{3}{4}$"

6 $\frac{3}{4}$" 6 $\frac{3}{4}$" Holes for Type J Lifting Pad Inner Face

Figure 2.20: Plan view of the Connection of the Lower Belt to FR166 Bulkhead Below Second Platform (1:6). There is a transverse watertight bulkhead at FR166 that is intercostal to the belt armor. This drawing shows how the belt forward of FR166 connects to the transverse bulkhead below second platform.

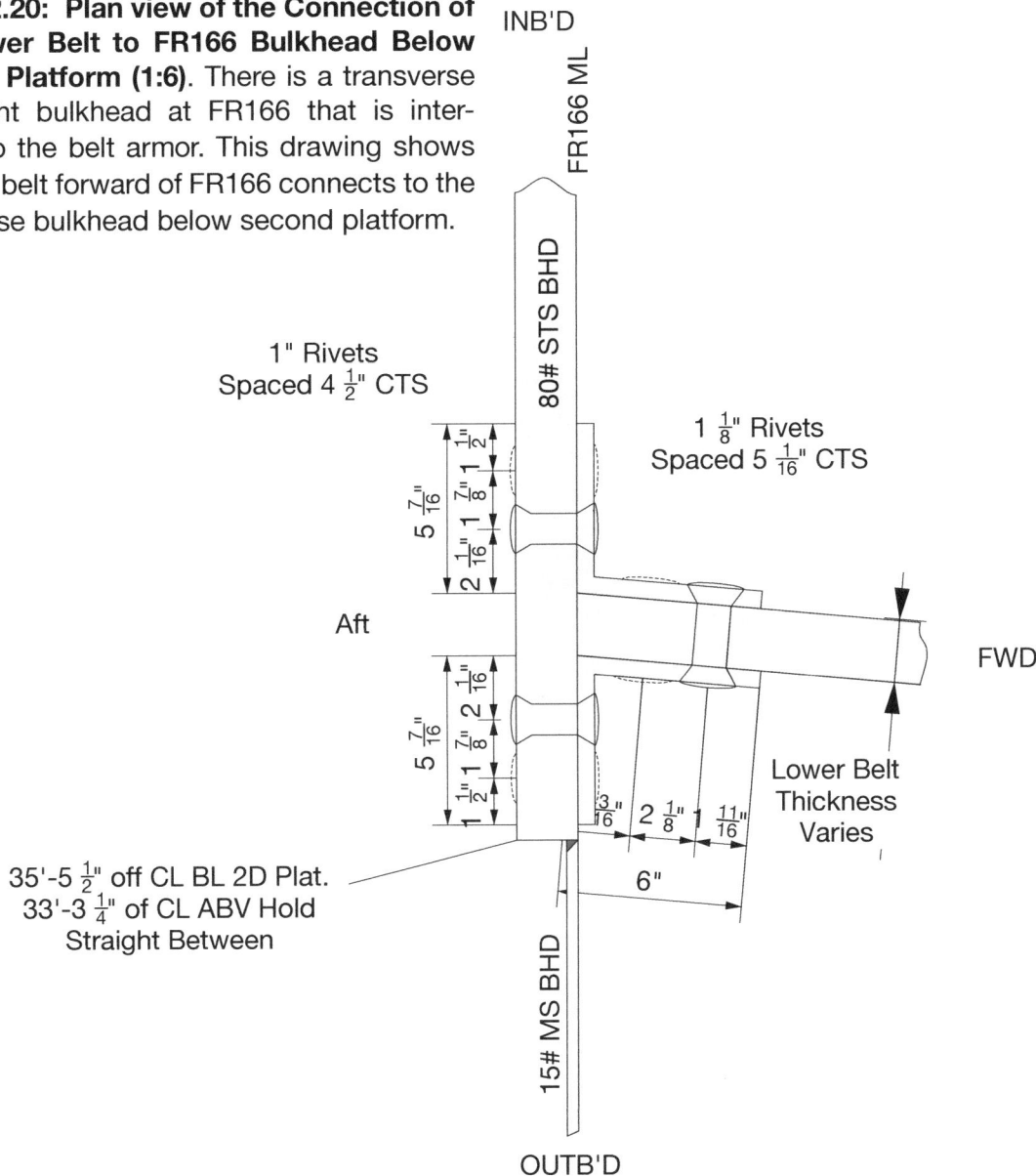

INB'D

FR166 ML

80# STS BHD

1" Rivets
Spaced 4 $\frac{1}{2}$" CTS

1 $\frac{1}{8}$" Rivets
Spaced 5 $\frac{1}{16}$" CTS

5 $\frac{7}{16}$" 1 $\frac{1}{2}$" 1 $\frac{7}{8}$" 1 $\frac{1}{16}$" 2 $\frac{1}{16}$"

Aft

FWD

5 $\frac{7}{16}$" 2 $\frac{1}{16}$" 1 $\frac{1}{16}$" 1 $\frac{7}{8}$" 1 $\frac{1}{2}$"

Lower Belt
Thickness
Varies

$\frac{3}{16}$" 2 $\frac{1}{8}$" 1" 1 $\frac{11}{16}$"

35'-5 $\frac{1}{2}$" off CL BL 2D Plat.
33'-3 $\frac{1}{4}$" of CL ABV Hold
Straight Between

6"

15# MS BHD

OUTB'D

Figure 2.21: Lifting Bolt Positions (1:96). This drawing shows how lifting pads were connected to lower belt plates. The lower belt plates are in groups of two or three aligned underneath an upper belt plate (see Figure 2.22). The lower belt plates were craned two groups at a time (4, 5, or 6 plates). The lifting pads were positioned relative to plate groups rather than individual plates. Three holes were drilled and tapped for each lifting pad. These holes were plugged after the plate group was in position.

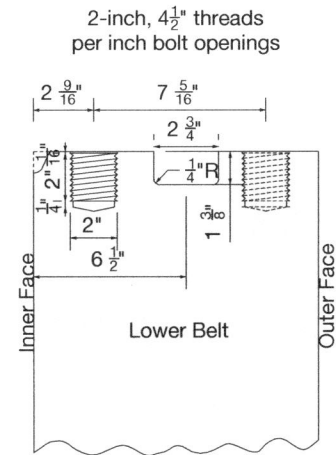

2-inch, 4 $\frac{1}{2}$" threads
per inch bolt openings

2 $\frac{9}{16}$" 7 $\frac{5}{16}$"

2 $\frac{3}{4}$"

$\frac{1}{4}$"R

2 $\frac{1}{16}$" $\frac{1}{4}$" 2"

2" 1 $\frac{3}{8}$"

6 $\frac{1}{2}$"

Inner Face Outer Face

Lower Belt

Figure 2.22: Lifting Bolt Holes (1:8). The holes for lifting pad were drilled on either side of the upper keyway. This drawing shows the position of the holes relative to the inside face of the plates.

SECT. AT FR. 78

SECT. AT FR. 57½

SECT. AT FR. 51

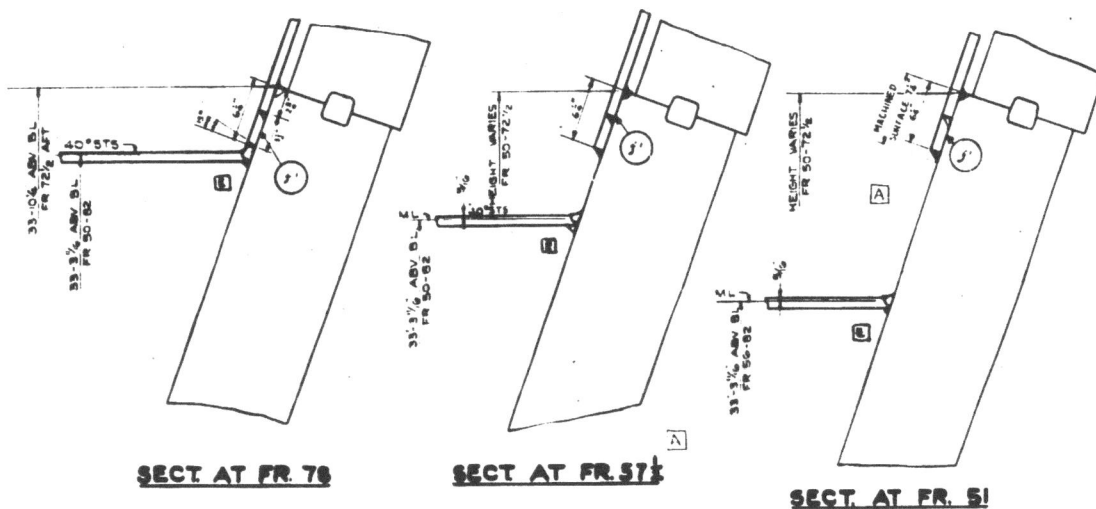

Plan 2.8: (1302CI). The elevation of third deck relative to the belt changes over the length of the ship due to deck sheer. These plan views show how third deck connected to the belt at various heights.

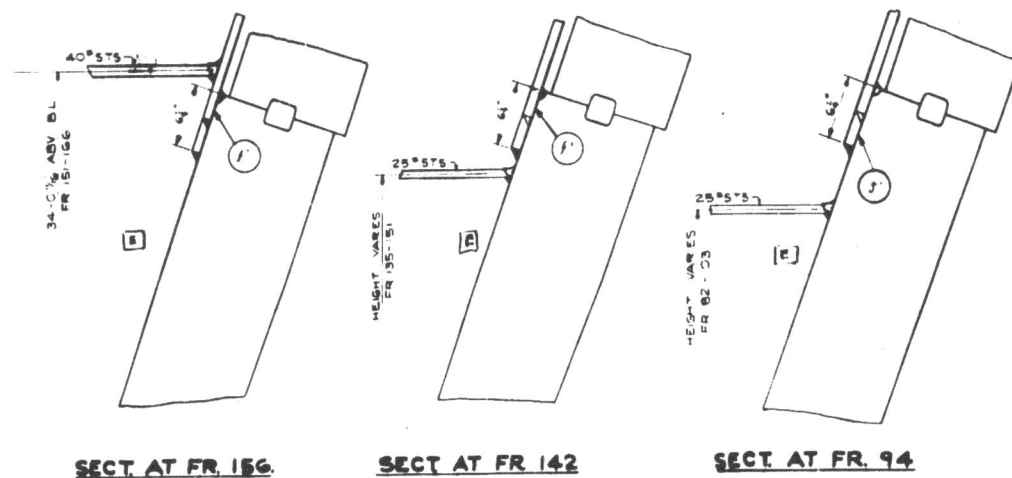

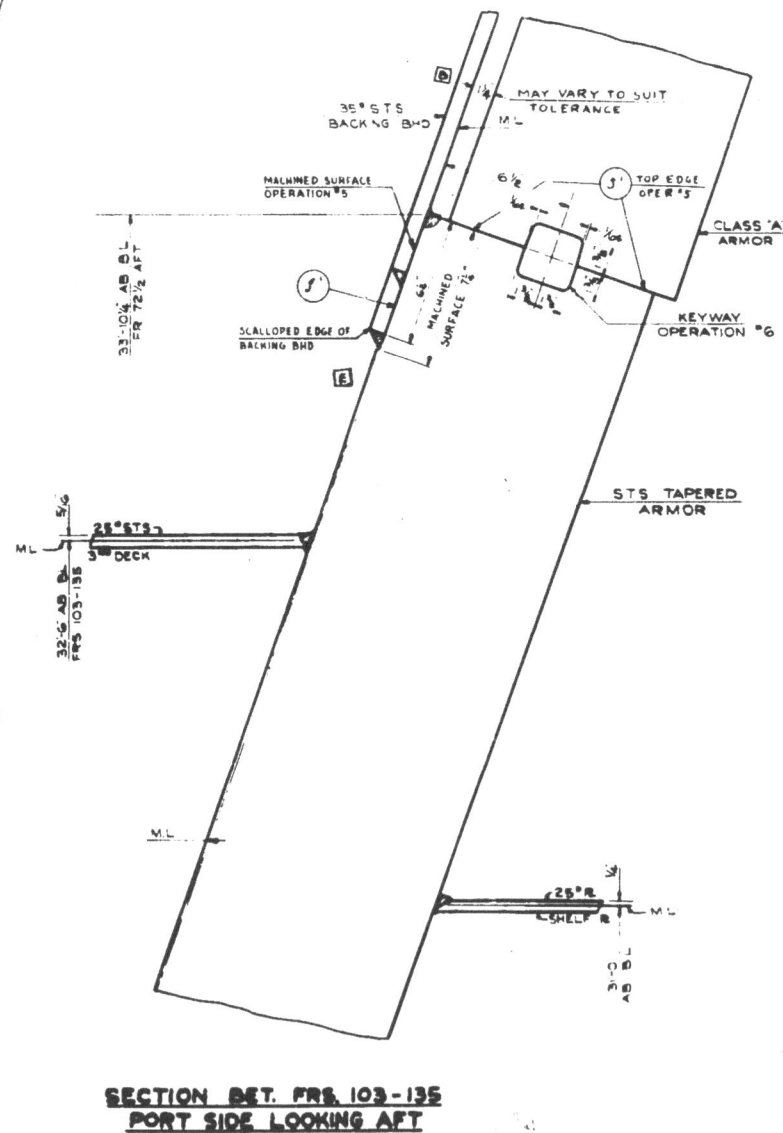

SECT. AT FR. 156.

SECT. AT FR 142

SECT. AT FR. 94

SECTION BET. FRS. 103-135
PORT SIDE LOOKING AFT

2.8. Frame 166 Transverse Armor

The top of the citadel drops from second deck to third deck at FR166. A transverse bulkhead extending between these decks closes the gap. This bulkhead is connected directly to the upper belt. Therefore, this chapter treats this transverse armor as part of the upper belt.

The transverse armor consists of three 11.3-inch Class A plates. These plates were aligned using a keyway and they were welded together along the inside face. The outer plates (B-166-2 and B-166-3) have a tongue. The aftmost upper belt plates (U-B-16-P, U-B-16-S) were ordered oversized in length. The yard had to machine a groove into the belt plates and the excess plating was ground flush with the aft face of the transverse plates. A cover plate was tap riveted on the outer face over the joint between the transverse and belt plates. On the inner face the transverse plates were welded to the belt armor backing plate along angled scallops placed in the corners.

The lower edge of the plates have a sloped tongue that sits in a groove in the third deck armor. The slope of the groove is oriented to cause the weight to push the plate forward. The transverse plate is secured from below by bolts. An angle with a scallop along its horizontal flange was welded to third deck against the aft face of the transverse plates.

The upper edges of the transverse plates were drilled and tapped for bolts that were driven though holes in the second deck armor. A scalloped kickplate and scalloped angle were welded over the inside joint between the underside of second deck and the transverse plates. An angle with the horizontal flange scalloped was welded

to the outside joint. The unscalloped flange was welded to a soft area of the armor face.

Plate	Weight (lbs)
B-166-1	122,766
B-166-2	108,093
B-166-3	108,093
Total	338,952

Table 2.5: Weights of FR166 Transverse Plates. Plates were supplied by Carnegie-Illinois Steel except for USS Wisconsin and USS Illinois which were supplied by The Midvale Company.

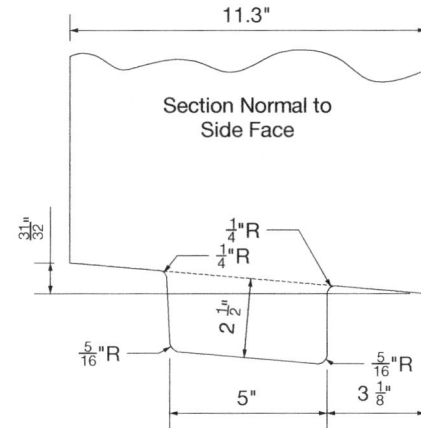

Figure 2.23: The transverse plates link to the upper belt plates with a tongue and groove joint. The tongues are on the outer transverse plates.

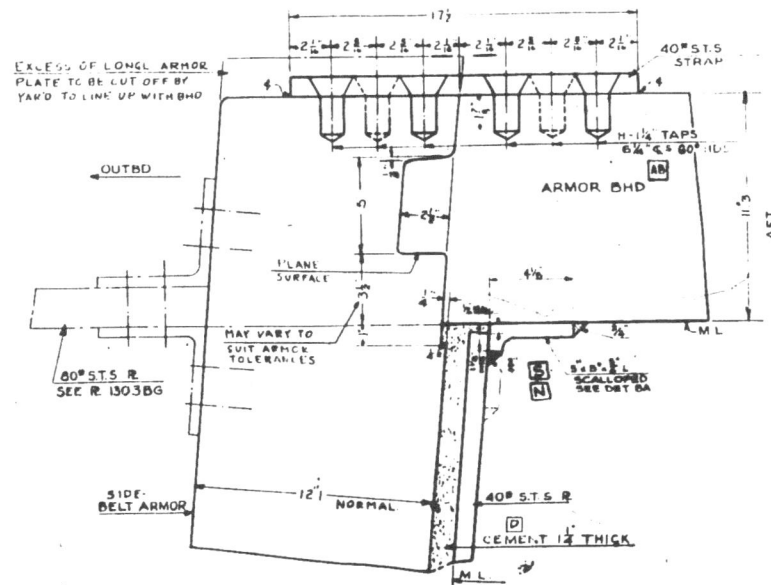

Plan 2.9: (1303AT). The connection of the transverse plates at FR166 to the upper belt and backing plate.

Top Plan

6 ½" | 6 ½"   6 ½" | 6 ½"

7 ⅞" 11 ¾" 11 ¾" 11 ¾" 11 ¾" 11 ¾" 12 ½" 12 ½" 12 ¼" 12 ¼" 12 ¼" 12 ¼" 12 ¼" 12 ¼" 12 ¼" 12 ¼" 12 ¼" 12 ¼" 12 ½" 11 ½" 11 ½" 11 ½" 11 ½" 11 ½" 9 ⅜"

Aft Face Smooth
Forward Face Smooth

2"

11.3"

6"

104 ⅜"   104 ²³⁄₃₂"

B-166-1
11.3" Class A

Aft Face Smooth

2"

Forward Face Elevation

Aft FWD

FWD Aft

7 ⅝" 15 ¼" 7 ⅝"

366"

Bottom Plan

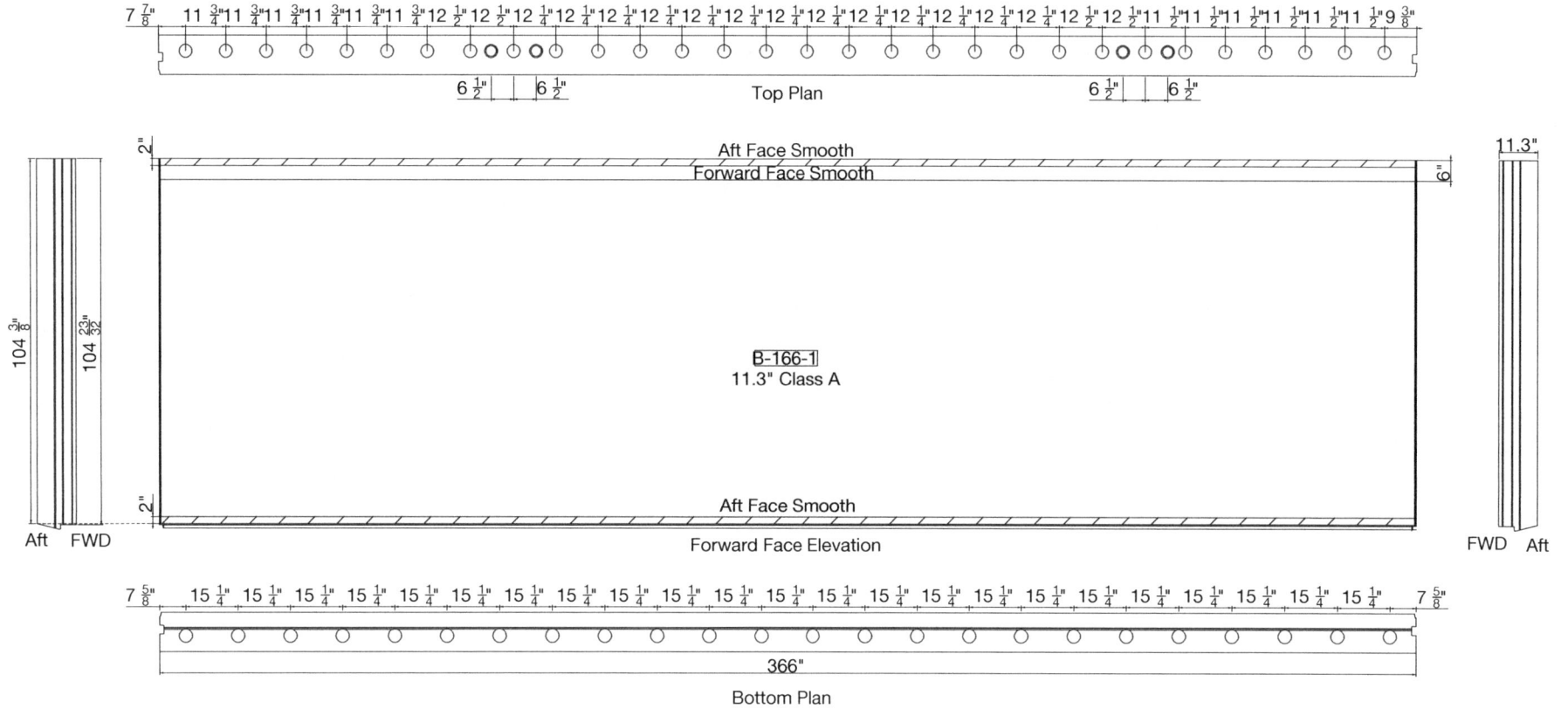

Figure 2.24: Plate B-166-1 (1:48)

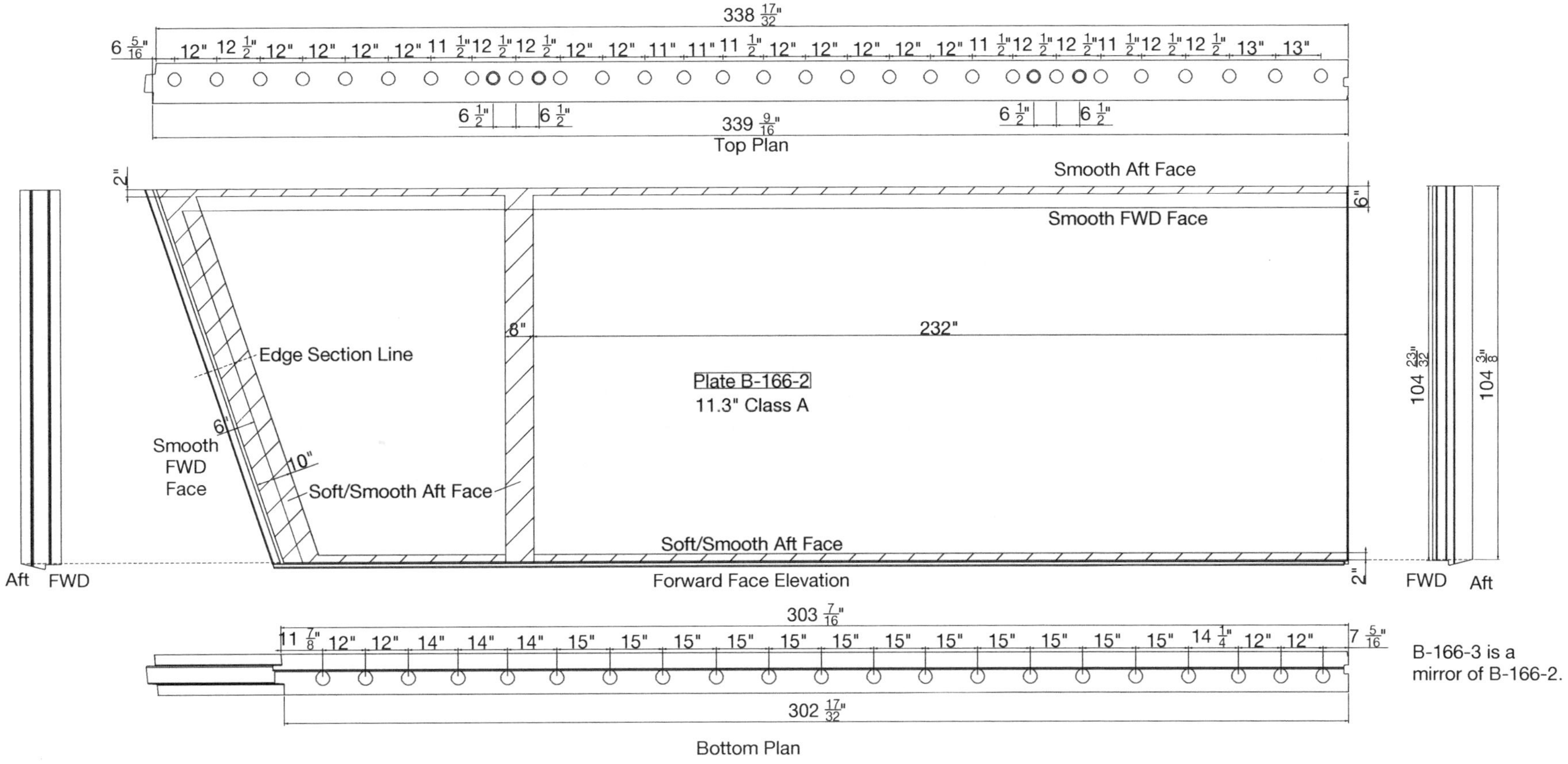

Top Plan

338 $\frac{17}{32}$"

6 $\frac{5}{16}$" | 12" | 12 $\frac{1}{2}$" | 12" | 12" | 12" | 12" | 11 $\frac{1}{2}$" | 12 $\frac{1}{2}$" | 12 $\frac{1}{2}$" | 12" | 12" | 11 $\frac{1}{2}$" | 11" | 11 $\frac{1}{2}$" | 12" | 12" | 12" | 12" | 12" | 11 $\frac{1}{2}$" | 12 $\frac{1}{2}$" | 12 $\frac{1}{2}$" | 11 $\frac{1}{2}$" | 12 $\frac{1}{2}$" | 12 $\frac{1}{2}$" | 13" | 13"

6 $\frac{1}{2}$" | 6 $\frac{1}{2}$"

339 $\frac{9}{16}$"

6 $\frac{1}{2}$" | 6 $\frac{1}{2}$"

Forward Face Elevation

Smooth Aft Face

Smooth FWD Face

2"

6"

8" — 232"

Plate B-166-2
11.3" Class A

Edge Section Line

Smooth FWD Face

6"

10"

Soft/Smooth Aft Face

Soft/Smooth Aft Face

104 $\frac{23}{32}$"

104 $\frac{3}{8}$"

Aft FWD

2"

FWD Aft

Bottom Plan

303 $\frac{7}{16}$"

11 $\frac{7}{8}$" | 12" | 12" | 14" | 14" | 14" | 15" | 15" | 15" | 15" | 15" | 15" | 15" | 15" | 15" | 15" | 15" | 15" | 14 $\frac{1}{4}$" | 12" | 12" | 7 $\frac{5}{16}$"

302 $\frac{17}{32}$"

B-166-3 is a mirror of B-166-2.

Figure 2.25: Plate B-166-2 (1:48). The port plate (B-166-3) is a mirror. Note that the tongue at the side is outside the measured molded widths.

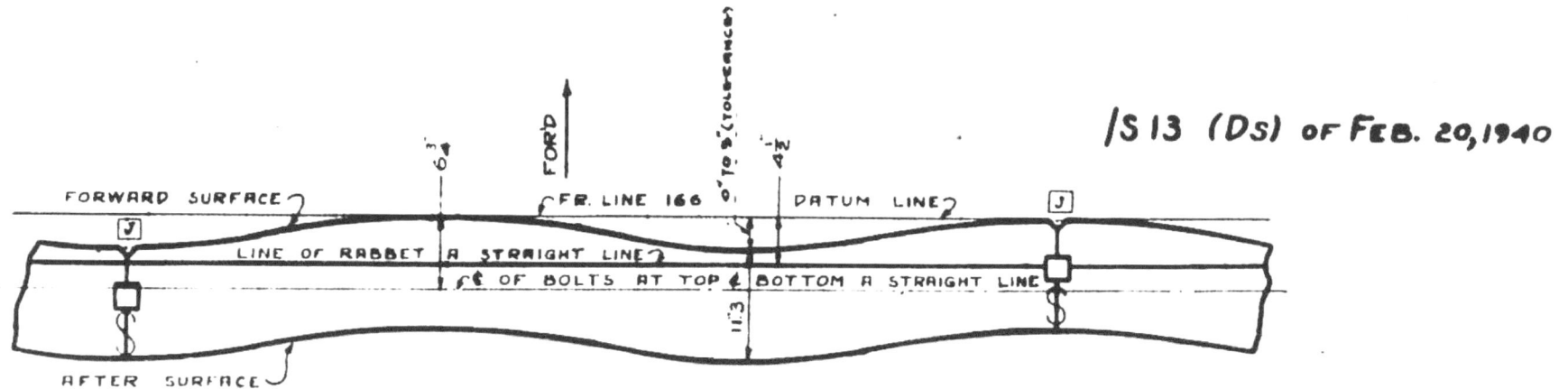

FORWARD SURFACE

FR. LINE 166

DATUM LINE

FORD

/S 13 (Ds) OF FEB. 20, 1940

LINE OF RABBET A STRAIGHT LINE

₡ OF BOLTS AT TOP & BOTTOM A STRAIGHT LINE

AFTER SURFACE

DIAGRAM SHOWING RABBET AND ₡ OF BOLTS AT TOP AND BOTTOM IN RELATION TO DATUM SURFACE AND ARMOR TOLERANCES.

Plan 2.10: (1303AH). The plans allowed the transverse face to vary up to 3 inches from the designed location.

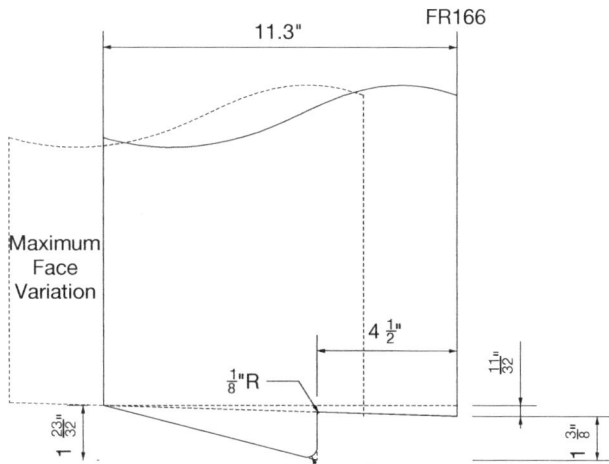

FR166

11.3"

Maximum Face Variation

$4\frac{1}{2}$"

$\frac{1}{8}$"R

$\frac{5}{16}$"R

The plans permitted the fore and aft faces to vary up to 3". The rabbet location remained fixed.

Figure 2.26: Lower Edge (1:6). The lower edges of the transverse plates were to be machined straight, ignoring any waviness in the plates.

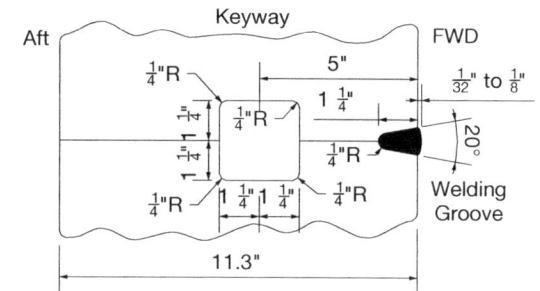

Aft   Keyway   FWD

$\frac{1}{4}$"R   5" $\frac{1}{32}$" to $\frac{1}{8}$"

$1\frac{1}{4}$

$\frac{1}{4}$"R

20°

Welding Groove

11.3"

Figure 2.27: Joint Between Plates (1:6). The transverse plates were aligned using a key within a keyway machined into the edges. The plates were welded together along a groove machined into the inside face.

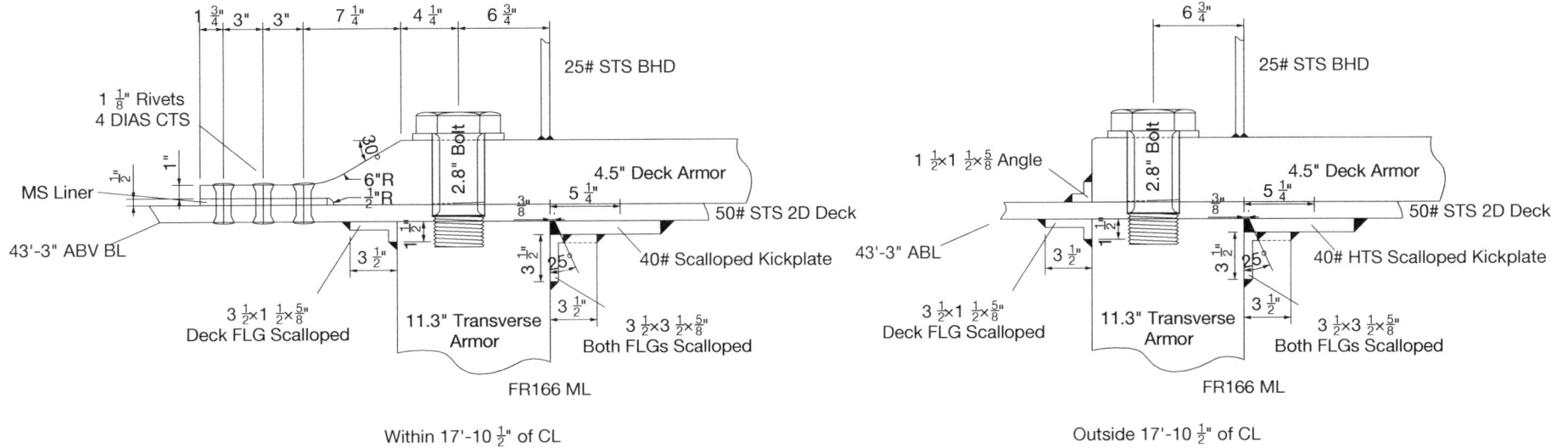

Figure 2.28: Connection to Second Deck (1:12). The connection to second deck is different at the centerline than at the edges. This figure shows both configurations

Photograph 2.1: Connection of Second Deck to the Transverse Armor. This is at the upper belt on USS New Jersey. The bolts are covered. Note the riveted extension to the right and the slope in the cover over the sloped belt armor.

Photograph 2.2: Connection of the Transverse Armor to Second Deck Near the centerline. This is USS New Jersey. This area lacks the riveted extension going aft.

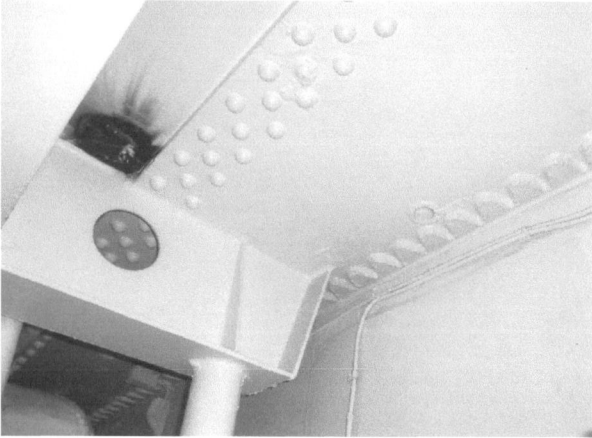

Photograph 2.3: Second deck viewed from below looking forward at FR166. The hardened face of the transverse armor is to the right. Note that the girder does not extend to the hardened face of the armor. Compare with Figure 2.28.

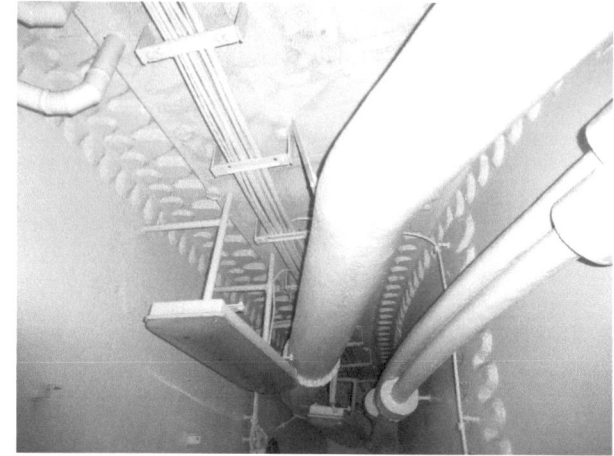

Photograph 2.4: Connection of Second Deck to the Transverse Armor at FR166. The transverse armor is to the left and the barbette support for turret no. 3 is to the right. The kickplate and angle create a double row scallops along second deck. The vertical flange of the angle is welded to the soft face of the transverse plates. The scalloped angle welded to the barbette support and second deck is at the upper right. Below is a scalloped seam-strap across the joint between two plates of the barbette support.

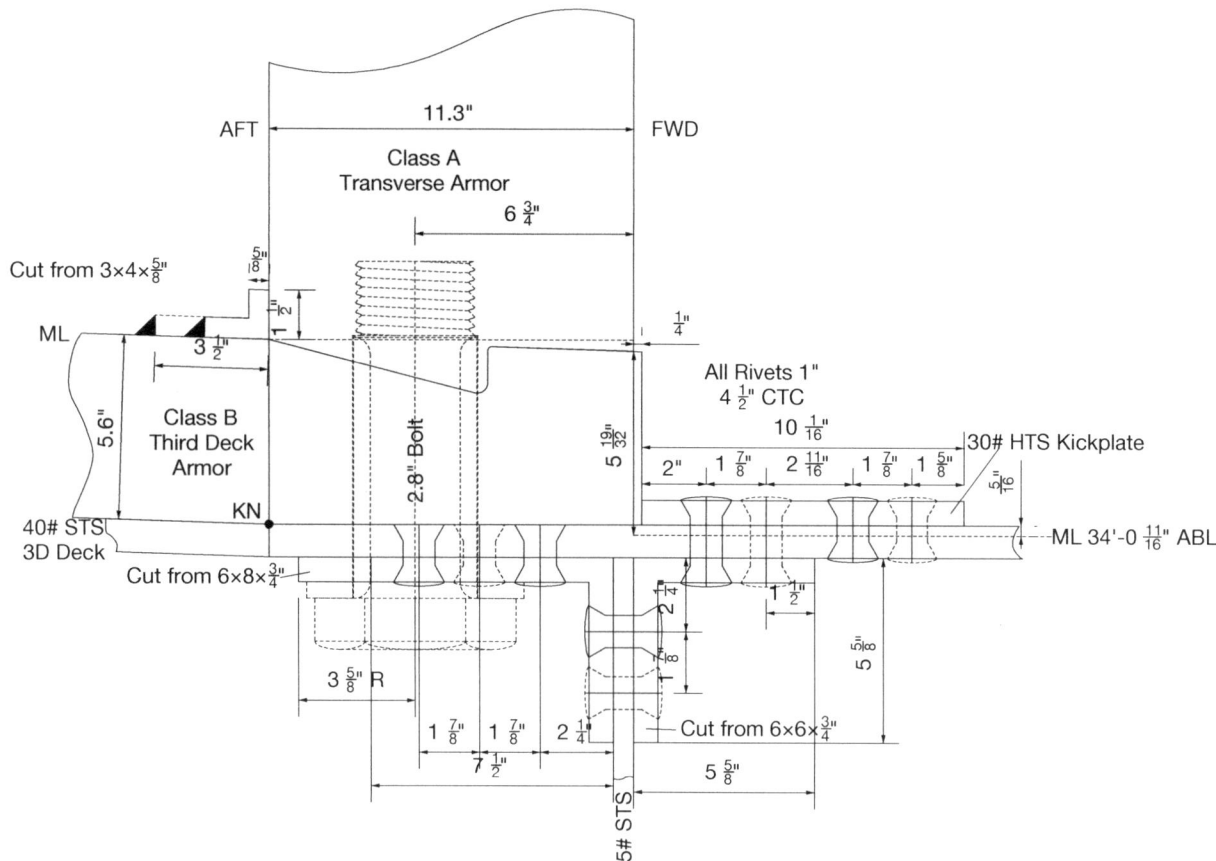

Figure 2.29: Connection to Third Deck (1:6). See also Figure 2.30.

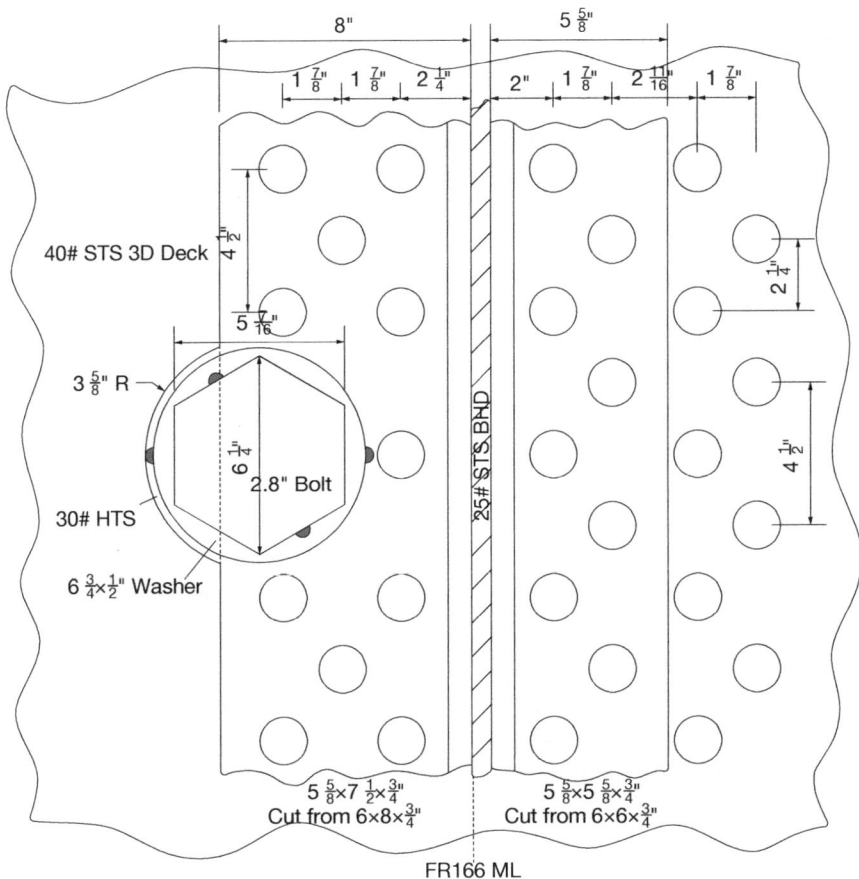

Figure 2.30: Bottom Plan of Connection to Third Deck (1:6). The connection of the transverse armor to third deck from below. See also Figure 2.29.

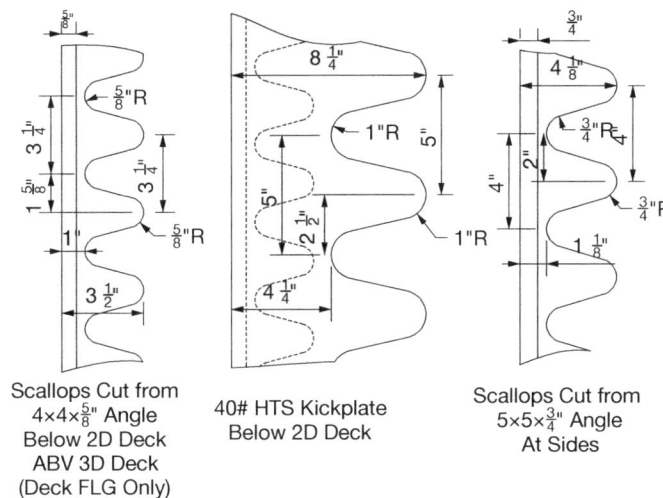

Figure 2.31: Scallops Used With Transverse Armor Connections (1:8).

Figure 2.32: Plan of Connection to Longitudinal Bulkhead (1:6). Two watertight longitudinal bulkheads attach to the transverse armor on the outer face. The bulkhead connect to a strap tap riveted to a soft area in the otherwise hardened face.

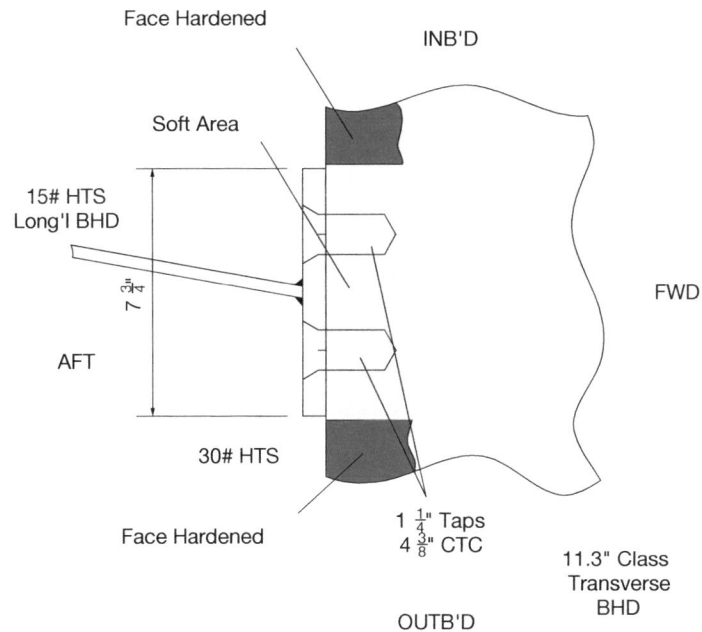

2.8.1. Upper Belt

The upper belt consists of sixteen 12.1" Class A armor plates on each side. The plate edges are oriented normal to the centerline. The plates forward are rhombuses (with sides parallel) while most of the plates are rectangular. The forward-most (U-B-1-P & S) and aftmost (U-B-16-P & S) plates were oversized in length except on USS *Iowa* where the mill supplied the plates to follow the profile of the transverse armor at FR50. The mill machined keyways into the bottom and sides of the plates for alignment. The mill also drilled and tapped bolt holes on the inside face that were used to install the plate,

2.8.1.1. Machining at the Yard

The yard had to trim the forwardmost plates (exept on USS *Iowa*) so the forward edge followed the profile of the transvers armor at FR50. The yard also had to trim aft end of the upper belt to match the edge of the of the transverse plates at FR166 and machine a groove into the inside face of the aftmost plates in order to receive the tongue at the edge of transverse armor plates.

Other than these end adjustments all of the upper belt machining was done at the mill. The finishing process at the mill included machining keyways along the bottom and sides for alignment as well as drilling and tapping holes along the upper face for lifting pads and on the inside face for securing the plate to the backing bulkhead.

The upper belt is the only non-structural region of the belt armor. In this area the belt plates are attached to a backing plate using bolts. The backing plate forms part of torpedo bulkhead 3. The upper belt plates were designed to

have a uniform thickness of 12.1". However, the face-hardening process tended to create irregular plate surfaces. In order to ensure the plates could be properly aligned, they were designed to sit $1\frac{1}{4}$" off the backing bulkhead. That way, any small bumps in the surface would sit in the gap rather than throw the plate out of alignment. The gap between the upper belt and backing plate was filled with cement so that a force applied to the outside face of the armor would be transmitted to the entire backing bulkhead.

2.8.1.2. Installing Upper Belt Plates

Installing the Upper Belt plates was a complex process. To start, the shipyard screwed special adapter plugs into each of the tapped holes on the inside of the plate. The adapter plugs were 3.2" diameter rings with threads on the outside and a $1\frac{3}{8}$" threaded opening in the center. These adapters served different purposes during the installation.

The upper belt plate had to be bolted to the torpedo bulkhead. This required bolt holes to be drilled into the backing bulkhead that exactly matched the bolt holes in the plate. To get the precision required, the shipyard created a template for each plate to locate the holes. Pointed marking plugs were screwed into the center openings of the adapters. Pressing the template over a plate caused the points to create small holes indicating the location of the opening in the plate. The shipyard then used the template to locate the positions for drilling two-inch pilot holes in the backing bulkhead matching the holes in the plate. The yard drilled and tapped four additional holes in the backing bulkhead for adjusting bolts at the corners. Once the template

was created, the marking plugs were removed from the adapters.

Several preparatory steps had to be taken before a plate could be moved into position. First, the yard had to install the key at the top of the lower belt that aligned the Upper Belt plate with the Lower Belt. Lifting pads had to be bolted to the top edge of the plate so that a crane could move the plate. At this stage the adapter plugs then served a second purpose. Wire ropes were screwed into the four adapters in the top row and four at the bottom row. The ropes were threaded through the corresponding bolts opening in the bulkhead.

Once these steps were completed, a crane could move the plate into position. The wire ropes screwed into the adapter plugs were tightened to pull the plate against the adjusting bolts threaded through the bulkhead. Once the plate was roughly in position draw bolts were screwed into the unused adaptors through the bulkhead. A nut and washer was screwed over the draw bolt. The wire ropes, draw bolts, and adjusting bolts were tightened and loosened as needed to position the plate.

After the plate was in its final position, the adapter plugs were replaced with attachment bolts that provided the permanent link to the ship. At this point in the installation each bolt hole had a 3.2 inch diameter adapter plug with either a wire rope or a draw bolt. Yet the pilot holes were only two inches in diameter. The shipyard had to install one bolt at a time. First, either the wire rope or draw bolt was removed. Second, the two-inch pilot holes were drilled out to $3\frac{5}{16}$", allowing the adapters to be unscrewed and removed through the bulkhead. Third, the adapter was replaced with a large attachment bolt. Fourth, a nut and washer were tightened over the bolt. Finally, the bolt was trimmed and a cover was welded over the bolt. When all the upper plates were in place, keys were hammered into the vertical keyways between the plates and the gap between the bulkhead and the plates was filled with cement. After the cement hardened, the adjusting bolts were removed and replaced with plugs.

	Frame	Lower		Upper	
		HT	HB	HT	HB
$13\frac{5}{8}$" Before	50	35' 0$\frac{3}{32}$"	23' 8$\frac{1}{32}$"	45' 7$\frac{23}{32}$"	27' 3$\frac{31}{32}$"
	50	34' 11$\frac{3}{4}$"	23' 10"	45' 7$\frac{3}{8}$"	27' 5$\frac{15}{16}$"
	57$\frac{1}{2}$	34' 2$\frac{3}{8}$"	28' 3$\frac{1}{16}$"	44' 10"	31' 11"
	65	34' 0$\frac{1}{8}$"	32' 10$\frac{9}{16}$"	44' $\frac{5}{8}$"	36' 4$\frac{1}{16}$"
	72$\frac{1}{2}$	33' 9$\frac{7}{8}$"	35' 9$\frac{5}{8}$"	43' 9"	39' 2$\frac{5}{8}$"
	80	33' 9$\frac{7}{8}$"	38' 9$\frac{7}{16}$"	43' 9"	42' 2$\frac{7}{16}$"
	87$\frac{1}{2}$	33' 9$\frac{7}{8}$"	41' 9$\frac{1}{4}$"	43' 9"	45' 2$\frac{1}{4}$"
	95	33' 9$\frac{7}{8}$"	42' 9$\frac{1}{4}$"	43' 9"	46' 2$\frac{1}{4}$"
	102$\frac{1}{2}$	33' 9$\frac{7}{8}$"	43' 9$\frac{1}{4}$"	43' 9"	47' 2$\frac{1}{4}$"
	110	33' 9$\frac{7}{8}$"	44' 9$\frac{1}{4}$"	43' 9"	48' 2$\frac{1}{4}$"
	117$\frac{1}{2}$	33' 9$\frac{7}{8}$"	45' 9$\frac{1}{4}$"	43' 9"	49' 2$\frac{1}{4}$"
	125	33' 9$\frac{7}{8}$"	45' 9$\frac{1}{4}$"	43' 9"	49' 2$\frac{1}{4}$"
	132$\frac{1}{2}$	33' 9$\frac{7}{8}$"	45' 9$\frac{1}{4}$"	43' 9"	49' 2$\frac{1}{4}$"
	140	33' 9$\frac{7}{8}$"	45' 9$\frac{1}{4}$"	43' 9"	49' 2$\frac{1}{4}$"
	145$\frac{1}{2}$	33' 9$\frac{7}{8}$"	45' 9$\frac{1}{4}$"	43' 9"	49' 2$\frac{1}{4}$"
	151	33' 9$\frac{7}{8}$"	45' 9$\frac{1}{4}$"	43' 9"	49' 2$\frac{1}{4}$"
	158$\frac{1}{2}$	33' 9$\frac{7}{8}$"	43' 0$\frac{1}{4}$"	43' 9"	46' 5$\frac{1}{4}$"
	166	33' 9$\frac{7}{8}$"	40' 3$\frac{1}{4}$"	43' 9"	43' 8$\frac{1}{4}$"
$13\frac{5}{8}$" After	166	33' 9$\frac{7}{8}$"	40' 2$\frac{1}{32}$"	43' 9"	43' 7$\frac{1}{32}$"

Table 2.6: Upper Belt Inner Plate Corners Given in Plans. While these same values are given in three different data sets, they are certainly in error because they position the plates too close to the backing bulkhead. These values are included for documentation purposes.

	Upper		Lower	
	HT	HB	HT	HB
50	34' 11$\frac{23}{32}$"	23' 10$\frac{7}{32}$"	45' 7$\frac{21}{64}$"	27' 6$\frac{3}{16}$"
57$\frac{1}{2}$	34' 2$\frac{11}{32}$"	28' 3$\frac{1}{4}$"	44' 9$\frac{61}{64}$"	31' 11$\frac{13}{64}$"
65	34' 0$\frac{3}{32}$"	32' 10$\frac{23}{32}$"	44' 7$\frac{45}{64}$"	36' 6$\frac{11}{16}$"
72$\frac{1}{2}$	33' 9$\frac{27}{32}$"	35' 9$\frac{3}{4}$"	43' 8$\frac{31}{32}$"	39' 2$\frac{25}{32}$"
80	33' 9$\frac{27}{32}$"	38' 9$\frac{9}{16}$"	43' 8$\frac{31}{32}$"	42' 2$\frac{19}{32}$"
87$\frac{1}{2}$	33' 9$\frac{27}{32}$"	41' 9$\frac{3}{8}$"	43' 8$\frac{31}{32}$"	45' 2$\frac{13}{32}$"
95	33' 9$\frac{27}{32}$"	42' 9$\frac{3}{8}$"	43' 8$\frac{31}{32}$"	46' 2$\frac{13}{32}$"
102$\frac{1}{2}$	33' 9$\frac{27}{32}$"	43' 9$\frac{3}{8}$"	43' 8$\frac{31}{32}$"	47' 2$\frac{13}{32}$"
110	33' 9$\frac{27}{32}$"	44' 9$\frac{3}{8}$"	43' 8$\frac{31}{32}$"	48' 2$\frac{13}{32}$"
117$\frac{1}{2}$	33' 9$\frac{27}{32}$"	45' 9$\frac{3}{8}$"	43' 8$\frac{31}{32}$"	49' 2$\frac{13}{32}$"
125	33' 9$\frac{27}{32}$"	45' 9$\frac{3}{8}$"	43' 8$\frac{31}{32}$"	49' 2$\frac{13}{32}$"
132$\frac{1}{2}$	33' 9$\frac{27}{32}$"	45' 9$\frac{3}{8}$"	43' 8$\frac{31}{32}$"	49' 2$\frac{13}{32}$"
140	33' 9$\frac{27}{32}$"	45' 9$\frac{3}{8}$"	43' 8$\frac{31}{32}$"	49' 2$\frac{13}{32}$"
145$\frac{1}{2}$	33' 9$\frac{27}{32}$"	45' 9$\frac{3}{8}$"	43' 8$\frac{31}{32}$"	49' 2$\frac{13}{32}$"
151	33' 9$\frac{27}{32}$"	45' 9$\frac{3}{8}$"	33' 9$\frac{27}{32}$"	45' 9$\frac{3}{8}$"
158$\frac{1}{2}$	33' 9$\frac{27}{32}$"	43' 0$\frac{3}{8}$"	43' 8$\frac{31}{32}$"	46' 5$\frac{13}{32}$"
166	33' 9$\frac{27}{32}$"	40' 3$\frac{3}{8}$"	43' 8$\frac{63}{64}$"	43' 8$\frac{29}{64}$"

Table 2.7: Calculated values for the locations of the inner faces of the upper belt plates if they are placed 1$\frac{1}{4}$" from the backing bulkhead as specified in the plans.

Plate	Weight (lbs.)
U-B-1	174,145
U-B-2	163,088
U-B-3	156,527
U-B-4	155,708
U-B-5	155,708
U-B-6	155,020
U-B-7	155,015
U-B-8	155,015
U-B-9	155,015
U-B-10	154,934
U-B-11	154,920
U-B-12	154,929
U-B-13	113,529
U-B-14	113,574
U-B-15	155,582
U-B-16	161,011
Total (one side)	2,433,720
Grand Total (both sides)	4,867,440

Table 2.8: Upper Belt Plate Weights

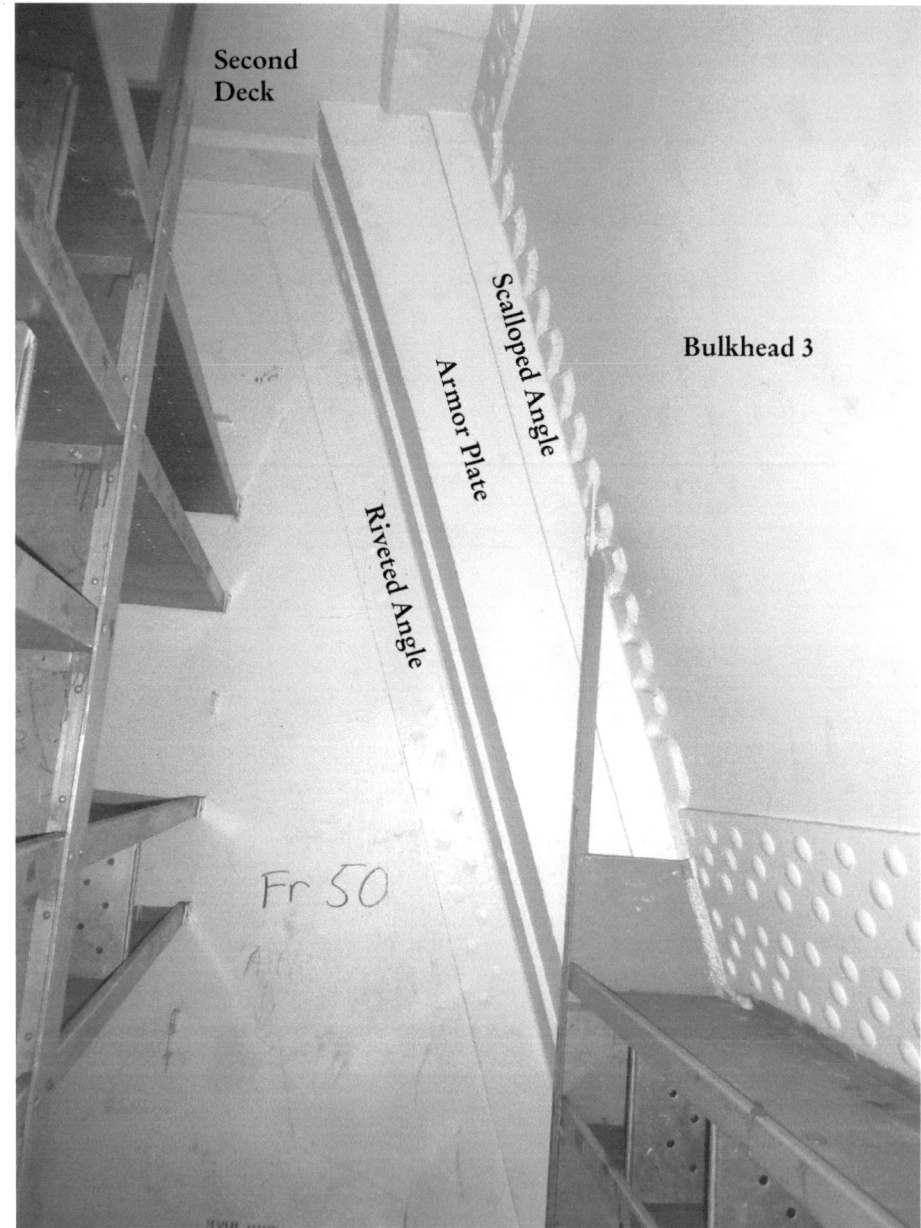

Photograph 2.5: Forward End of Upper Belt. This is the starboard side of USS New Jersey. Bulkhead 3 is at the right and the Frame 50 transverse bulkhead is to the left. See Figure 3.21 for the connections.

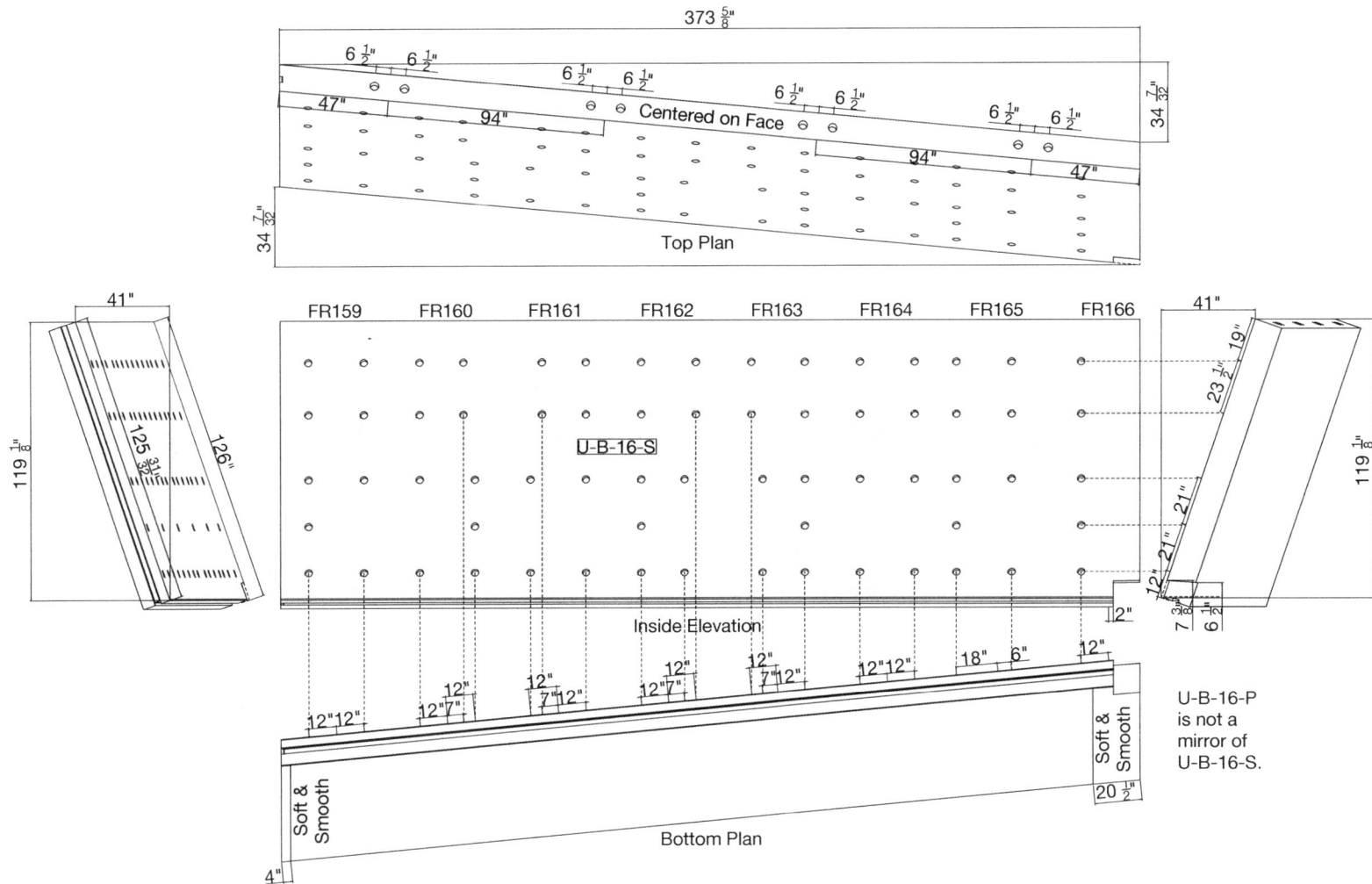

Figure 2.33: Plate U-B-16-S (1:72).

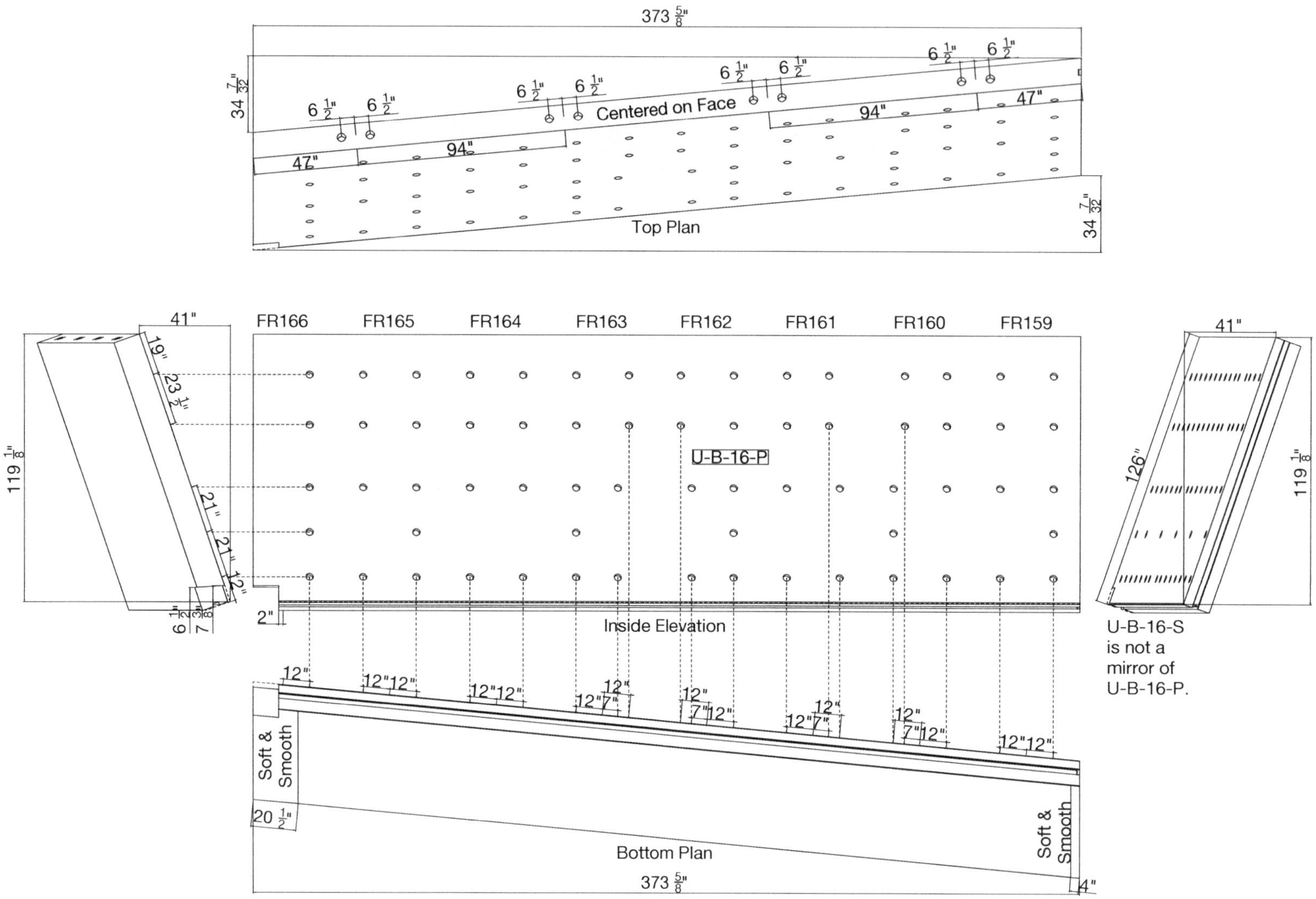

Figure 2.34: Plate U-B-16-P (1:72)

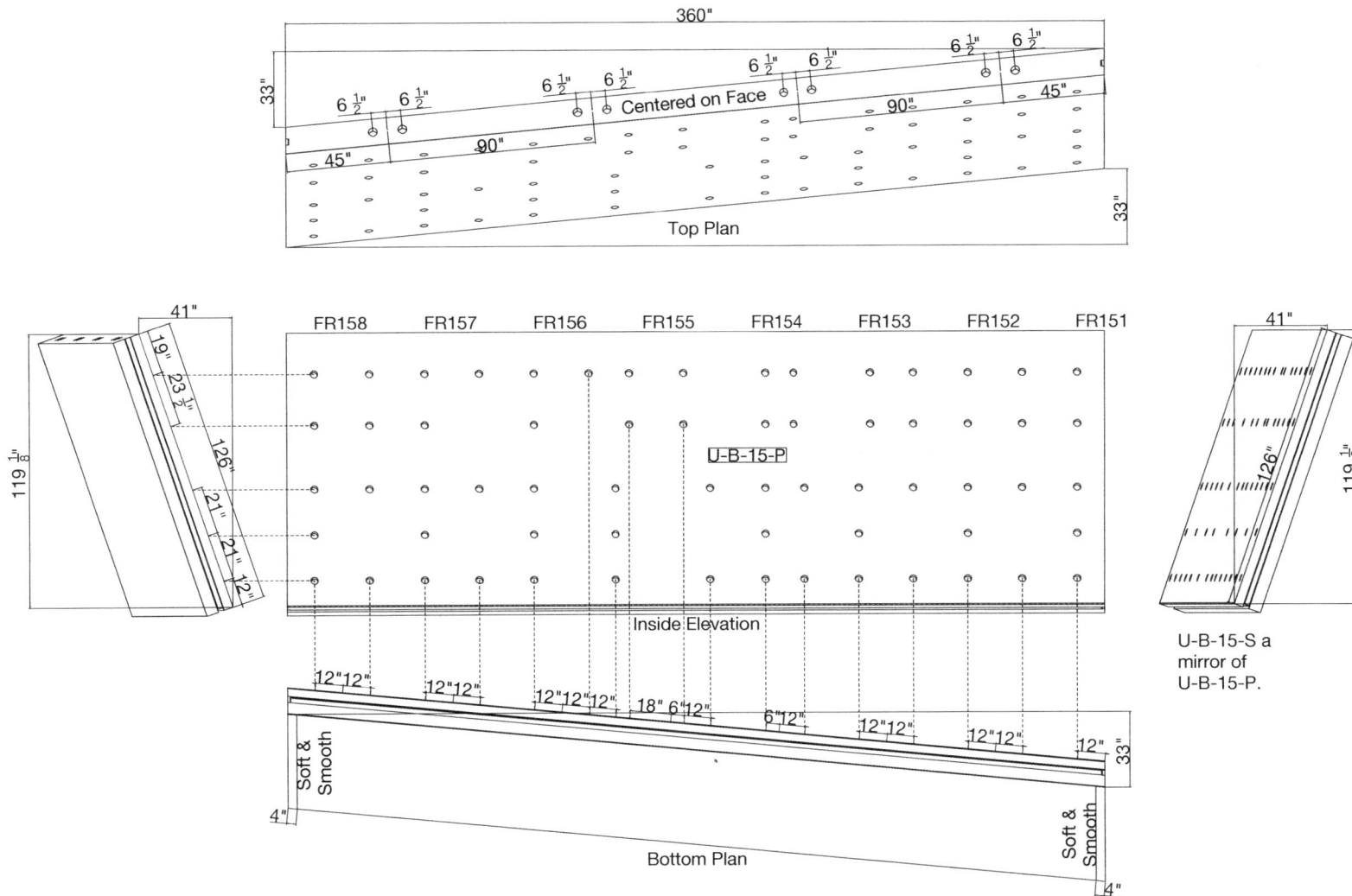

Figure 2.35: Plate U-B-15-P (1:72).

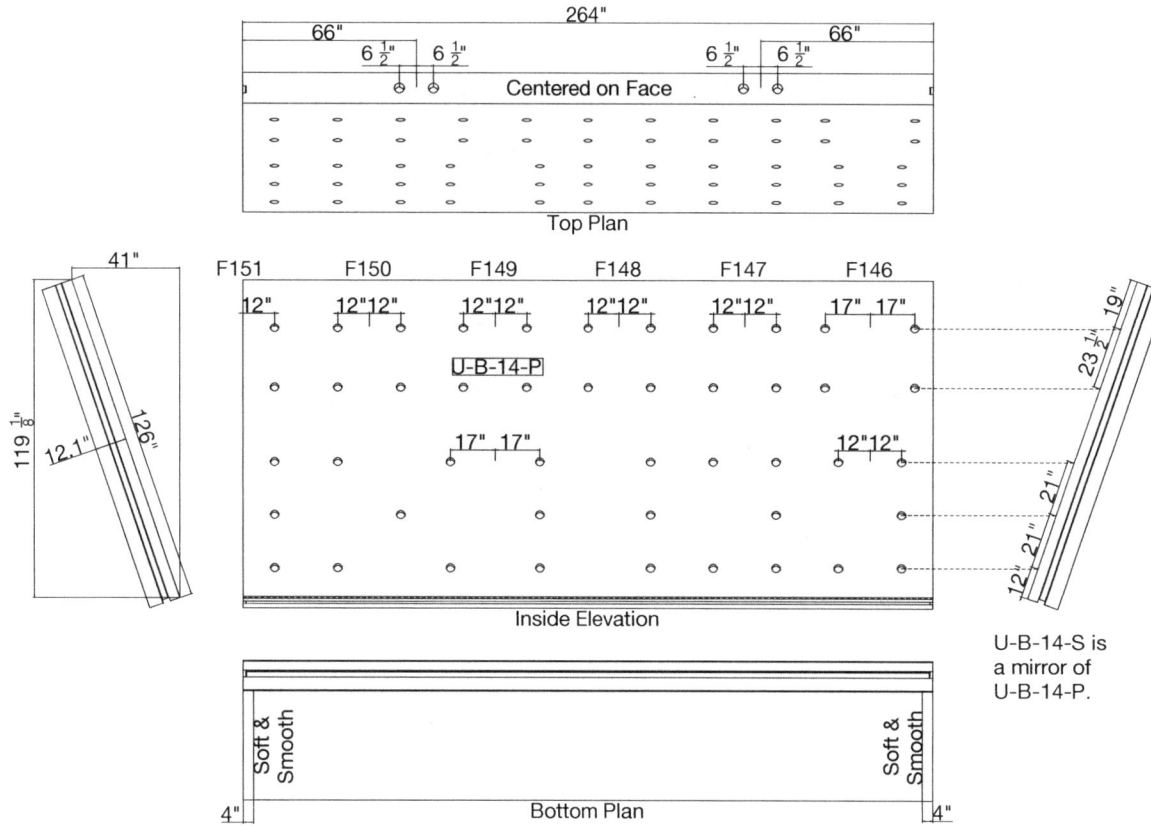

Figure 2.36: Plate U-B-14-P (1:72).

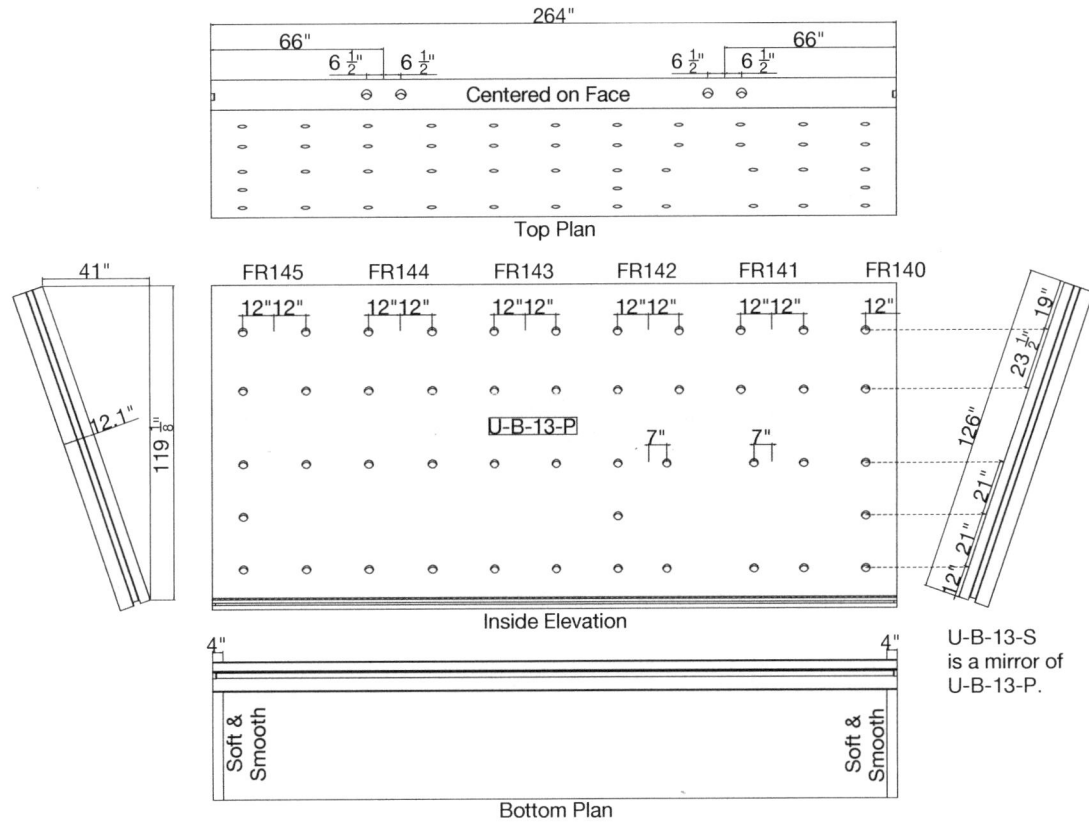

Figure 2.37: Plate U-B-13-P (1:72).

Top Plan

360"

45" 90" 90" 45"

6 ½" 6 ½" 6 ½" 6 ½" 6 ½" 6 ½" 6 ½" 6 ½"

Centered on Face

Inside Elevation

41"

FR140 FR139 FR138 FR137 FR136 FR135 FR134 FR133

12" 17" 17" 12" 12" 12" 12" 12" 12" 12" 12" 12" 12" 12" 12"

12" 12"

U-B-12-P

119 ⅛"

12.1"

23 ½" 19"

126"

21" 21"

12" 21"

+ Holes drilled by mistake on USS Missouri were filled

U-B-12-S is a mirror of U-B-12-P.

Bottom Plan

Soft & Smooth

Soft & Smooth

4" 4"

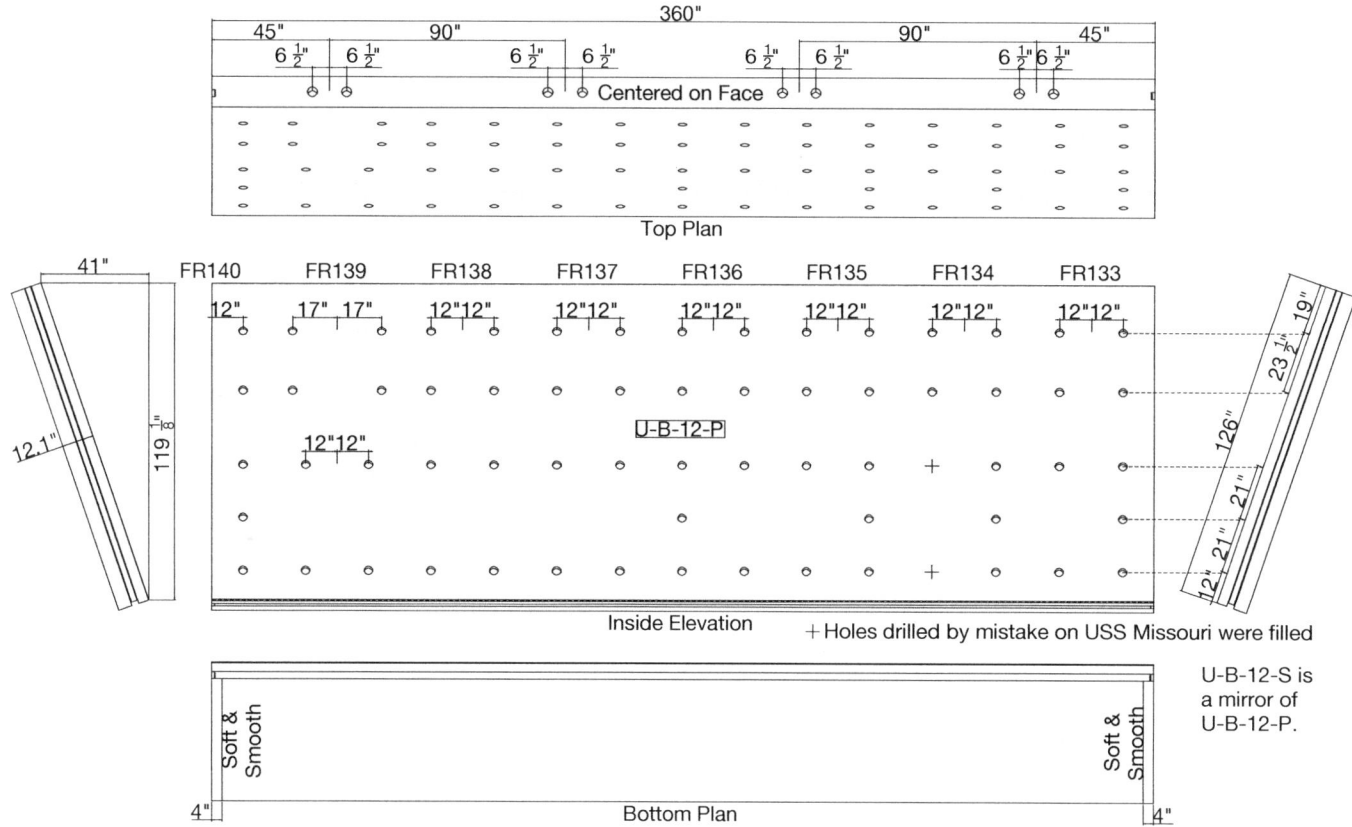

Figure 2.38: Plate U-B-12-P (1:72).

360"

45" | 90" | 90" | 45"

$6\frac{1}{2}$" | $6\frac{1}{2}$" | $6\frac{1}{2}$" | $6\frac{1}{2}$" | $6\frac{1}{2}$" | $6\frac{1}{2}$" | $6\frac{1}{2}$" | $6\frac{1}{2}$"

Centered on Face

Top Plan

41"

FR132   FR131   FR130   FR129   FR128   FR127   FR126   FR125

12"7"   7"12"   12"12"   12"12"   12"12"   12"12"   12"12"   12"

$119\frac{1}{8}$"

12.1"

U-B-11-P

12"   12"   7"   7"

$23\frac{1}{2}$"   19"

126"

21"   21"

12"   21"

Inside Elevation

U-B-11-S is
a mirror of
U-B-11-P.

Soft & Smooth

Soft & Smooth

4"   Bottom Plan   4"

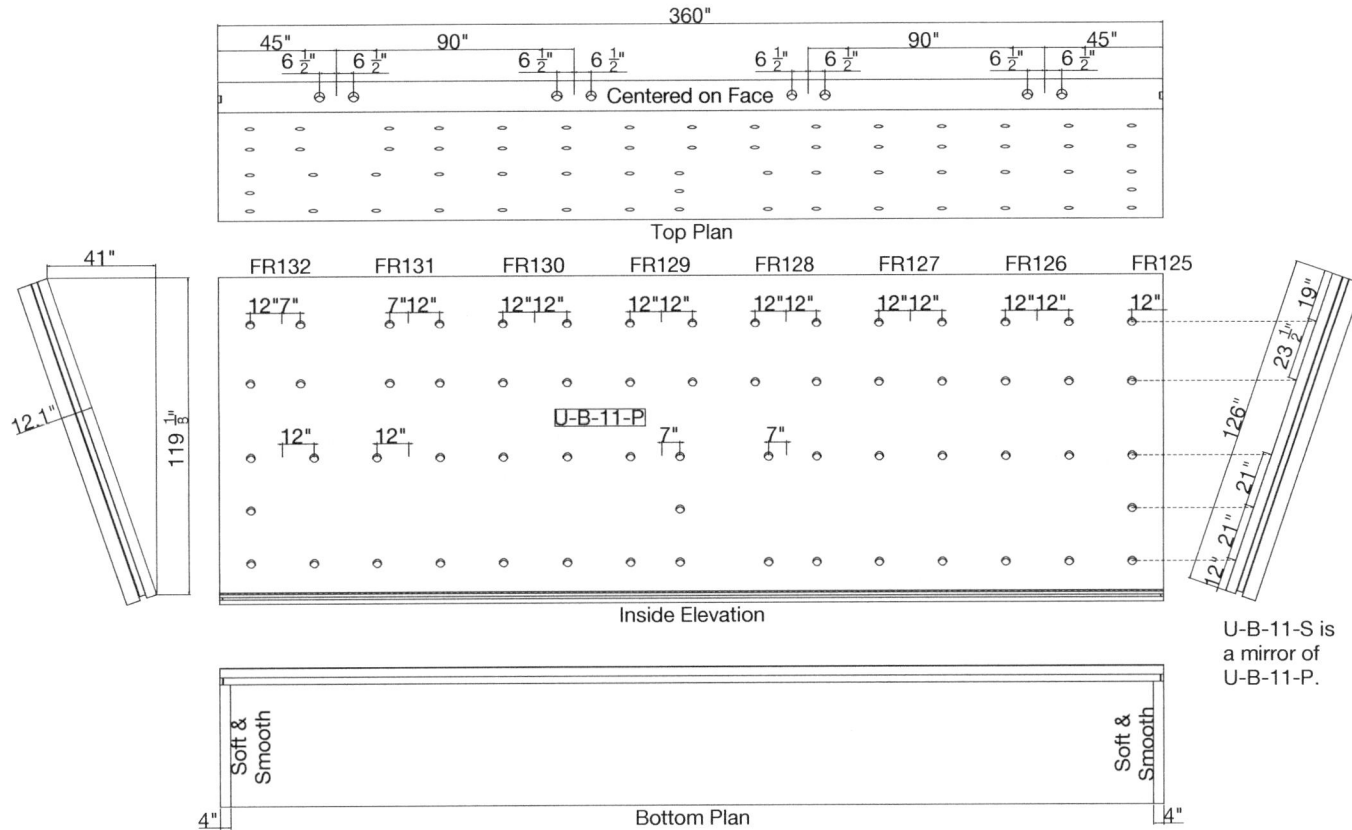

Figure 2.39: Plate U-B-11-P (1:72).

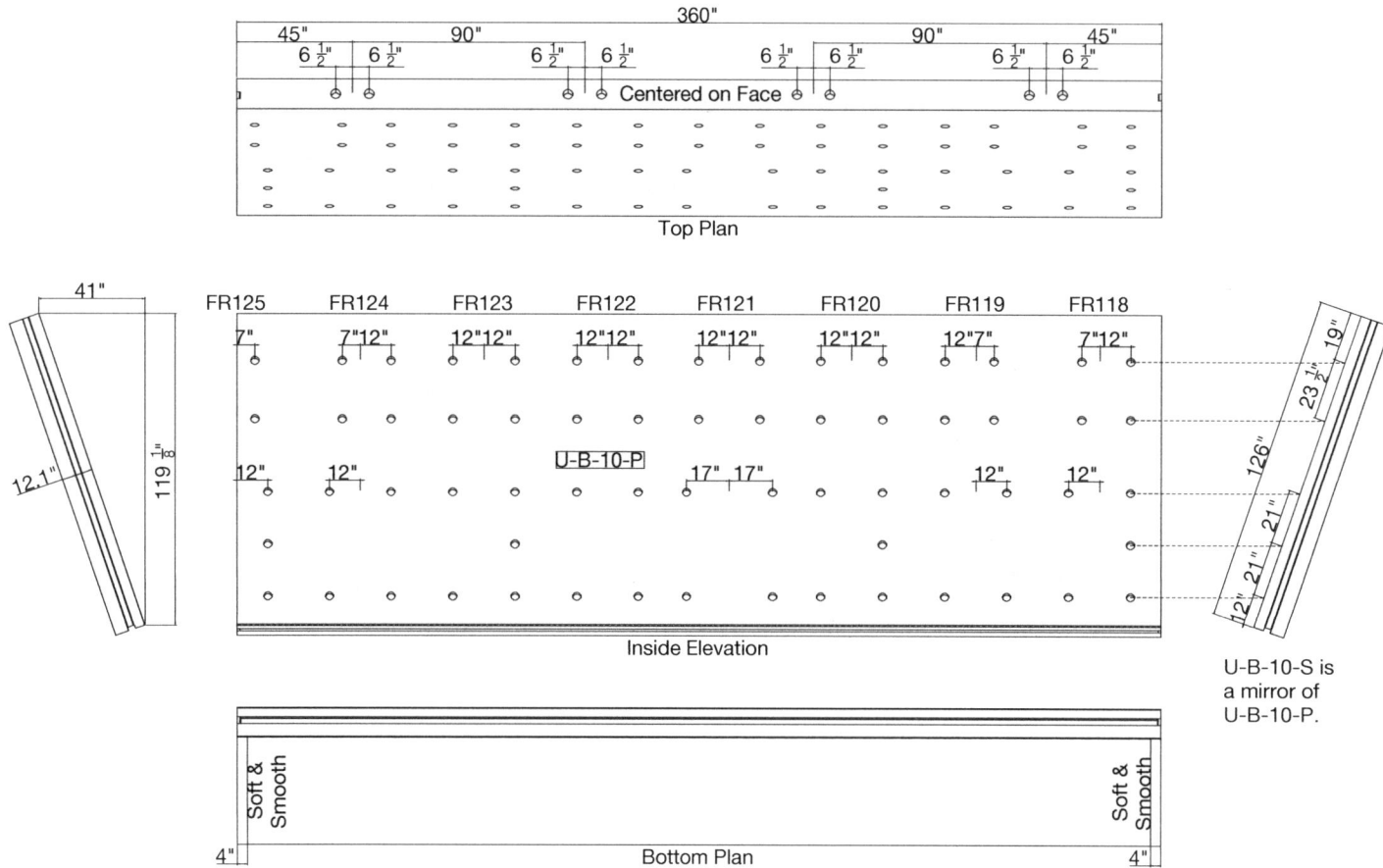

Figure 2.40: Plate U-B-10-P (1:72).

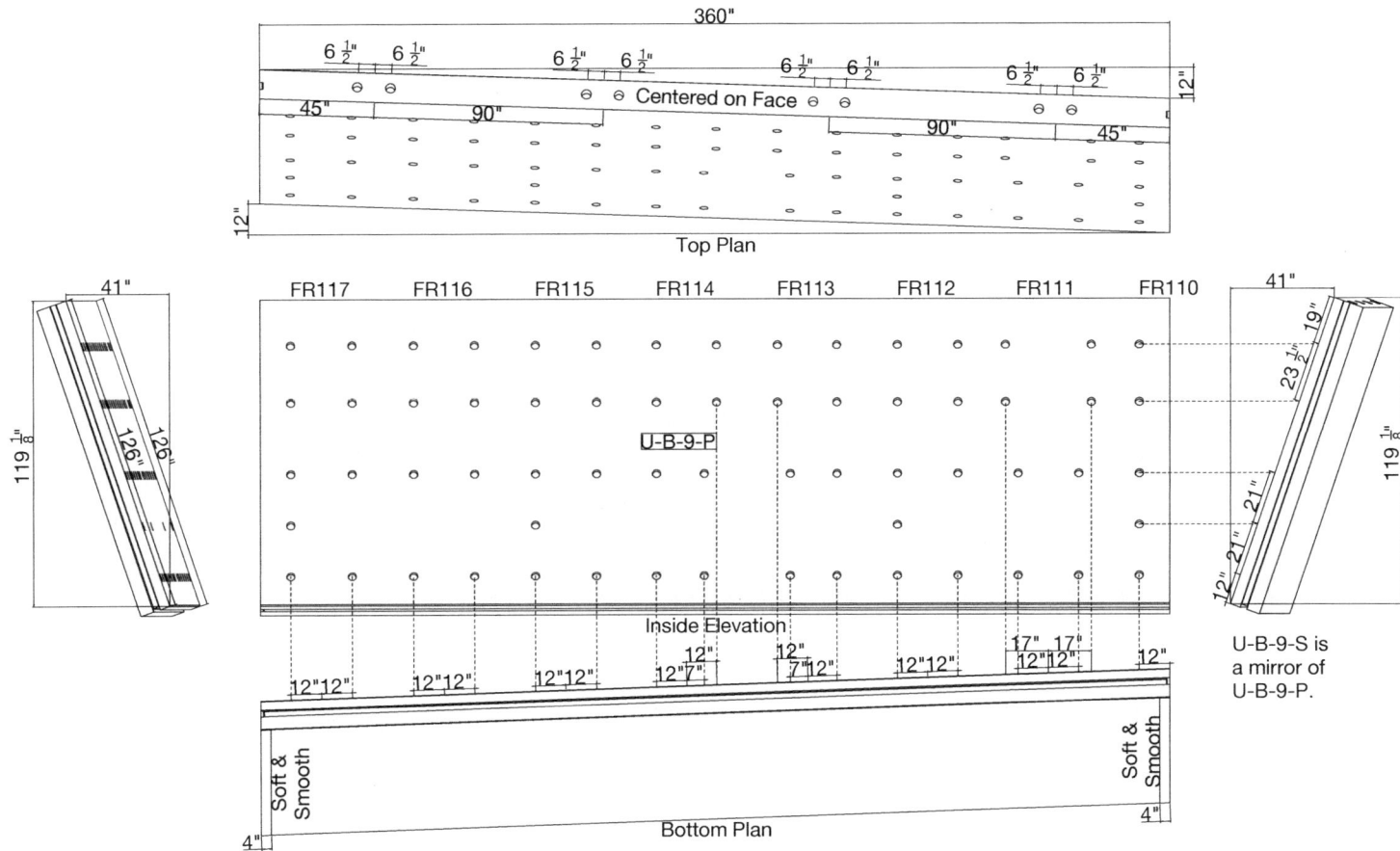

Figure 2.41: Plate U-B-9-P (1:72).

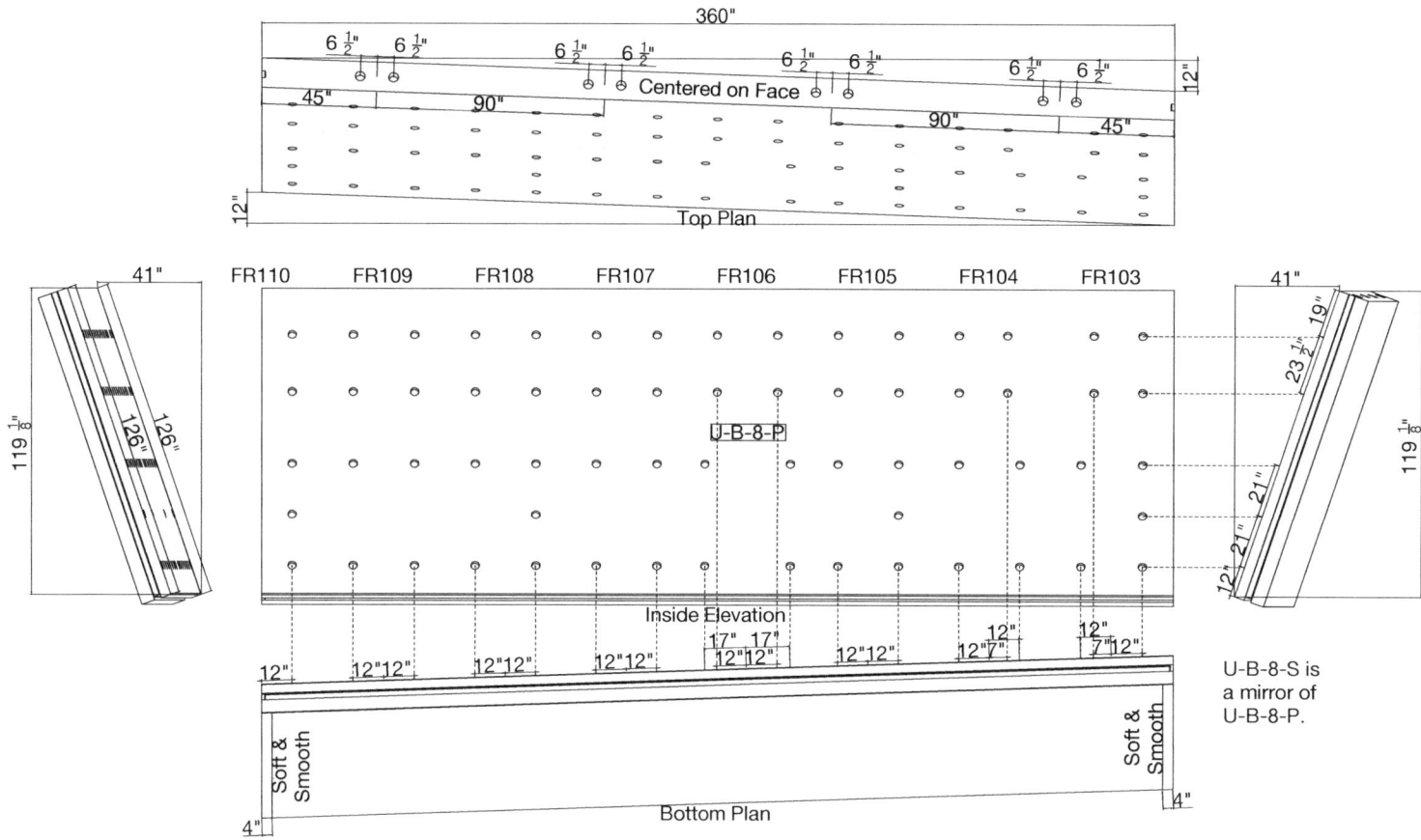

Figure 2.42: Plate U-B-8-P (1:72).

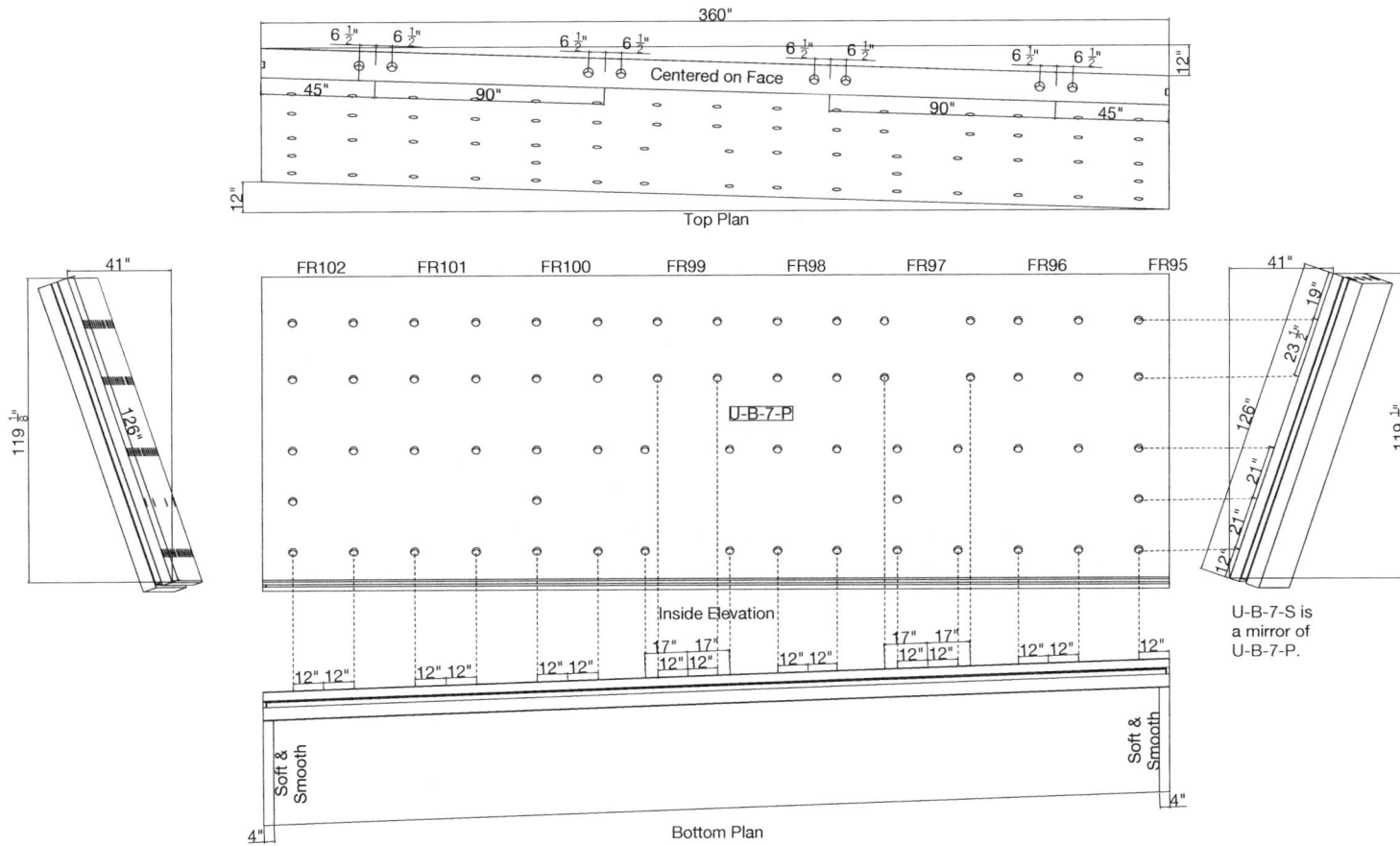

Figure 2.43: Plate U-B-7-P (1:72).

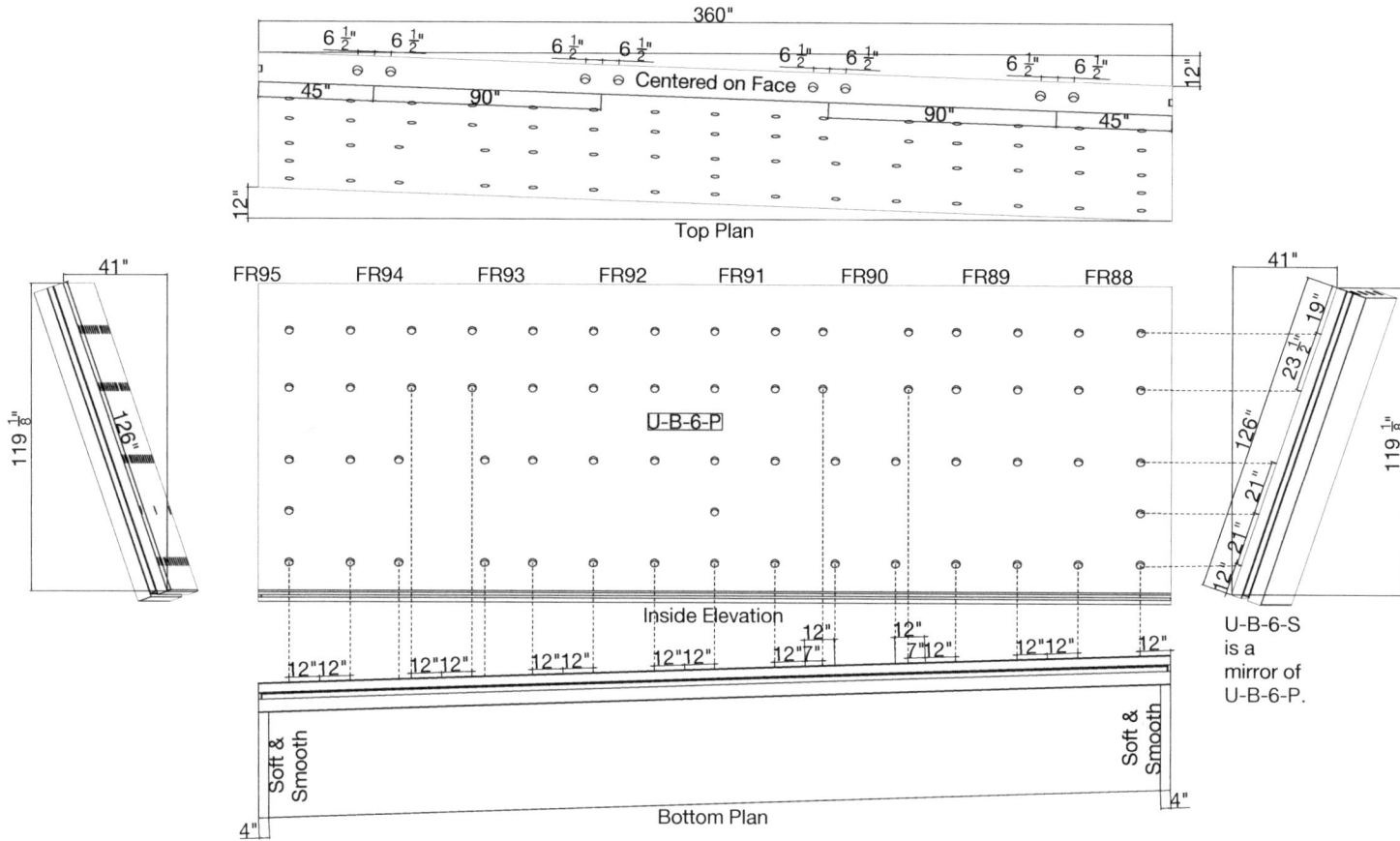

Figure 2.44: Plate U-B-6-P (1:72).

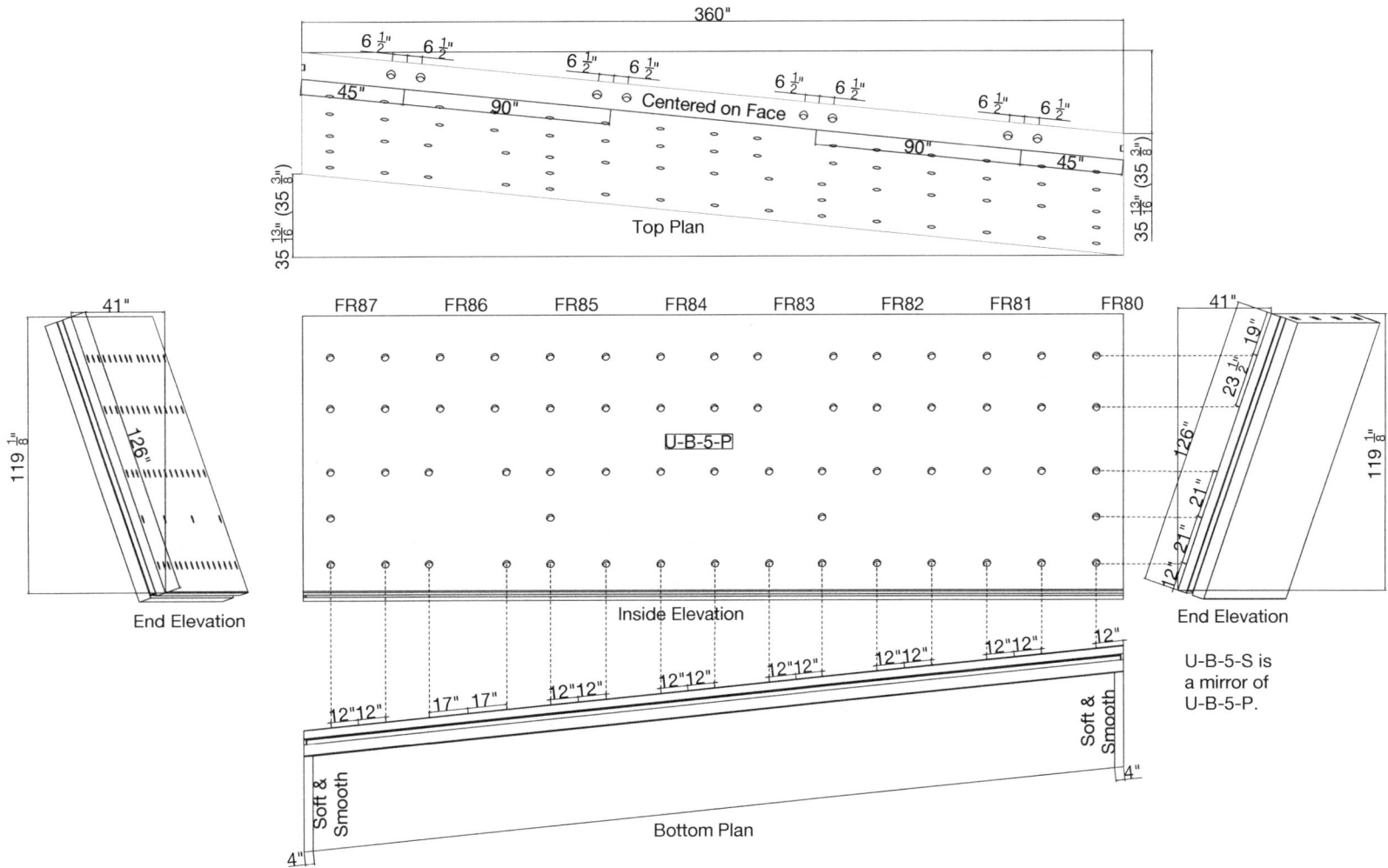

Figure 2.45: Plate U-B-5-P (1:72).

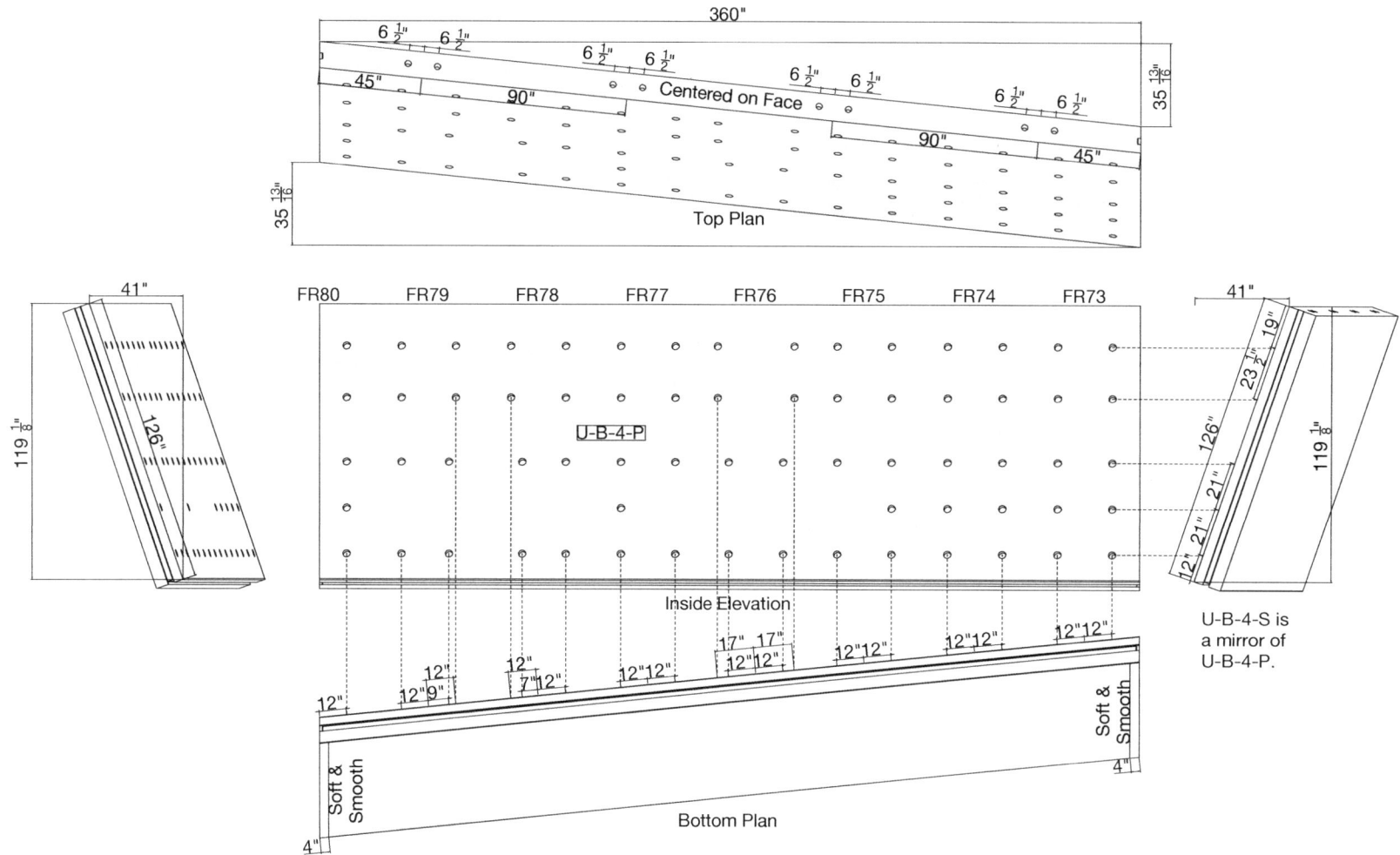

Figure 2.46: Plate U-B-4-P (1:72). The starboard side plate U-B-4-S is a mirror.

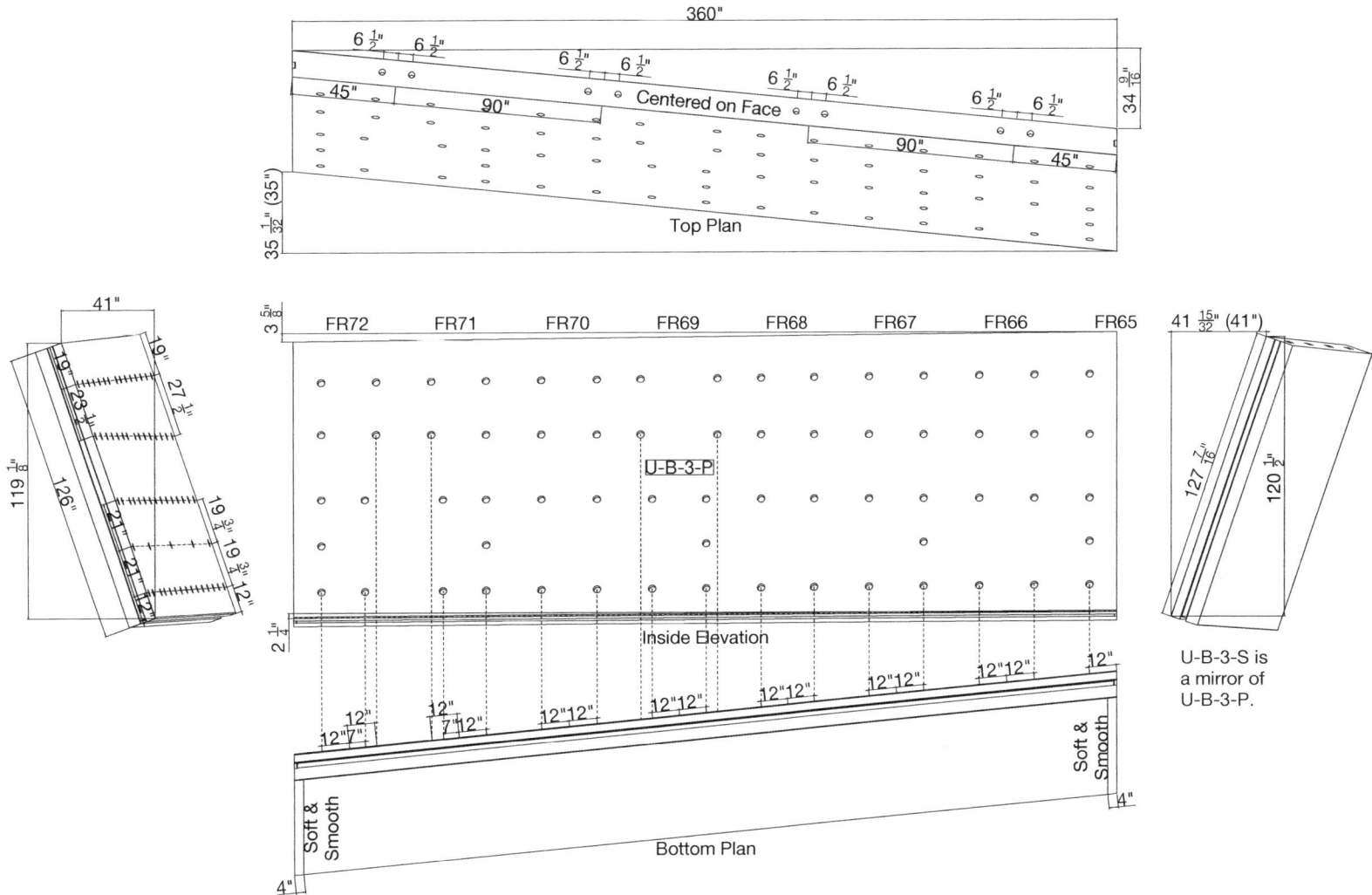

Figure 2.47: Plate U-B-3-P (1:72). The starboard side plate U-B-3-S is a mirror.

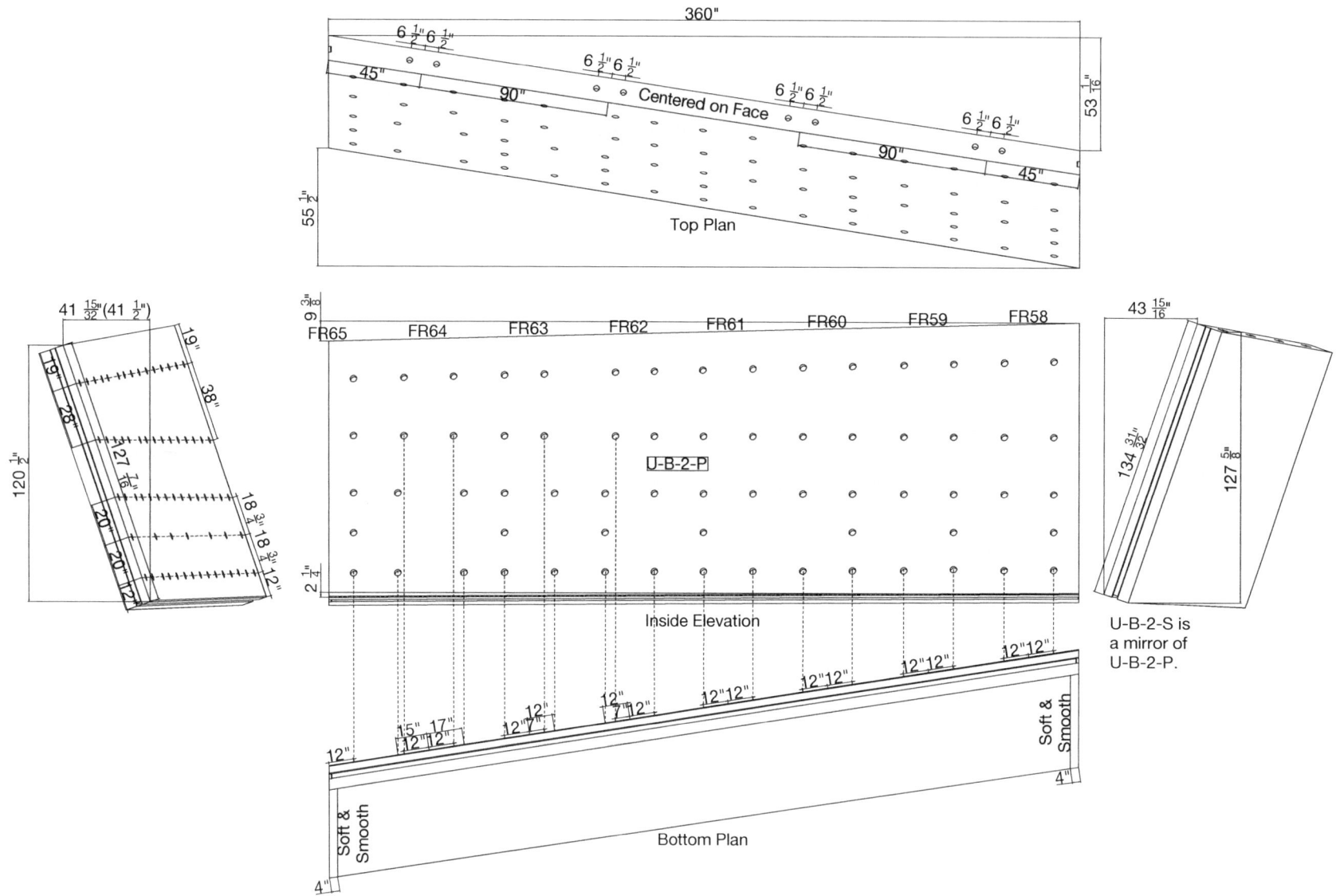

Figure 2.48: Plate U-B-2-P (1:72). The starboard side plate U-B-2-S is a mirror.

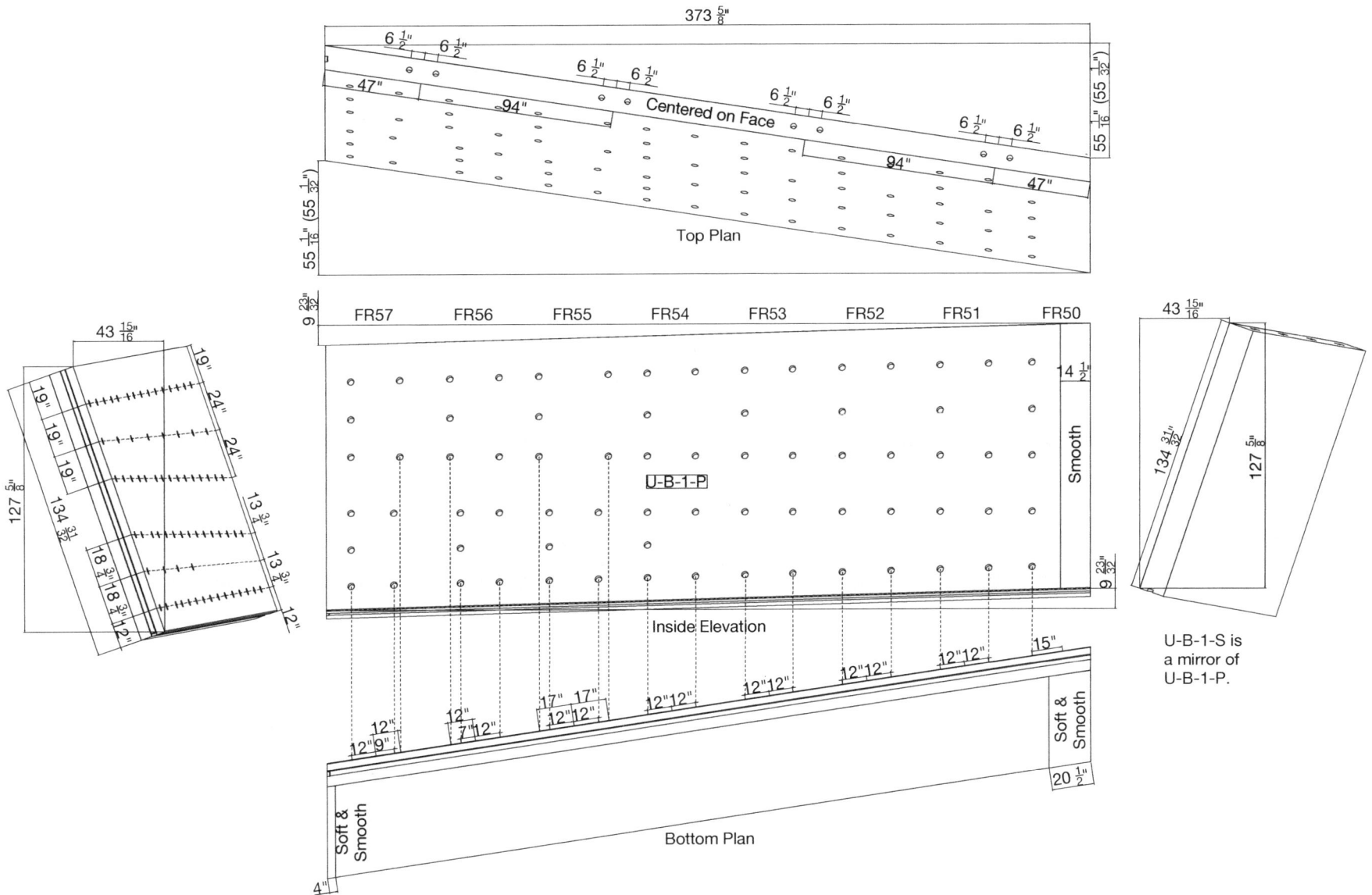

Figure 2.49: Plate U-B-1-P (1:72). The starboard side plate U-B-1-S is a mirror.

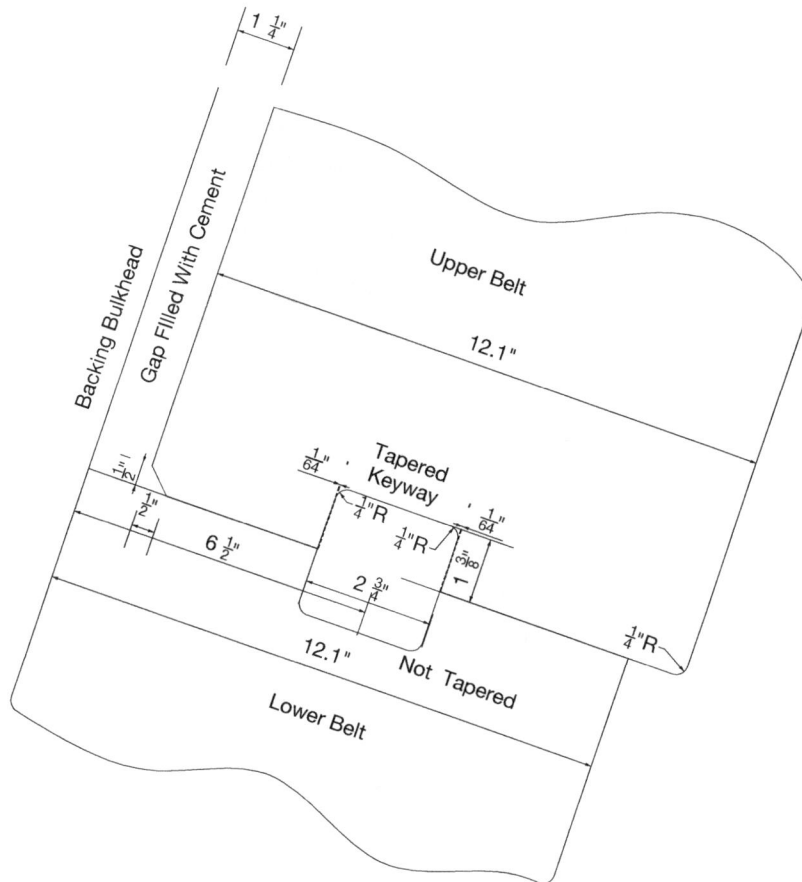

Figure 2.50: Elevation of the Upper Belt Lower Edge (1:4). The upper belt and the upper zone of the lower belt have the same thickness. However, their faces are not aligned. A key aligns the upper belt to the lower belt.

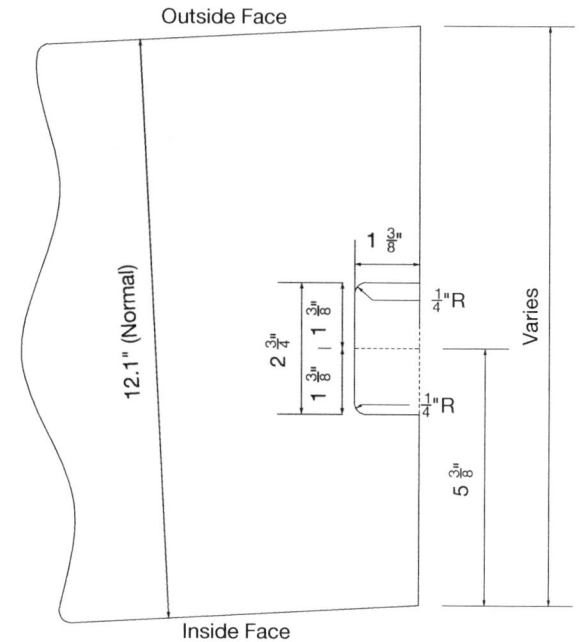

Figure 2.51: Section the Upper Belt Plate Normal to Faces (1:4). A key aligns the upper belt plates along the side.

FR166

13 $\frac{5}{8}$"

Upper Belt

Excess Trimmed by Navy Yard to Be Flush With Transverse Armor

Rabbet machined by Navy Yard

$\frac{1}{8}$"

$\frac{1}{4}$"R

$\frac{1}{4}$"R

$\frac{1}{4}$"R

To Fit

$\frac{5}{16}$"R

$\frac{5}{16}$"R

2 $\frac{1}{2}$"

1"

$\frac{1}{4}$

5"

3 $\frac{1}{8}$"

11.3"

Transverse Armor

Section Normal to Transverse Armor Side

Transverse

Section

Belt

Aft Reference View for Section

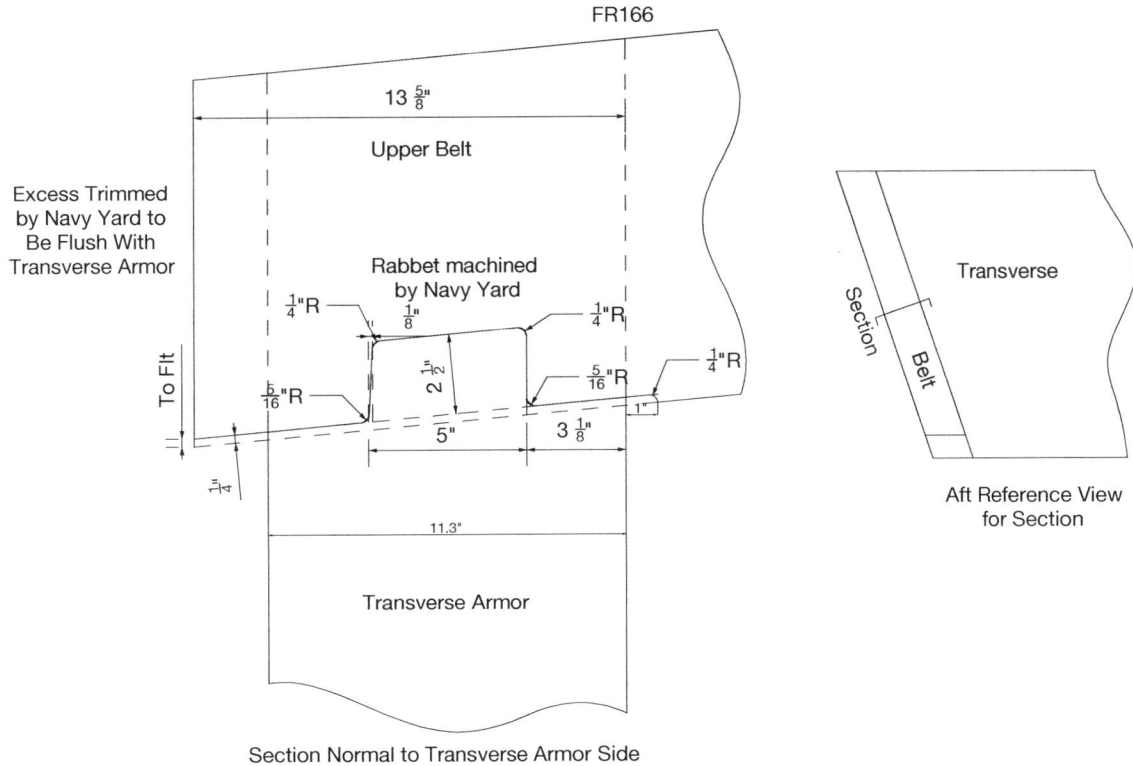

Figure 2.52: Section Through Joint Between the Upper Belt and Transverse Armor at FR166 (1:6). The lower belt rises to third deck aft of FR166. There is a notch in the lower aftmost upper belt plates (U-B-16-P and U-B-16-S) to allow those plates to fit over that rise. The mill machined an undersized starter notch in the plates. The yard had to machine the notch to fit.

Figure 2.53: Elevations of the Aft Corner of the Starboard Upper Belt at FR166 (1:24). The belt plates were ordered oversized in length at the ends to create an adjustment allowance as the belt was installed from the center outward. The yard had to machine a tongue in the aftmost plates (U-B-16-P and U-B-16-S) to accept a tongue the mill machined in the transverse plates at FR166. The yard also has to trim the plates to length to match the transverse armor

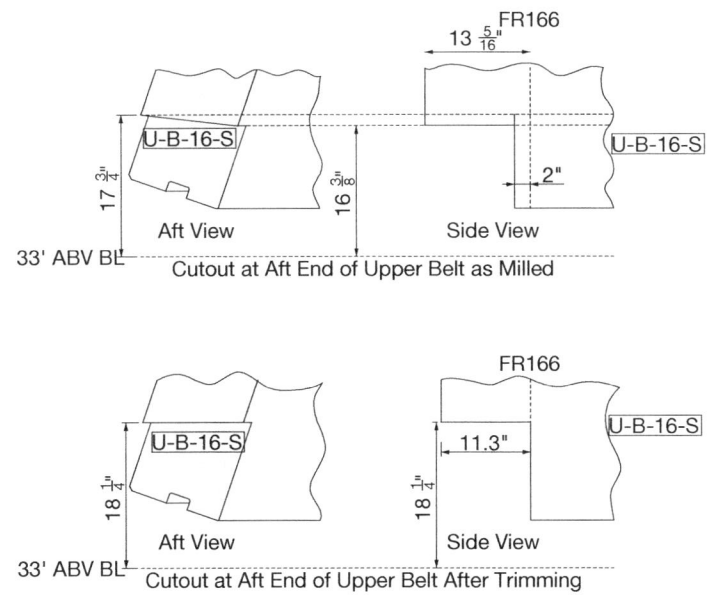

FR166

13 $\frac{5}{16}$"

U-B-16-S

U-B-16-S

17 $\frac{3}{4}$"

16 $\frac{3}{8}$"

2"

33' ABV BL

Aft View

Side View

Cutout at Aft End of Upper Belt as Milled

FR166

U-B-16-S

U-B-16-S

18 $\frac{1}{4}$"

18 $\frac{1}{4}$"

11.3"

33' ABV BL

Aft View

Side View

Cutout at Aft End of Upper Belt After Trimming

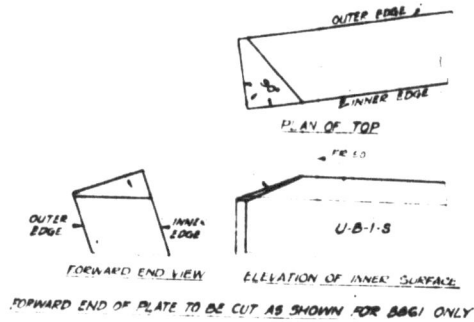

OUTER EDGE
INNER EDGE
PLAN OF TOP
FR 50
OUTER EDGE
INNER EDGE
U·B·1·S
FORWARD END VIEW   ELEVATION OF INNER SURFACE
FORWARD END OF PLATE TO BE CUT AS SHOWN FOR 8861 ONLY

Plan 2.6: (1302Al). Shows how the upper forward corners of the upper belt were beveled on USS Iowa.

Bulkhead 3   FR50

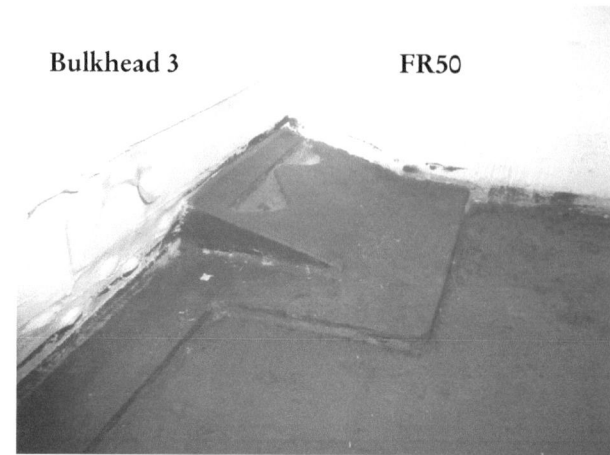

Photograph 2.7: The Start of the Port Upper Belt on USS New Jersey at Second Deck, FR50. The support for second deck shelf covers most of the armor. However, part of the upper face is visible between the scalloped edge. The front face is not beveled as on USS Iowa.

FR50

Bulkhead 3

Photograph 2.8: The Start of the Port Upper Belt on USS Iowa at Second Deck, FR50. The only part of the armor visible is a small area at the corner at the upper left. Uniquely, the forward edge of the belt is sloped. (Jim Kurrasch)

2.9. Lower Belt Extension

The top of the citadel drops from second deck to third deck at FR166. From FR166 to FR189 the belt extends from second platform to third deck to account for the lower citadel and the rise of the hull at the stern. The armor belt consists of ten plates on each side over this range. On USS *Iowa* and USS *New Jersey* the plates are STS and on the later ships they were Class B. Like the belt plates forward of FR166, these plates taper at knuckles from top to bottom. Unlike those of the forward plates, the knuckles over this range slope downwards towards the aft end. From FR166 to FR174 the upper thickness of the plates is 13 inches. That increases to 13.5 inches between FR174 and FR189. The plates were ordered $\frac{3}{8}$" oversized at the sides and bottom.

2.9.1. Machining at the Yard

The mold loft directed the machining of the oversized plates to fit their assigned locations. Except at FR166 and FR189, the joints between the plates were cut so that the inside edge was aligned with their frame or half frame location and that the butts are normal to the upper edge at the third deck molded line. At FR166 and FR189 the butts were cut normal to the centerline.

The yard machined a keyed hook joint along the upper edge of the plates where the belt was later joined to third deck armor. The yard also drilled and tapped holes for lifting bolts in the tongue of the joint.

To align the plates, the yard machined keyways at the plate edges. Except at FR189, the keyways are rectangular. The joint at FR189 uses an hourglass keyway. The yard machined a welding groove on the inside face at FR189.

2.9.2. Plate Installation

After the plates were trimmed to fit, the knuckle lines of adjoining plates were no longer aligned. The outside edge of the plates had to be grinded to make them flush so that scalloped cover plates could be welded over both sides of the joint. At FR174, where the plate changes thickness, the outer corner of the thicker aft plate was grinded to a 45 degree angle. The outer cover plate was made thicker on the forward side to account for the different in plate thickness. At FR166, the machining of the outer joint was more complex. The upper part of the joint was planed so the faces of

both plates matched. The lower part of the joint was machined at a 45 degree angle. The scalloped cover plate was machined to fit the outer joint.

At FR189, the yard hammered an hourglass key into the joint that both aligned and secured the two plates. The inside face of the joint was welded along the inside face. A rectangular cover plate was attached over the outside face of the joint using tap rivets.

Frame	Lower		Upper	
	HT	HB	HT	HB
FR166	17'-6"	34'-6 $\frac{5}{8}$"	34'-6 $\frac{9}{32}$"	40'-4 $\frac{15}{16}$"
FR189	21'-9 $\frac{3}{4}$"	15'-2 $\frac{5}{8}$"	37'-2 $\frac{5}{8}$"	20'-6 $\frac{5}{16}$"

Table 2.9: Molded locations for the insider corners of the belt. The belt in this area extends from second platform to third deck. The top of the second platform plating is $\frac{5}{16}$" above the molded line. The plates had to be trimmed at the bottom edge to account for the deck plating. The molded line of third deck is the bottom of the plating.

Plate	Weight (lbs)
L-B-47 (-P or -S)	55,941
L-B-48	55,747
L-B-49	56,818
L-B-50	61,301
L-B-51	77,981
L-B-52	79,409
L-B-53	80,695
L-B-54	81,840
L-B-55	82,842
L-B-56	84,086
Total Per Side	716,660
Grand Total	1,433,320

Table 2.10: Lower Belt Plate Weights FR166–FR189

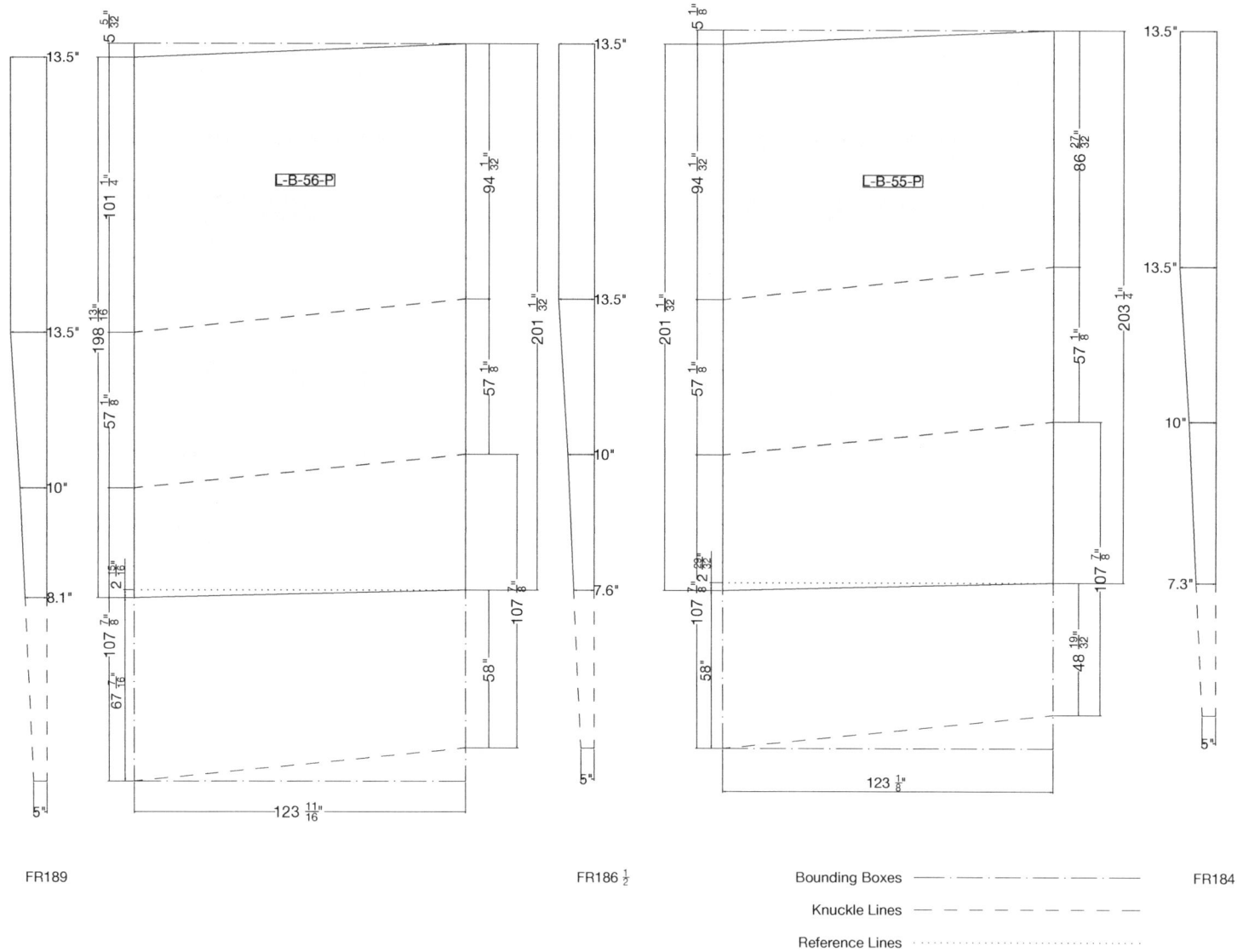

Figure 2.54: Mill sizes for the port lower belt plates between FR184 and FR189. The starboard plates are mirrors. The yard had to trim the plates under the direction of the mold loft.

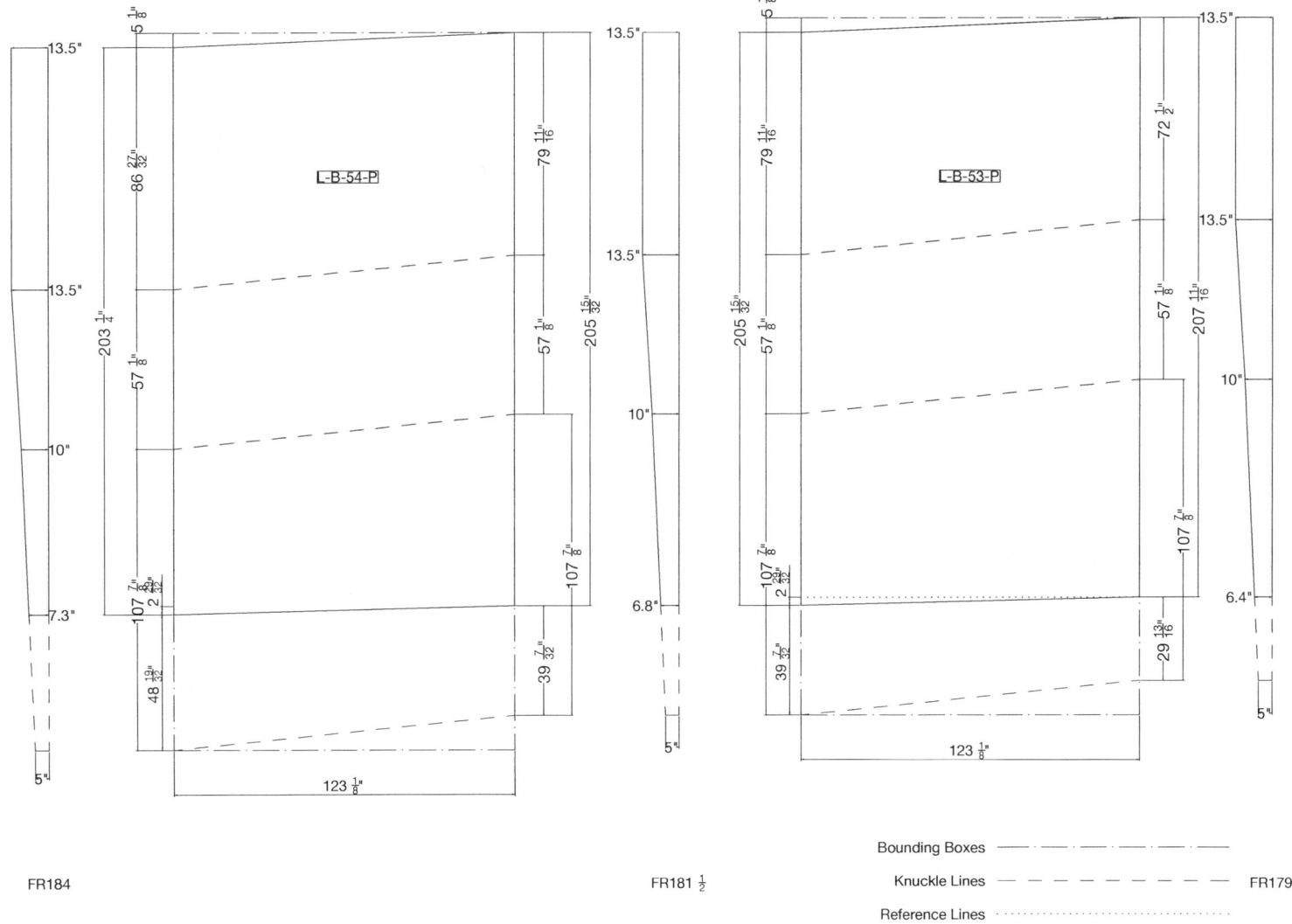

Figure 2.55: Mill sizes for the port lower belt plates between FR179 and FR184. The starboard plates are mirrors.

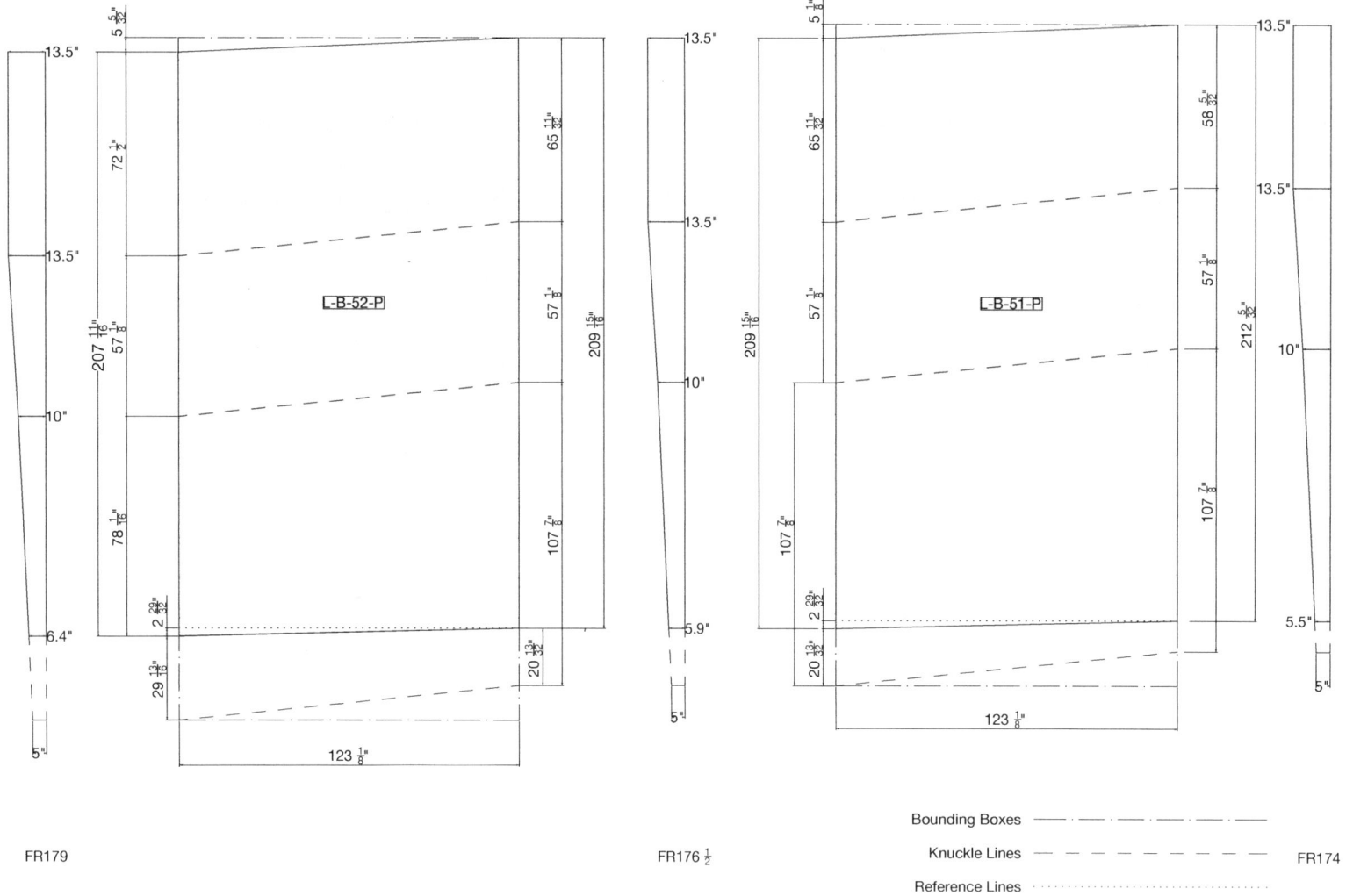

Figure 2.56: Mill sizes for the port lower belt plates between FR174 and FR179. The starboard plates are mirrors.

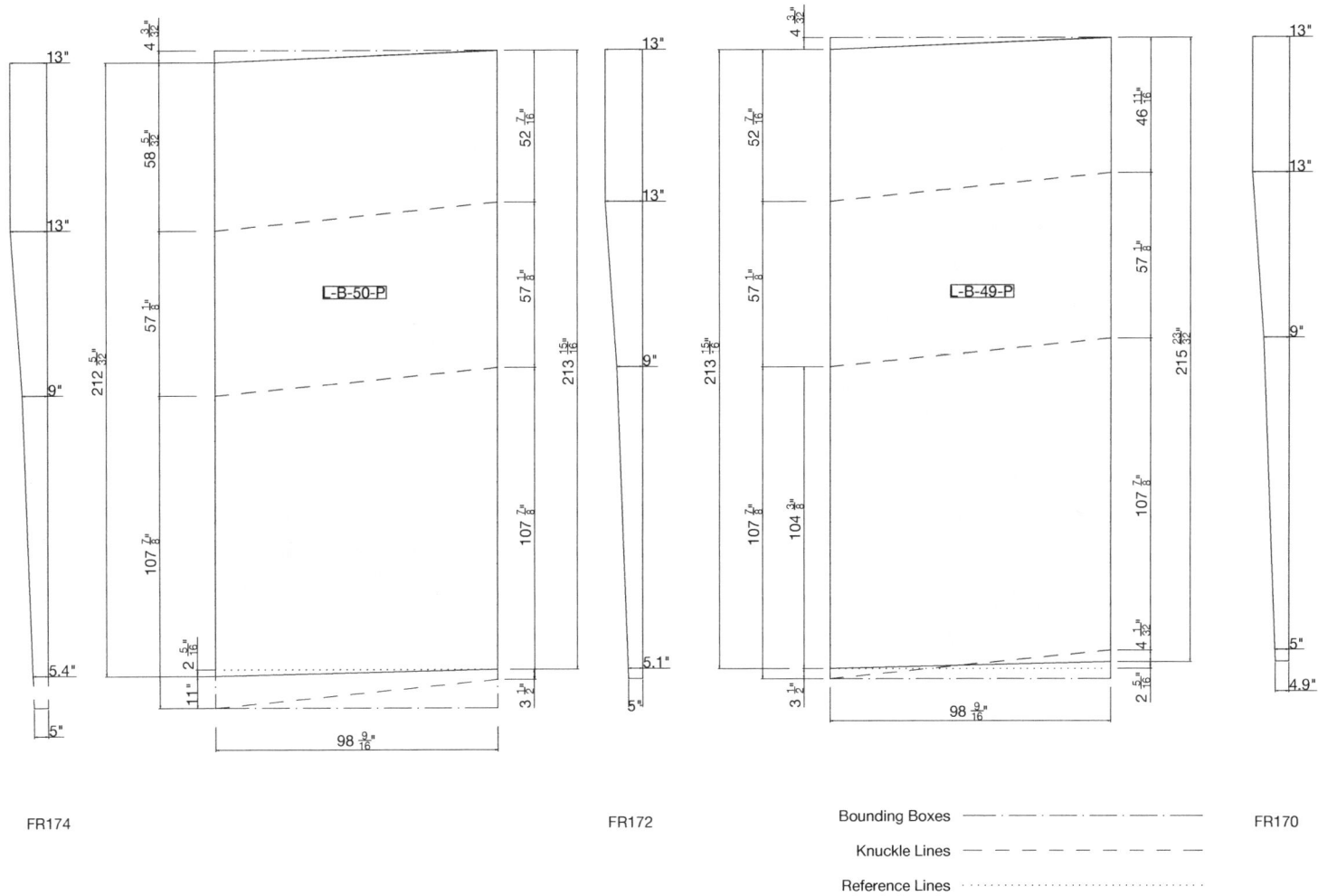

Figure 2.57: Mill sizes for the port lower belt plates between FR170 and FR174. The starboard plates are mirrors. These plates are thinner than those aft of FR174.

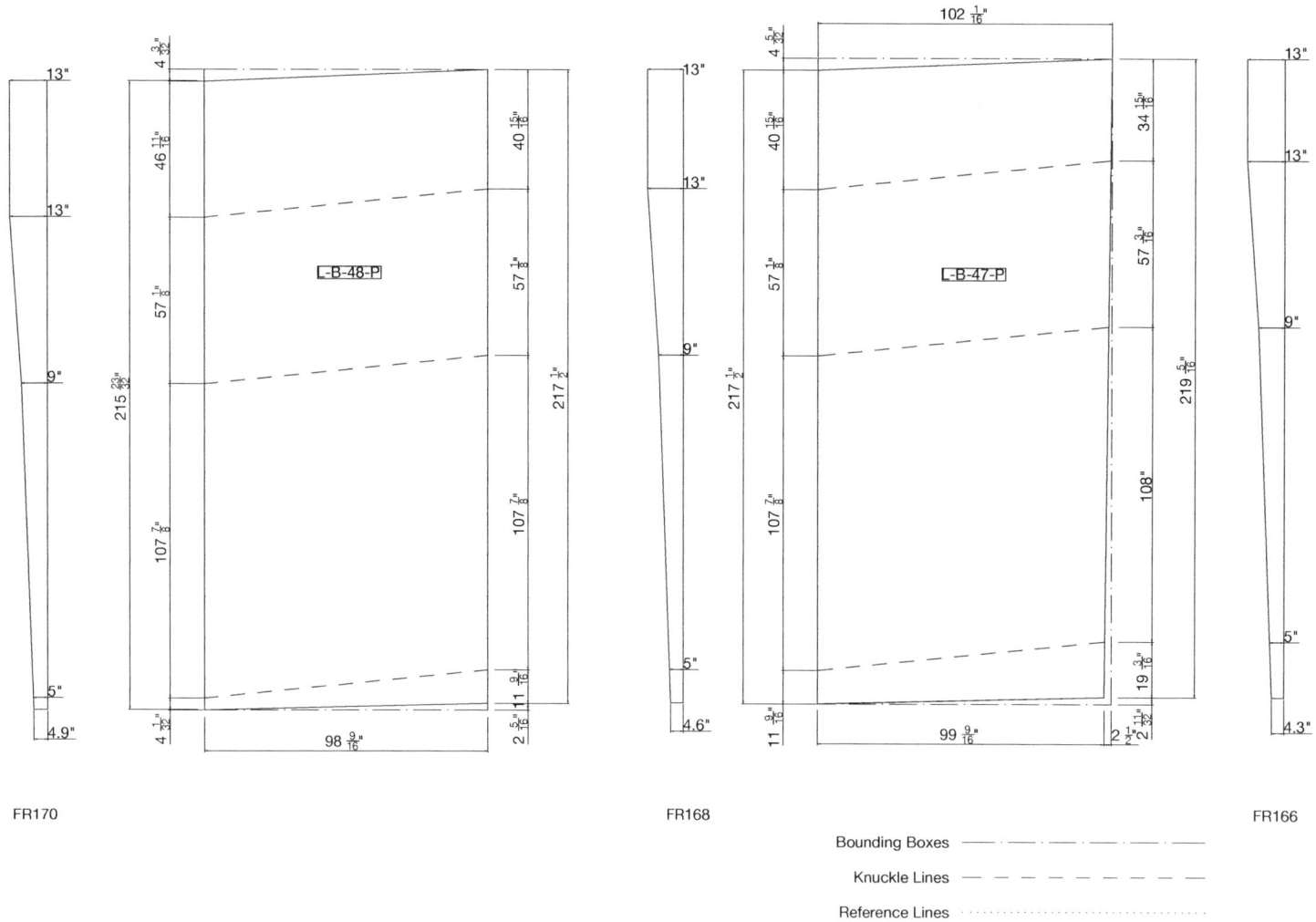

Figure 2.58: Mill sizes for the port lower belt plates between FR166 and FR170. The starboard plates are mirrors.

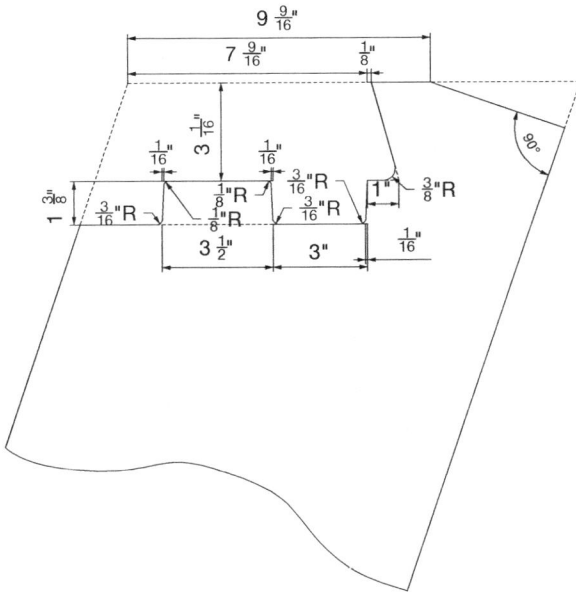

Figure 2.59: Upper Edge of the Belt (1:6). The yard machined a keyed hook joint into the upper edge of the lower belt plates. The Third Deck Armor section describes how that deck connected to the belt.

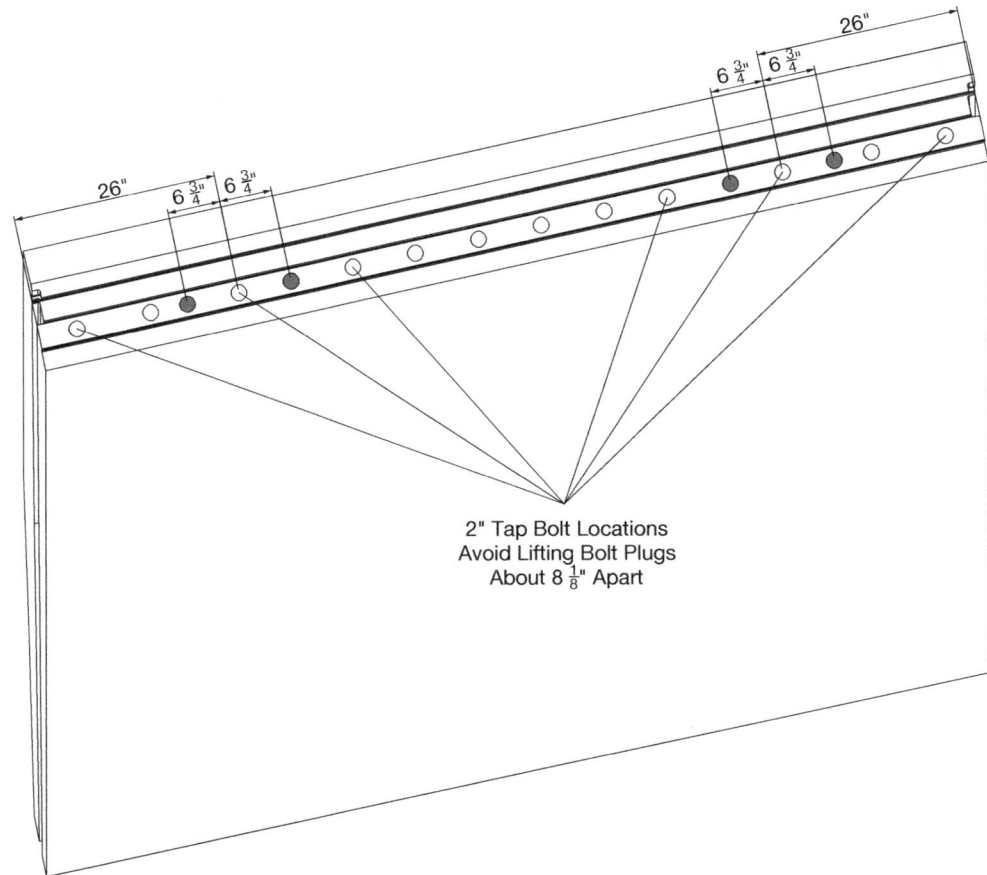

2" Tap Bolt Locations
Avoid Lifting Bolt Plugs
About 8 $\frac{1}{8}$" Apart

Figure 2.60: Location of Lifting Bolt Holes (1:24). The yard drilled and tapped holes for lifting bolts and tap bolts used to secure third deck into the tongue of the hook joint. The lifting bolt holes were plugged after the plate was in position. Presumably, the lifting bolt hole locations were marked on the intact area of the upper face and the tap bolt holes were drilled after third deck was in place to avoid those locations.

37'-2 $\frac{5}{8}$" ABL

Welding Groove

28'-11 $\frac{5}{8}$" ABL

Photograph 2.9: The seamstrap at FR196 starboard on USS New Jersey. This is at first platform. The belt armor is at the right and the transverse bulkhead to the left is welded to the seamstrap.

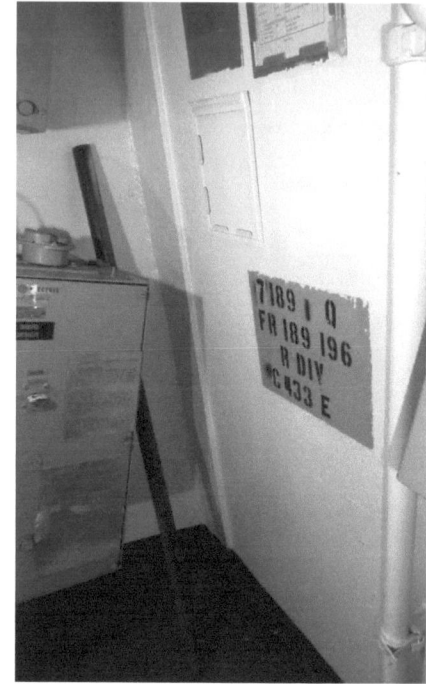

25# STS

20°

$\frac{1}{4}$"R

6 $\frac{1}{8}$"

Inboard

3 $\frac{7}{16}$"

3"

$\frac{1}{8}$"R $\frac{1}{8}$"R

3"

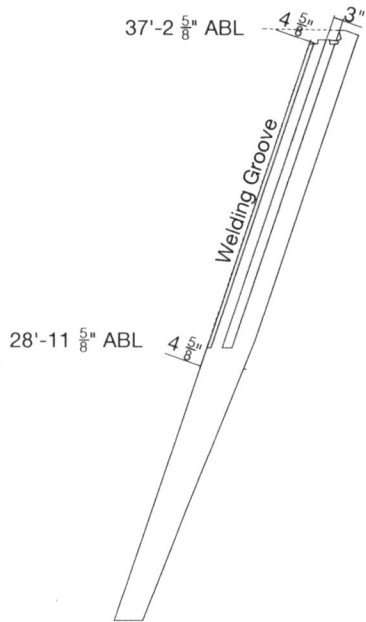

Figure 2.61: Elevation of Keyway at FR189 Looking Forward (1:60). The joint at FR189 connects the lower belt extension to the steering gear armor. Forward of FR189 the lower belt extends down to second platform while aft its only extends down to first platform. The keyway and welding groove only extends down to first platform at this location.

13.5" Class A

Class B
Thickness Varies

Shim   Outboard

18 $\frac{1}{2}$" 30# STS
Seamstrap

15# HTS

Figure 2.62: Section Through Keyway at FR189 Normal to Slope from Centerline (1:6). The joint at FR189 uses an hourglass key and keyway. The key was hammered into the keyway from the top after the plates were in place. The joint is welded on the inboard side and a seamstrap was tap riveted over the outboard side. The layout of the seamstrap is shown in Figure 2.92. Bulkheads were welded to either side of the armor.

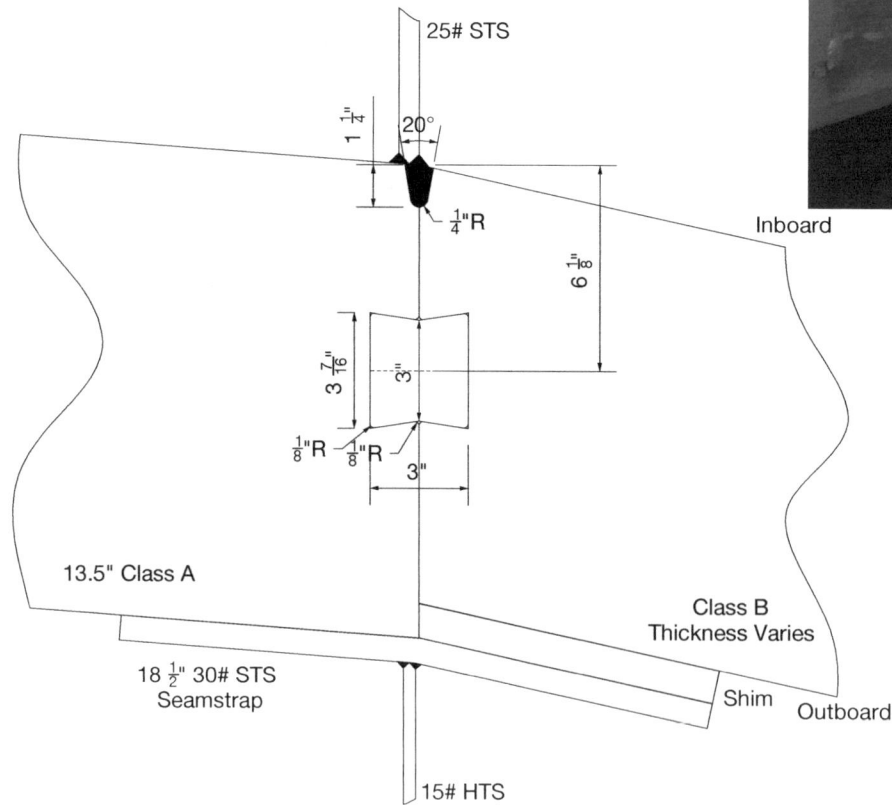

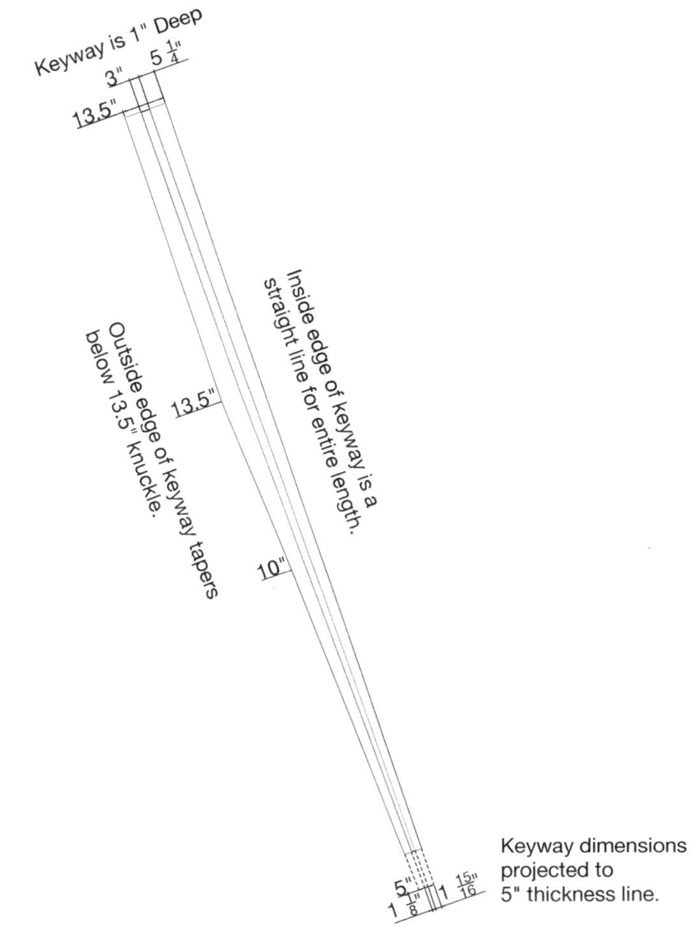

Keyway is 1" Deep
3" 5 1/4"
13.5"

Inside edge of keyway is a
straight line for entire length.

Outside edge of keyway tapers
below 13.5" knuckle.

13.5"

10"

5" 1 15/16"
1 1/8"

Keyway dimensions
projected to
5" thickness line.

View Parallel to Inside Face

At 5" Thickness   At Top
1 1/8"   3"

Inboard   Outboard

2"   2"
1 15/16"
1/4"R   1/4"R
5 1/4"

5/16"R
1 15/16"   1 15/16"   Key
1 1/16"   2 15/16"   5/16"R

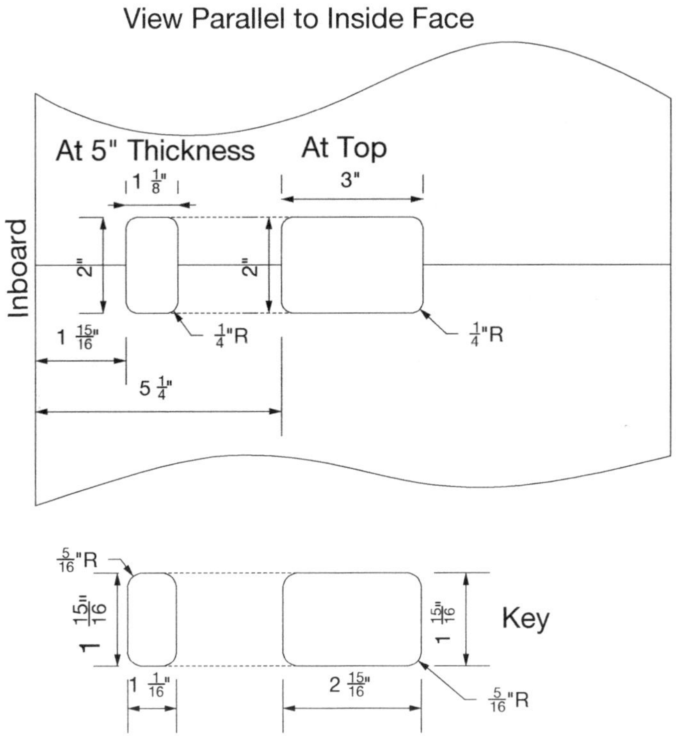

Figure 2.64: Section Through Keyways FR174-FR186 Normal to Inside Face (1:4). The keys are slightly smaller and their radii are slightly larger than the keyway.

Figure 2.63: Elevation of Keyways FR174–FR186 (1:60). The joints within the lower belt extension were aligned with a tapered key. The widths of the keyways in this range are measured from the top of the plate before machining the hook joint to the theoretical 5" thickness line that is located below the bottom of the plate edge.

Keyways are 1" Deep

2 7/8"   5 1/16"
13"

2"   7/8"

13"

Outside edge of keyway tapers
below 13" knuckle.

9"

5"
1 1/8"   1 15/16"

At FR172 the keyway
dimensions project past the
bottom of the plate to the
5" thickness line

2 7/8"   5 1/16"
13"

2"   7/8"
18"

Inside edge of keyway is
straight for entire length.

9"

5"
1 1/8"   1 15/16"   Measured at 5" Thickness Line

At FR168 and FR170 the
keyway extends past the
5" thickness line.

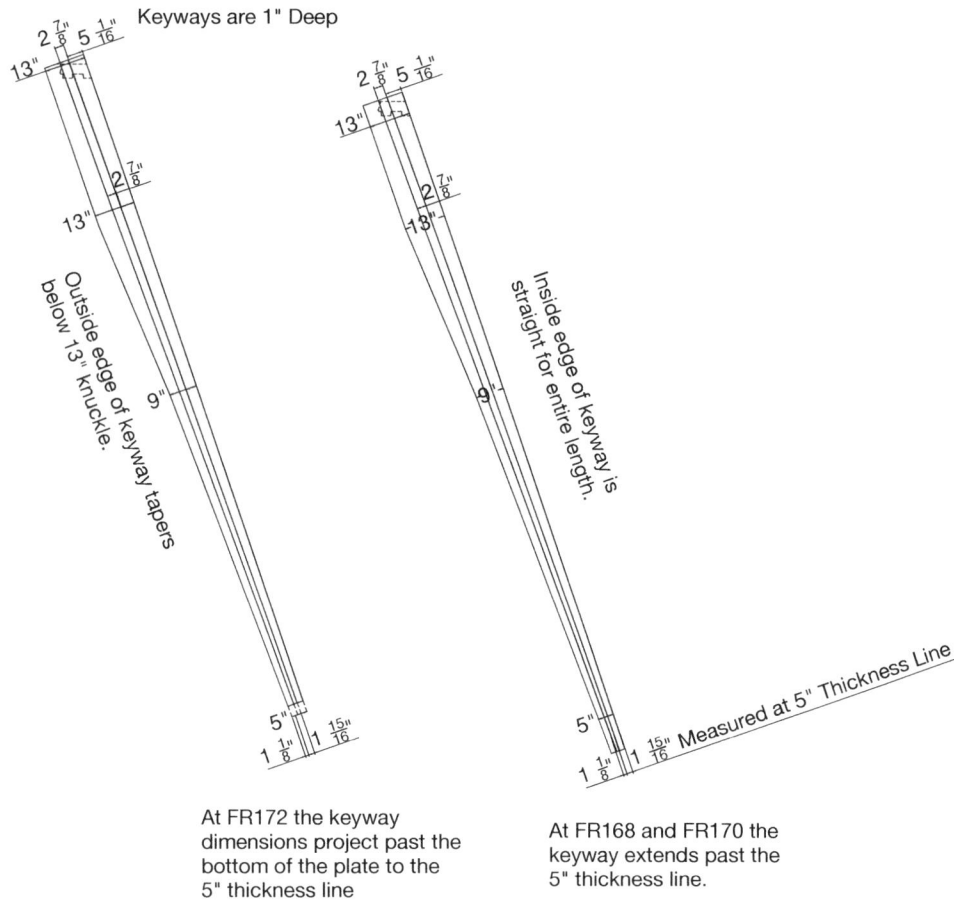

Figure 2.65: Elevation of Keyways FR168–FR172 (1:60). The plates forward of FR172 are thinner than those aft so the keyway location is adjusted to compensate. The keyway was measured at the top of the plate and at the 5" thickness line. At FR172 this is a theoretical location below the bottom of the plate.

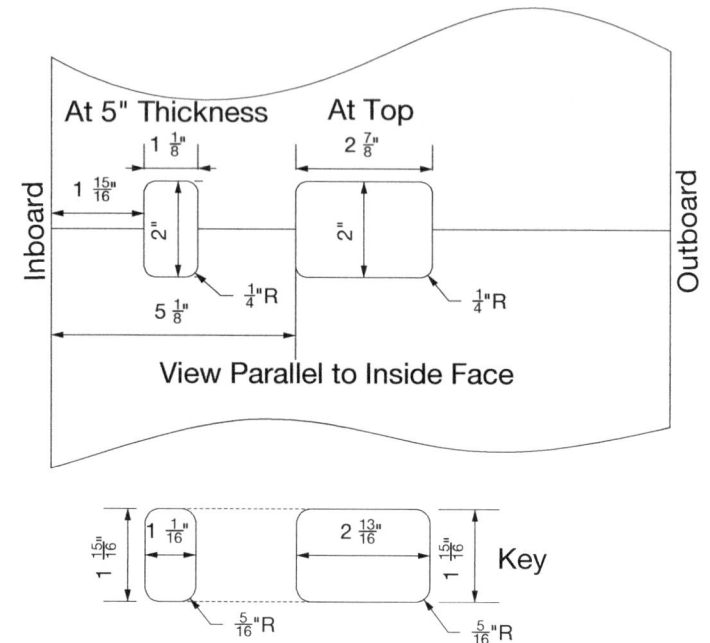

Inboard   Outboard

At 5" Thickness   At Top

1 1/8"   2 7/8"

1 15/16"

2"   2"

1/4"R   1/4"R

5 1/8"

View Parallel to Inside Face

1 15/16"   1 1/16"   2 13/16"   1 15/16"

1 1/16"   5/16"R   5/16"R   Key

Figure 2.66: Section Through Keyways FR168–FR172 Normal to Inside Face (1:4). The keyways were measured from the top of the plate before machining the hook joint to the 5" thickness line. At FR168 and FR170 the plates extend below this location so the bottom of the keyway is narrower than the measurement shown.

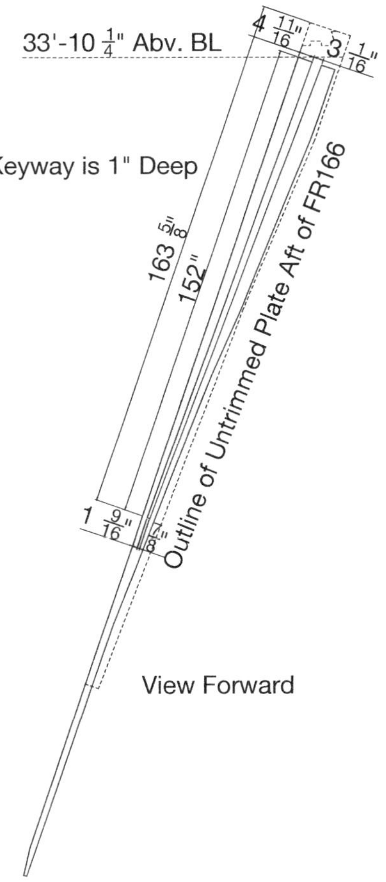
33'-10 1/4" Abv. BL
4 11/16"
3 1/16"
Keyway is 1" Deep
163 5/8"
152"
Outline of Untrimmed Plate Aft of FR166
1 9/16"
7/8"
View Forward

Figure 2.67: Elevation of Keyway FR166 (1:60). The forward plate is shown shaded and the outline of the aft plate is the dashed line. This keyway does not extend the full length of the shorter aft plate because the forward plate becomes too thin at the bottom and is too short at the top.

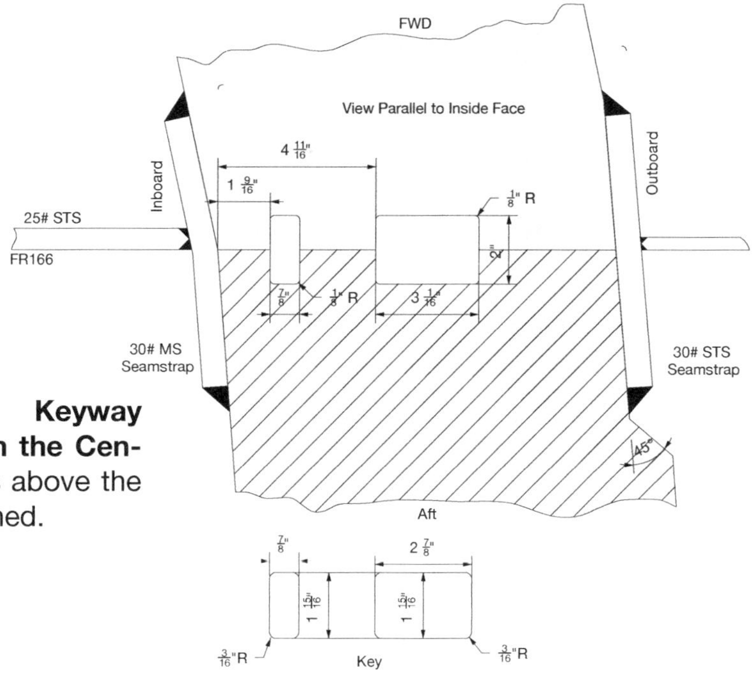
FWD
View Parallel to Inside Face
Inboard
Outboard
4 11/16"
1 9/16"
1/8" R
25# STS
FR166
1/2"
7/8"
1/8" R
3 1/16"
30# MS Seamstrap
30# STS Seamstrap
45°
Aft
7/8"
2 7/8"
1 15/16"
1 15/16"
3/16"R
Key
3/16"R

Figure 2.68: Section Through Keyway FR166 Normal to the Slope from the Centerline (1:4). The aft plate extends above the forward plate so it is shown sectioned.

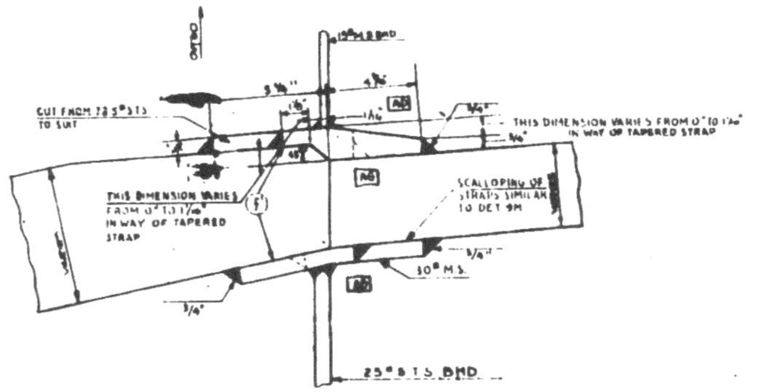

Plan 2.10: (1302BU) Towards the bottom of the joint at FR166 the difference in plate thickness required the seamstrap to be machined thicker on the forward side of the joint.

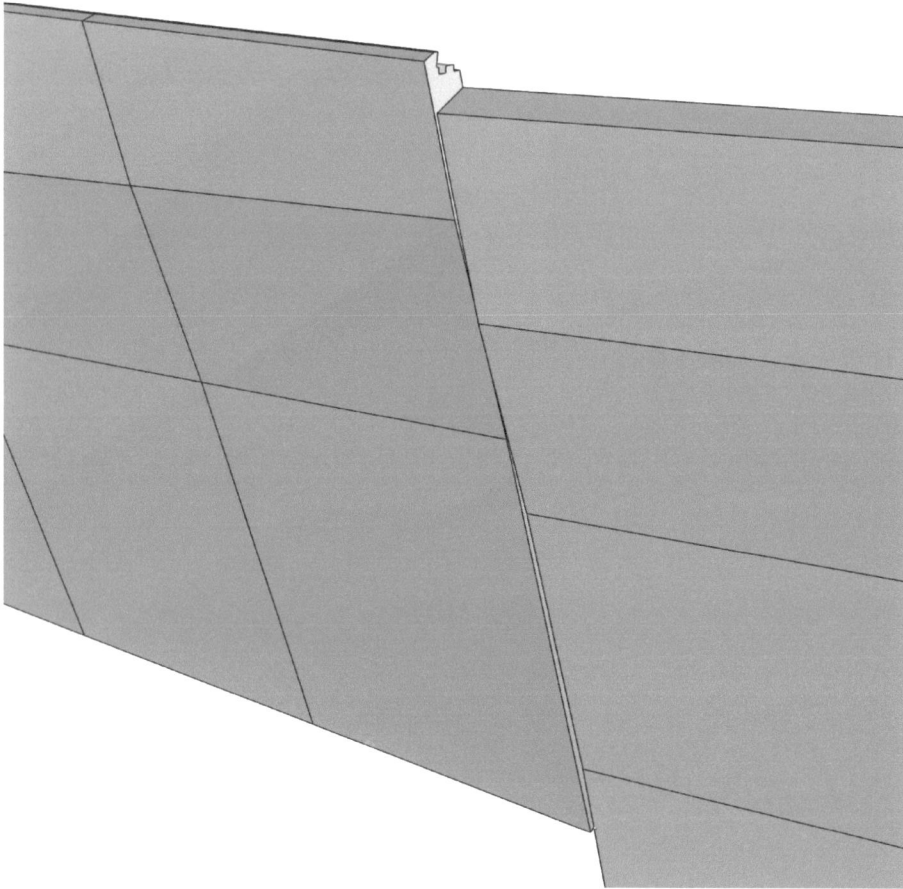

Figure 2.69: Rendering of the Outside Joint at FR166 Before Planing. At FR166 the configuration of the lower belt plates changes. This creates discontinuities along the outer face. At the top and bottom the aft plate is thicker and near the middle the forward plate is thicker.

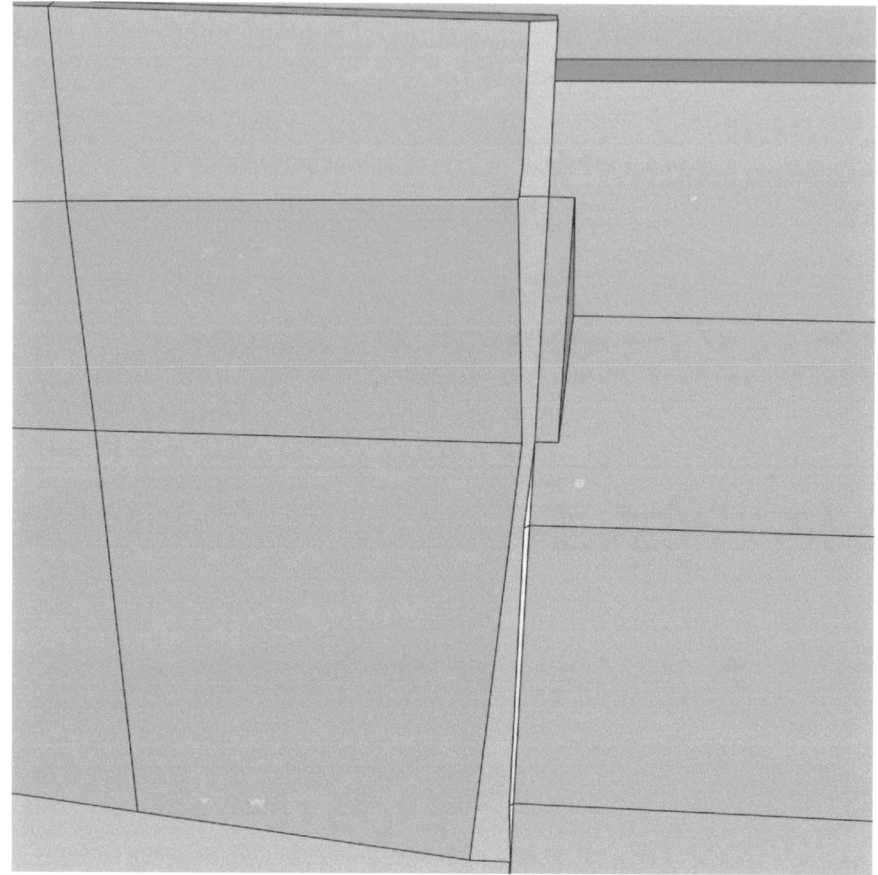

Figure 2.70: Rendering of the Outside Joint at FR166 After Planing. The joint at FR166 was planed to allow seamstraps to fit over the joint. The upper part of the joint was planed even while the lower part of the joint has a 45 degree angle. The outer corners of the aft plates at FR172 (where the plate thickness changes) were cut at a 45 angle along their entire length to mate with outer faces of the forward plates.

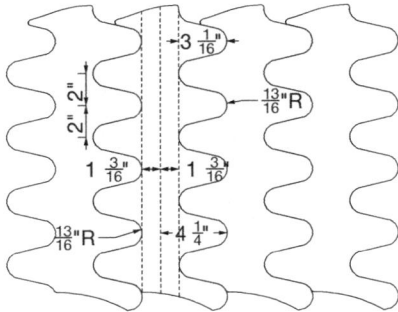

Figure 2.71: Seamstraps (1:12). This is the pattern for cutting the seamstraps welded over the plate joints. Multiple seamstraps were normally cut from a single plate as shown. The cutting torch would create lossage so the seamstraps would be slighly smaller than indicated in the plans.

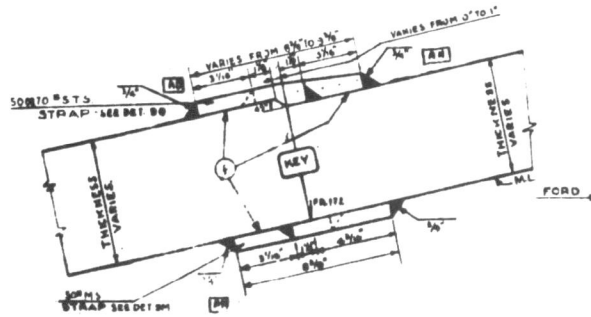

Plan 2.11: (1302BU) At the lower part of the joint at FR166 and the joint at FR172 the difference in plate thicknesses on either side of the joint precluded planing the entire outer face of the joint flush. In such areas the scalloped seamstrap was made thicker on one side of the joint as shown in this plan.

Figure 2.72: Connection of Belt Armor to Second Platform. The lower belt was attached to second platform using a scalloped angle welded over a scalloped kick plate. The outer face was caulk welded to the deck to make the joint watertight.

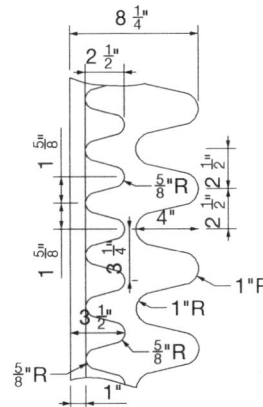

Scallop Pattern for
4×4×⅝" Angle
on top of
40# Kickplate (Left)
and for the Kickplate (Right)

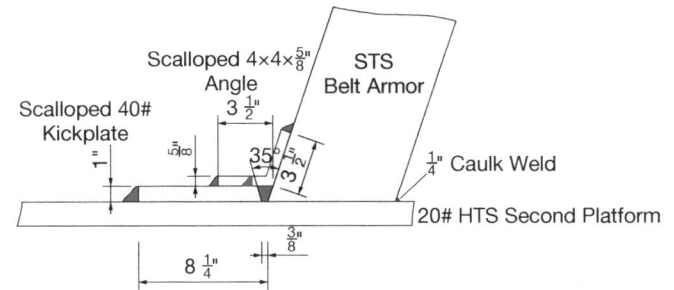

Belt Armor

60# STS First Platform

$\frac{3}{4}"$

$2\frac{1}{2}"$

$3\frac{1}{2}"$

$\frac{5}{8}"R$

Outboard

$3\frac{1}{4}"$

$1\frac{3}{8}"$

Inboard

25# STS BHD

25# HTS BHD

20# HTS Second Platform

25# HTS LONGL #4

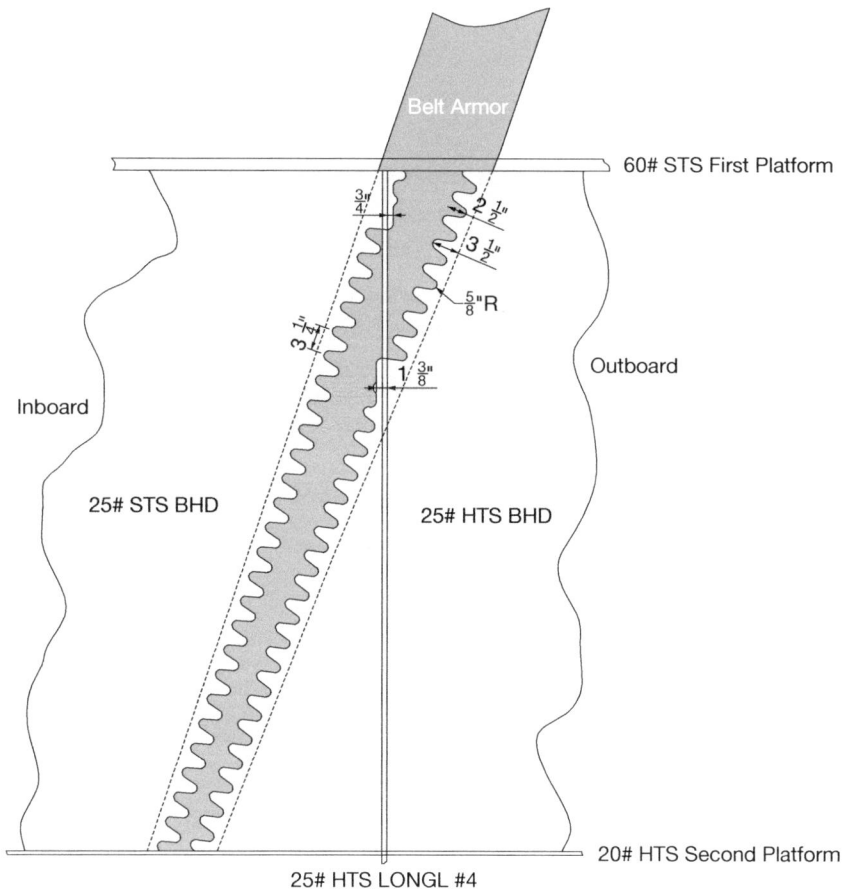

Figure 2.73: Attachment of Belt to Transverse Bulkhead at FR189 Looking Foward. Second platform ends at a transverse bulkhead at FR189. The lower edge of the belt jumps up to first platform at this location. The designers used the belt armor as a transition point in the bulkhead. The part of the bulkhead that closes off the end of the citadel is special treatment steel while the part outside the citadel is less expensive high tensile steel. The edges of the bulkhead plates were scalloped and welded to the belt plate forward of FR189. Longitudinal 4 attaches to the bulkhead in this area. The scallops were truncated where the longitudinal meets the belt plate to make trimming the longitudinal simpler in the vicinity of the scallops.

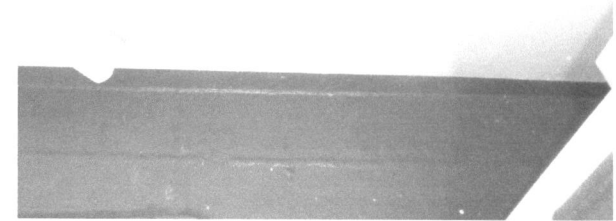

Photograph 2.12: The caulk weld on the outside of the belt around the FR 188 on USS New Jersey.

Photograph 2.13: Second platform on USS Iowa looking forward at FR173 port side from inside the citadel. The belt armor is to the left. The scalloped cover plates linking the armored plates are clearly visible. The scalloped kick plate and angle attaching the belt to the deck is also visible. The beams supporting first platform are welded directly to the armor or, in the case of the beam in the background at FR174, welded over the scalloped cover plates. (Jim Kurrasch)

2.10. Steering Gear Armor

The steering gear armor extends between FR189–FR203 longitudinally and from first platform to third deck vertically. Transverse armor at FR203 links the two sides of the belt armor.

2.11. Armor Plates

The steering gear belt armor consists of two 13.5-inch Class A plates per side (S-G-1-S and S-G-1-P forward and S-G-2-S and S-G-2-P aft). The belt plates are sloped 19° (designed 18° 59' 10) from the centerline and the belt plates at each side meet at a knuckle with an angle less than one degree. The molded lines of the inner and outer faces are rhomboidal with the sides being parallel and vertical. The upper and lower edges are sloped to conform to the angles of third deck above and first platform below.

The transverse armor consists of single 11.3-inch Class A plate (S-G-203) whose fore and aft faces are vertical and sides are angled to match the belt plates. The molded lines of all the faces are trapezoidal. Because of the deck slopes above and below, the upper and lower edges are higher at the aft face than the inner face.

The specified dimensions of the transverse

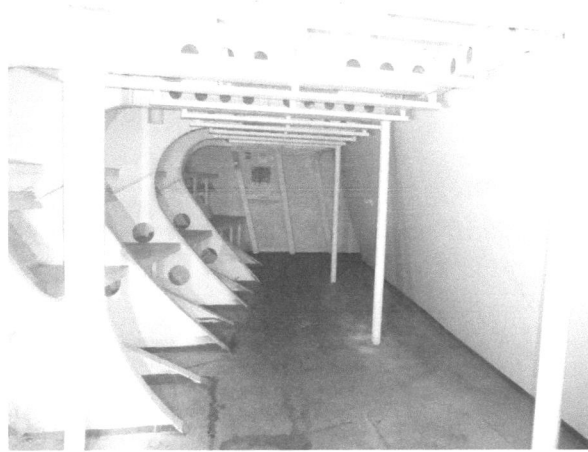

Photograph 2.14: The first platform storeroom at FR189 looking aft. The steering gear armor (plate S-G-2-S) runs along the right side of the photograph. The beams supporting third deck do not extend to the armor. The beams are supported by the stanchions. The bulkhead in the background is part of an aviation fuel tank.

plate are unusual in that the plate extends beyond its molded lines. The molded lines conform to the pre-machining shape of the adjacent belt plates. The plans define a tongue on either side that extends beyond the molded lines. The physical dimensions of the transverse armor plate are the molded lines plus the tongue dimensions (see below).

2.12. Machining at the Mill

Because the steering gear armor consists of Class A plates, most machining of the plates took place at the mill. The upper edges of the side plates have a facet that slopes downwards and a keyed hook joint machined into their upper edge. The hook joint flows around the angle between the aft belt plates to the transverse plate. Along the transverse plate, the hook joint was machined at a slight downward slope in order to match the slope of the upper edge of the belt plates. The joint is also slightly narrower at the transverse plate than on the belt plates to compensate for its thinner armor.

The steering gear plates were kept in alignment in two different ways. First, a tongue and groove joint aligns the aft belt plates (groove) and the transverse plate (tongue). Second, hourglass keyways were machined into the edges between the belt plates. In addition, the inside (soft) edges of these joints were machined with a

	Frame	Lower		Upper	
		HT	HB	HT	HB
	189	28 $11\frac{5}{16}$"	17 $8\frac{1}{8}$"	37 $2\frac{5}{8}$"	20 $6\frac{5}{16}$"
	196	30 $1\frac{5}{16}$"	15 $9\frac{3}{8}$"	38 $\frac{1}{2}$"	18 $6\frac{1}{8}$"
	203	31 $3\frac{5}{16}$"	13 $5\frac{3}{8}$"	38 $10\frac{5}{16}$"	16 $\frac{11}{16}$"
11.3" Aft of	203	31 $3\frac{25}{32}$"	13 $4\frac{7}{16}$"	38 $10\frac{21}{32}$"	15 $\frac{11}{16}$"

Table 2.11: The locations of inner corners of the molded lines of the belt plates of the steering gear armor. These also give the molded locations of the eight corners of the molded lines of the transverse armor plate at FR203,

groove for welding.

Each plate had eight holes drilled and tapped at the top for attaching lifting pads. These holes were plugged after the armor was in place. The steering gear plates used K type lifting pads. When lifting the transverse plate, liners were added to either side of the tongue below the pad in order to compensate for it being narrower than on the transverse plates.

2.13. Plate Installation.

2.13.1. Belt to Transverse Armor Connections

The transverse to belt armor joints have a tongue and groove joint to preserve alignment. A diagonal 40# STS seamstrap was tap riveted over the outside of the joint. This area was not face-hardened. At the inside, the armor plates were welded to a scalloped 5×5×$\frac{1}{4}$" angle placed over the joints.

2.13.2. Belt Plate Connections

The two Class A belt plates at each side meet at a knuckle. While the plates have the same slope normal to the centerline, they have different

Plate	Weight (lbs.)
S-G-1-S	127,524
S-G-1-P	127,524
S-G-2-S	125,660
S-G-2-P	125,660
S-G-203	99,070
Total	605,438

Table 2.13: Steering Gear Armor Plate Weights

Ship	Supplier
USS *Iowa*	Carnegie-Illinois Steel
USS *New Jersey*	Carnegie-Illinois Steel
USS *Missouri*	The Midvale Co.
USS *Wisconsin*	Carnegie-Illinois Steel
USS *Illinois*	The Midvale Co.
USS *Kentucky*	Carnegie-Illinois Steel

Table 2.12: Suppliers of Steering Gear Armor

slopes normal to their inside edges. Therefore, when the plates are aligned at their inside edge, the outside edges are not aligned. Consequently, the outer faces of the aft belt plates had to be planed to match the surface of the plate forward. This area of the plate was not face-hardened. The two belt plates at each side of the joint were joined on the outside using a 30# STS seamstrap tap riveted over the joint. The plates were welded along the groove machined into the inside faces and an hourglass key was hammered into the keyway that had been machined into the sides of the plates.

The connections to the Class B belt plates forward are nearly identical using a key, seamstrap, and weld. A 30# STS seamstrap was tap riveted over the outer joint. Unlike the joints between the Class A plates, the seamstrap was bent to follow the angle between the plates and was shimmed at the aft side of the joint using a liner that would have had to be about $\frac{1}{4}$" thick. The inner faces of the plates were welded along the groove machined into the plates and an hour-glass key was hammered into the keyway machined into the side of the joint.

2.13.3. Transverse Armor to First Platform

The aft edge of the transverse armor was aligned to first plate form using an 5×3×$\frac{3}{4}$" angle that was cut with 4$\frac{1}{4}$" scallops on one face and cut down to 1$\frac{1}{2}$" on the other face. The scalloped face was welded to the deck with the other face pressed against the Class A armor plate.

A double layer of material was used on the inside face over most of the joint. A 40# kick-plate was cut with scallops. It was then welded to the deck along the scallops and to the inside face of the Class A armor along the straight edge. Scallops were cut into a 4×4×$\frac{5}{8}$" angle that was then welded on top of the kickplate and to the inside face of the transverse armor plate.

2.13.4. Belt Armor to First Platform

The plans specified two alternatives at outside joint of the Class A belt and first platform. Either the joint could be caulk welded or a 1$\frac{1}{2}$" flat bar would be welded to the deck upright against the armor with the upper joint caulked.

Two patterns were followed to secure the inside face of the Class A belt armor to first platform. The first was used between FR199$\frac{1}{2}$

and FR201 $\frac{1}{2}$ above the rudder bearing assembly. Here, a tapered 60# STS plate was cut with scallops that were welded to the deck and the edge was welded to the inside face of the Class A plates. Brackets were welded to the deck and armor plate along the inside face every two feet. Between each bracket location, the armor plate was secured by four tap rivets that were driven through first platform from below into the belt plates.

The second pattern was to use a double layer of scalloped bars. Scallops were cut into one edge of a 40# kickplate that was placed against the inside face of the armor. The scalloped edge was welded to the deck and the flat edge was welded to the base of the inside face of the armor. A $4\times4\times\frac{5}{8}"$ angle was scalloped on both faces then welded on top of the bracket and to the armor plate. These joints were the similar to those used between the transverse armor and first platform with the only difference being the belt plate joints had to take into account the slope of the inside face.

2.13.5. Transverse Armor Bulkheads

Bulkheads within the steering gear armor were welded directly to the soft face of the armor plates. Transverse bulkheads at FR189 and FR196 were welded to the seamstraps joining belt plates at those locations. A centerline bulkhead joins the aft face of the Transverse plate. Here a 30# high tensile steel plate was tap rived to the armor plate which was specified not to be face hardened in this area. The bulkhead was welded to that plate.

2.13.6. Connection to Third Deck Armor

See the section on Third Deck armor.

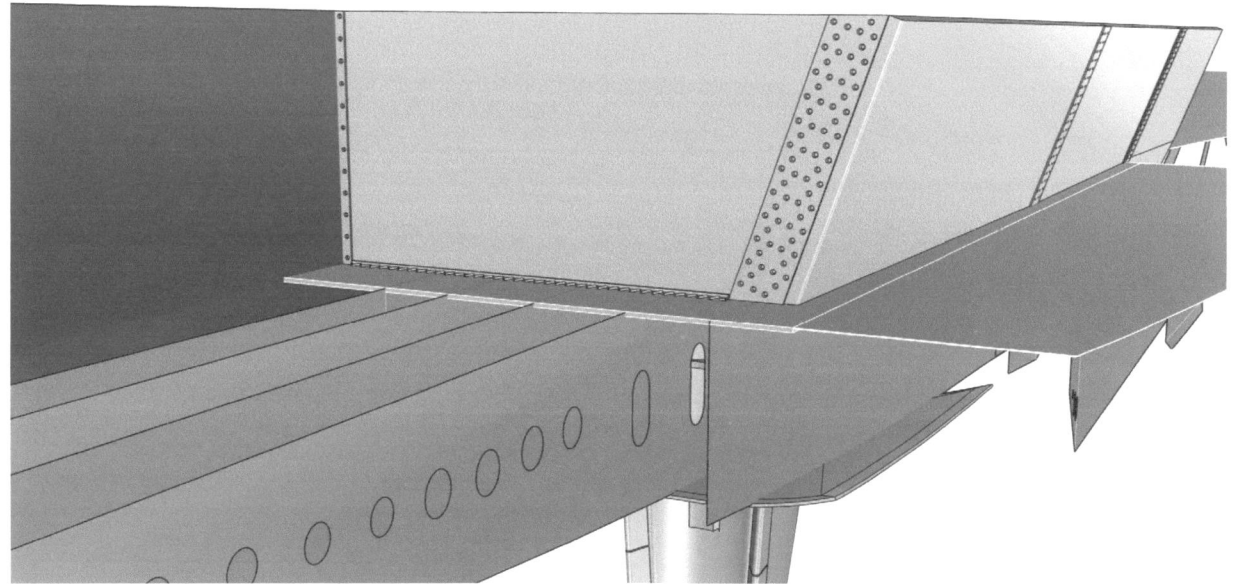

Figure 2.74: Rendering shows the steering gear armor viewed from the starboard side looking forward. The bulkheads and framing have been removed. First platform ends as shown. The area aft of the end of the deck is where the aviation fuel tanks are located. There is a transverse bulkhead (not shown) at the end of the deck that creates the fuel tank compartment. An additional transverse bulkhead connects to the seamstraps at the side and is not shown. To the left vertical keel rises to third deck. The vertical keel originally formed a non-tight bulkhead in this area and had openings to move between the fuel tanks on either side. In the 1980's the fuel tanks on the starboard side were removed, the bulkhead was made tight and the entire side became a larger fuel tank.

Lifting Pad Bolt Holes

$74\frac{3}{4}"$ $43\frac{1}{2}"$

$30\frac{7}{16}"$

$1"R$ $43\frac{1}{2}"$ $76\frac{1}{2}"$

$6\frac{3}{4}"$ $6\frac{3}{4}"$ $6\frac{3}{4}"$ $6\frac{3}{4}"$

$6\frac{3}{4}"$ $6\frac{3}{4}"$ $6\frac{3}{4}"$ $6\frac{3}{4}"$

Top Plan

Soft and Smooth $31\frac{9}{16}"$

9

$96\frac{3}{32}"$

$90\frac{7}{8}"$

S-G-2-P

Smooth

$6"$

$14\frac{15}{32}"$

End Elevation   Inside Elevation

$32\frac{3}{4}"$

$10\frac{1}{8}"$

$95\frac{3}{16}"$

$100\frac{11}{16}"$

End Elevation

$3\frac{1}{4}"$ $3\frac{1}{4}"$ $3\frac{1}{4}"$ $3\frac{1}{4}"$

$3\frac{1}{2}"$

$1"R$ $4\frac{1}{4}"$ $4\frac{1}{4}"$ $4\frac{1}{4}"$ $4\frac{1}{4}"$

$28\frac{15}{16}"$

Tap Offsets from Half Frames
Other Taps are Spaced $5\frac{1}{2}"$

1 $1.3"$ $6"$

Soft and Smooth

$10"$

Bottom Plan

FR203   FR202   FR201   FR200   FR199   FR198   FR197   FR196

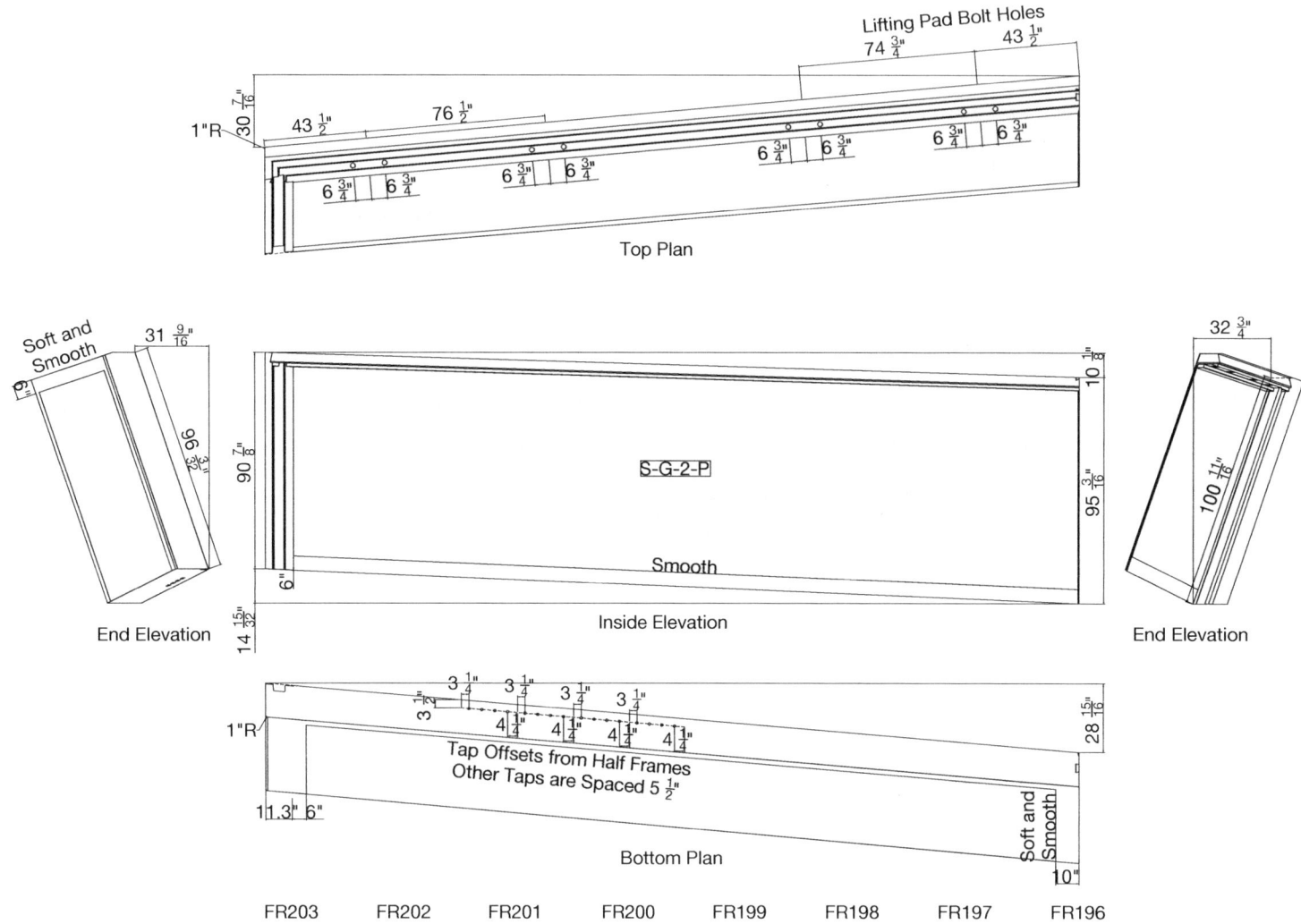

Figure 2.75: Steering Gear Plate S-G-2-P
(1:72). The starboard plate S-G-2-S is a mirror.

Lifting Pad Bolt Holes

$74 \frac{3}{4}"$ $43 \frac{1}{2}"$

$24 \frac{3}{16}"$ $43 \frac{1}{2}"$ $76 \frac{1}{2}"$

Plan of Top

Soft and Smooth $32 \frac{3}{4}"$

$6 \frac{7}{8}"$

$9"$

$100 \frac{11}{16}"$

$95 \frac{3}{16}"$

S-G-1-P

$99 \frac{5}{16}"$

Smooth

$14" \ 6"$ $6"$

End Elevation

Elevation of Inner Surface

$34 \frac{3}{16}"$

End Elevation

$22 \frac{3}{4}"$

$10"$ Soft and Smooth

Plan of Bottom

Soft and Smooth $10"$

FR196

FR189

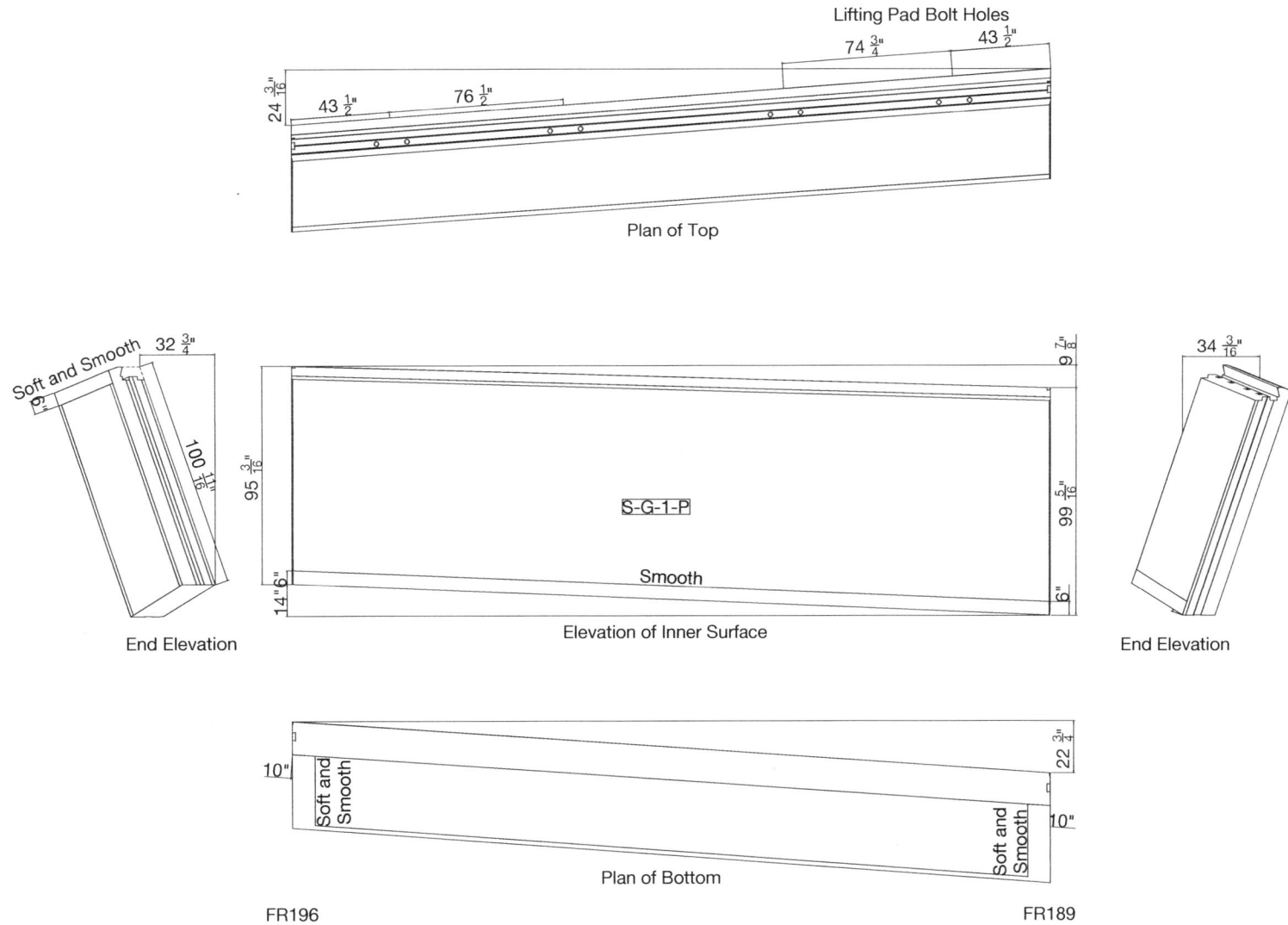

Figure 2.76: Steering Gear Plate S-G-1-P
(1:72). The starboard plate S-G-1-S is a mirror.

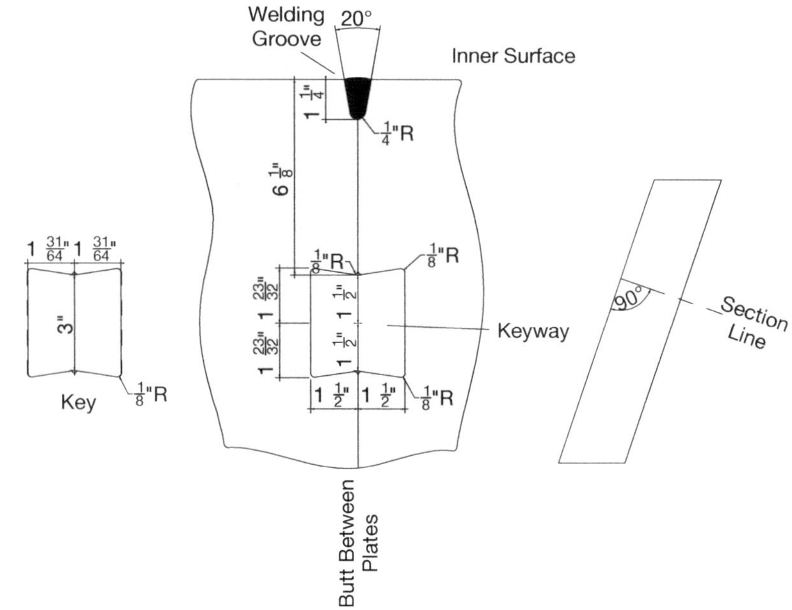

Figure 2.77: Keyway Between Steering Gear Belt Plates (1:6). The steering gear belt plates are linked to each other and the belt plate forward of FR189 with an hourglass key hammered into the keyway and at the inside edge by a weld. This type of joining was used at FR196 (Class A to Class A) and at FR189 (Class A to Class B). The parallel edges of the key have a $\frac{1}{64}$" gap from the keyway. The angles edges of the key are at the edge of the keyway.

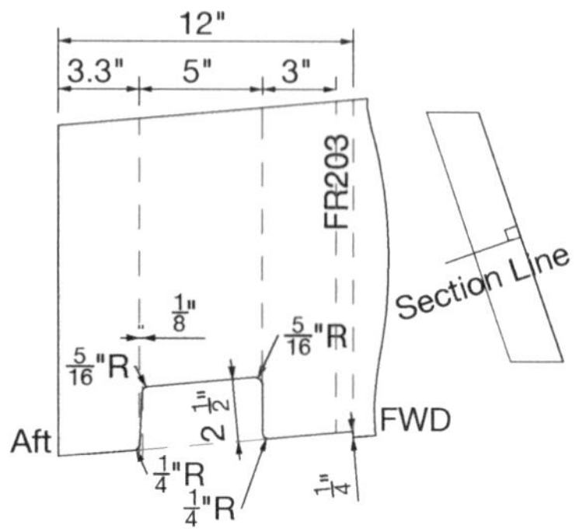

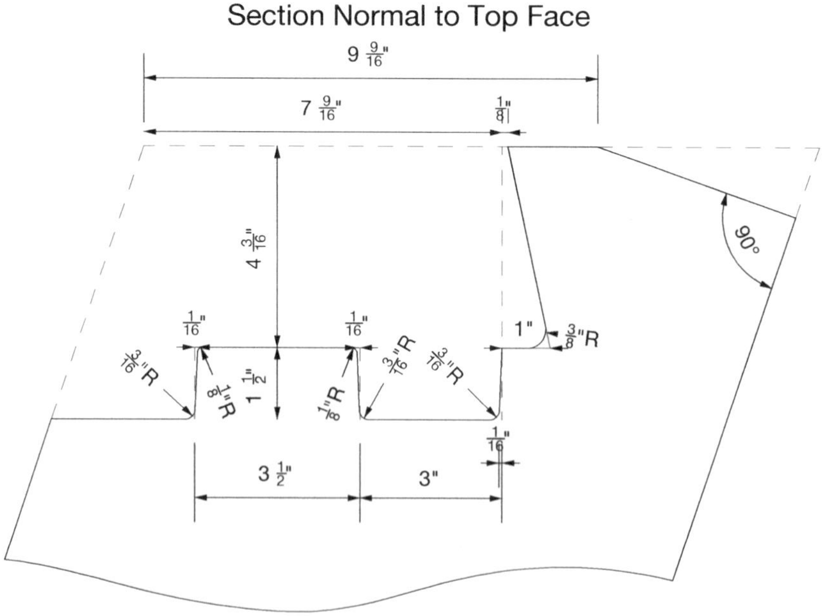

Figure 2.79: Upper Edge of Belt Plates (1:4). The mill machined a keyed hook joint into the upper edge of the steering gear belt plates. This joint is similar to, but deeper, than the joint between FR166 and FR189. Holes for lifting pad bolts were drilled centered on the tongue. The third deck armor plates were machined to match this joint.

Figure 2.78: Aft End of Belt Plate (1:12). The machining of the joint at the aft end of the steering gear belt (S-G-2-P looking down or S-G-2-S looking up) that mates with the transverse armor plate (S-G-203). The section here is rotated 19° from horizontal along the centerline.

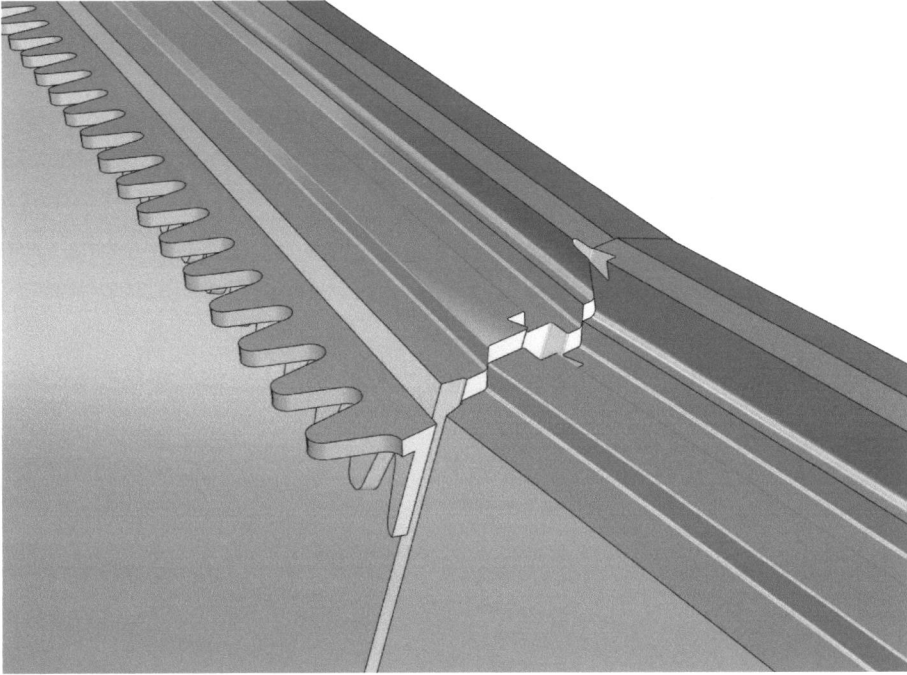

Figure 2.80: Belt at Frame 189. Forward is to the left. The transverse bulkheads and deck have been removed. The hook joint is shallower forward because of the thinner deck armor.

Figure 2.81: Rendering showing the keyway and welding groove machined into the sides of the belt plates. The keyway had to extend all the way to the top of the plate in order for the key to fit.

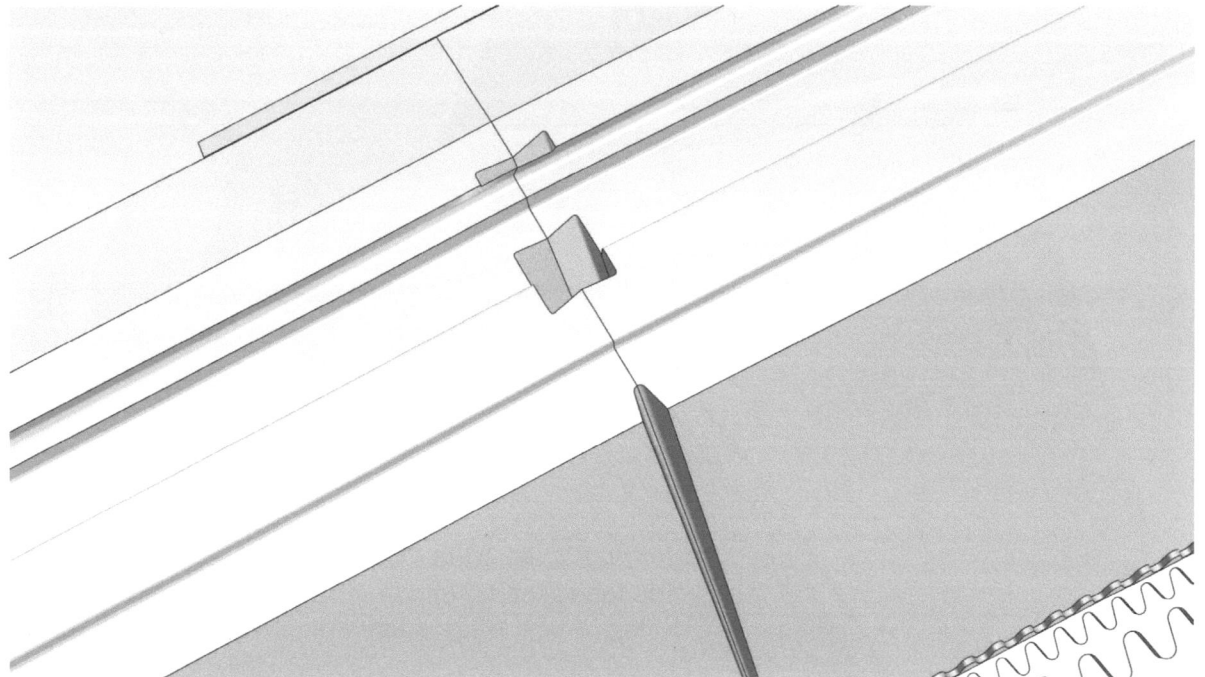

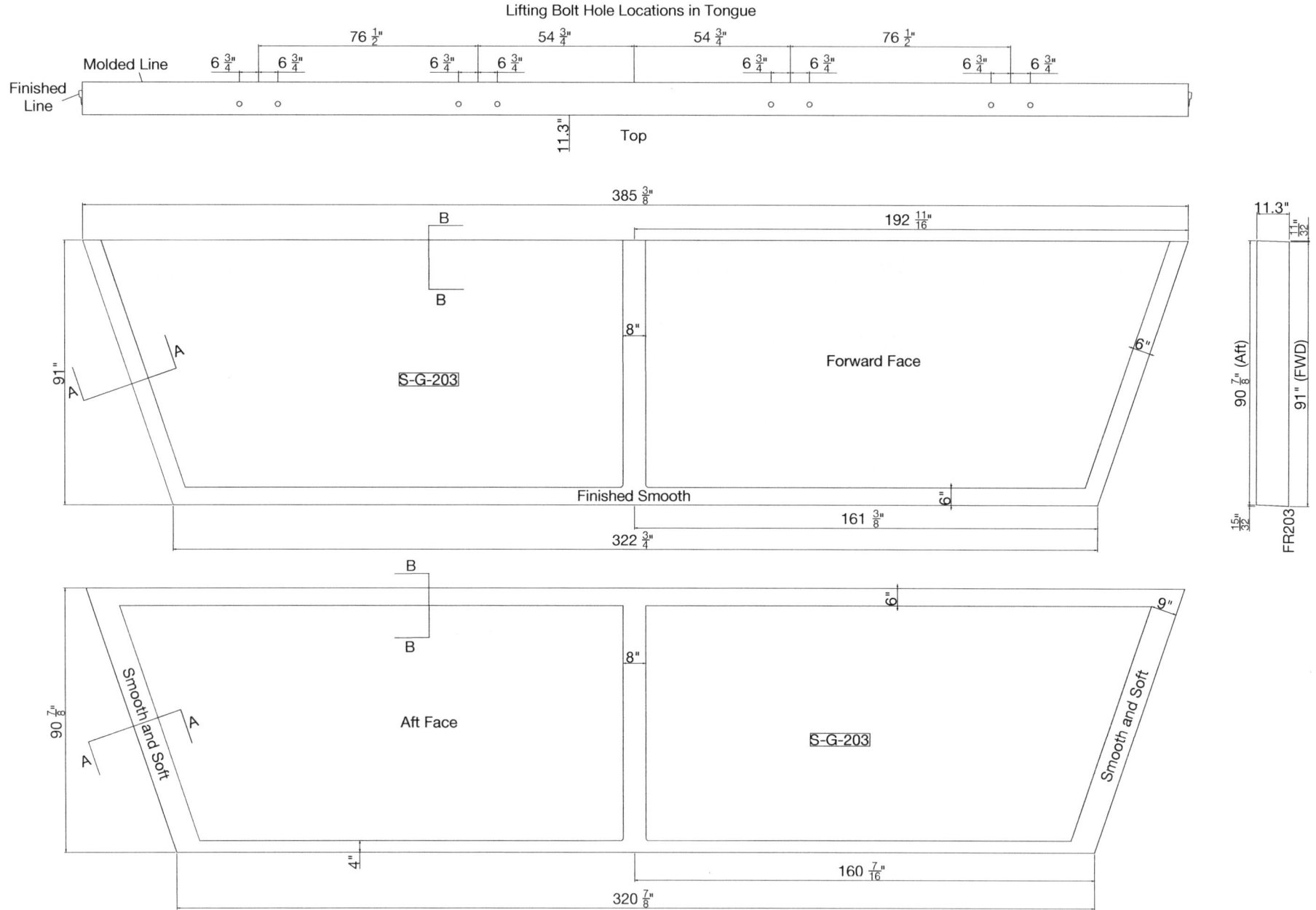

Figure 2.82: Molded Lines of Transverse Plate S-G-203 (1:48). The inside face is at the top and the outside face is at the bottom. Note that the tongue at the sides extends beyond the molded line.

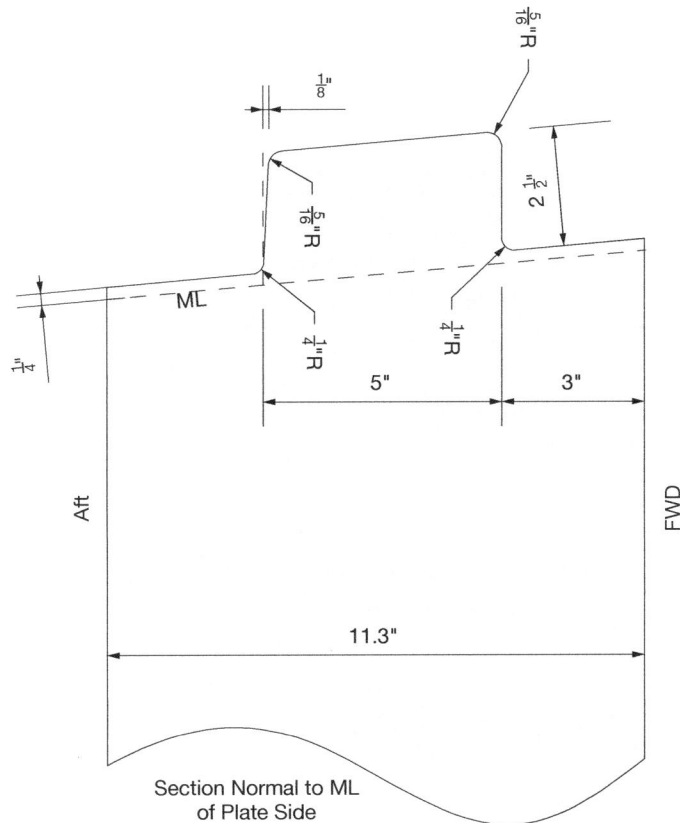

Figure 2.83: **Section A–Tongue at Side of Plate S-G-203 (1:4)**. The section shown is rotated 19° along the centerline to match the slope of the molded edge of the plate. This view is both the port side looking down or starboard, side looking down). The tongue is added to the molded dimensions of the plate. Specifying the dimensions as angled extensions relative to molded lines allowed the measurements to use $\frac{1}{32}$ " precision and have a good fit. If the measurements of the tongue had been specified as absolute horizontal, vertical, and longitudinal terms, $\frac{1}{128}$ " precision would have been required to get the same fit.

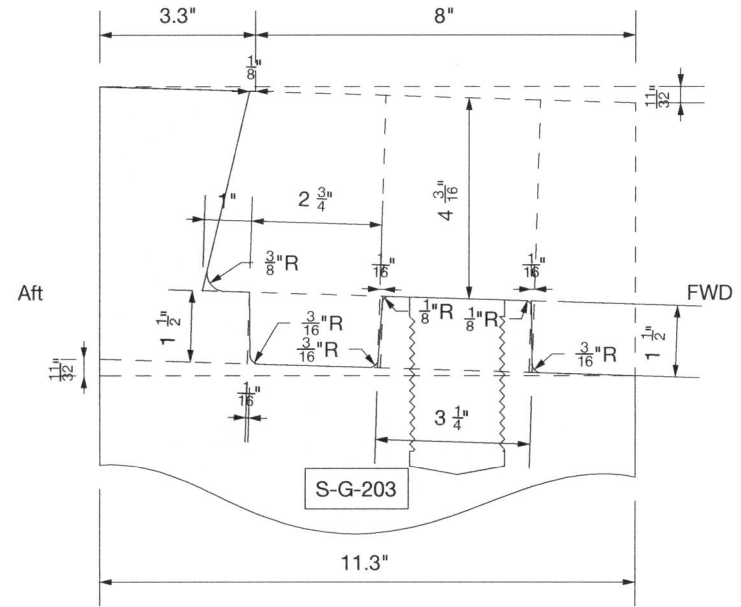

Figure 2.84: **Section B–Top Edge of Plate S-G-203 (1:4)**. The keyed hook joint is very similar to that of the belt plates ("Figure 2.79: Upper Edge of Belt Plates (1:4). The mill machined a keyed hook joint into the upper edge of the steering gear belt plates. This joint is similar to, but deeper, than the joint between FR166 and FR189. Holes for lifting pad bolts were drilled centered on the tongue. The third deck armor plates were machined to match this joint." on page 86 except that it is angled to match the slope of the deck.

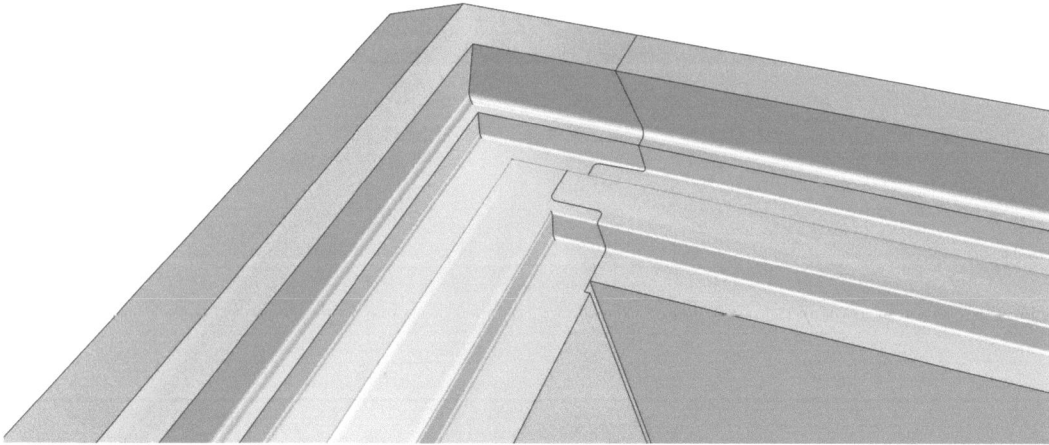

Figure 2.85: Rendering showing how the keyed hook joint wraps around the aft starboard corner of the steering gear armor.

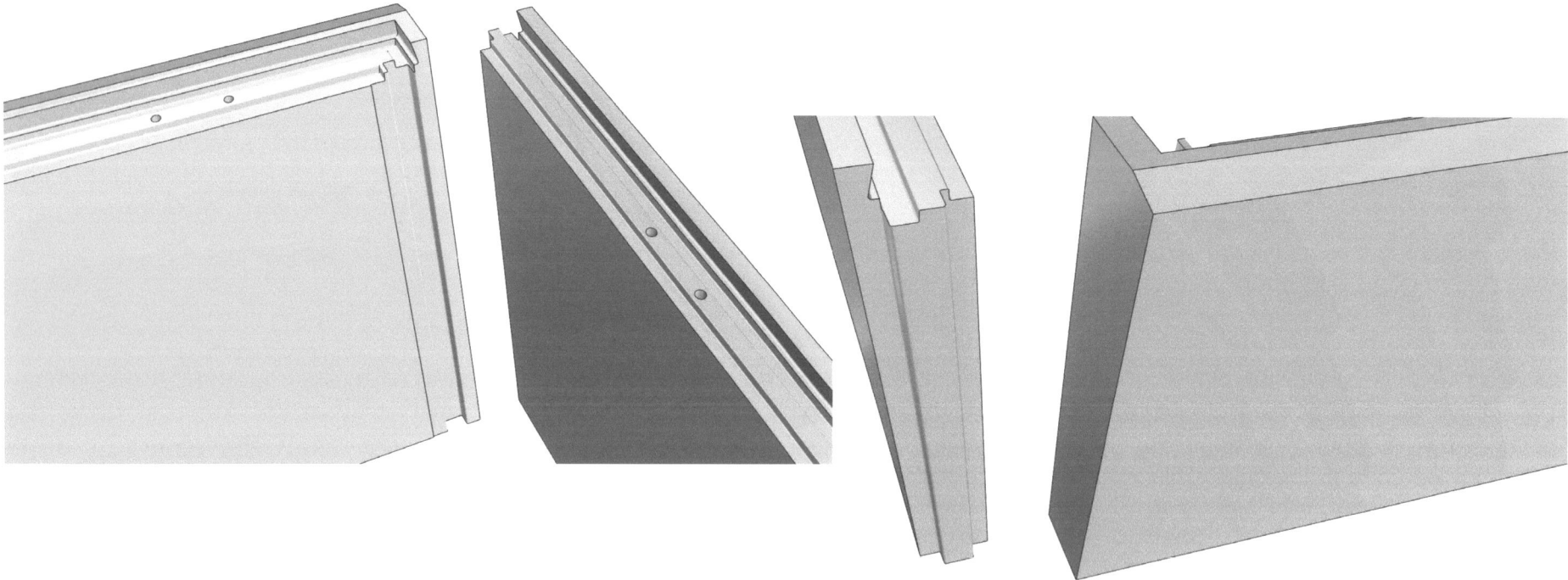

Figure 2.86: Renderings showing the tongue and groove joint between the starboard belt and the transverse plate.

Figure 2.88: Rendering of the steering gear armor looking towards the aft starboard corner. The circles on the deck are the rudder stock locations.

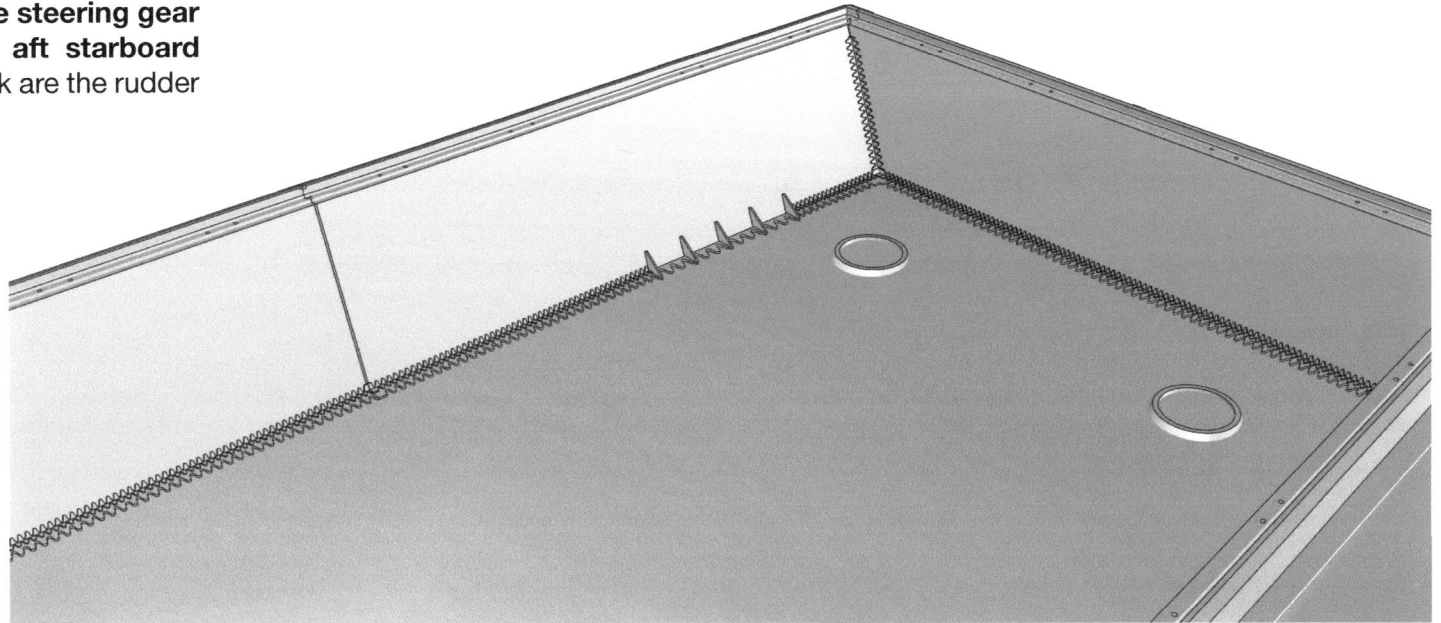

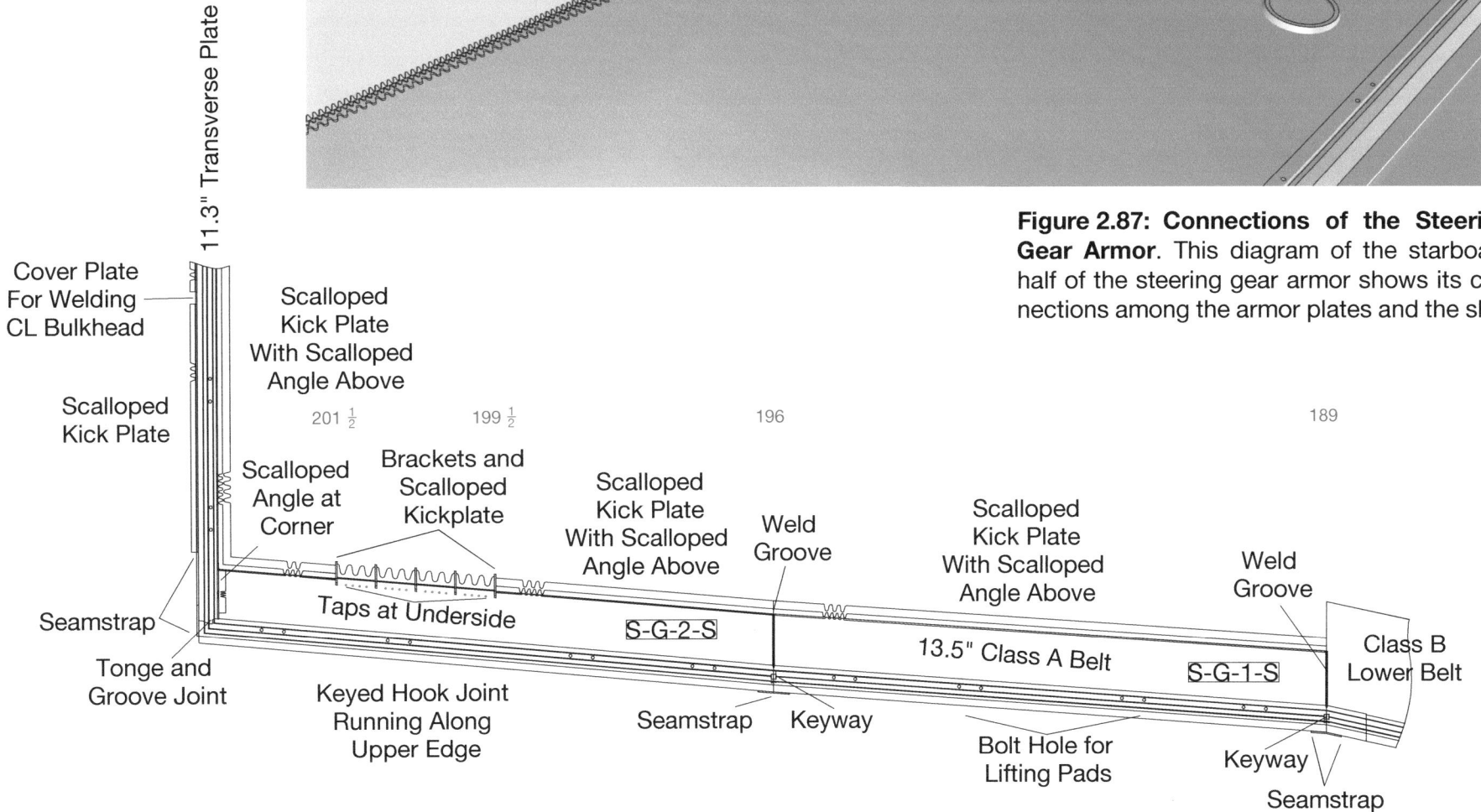

Figure 2.87: Connections of the Steering Gear Armor. This diagram of the starboard half of the steering gear armor shows its connections among the armor plates and the ship.

11.3" Transverse Plate

Cover Plate For Welding CL Bulkhead

Scalloped Kick Plate With Scalloped Angle Above

Scalloped Kick Plate

201 ½   199 ½   196   189

Scalloped Angle at Corner

Brackets and Scalloped Kickplate

Scalloped Kick Plate With Scalloped Angle Above

Weld Groove

Scalloped Kick Plate With Scalloped Angle Above

Weld Groove

Taps at Underside

S-G-2-S

13.5" Class A Belt

S-G-1-S

Class B Lower Belt

Seamstrap

Tonge and Groove Joint

Keyed Hook Joint Running Along Upper Edge

Seamstrap   Keyway

Bolt Hole for Lifting Pads

Keyway

Seamstrap

$4\frac{1}{8}$"

$\frac{3}{4}$"R

$1\frac{1}{8}$" $\frac{3}{4}$"R

2"

3"

Scallop Used for Aft Inside Corners and Aft Base of Transverse Steering Gear Armor Cut From 5x5x$\frac{3}{4}$" Angle Bar

Horizonal Cross Section of Scalloped $1\frac{1}{2}$×4 $\frac{1}{8}$×$\frac{3}{4}$" Angle at Inside Aft Corners

Alternative for Caulk Welding

Caulk
1×$\frac{1}{2}$" Flat Bar
Weld

Transverse Armor
11.3"

Scalloped 5×5×$\frac{3}{4}$" Angle

Scalloped 4×4×$\frac{5}{8}$" Angle

$3\frac{1}{2}$"

$\frac{5}{8}$"

Scalloped 40# Kickplate

S-G-203

$\frac{3}{4}$" $1\frac{1}{2}$"

$3\frac{1}{2}$" $1\frac{1}{8}$"

$\frac{3}{4}$"

$\frac{5}{8}$"   1"

$\frac{5}{8}$"

$4\frac{1}{8}$" $8\frac{1}{4}$"

1"

FR203

$8\frac{1}{4}$"

$2\frac{1}{2}$"

$1\frac{5}{8}$"

$3\frac{1}{4}$"

$1\frac{5}{8}$"

$\frac{5}{8}$"R

$2\frac{1}{2}$" $2\frac{1}{2}$"

$\frac{5}{8}$"R

4"

1"R

1"R

$3\frac{1}{2}$"

$\frac{5}{8}$"R

$\frac{5}{8}$"R

1"

Scallop Pattern for 4×4×$\frac{5}{8}$" Angle on top of 40# Kickplate (Left) and for the Kickplate (Right)

Scalloped 4×4×$\frac{5}{8}$" Angle

13.5" Belt Armor

$3\frac{1}{2}$"

Scalloped 40# Kickplate

1"

$\frac{5}{8}$"

35°

$3\frac{1}{2}$"

$\frac{3}{8}$"

$\frac{1}{4}$" Caulk Weld

First Platform

$8\frac{1}{4}$"

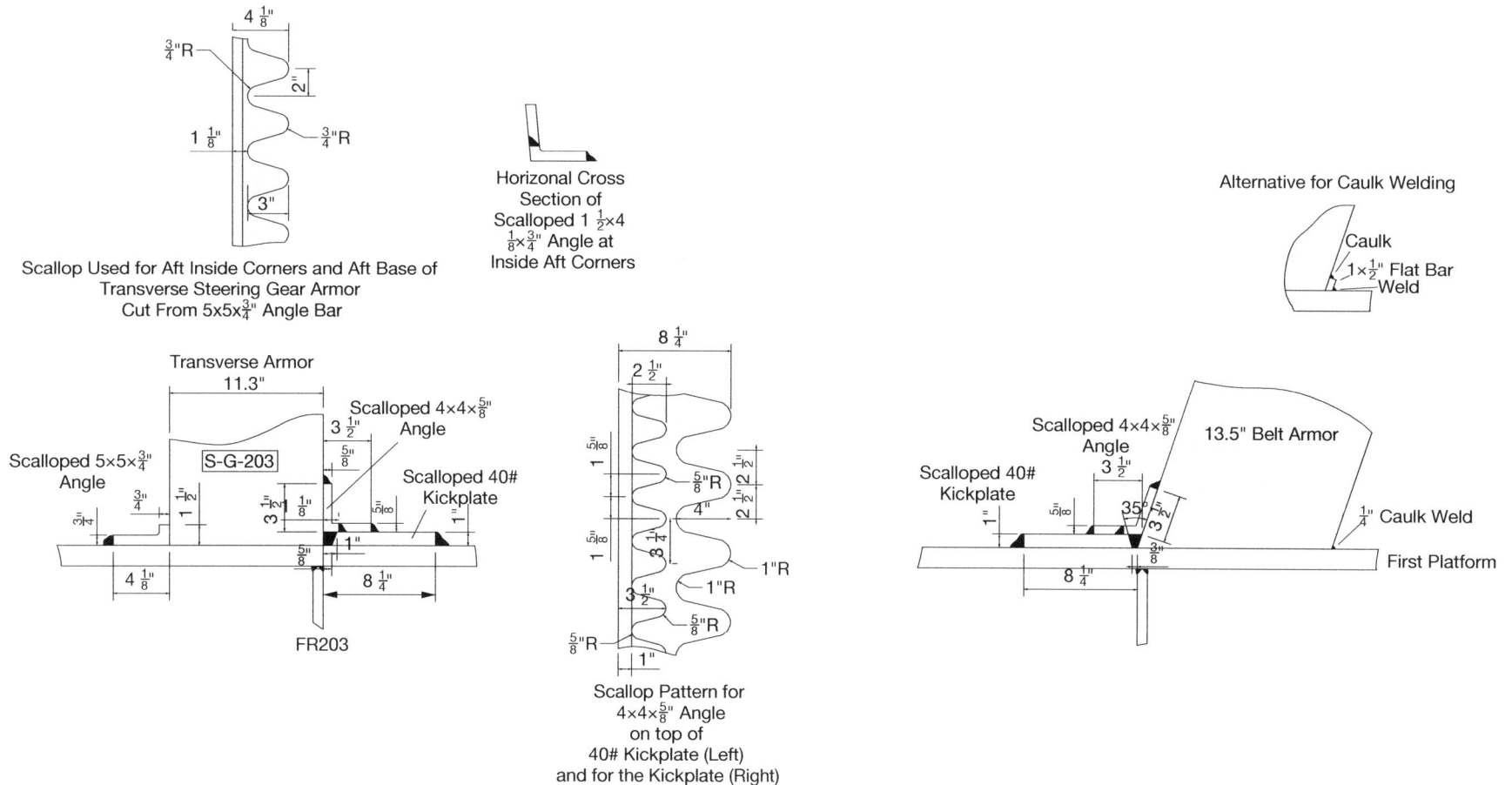

Figure 2.89: Steering Gear Scallops (1:12). At the upper left is the scallop pattern used for 5×5×$\frac{3}{4}$" angles. This sized angle was used in the corners between the transverse and belt armor and along the deck at the aft face of the transverse armor. Scallops were cut into both sides of the angle for the corners but only the deck side for the angle against the deck. In all cases, scallops cut into both sides of angles were staggered in order to not create a narrow spot. The diagram at the lower center shows the scallop pattern used for 4×4×$\frac{3}{8}$" angles and 40# kickplates. The kickplates run along the deck at the inside face of the steering gear plates. The angles run along the top of the kickplates as shown in the diagrams at the left and right. The plans do not specify what was to be done at the ends, where multiple scalloped bars came together. In general practice, the ends were not scalloped so the bars could be joined at flat edges.

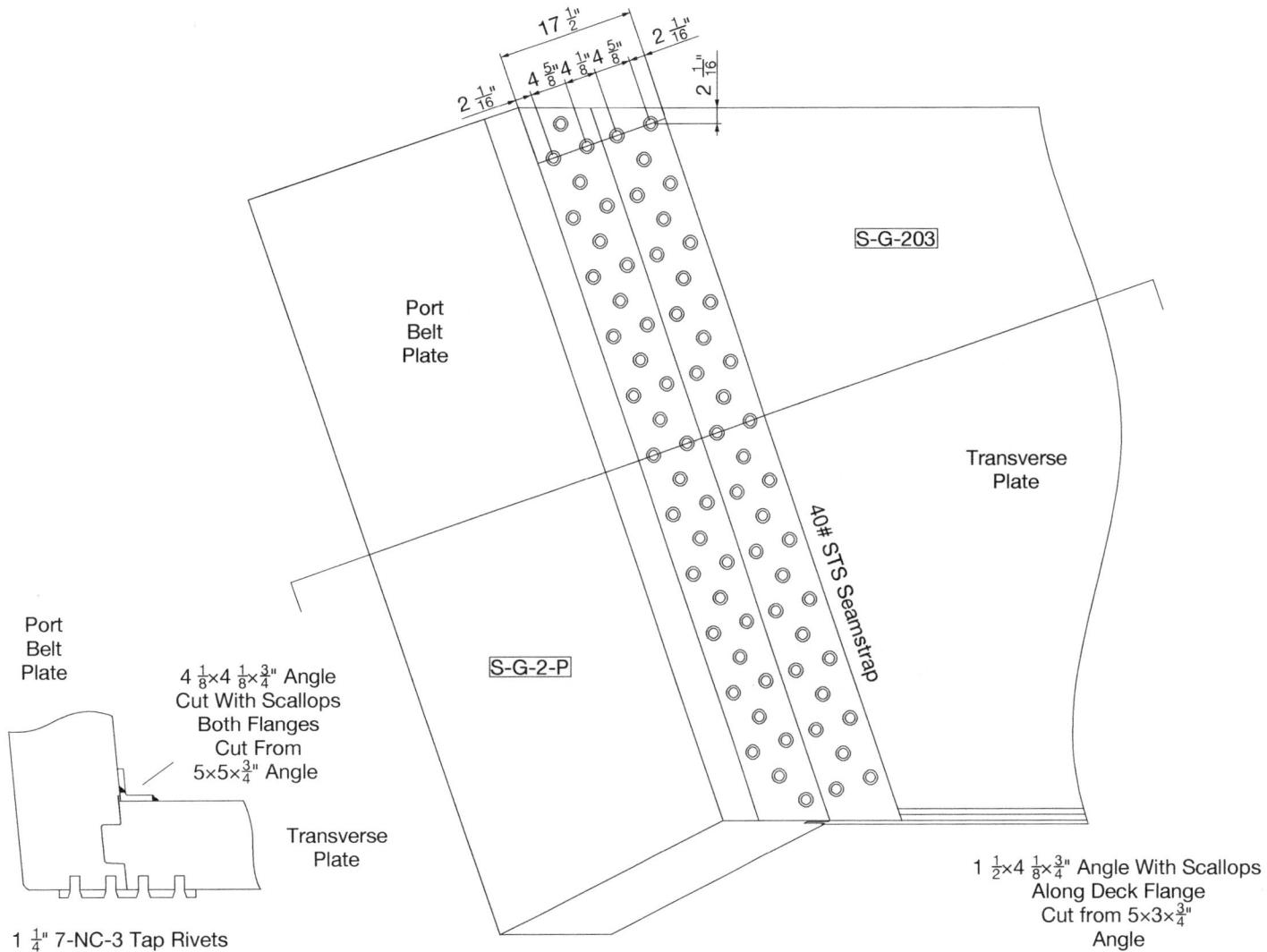

Figure 2.90: Connection Between Belt and Transverse Plates (1:24). A scalloped plate was welded over the joint along the inside corner (see Plan 2.12). A 30# STS cover plate was tap riveted of the outside of the joint. This area was not face hardened.

Photograph 2.15: The scalloped angle at the base of the aft edge of the FR203 transverse plate S-G-2-S on USS New Jersey. The angle is not welded to the plate so only the deck flange side is scalloped. The other flange was cut simply cut down.

Photograph 2.16: The aft, port inside corner of the steering gear armor at FR203 on USS New Jersey. The transverse plate (S-G-203) is to the right and belt plate is at the left (S-G-2-S). A scalloped kickplate runs along the deck. A scalloped angle was welded on top of the kickplate creating a double row of scallops at the deck. Another scalloped angle runs along the joint between the transverse and belt plates.

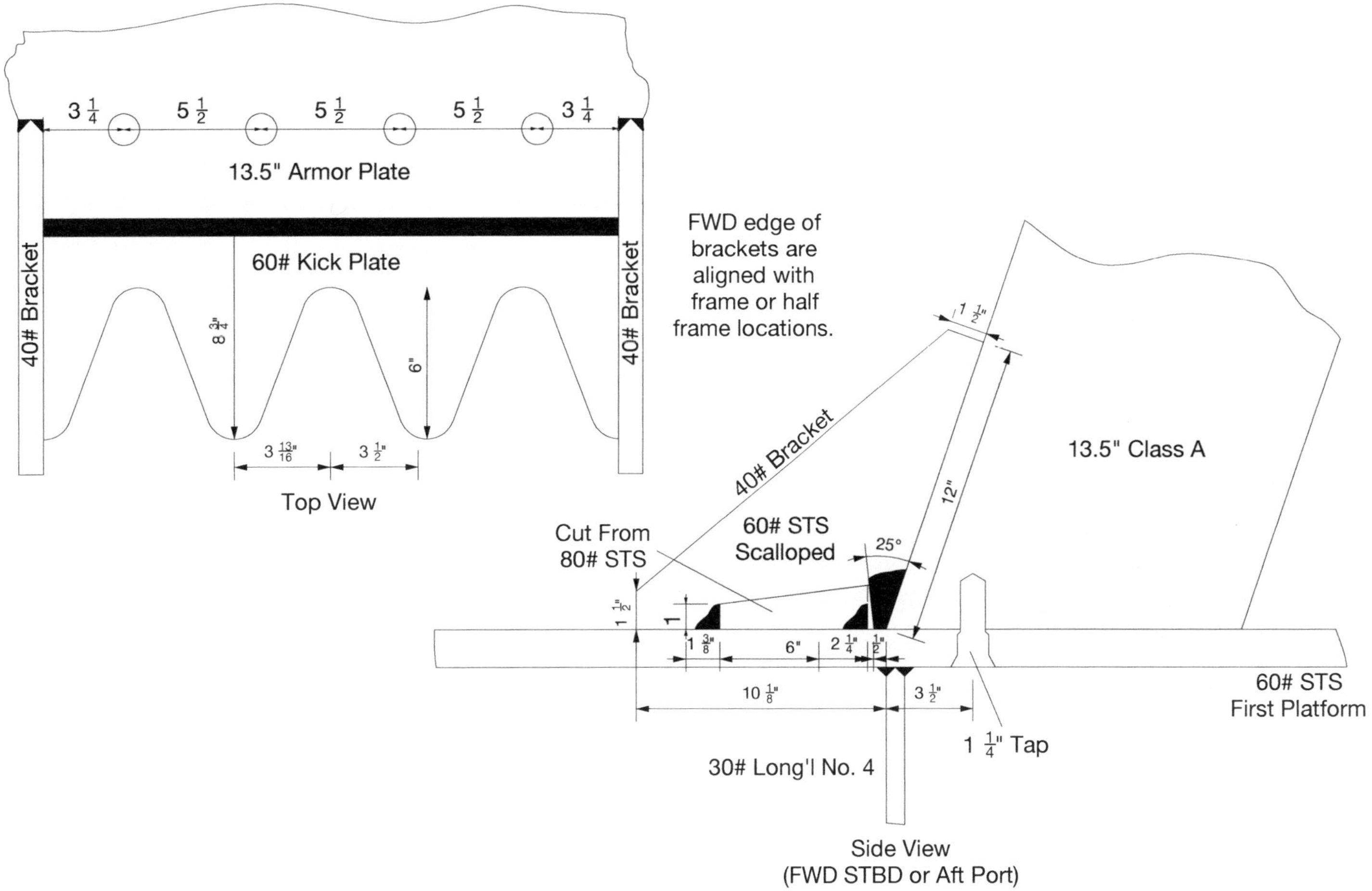

3 $\frac{1}{4}$ 5 $\frac{1}{2}$ 5 $\frac{1}{2}$ 5 $\frac{1}{2}$ 3 $\frac{1}{4}$

13.5" Armor Plate

40# Bracket

60# Kick Plate

40# Bracket

8 $\frac{3}{4}$

6"

3 $\frac{13}{16}$"   3 $\frac{1}{2}$"

Top View

FWD edge of brackets are aligned with frame or half frame locations.

1 $\frac{1}{2}$"

40# Bracket

12"

13.5" Class A

Cut From 80# STS

60# STS Scalloped

25°

1 $\frac{1}{2}$"

1

1 $\frac{3}{8}$"   6"   2 $\frac{1}{4}$"   1 $\frac{1}{2}$"

60# STS First Platform

10 $\frac{1}{8}$"   3 $\frac{1}{2}$"

1 $\frac{1}{4}$" Tap

30# Long'l No. 4

Side View
(FWD STBD or Aft Port)

Figure 2.91: Attachment of Belt Plates Near Rudder Stock (1:6). These diagrams show how the belt plates were connected to first platform above the rudder bearing assembly between FR199 $\frac{1}{2}$ and FR201 $\frac{1}{2}$. One of four 2-foot segments per side is shown here.

Photograph 2.17: The angled buttstrap at the starboard side looking inboard and forward on USS New Jersey. To the left is the FR203 transverse plate and to the right is the belt plate. The angle at the aft edge ends at the seamstrap. The lower edge of the belt plate is caulk welded to the deck. This weld is solely for watertightness.

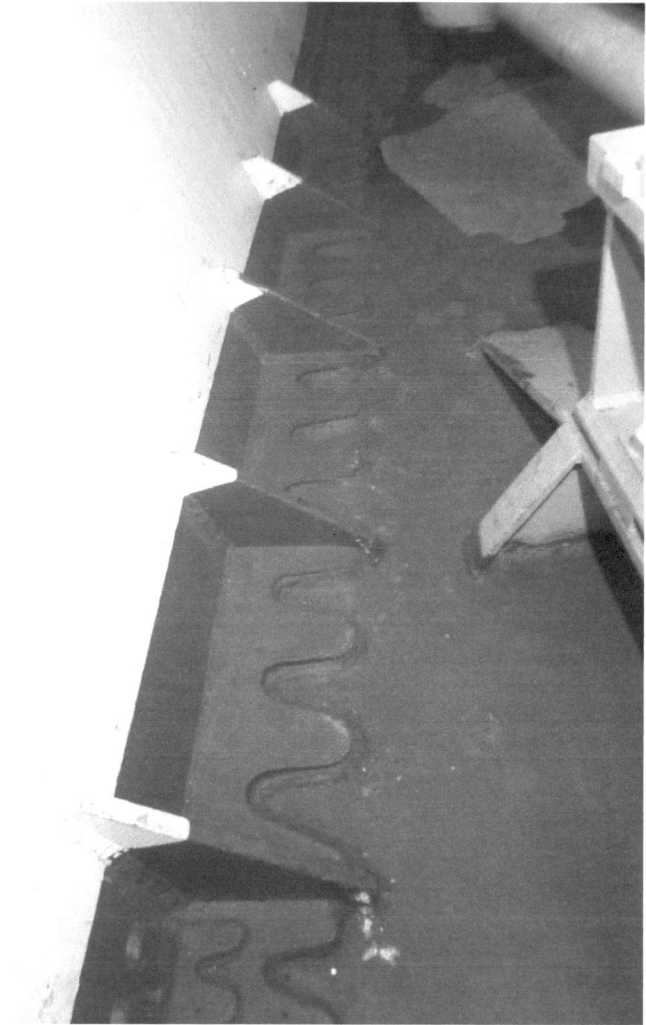

Photograph 2.18: The starboard belt at the rudder bearing assembly on USS New Jersey. The kickplate in this area is thicker and tapered and there is no angle welded on top.

9 $\frac{1}{8}$"

1 $\frac{7}{8}$" 2" $\frac{3}{8}$" 2" $\frac{7}{16}$" 2" $\frac{7}{16}$"

Inside

Outside

2 $\frac{1}{2}$" 2" $\frac{3}{8}$" 2" $\frac{3}{8}$" $\frac{7}{8}$"

9 $\frac{1}{8}$"

7 $\frac{5}{16}$"

1 $\frac{11}{16}$" 3 $\frac{15}{16}$" 1 $\frac{11}{16}$"

18 $\frac{1}{4}$"

1 $\frac{7}{8}$" 2" $\frac{3}{8}$" 2" $\frac{3}{8}$" 2" $\frac{1}{2}$" 2" $\frac{1}{2}$" 2" $\frac{3}{8}$" 2" $\frac{3}{8}$" 1" $\frac{7}{8}$"

30# HTS
Cover Plate
Centerline
FR203+11.3"
1 $\frac{1}{8}$" SPEC Taps

30# STS
Buttstrap
FR196
1 $\frac{1}{4}$" SPEC Taps

30# STS
Buttstrap
FR189
1 $\frac{1}{4}$" SPEC Taps
Plans state a liner
may be fitted.

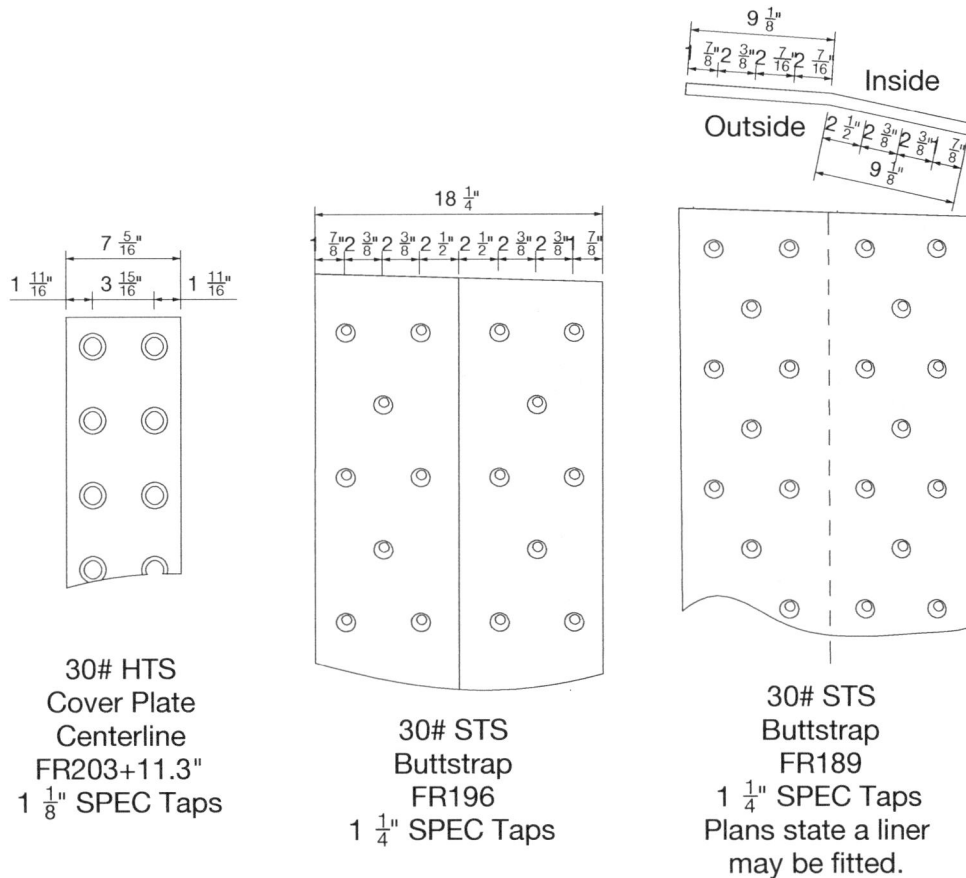

Figure 2.92: Steering Gear Armor Buttstraps (1:12). At the left is the strap that is tap riveted to the aft face of the transverse plate at the centerline. This serves as an attachment point for welding the centerline bulkhead to the face-hardened armor plate. There is no joint at this location. The center illustration shows the buttstraps that were tap riveted over the joint between the Class A belt plates at FR196. The aft plates had to be planed to make this joint flat. This area of the plate was not face hardened. The illustration at the right shows the buttstraps that were tap riveted over the joint between the Class A and STS (or Class B) belt plates at FR189. These plates meet at an angle (over 8°) that created a gap where the edge of the forward plate extended about $\frac{1}{4}$" outboard of the edge of the aft plate. The plans indicate that a liner could have been fitted here "if required." A liner was certainly required to fill such a gap. The ships' plans specify the horizontal spacing of the taps but not their vertical spacing.

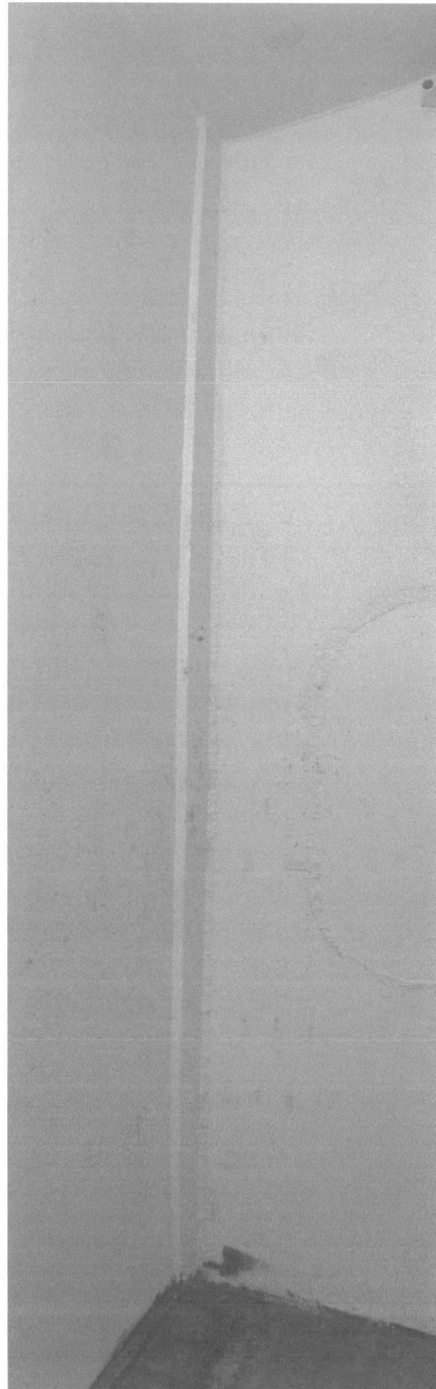

Photograph 2.19: The centerline strap at the aft face of the FR203 transverse plate on USS New Jersey. The transverse armor is to the left and the longitudinal bulkhead is at the right. The strap ends at the scalloped angle that passes underneath. Third deck is welded directly to the transverse armor in this area. An opening in the bulhead has been welded closed. The transverse bulkhead acts as an structural extension of the keel to third deck.

Photograph 2.20: The connection of the port steering gear belt plates on USS New Jersey at FR189. The steering gear belt plate (S-G-2-1) is at the left. The extension belt plate (L-B-56-P) is obscured by the bulkhead to the left.

Chapter 3: Frame 50 Transverse Armor

The transverse armor at FR50 refutes the frequently made claim that Class A armor could not be used structurally. The transverse armor forms a massive watertight bulkhead between torpedo bulkhead 3 at the sides and the third skin and second deck.

The surviving plans for the transverse armor are for USS *Iowa* and USS *New Jersey*. These plans have notations indicating the changes for the later *Iowa*-class battleships. As described below, the plates for USS *Iowa* did not follow the plans. The illustrations in this chapter for the later ships show measurements than can be inferred from other dimensions shown in the updated plans.

3.14. Armor Plates

The FR50 armor consists of six Class A plates. Third platform passes through the transverse armor and emerges as second platform on the aft side. Five plates (B-50-1, B-50-2, B-50-3, B-50-4, and B-50-5) form an assembly above this deck. The last plate (B-50-6) is below the deck and has no connections with the other transverse plates.

On USS *Iowa* and USS *New Jersey* the transverse armor is 11.3" from the top to a knuckle twenty feet above the baseline where it tapers to 8.5" at the third skin. On the later *Iowa*-class battleships, the armor was increased to 14.5" tapering to 11.7". The upper face of the bottom plate has the same width as the lower faces of the upper plates so there is no theoretical taper across the deck gap.

The same plans were used for the transverse armor for USS *Iowa* and USS *New Jersey*. However, one of the sheets contain a note that the armor for USS *Iowa* was built with the knuckle on the aft side rather than as specified on the forward side. This suggests that the mill made a mistake and the Navy accepted the plates as-is rather than delay construction for months for new plates. If this theory be correct, the alteration would have only forced relatively minor changes to the rest of the ship.

The belt armor for USS *Iowa* and USS *New Jersey* extend well forward of the transverse armor. The plans instruct that the belt plates were to be trimmed flush with the transverse armor. The existence of this extra length for the belt suggests that thicker transverse armor was considered for USS *Iowa* and USS *New Jersey* as well, but this had to be reduced to comply with treaty limitations that did not apply to the later ships of the *Iowa*-class.

3.15. Machining at the Mill

The FR50 transverse plates were fully shaped at the mill. The joints between the five upper plates were machined with a welding groove along the aft (soft) edge and rectangular keyways were machined along the side. The keyways have a constant width to the plate knuckle and taper below the knuckle location. The upper plates were drilled and tapped for bolts to secure them to second deck. All of the plates were drilled and tapped for lifting bolt holes.

3.16. Plate Installation

3.16.1. Plate to plate connections

The five upper plates were joined together to form a single assembly. Keys were inserted into the keyways to align the plates. The plates were joined by welding along the groove on the inside face.

3.16.2. Connection to Third Skin

The lower plate (B-50-6) rests on the third skin. A 40# scalloped kick plate was welded to the deck on the aft side. A $3\frac{1}{2} \times 3\frac{1}{2} \times \frac{5}{8}$" scalloped angle was welded to the top of the kick plate and the aft face of the armor. For the outside face an angle was cut down to $1\frac{1}{2} \times 3\frac{1}{2} \times \frac{5}{8}$" with a scallop on the long side. The angle was pressed against the forward face and welded to the third skin.

3.16.3. Connection to Second/Third Platform

The connections to the second/third platform use the same scallop pattern. The only significant difference is that the scallops were welded to both the upper and lower faces of the deck.

3.16.4. Forward Connection to First Platform, Third Deck, and Second Deck

Each joint in the forward face of the transverse armor contains slots for deck support brackets. Before the upper sections were assembled, supports were inserted into the slots and were secured to the plate using tap bolts. A beam connects the brackets.

3.16.5. Aft connection to First Platform

First platform was welded directly to the aft face of the transverse plates.

3.16.6. Aft Connection to Third Deck

Third deck was indirectly welded to the aft face. A 40-inch wide 40# HTS plate was welded to the aft face of each upper plate centered at the location of third deck. These plates were perforated to create welding locations across the width of the plate. Third deck was welded to this plate rather than directly to the armor plates. The plans give no explanation why this extra step was employed.

3.16.7. Aft Connection to Second Deck

Second deck runs over the FR50 transverse armor. The connection to the aft face uses the same method as lower plate (B-50-6) to second platform. A 40# scalloped kick plate was welded to the underside of second deck and against the inside face of the armor plates. A $3\frac{1}{2}\times3\frac{1}{2}\times\frac{5}{8}$" scalloped angle was welded to the kick plate and the armor. The second deck armor was bolted to the top face of the transverse armor.

3.16.8. Connection to Torpedo Bulkhead and Belt Armor

Torpedo bulkhead 4 was welded directly to the inside face of the transverse armor. Torpedo bulkhead 3 passes between the belt and the transverse armor so there is no direct connection between them. The bulkhead resumes with normal plating at the forward end of the lower belt plates, aft of the transverse armor. The aft edge of the bulkhead plating here is scalloped and welded to the lower belt plates. The joint along the corners uses a kick plate and angle with the same dimension as used for the connection with third skin and third platform and second platform. At the lower belt, there are three tiers of scallops.

Photograph 3.1: USS New Jersey under construction in Oct. 1941. Transverse plate 50-B-6 is in place with second platform above it. Torpedo bulkhead 4 is visible to the right of the fixed turrets. The belt is under construction to the right of the torpedo bulkhead. The transition from triple bottom to double bottom is at the lower right of the photograph. (Nat'l Archives)

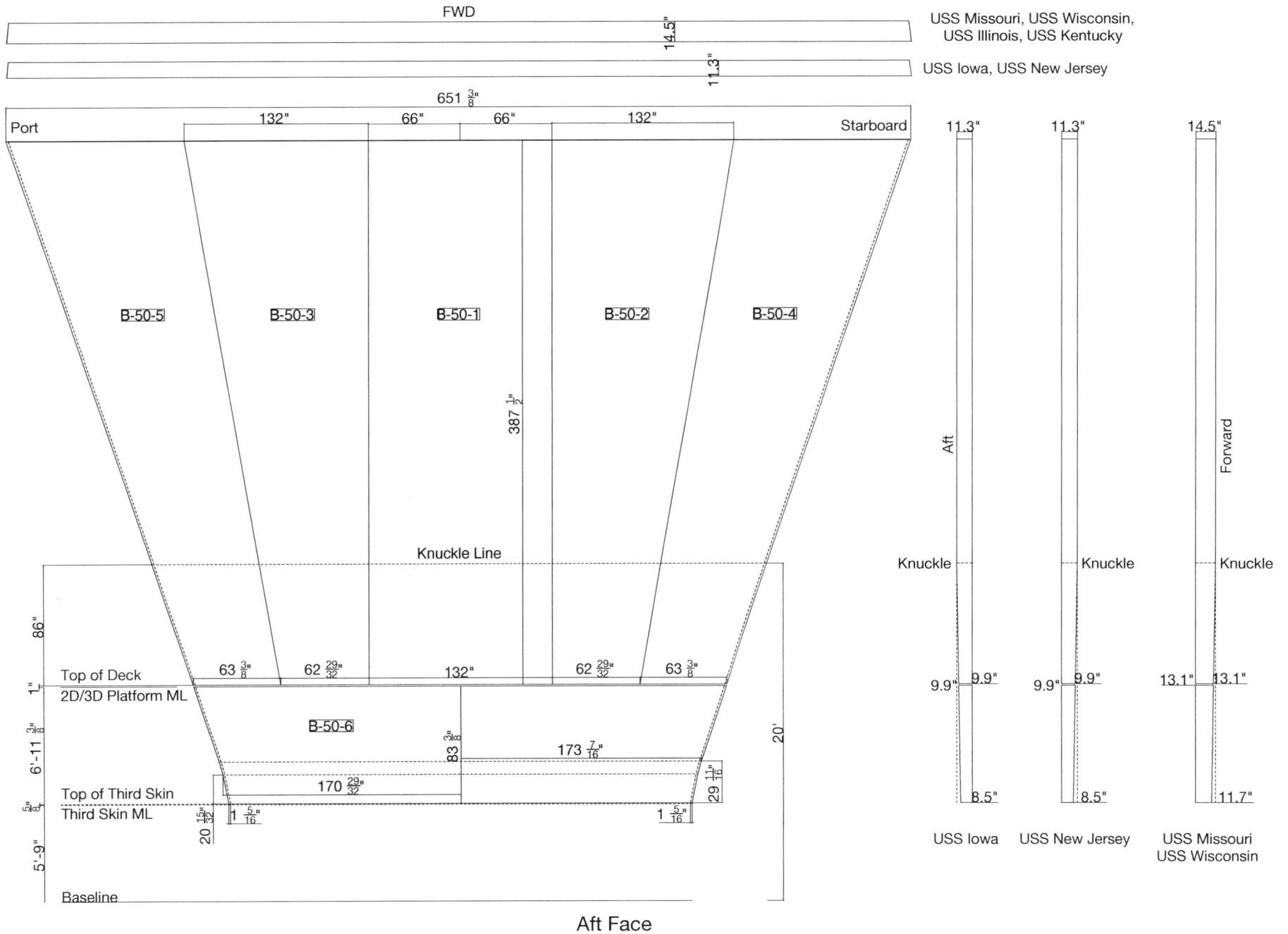

Figure 3.1: Arrangement of transverse plates (1:96).

Bolt Holes

Weld Groove

Lifting Bolt Holes

Top Plan

Weld Groove

Smooth

Smooth

B-50-1

Smooth

Aft Surface Elevation

Normal End Elevation

Normal End Elevation

Keyway Knuckle on its Outer Face

Weld Grooves

132"

Bottom Plan

Smooth

Smooth

Smooth

Knuckle

Smooth

Smooth

Forward Elevation

USS New Jersey Only
USS Iowa has knuckle on aft face.

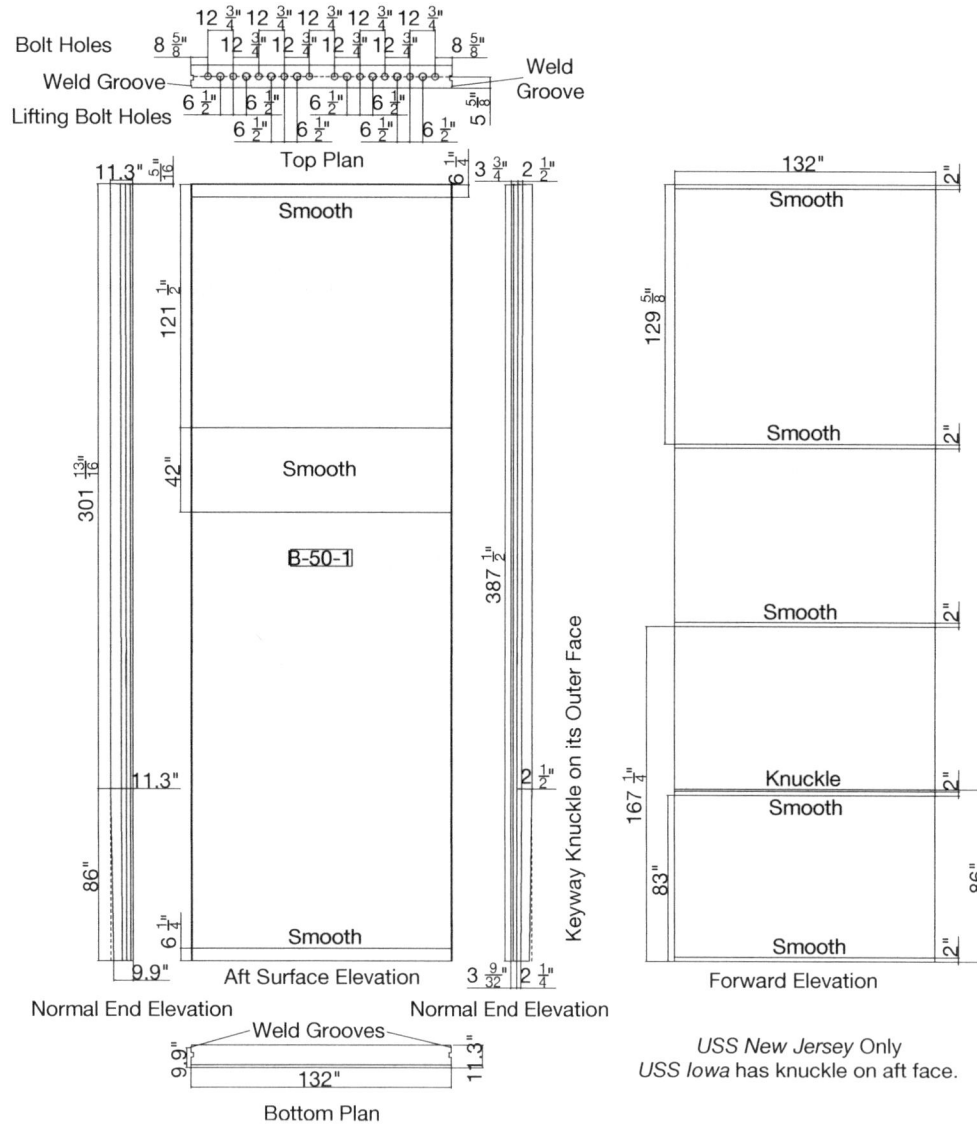

Figure 3.2: Plate B-50-1 on USS New Jersey (1:96)

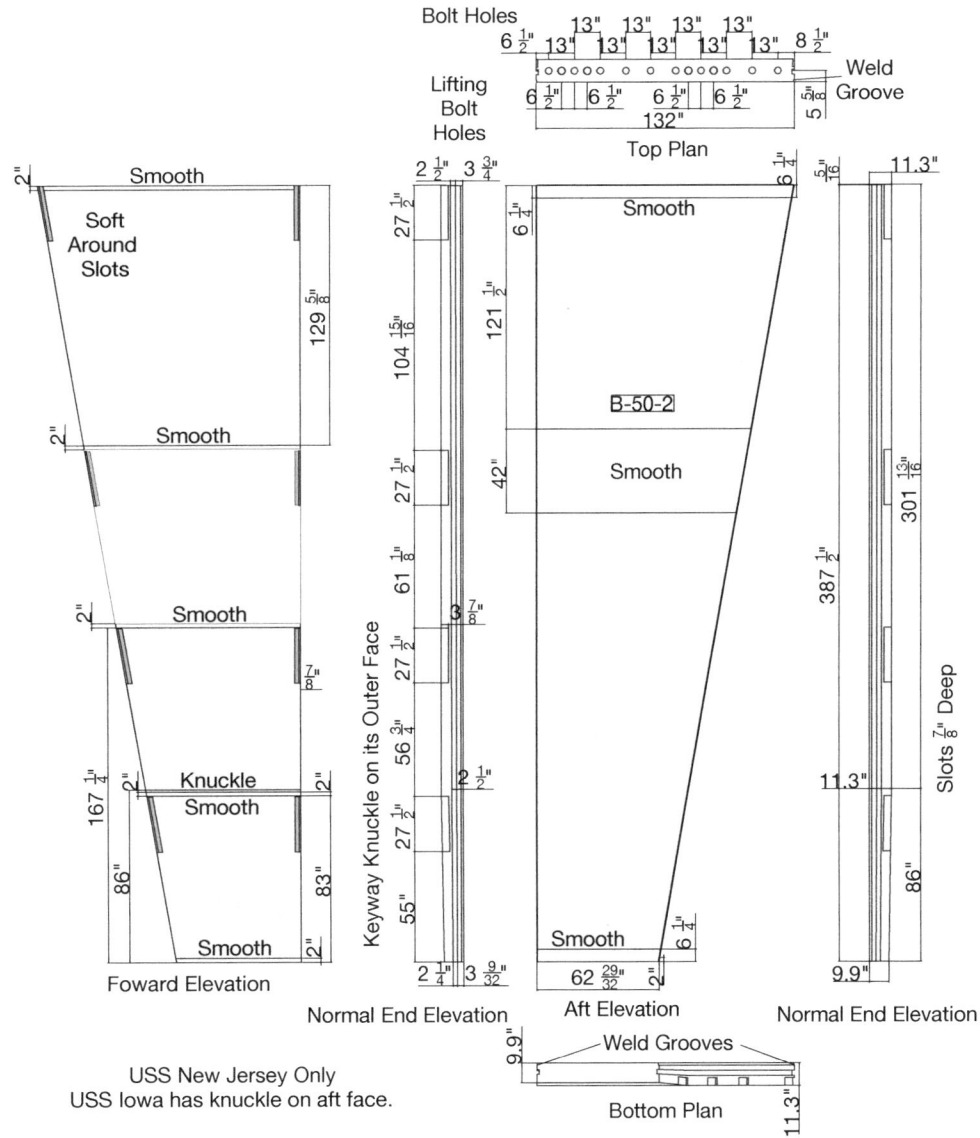

Figure 3.3: Plate B-50-2 on USS New Jersey (1:96)

Bolt Holes
Weld Groove
Lifting
Bolt Holes

8 1/2"   13"   13"   13"   13"   6 1/2"
      13"   13"   13"   13"   13"
      6 1/2"   6 1/2"   6 1/2"   6 1/2"
             132"   5 5/8"
                Top Plan

Weld
Groove

Smooth

B-50-3

Smooth

Smooth

Aft Elevation

Slots 7/8" Deep

Normal End Elevation

11.3"   5 5/16"   6 1/4"
301 13/16"
387 1/2"
86"
11.3"
9.9"
62 29/32"
2 7/8"
6 1/4"
2"

2"
121 1/2"
42"

3 3/4"   2 1/2"
27 1/2"
104 15/16"
27 1/2"
61 1/8"
3 7/8"
27 1/2"
56 3/4"
2 1/2"
27 1/2"
55"
3 9/32"   2 1/4"

Keyway Knuckle on its Outer Face

Normal End Elevation

Smooth
Soft
Around
Slots

Smooth

Smooth

Knuckle
Smooth

Smooth

Foward Elevation

129 5/8"
2"
2"
7/8"
167 1/4"
83"
2"
86"

Weld Grooves
9.9"
11.3"
Bottom Plan

USS New Jersey Only
USS Iowa has knuckle on aft face.

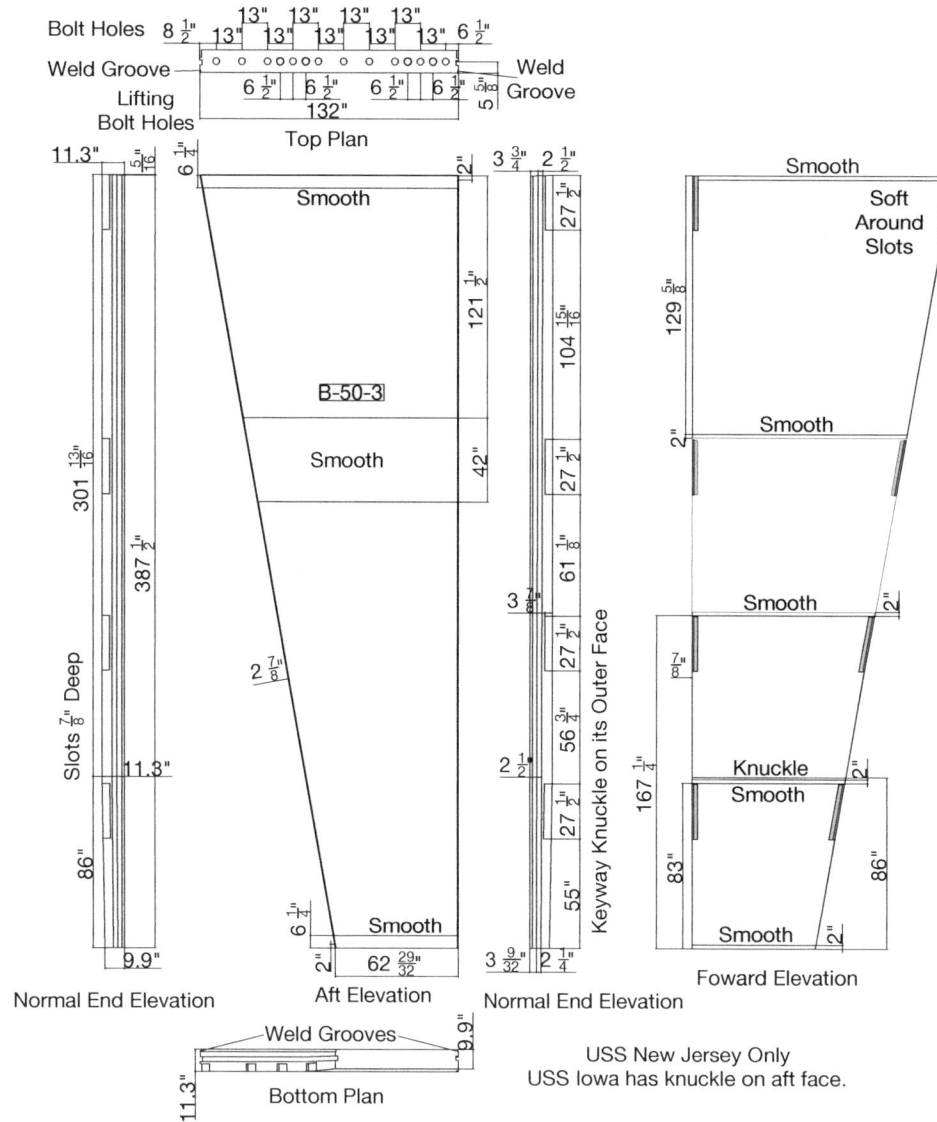

Figure 3.4: Plate B-50-3 on USS New Jersey (1:96)

Bolt Holes

12" 12 1/4" 12" 12"

Weld Groove

12" 12" 12 3/4" 12" 12" 6 11/16"

Lifting
Bolt Holes

6 1/2" 6 1/2" 6 1/2" 6 1/2"

127 11/16"

5 5/8"

Top Plan

Smooth

2"

129 5/8"

Soft and Smooth

Smooth

2"

2 1/2" 3 3/4"

6 1/4"

B-50-4

121 1/2"

Smooth

42"

387 1/2"

2"

167 1/4"

Knuckle

Smooth

86"

5 1/4"

83"

2"

2 1/2"

6 1/4"

Smooth

2"

Forward Elevation

2 1/4" 3 9/32"

6 1/4"

Aft Elevation

Normal End Elevation

USS New Jersey Only
USS Iowa has knuckle on aft face.

5/16"

11.3"

301 13/16"

11.3"

86"

9.9"

Normal End Elevation

Weld Groove

1 17/32"

9.9"

Bottom Plan

11.3"

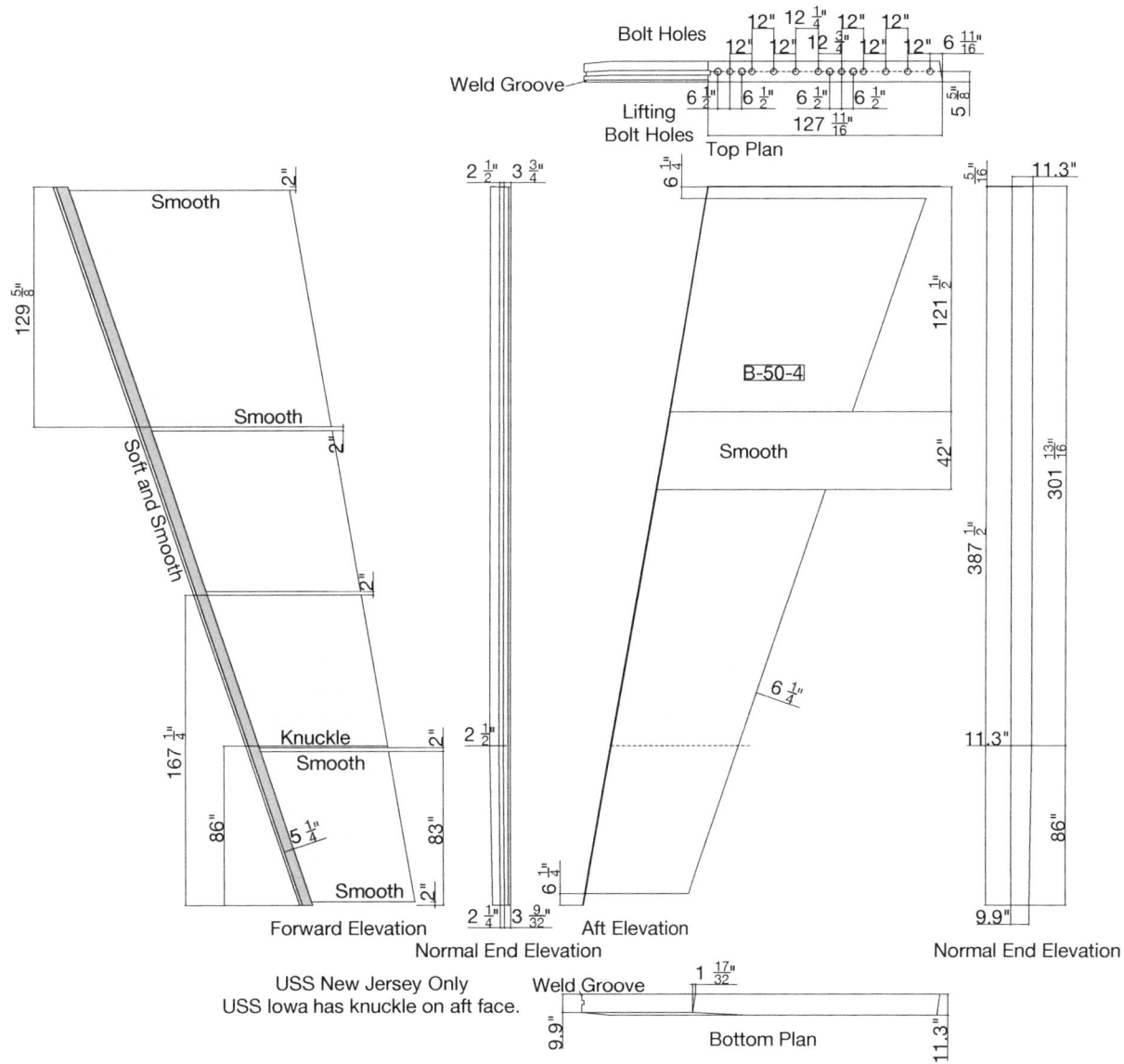

Figure 3.5: Plate B-50-4 on USS New Jersey (1:96)

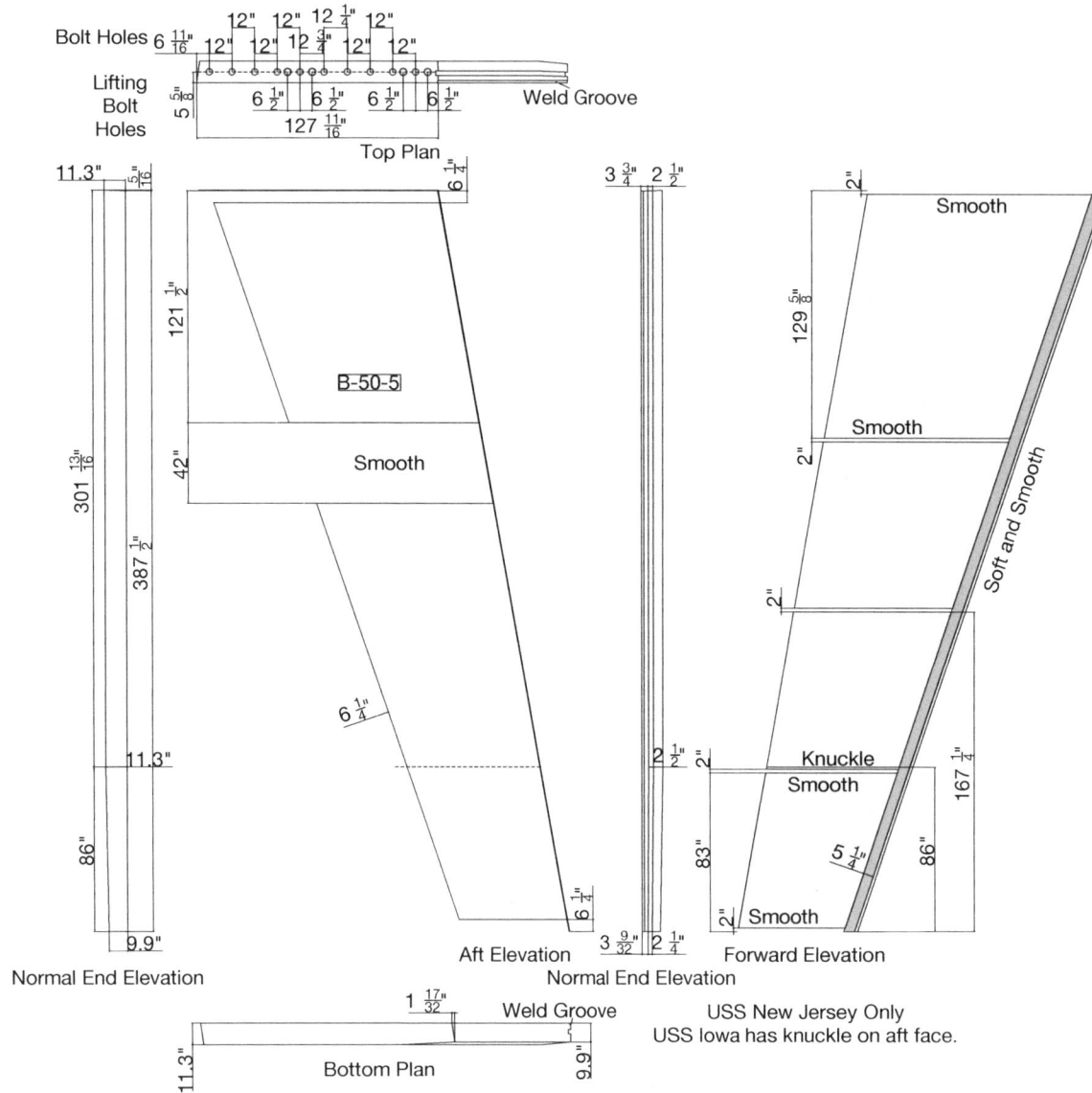

Figure 3.6: Plate B-50-5 on USS New Jersey (1:96).

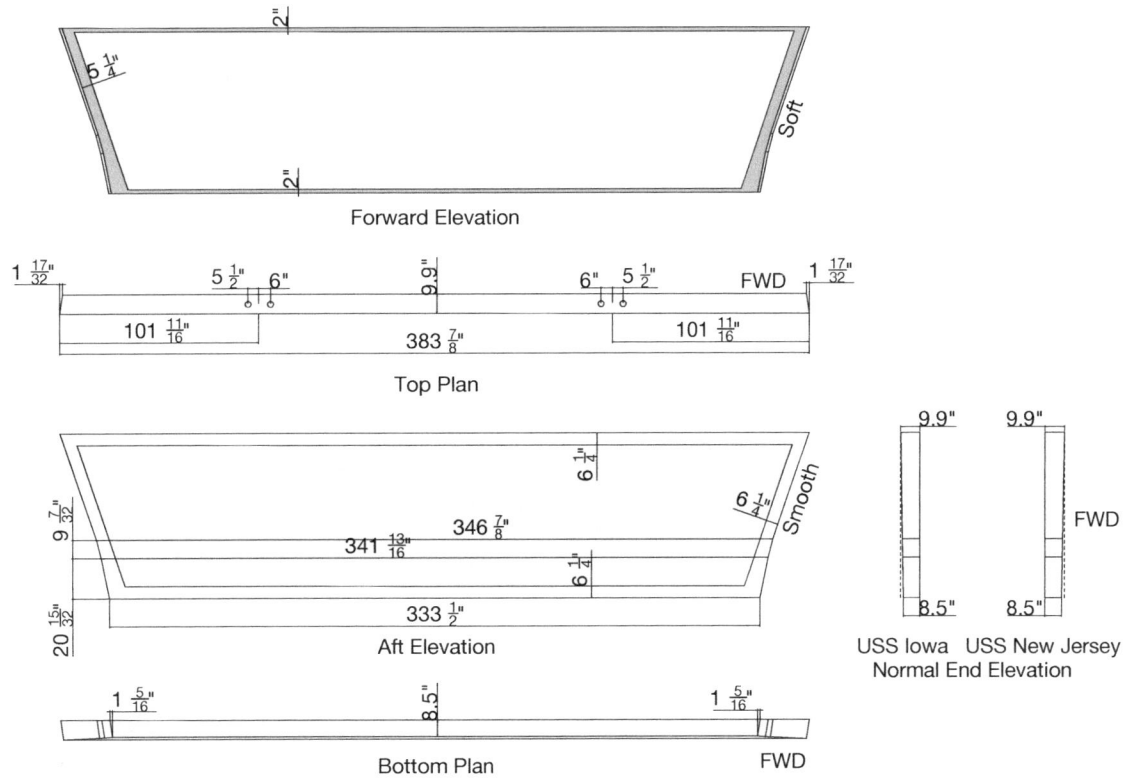

Figure 3.7: Plate B-50-6 on USS New Jersey (1:96)

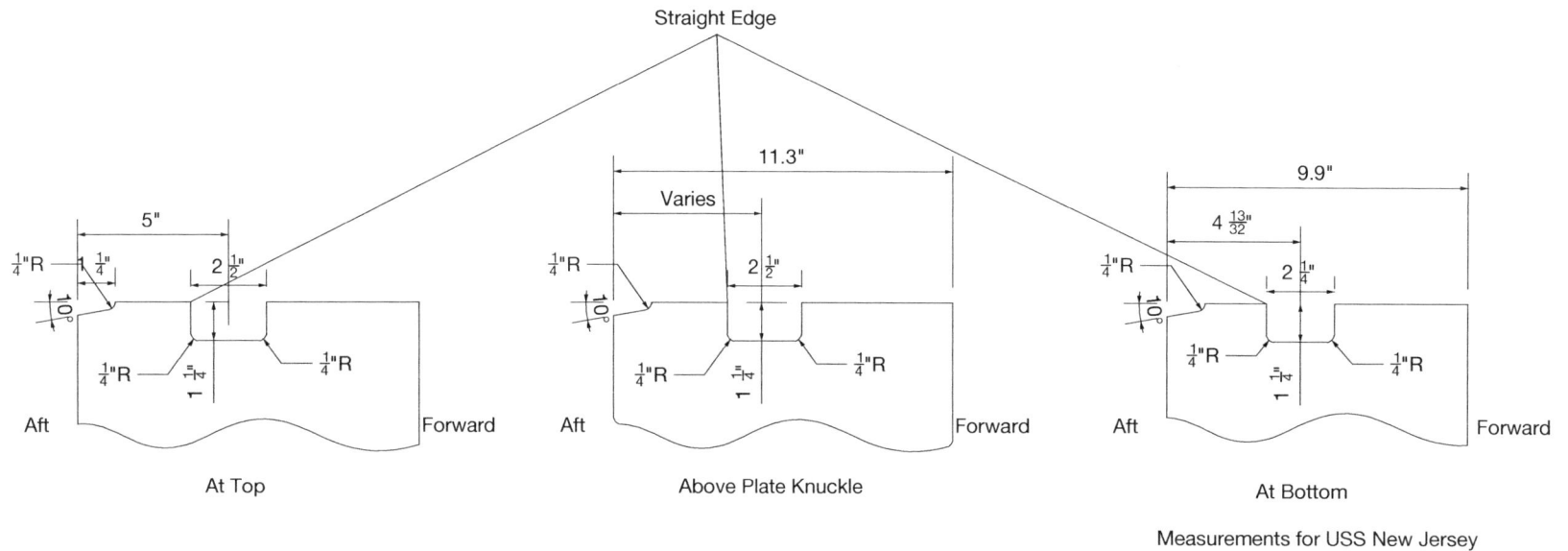

Figure 3.8: Sections Normal to Edge of Transverse Plates (1:6). The mill machined a welding groove and a keyway for alignment on the inside face of each plate joint. The keyway tapers from the plate thickness knuckle to the bottom.

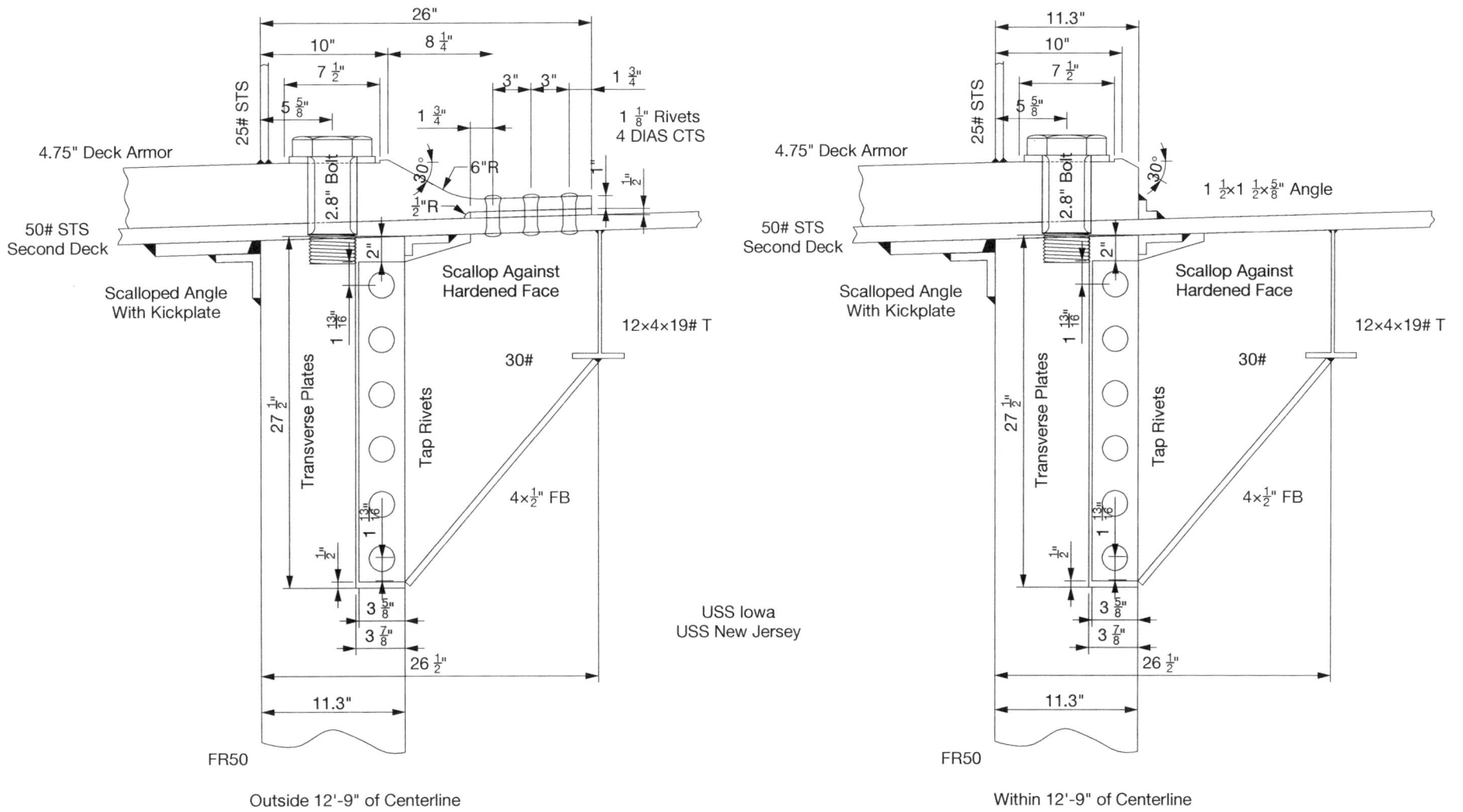

Figure 3.9: Connection to Second Deck (1:12). The attachment of the transverse armor to second deck has the two configurations shown above. Brackets were inserted into slots in the armor and tap riveted in place before the armor plates were assembled.

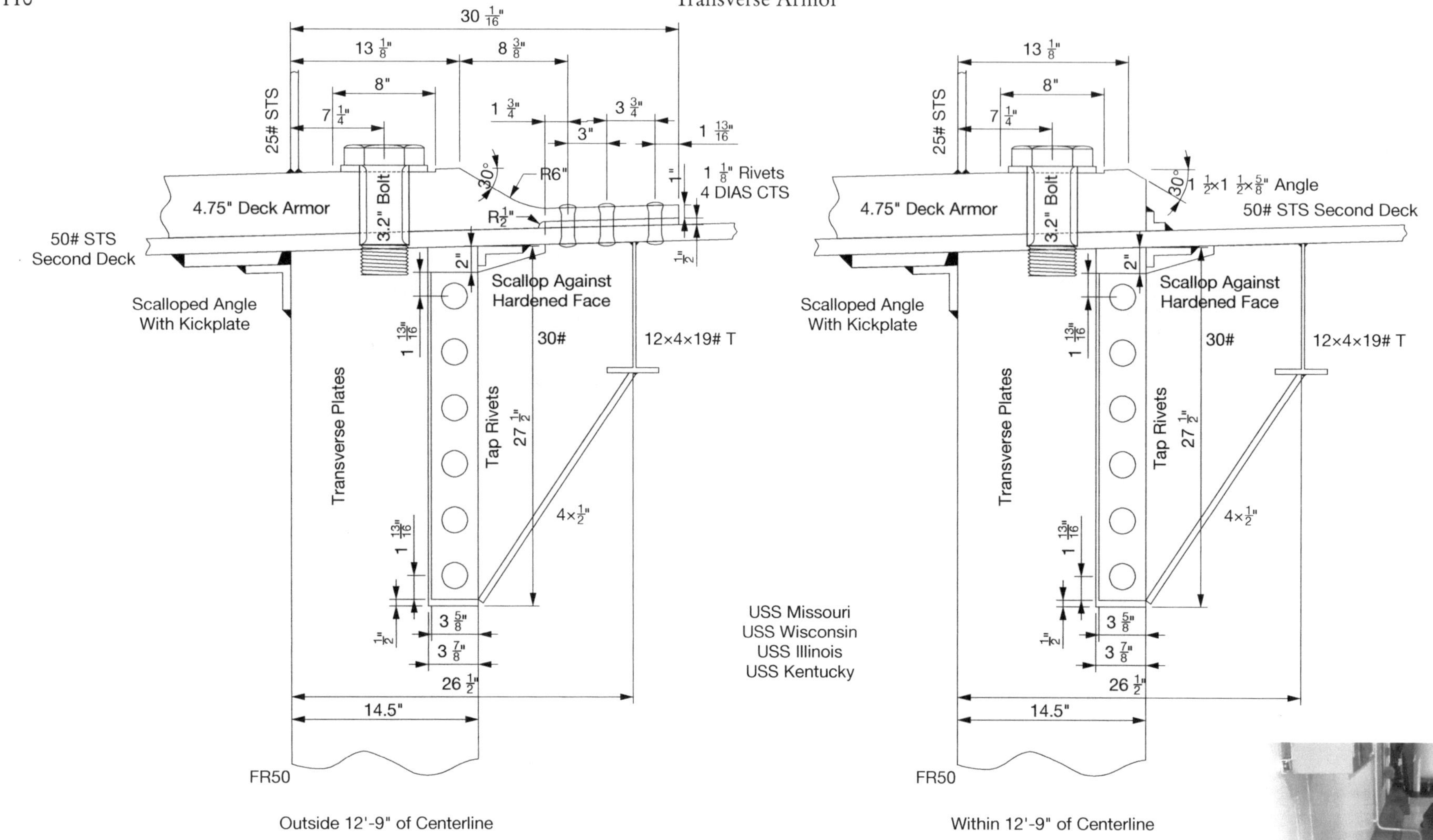

Figure 3.10: Connection to Second Deck (1:12). The attachment of the transverse armor to second deck on the later ships was modified to reflect the thicker transverse armor.

Photograph 3.1: Connection to Second Deck on USS New Jersey. This view is looking to port forward of FR50. A cover was welded over the bolts. The transition between the two types of connections shown above is as the bottom of the photograph.

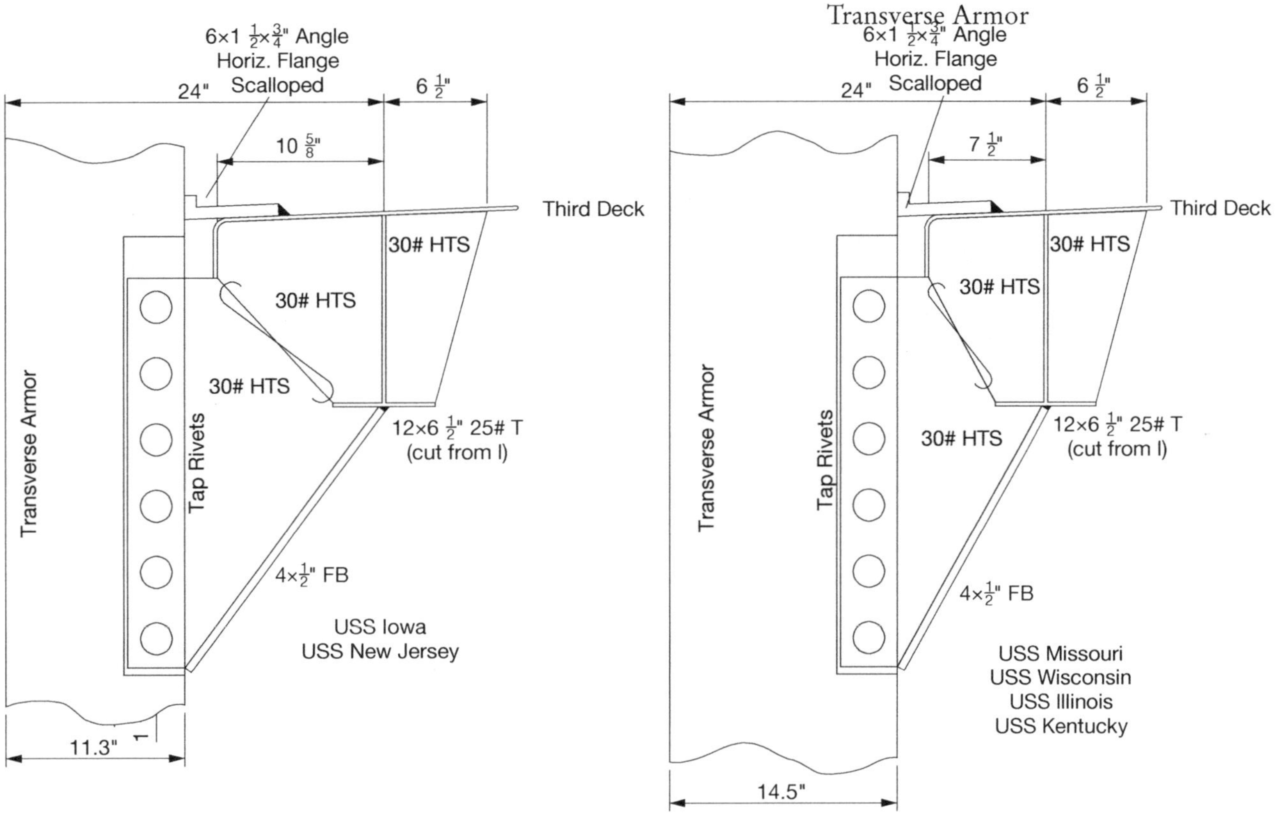

Transverse Armor

6×1 $\frac{1}{2}$×$\frac{3}{4}$" Angle
Horiz. Flange
Scalloped

24"

6 $\frac{1}{2}$"

10 $\frac{5}{8}$"

Third Deck

30# HTS

30# HTS

30# HTS

Transverse Armor

Tap Rivets

12×6 $\frac{1}{2}$" 25# T
(cut from I)

4×$\frac{1}{2}$" FB

USS Iowa
USS New Jersey

11.3"

1

Figure 3.11: Connection to Third Deck (1:12). On the outside face, third deck is supported entirely by brackets tap riveted into slots in the armor plates.

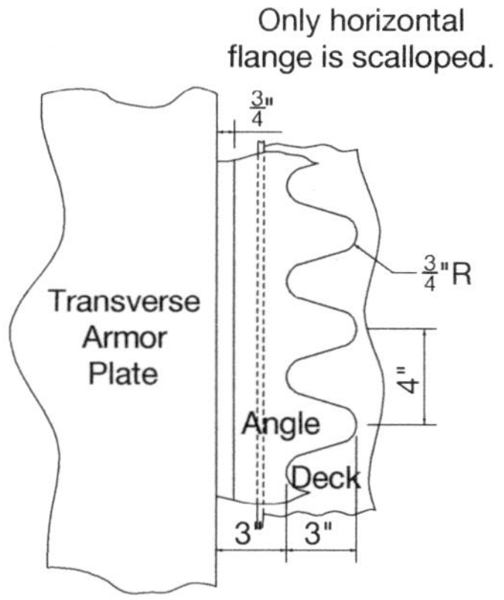

6×1 $\frac{1}{2}$×$\frac{3}{4}$" Angle
Horiz. Flange
Scalloped

24"

6 $\frac{1}{2}$"

7 $\frac{1}{2}$"

Third Deck

30# HTS

30# HTS

30# HTS

Transverse Armor

Tap Rivets

12×6 $\frac{1}{2}$" 25# T
(cut from I)

4×$\frac{1}{2}$" FB

USS Missouri
USS Wisconsin
USS Illinois
USS Kentucky

14.5"

Only horizontal flange is scalloped.

$\frac{3}{4}$"

$\frac{3}{4}$"R

4"

Transverse Armor Plate

Angle

Deck

3" 3"

Figure 3.12: Plan of 6-Inch Angle at Third Deck, First Platform, and Second Platform (1:12). An angle with the horizontal flange scalloped was welded to decks against the hard face of the transverse armor.

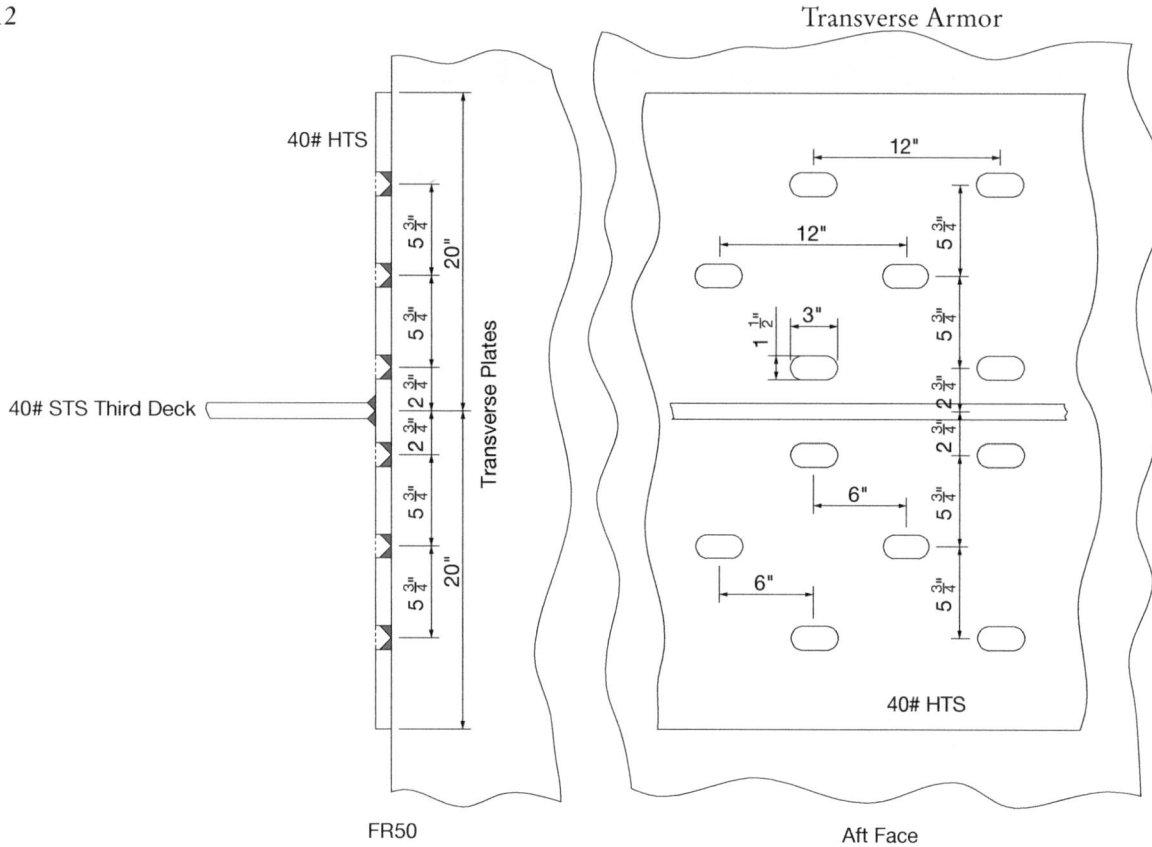

Figure 3.13: Aft Connection to Third Deck (1:12). Third deck was not welded directly to the inside face of the transverse plates as was first platform. Instead 40-inch tall 40# high tensile steel plates were welded to each armor plate at the deck level and third deck was welded to these plates. The plates have oval holes to create welding surfaces in addition to the edges. The hole pattern shown is from the plans but it was not followed exactly. The plans do not indicate why this was done for third deck and not first platform.

Figure 3.14: The connection between third deck and the transverse armor on USS New Jersey viewed from below. The seams in the 40# plates follow the joints in the armor plates. A weld between armor plates is visible below the weld between the two 40# plates shown.

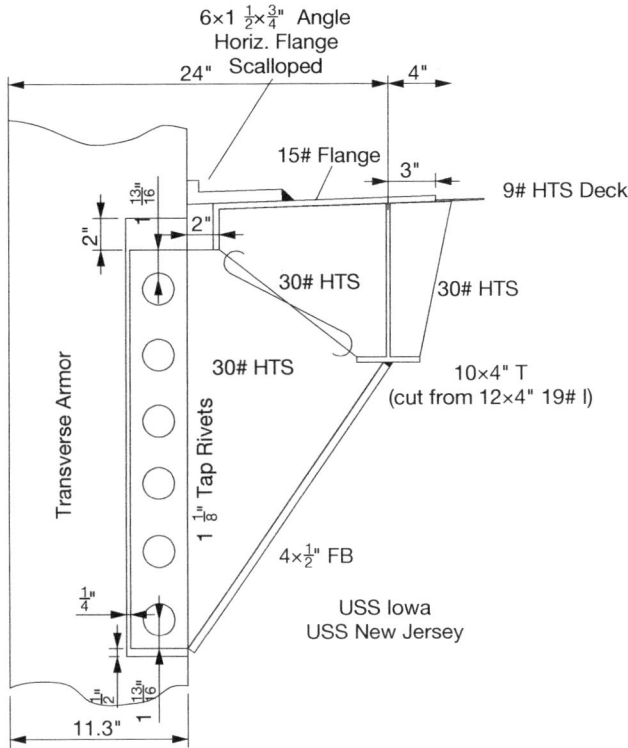

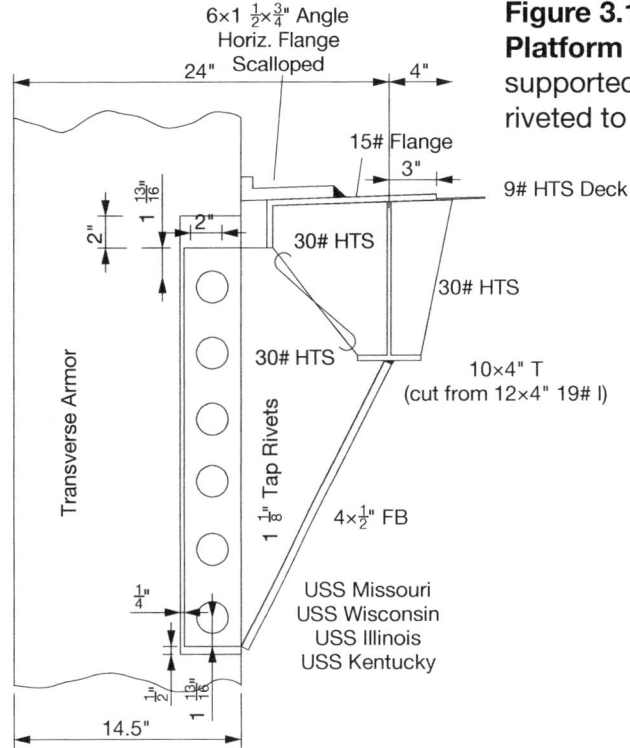

Figure 3.15: Forward Connection to First Platform (1:12). First platform is also entirely supported on the outside face by brackets tap riveted to slots in the armor plates.

Left diagram labels:

6×1 $\frac{1}{2}$×$\frac{3}{4}$" Angle
Horiz. Flange
Scalloped
24"
4"
15# Flange
3"
9# HTS Deck
1 $\frac{13}{16}$
2"
2"
Transverse Armor
30# HTS
30# HTS
1 $\frac{1}{8}$" Tap Rivets
30# HTS
10×4" T
(cut from 12×4" 19# I)
$\frac{1}{4}$
4×$\frac{1}{2}$" FB
USS Iowa
USS New Jersey
$\frac{1}{2}$
1 $\frac{13}{16}$
11.3"

Right diagram labels:

6×1 $\frac{1}{2}$×$\frac{3}{4}$" Angle
Horiz. Flange
Scalloped
24"
4"
15# Flange
3"
9# HTS Deck
1 $\frac{13}{16}$
2"
2"
30# HTS
Transverse Armor
30# HTS
30# HTS
1 $\frac{1}{8}$" Tap Rivets
10×4" T
(cut from 12×4" 19# I)
4×$\frac{1}{2}$" FB
$\frac{1}{4}$
USS Missouri
USS Wisconsin
USS Illinois
USS Kentucky
$\frac{1}{2}$
1 $\frac{13}{16}$
14.5"

Photograph 3.2: Connection of First Platform to FR50 Transverse Armor. This view is from second platform on USS New Jersey looking aft. Two brackets connected to the armor in slots are visible. They support a beam running among the brackets. The large pipe runs to a CHT tank on third platform.

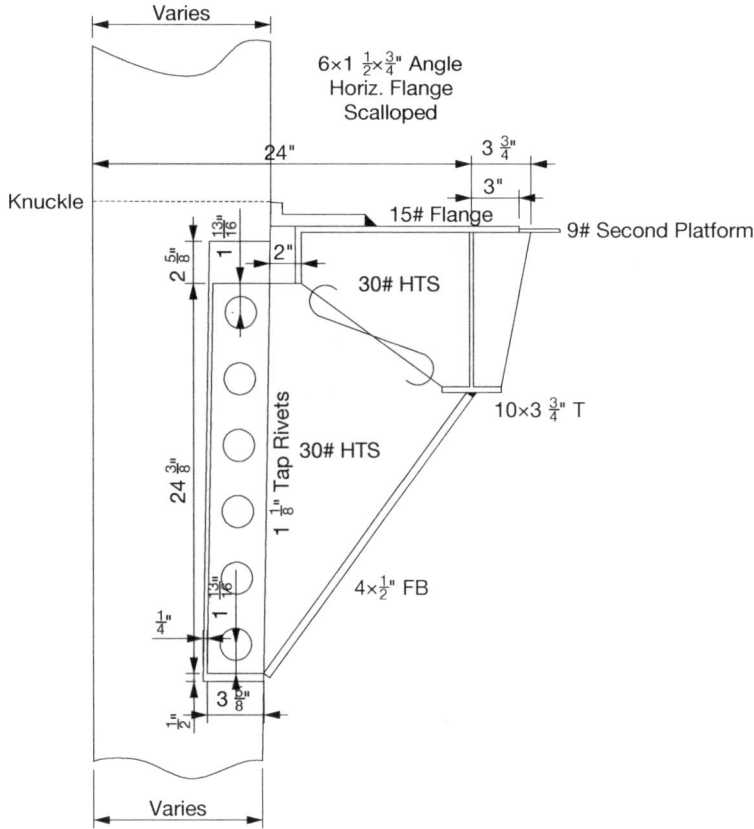

Figure 3.16: Forward Connection to Second Platform (1:12). Brackets tap riveted to slots in the armor pates support second deck on the outside face of the transverse armor. Structurally, second deck ends at the transverse armor and third platform becomes second platform.

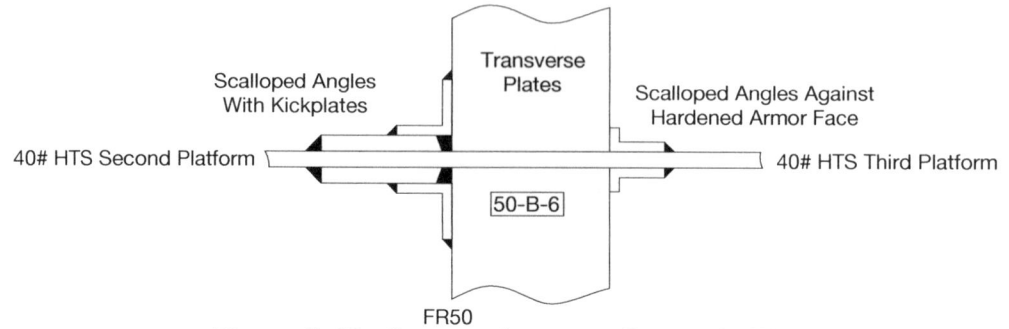

Figure 3.17: Connection to Third Skin (1:12). The bottom of the transverse armor rests on the third skin which is also the hold at this location.

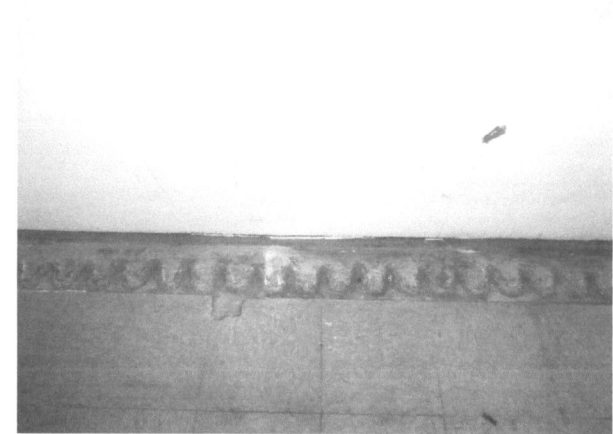

Figure 3.18: Connection to Second Platform and Third Platform (1:12). Third platform becomes second platform aft of FR50 The deck passes through the transverse armor so the lower plate (50-B-6) does not have any connection to the upper plates of the transverse armor.

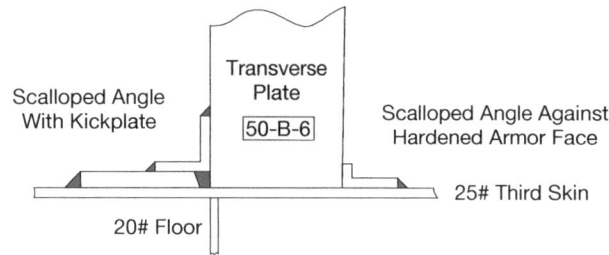

Photograph 3.3: Scalloped Angle Bracing Second Platform Against Transverse Armor. This USS New Jersey looking aft at FR50. The scalloped flange of the angle is welded to the deck. The vertical flange is pressed against the armor and caulked.

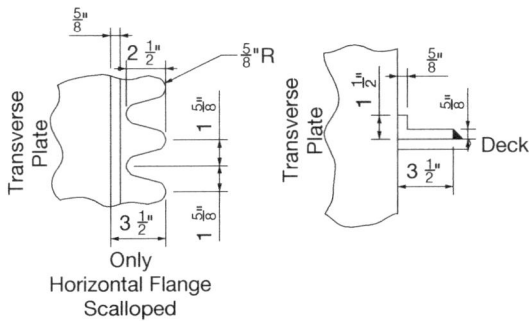

Figure 3.19: Scalloped Angle at Deck Against a Hardened Face (1:12). The hardened outer face of the transverse plates could not be welded. The deck connections to the out face incorporate angles that were scalloped on one flange. The scalloped flange was welded to the top or underside of the deck. The other flange served as a brace pressed up against the armor plate.

Photograph 3.4: The forward face of the FR50 transverse armor on USS New Jersey at the hold. The transverse armor is to the right and torpedo bulkhead 3 is to the left. Angles with one flange scalloped are visible along the torpedo bulkhead and the underside of third platform. The unscalloped flange along the torpedo bulkhead is tap riveted to the armor plate. The similar flange along the deck is just pressed against the armor.

Figure 3.20: Scalloped Angle with Kickplate (1:12). The connections to torpedo bulkhead 3, the hold at the third skin, to second platform, and second deck on the inside (soft) face of the transverse plates use a double layer of scallops. A scalloped kickplate was welded to the deck on either or both sides. The kick plate established a straight line for positioning the armor plate. After the armor was in position, a scalloped angle was welded to the kickplate and the armor plate.

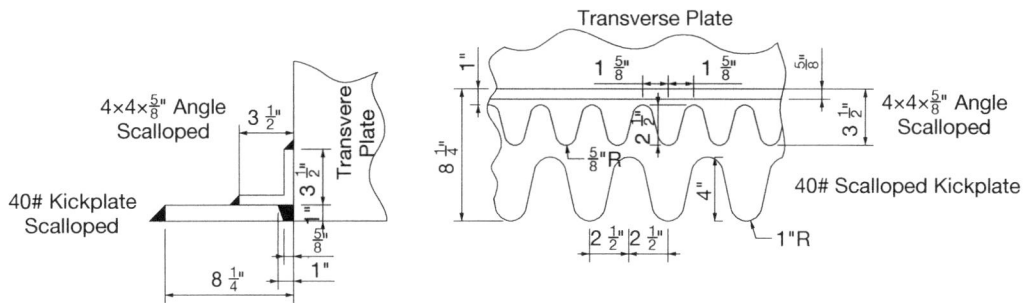

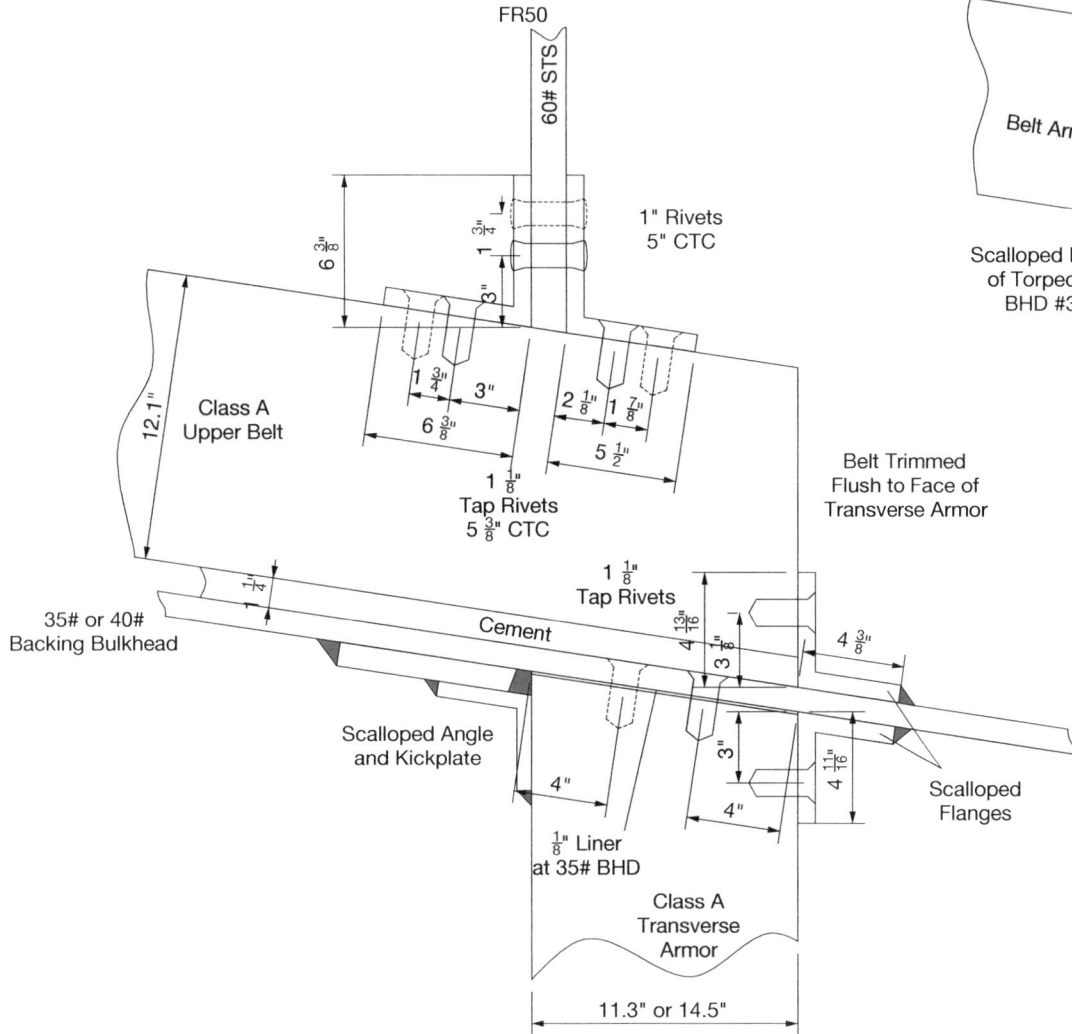

Figure 3.21: Connection to the Upper Belt (1:8). The connection between the transverse armor and torpedo bulkhead 3 changes vertically along with the configuration of the belt armor. There is no direct connection between the belt and transverse armor. This is the configuration at the upper belt.

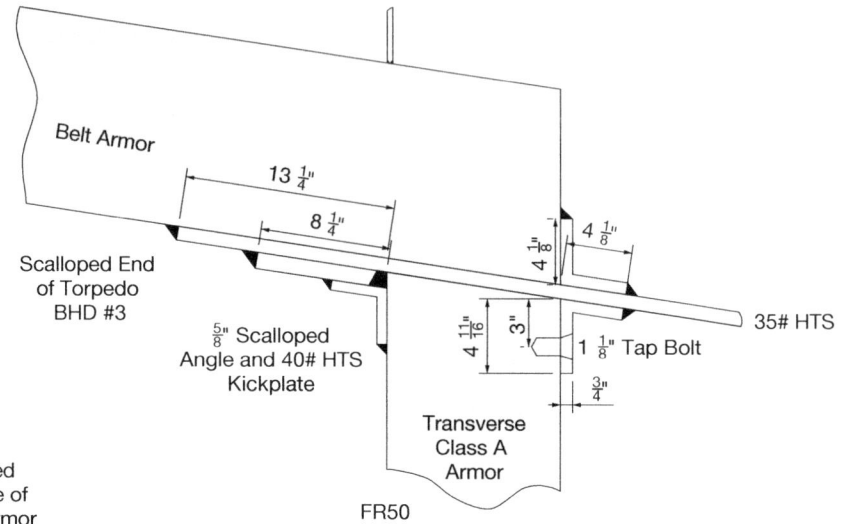

Figure 3.22: Connection to the Lower Belt Above First Platform (1:12). This is the configuration of the connection to the belt above first platform at the top of the lower belt.

Photograph 3.5: Attachment of Transverse Armor to Torpedo Bulkhead 3. An angle runs along the joint. The torpedo bulkhead is to the left and armor is to the right. An angled bar covers the joint. The flange against the torpedo bulkhead is scalloped and welded. The flange against the armor is flat and tap riveted.

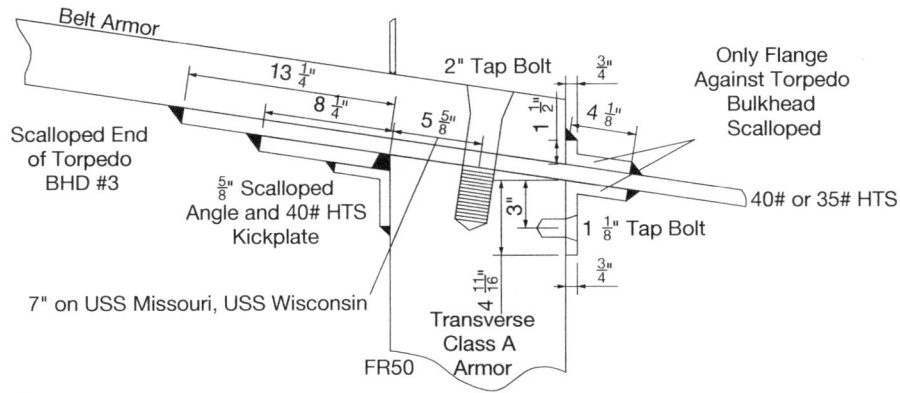

Figure 3.23: Connection to Lower Belt Between First Platform and Second Platform (1:12). Between second platform and first platform the lower belt becomes thin enough to use tap bolts to attach to the transverse armor. These tap bolts vary in length so they get shorter as the belt armor becomes thinner.

Plate	Weight (lbs.)
B-50-1	161,125
B-50-2	118,984
B-50-3	118,984
B-50-4	116,106
B-50-5	116,106
B-50-6	77,492
Total	708,797

Table 3.1: Frame 50 Transverse Plate Weights

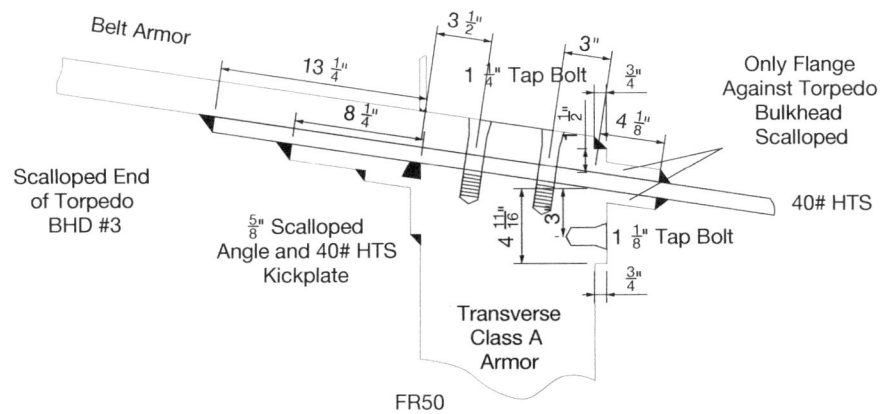

Figure 3.24: Connection to Lower Belt Below Second Platform (1:12). Below second platform two smaller tap bolts are used to connect the lower belt to the transverse armor. Here as well, the tap bolts become shorter towards the bottom to compensate for the narrowing of the belt armor.

Backing Bulkhead for
Upper Belt Armor

30# STS

Transverse Armor

Location of
Lower Belt

35# HTS

40# HTS

Inner Seamstrap

40# STS

40# STS   Third Skin

FR50

Location of
Lower Strake
Torpedo BHD 3

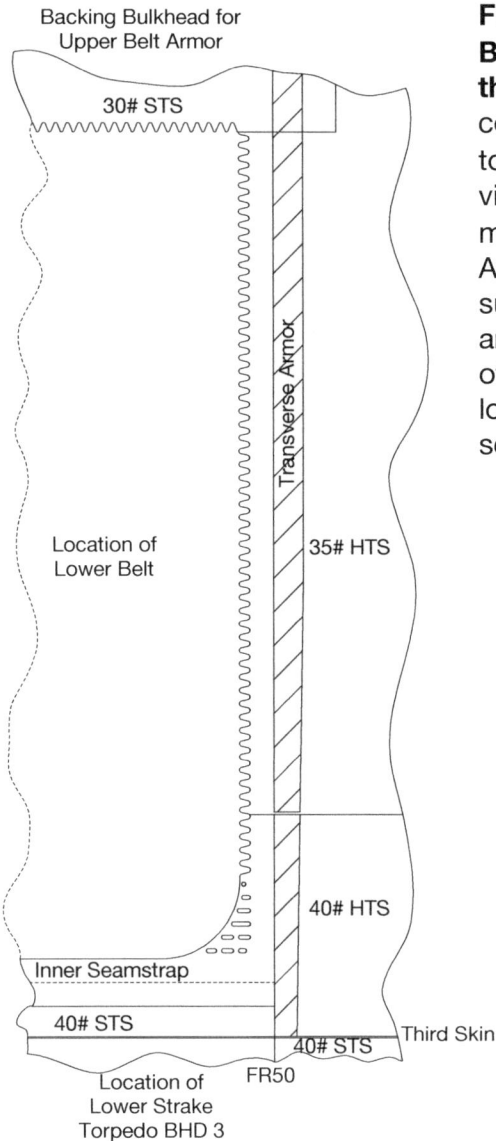

Figure 3.25: Elevation of the Connection Between the Lower Belt and Bulhead 3 at the FR50 Transverse Bulkhead (1:72). The connection between the transverse armor and torpedo bulkhead 3 is rather complex. The view in this figure is inboard with the belt armor and lower strake of bulkhead 3 removed. At FR50 the torpedo bulkhead shifts into a support for the armor belt. The scallops shown are for welding the lower belt plates to the rest of the building. A scalloped kickplate and scalloped angle are welded inboard of the vertical scallop, creating three layers of scallops.

Photograph 3.6: USS Kentucky Under Construction. The transverse armor is in place. The fixed turret part of turret 2 are visible behind. Aft of turret 2 work is taking place at second deck. (Naval History and Heritage Command)

Chapter 4: Second Deck Armor

The second deck armor consists of 169 plates that extend from FR50 to FR166 and sit within torpedo bulkhead no. 3 at the sides. The deck armor plates are 4.75-inches thick and rests on the 50# STS structural second deck. The combined second deck protection is six inches. The deck armor plates are positioned so that their seams do not align with the seams between deck plates in order avoid creating weak points. In this frame range, the bottom of 50# STS structural second deck is aligned to the molded line. The armor resting on top creates a noticeable step at FR50 and FR166 where the deck armor starts and ends.

The deck armor plates required extensive machining at the yard. They were ordered oversized from the mill and came as quadrilaterals or irregular pentagons. The yard had to cut all the curves, openings, and grind the plates to fit.

The yard also had to machine the joints between adjoining plates. Longitudinal edges have keyed joints The transverse plates edge were beveled so that the adjoining plates were welded together and to the deck.

There are eight removable sections (two over each engine room) in the deck that deviate from the generate arrangement. The removable sections consist of either two or three rectangular plates. The sections have a scarf joints oriented so the plates can be lifted upward along their transverse edges. The two of the adjoining removable plates also have a scarf joint. The other longitudinal joints are keyed joints. Under this arrangement, the upper plate at the scarf joint

would be pulled up first. As this plate was pulled away, the keyed joints would come loose. The opening in the 50# STS second deck plates was smaller than the opening in the armor to avoid having armor plates fall through the deck as they were being removed.

The removable plates were only intended to assist in repairing serious damage. They did not open like doors. Bulkheads and equipment sit over the removal section and would have to be removed first. Once the area above the removable section was cleared, rivets had to be drilled out. Similar plates would have to be removed at the 01 level, main deck, and third deck to create a usable opening.

The armor plates are secured to second deck using quilting pins. The quilting pins are similar to rivets. Unlike rivets, the pins were inserted, without being heated, from below then welded to the upper face of the armor. The pins were spaced about every 12 inches around plate edges and on either side of structural bulkheads. In other areas of the plate they were spaced about 24 inches.

The connections between the second deck armor with the upper belt and the transverse armor are described in the chapters covering those other sections.

4.17. Splinter Deck

The splinter deck provides additional protection below second deck. The purpose of the splinter deck was to catch parts of second deck (such as the heads of quilting pins) that might break off under an impact. The splinter deck extends between turret no. 2 and turret no. 3. The top of the splinter deck is 30 inches below the molded

Ship	Suppliers
USS *Iowa*	Midvale, Carnegie-Illinois, Bethlehem
USS *New Jersey*	Midvale, Carnegie-Illinois, Bethlehem
USS *Missouri*	Midvale
USS *Wisconsin*	Carnegie-Illinois
USS *Illinois*	Midvale, Carnegie-Illinois
USS *Kentucky*	Carnegie-Illinois

Table 4.1: Manufacturers of Second Deck Armor Plates

line of second deck until FR151 where it angles upward. From FR151 $\frac{1}{4}$ the splinter deck is 16 inches below second deck. The splinter deck creates a crawl space for wiring. The splinter deck layout was covered in Volume 1. The splinter deck is 25# on USS *Iowa* and USS *New Jersey* and 30# on USS *Missouri* and USS *Wisconsin*.

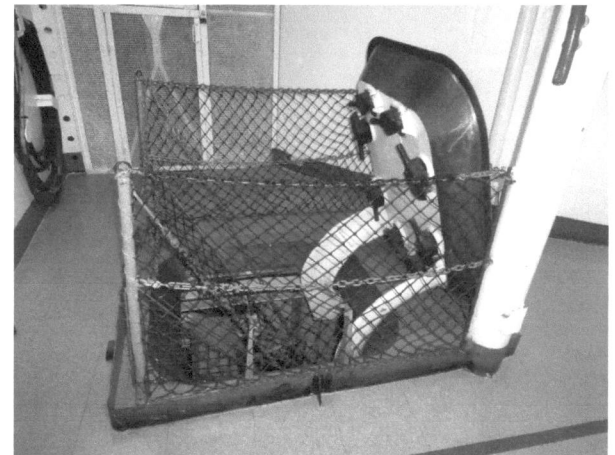

Photograph 4.1: Hatch in Second Deck. The second deck hatches make the armor thickness apparent. This is USS New Jersey.

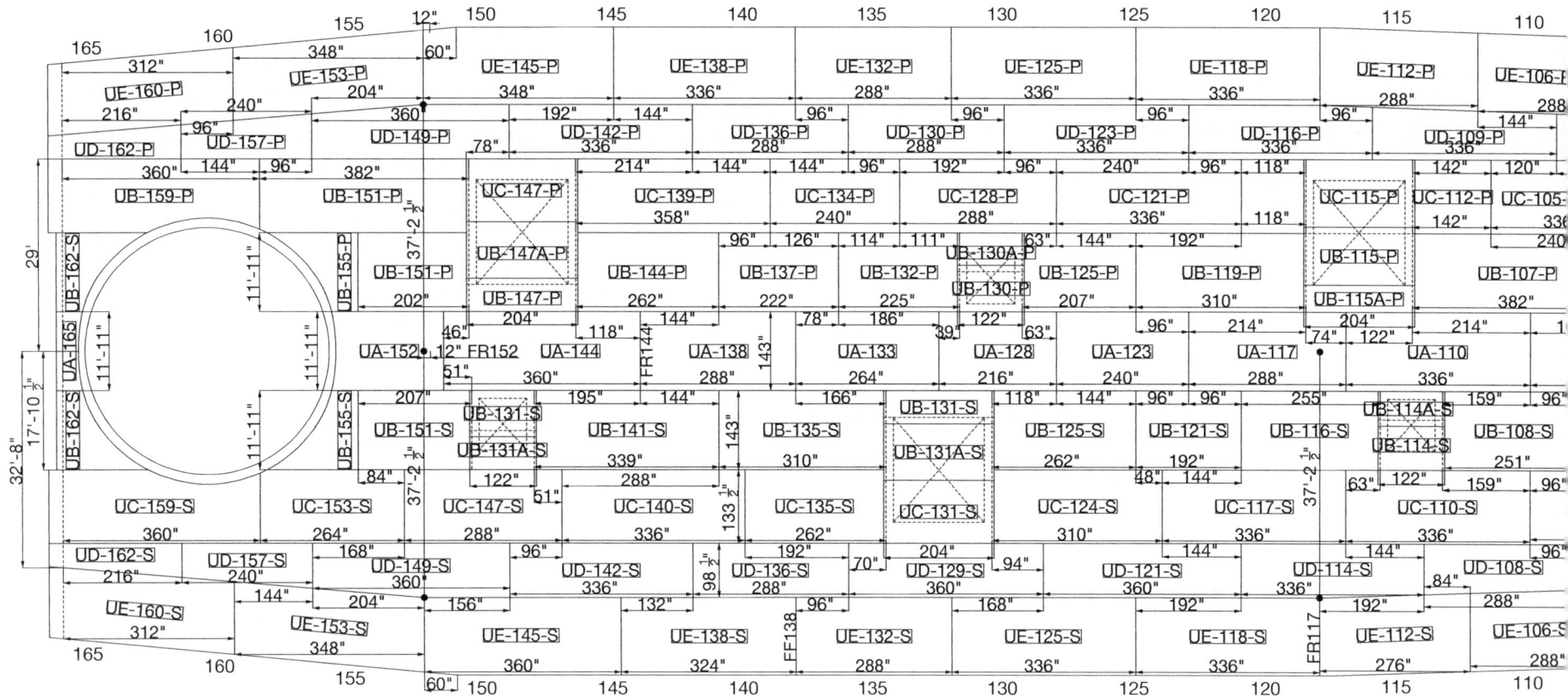

Figure 4.1: Second Deck Armored Plates (1:288). This diagram shows the plates trimmed and in position. Each engine room has two areas of removable plates. The clear openings are marked with an X.

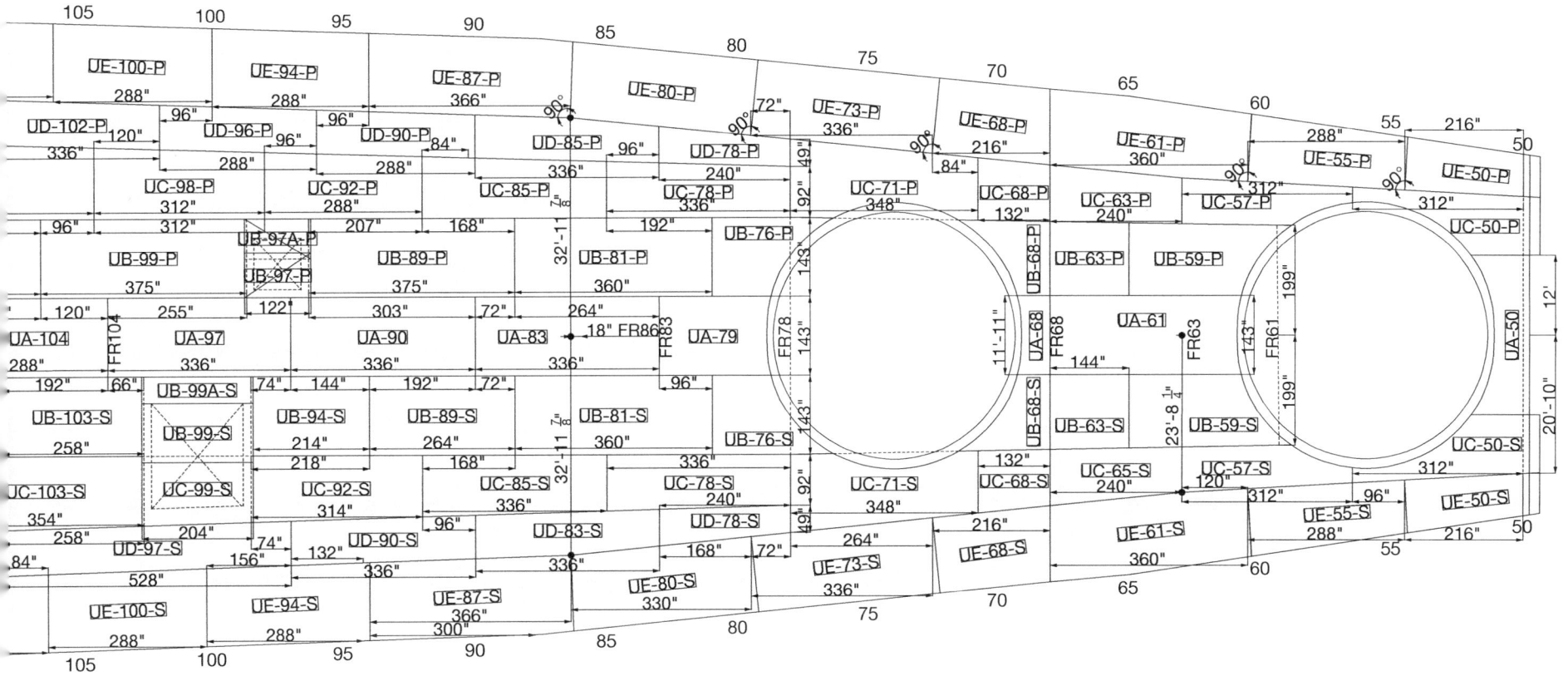

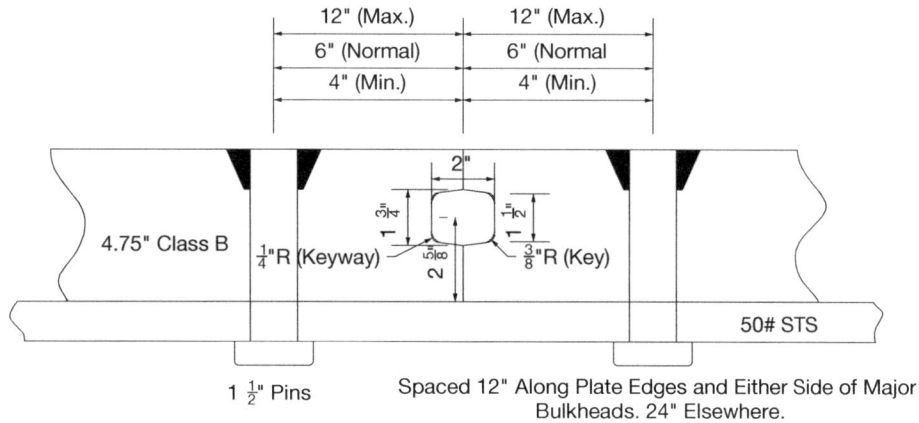

Figure 4.2: Longitudinal Joints Between Second Deck Plates (1:6). Longitudinal joints were keyed for alignment,

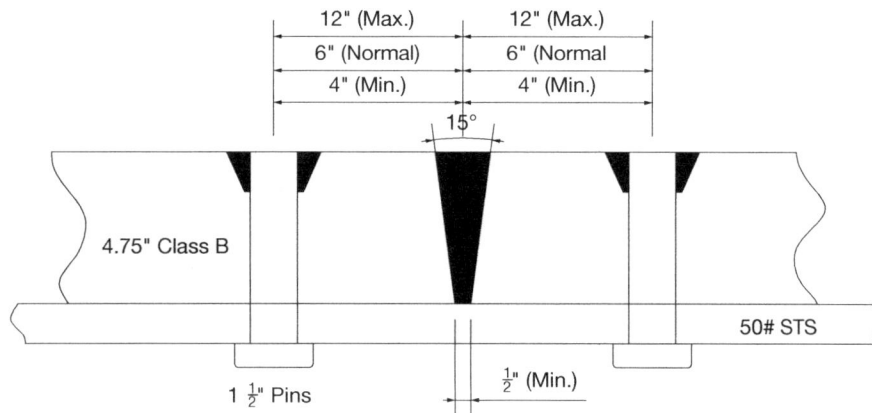

Figure 4.3: Transverse Joints Between Second Deck Armor Plates (1:6). Transverse joints were welded to second deck.

Photograph 4.2: Second deck in the carpenter's shop of USS New Jersey. The welded ends of quilting points are visible. The pins along the seam between armor plates are more closely spaced than those away from the edge.

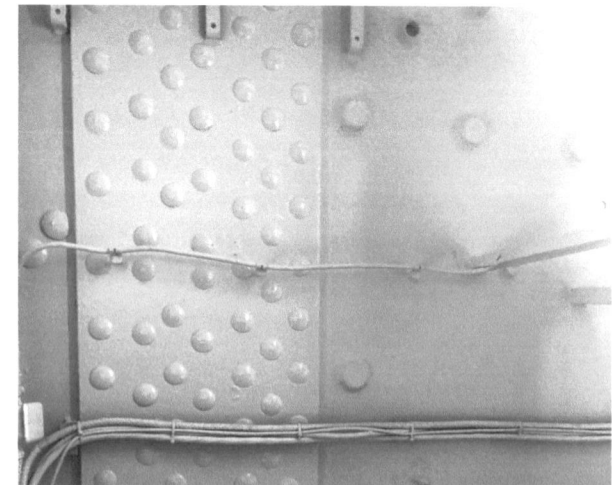

Photograph 4.3: The overhead in the X-Ray room of USS New Jersey. The riveted butt-strap connects second deck plates. The heads of quilting pins are visible to either side.

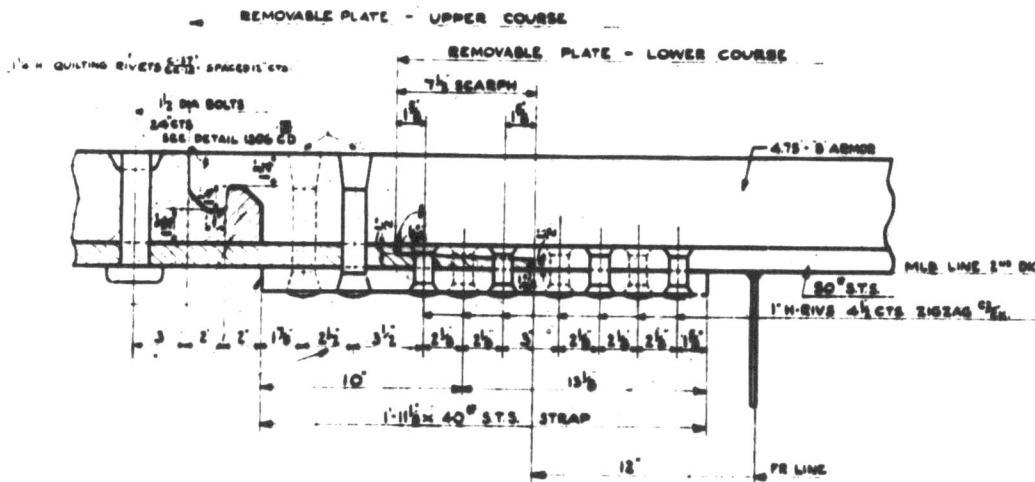

Plan 4.4: (1106BH). A plan showing the transverse joint of a removable plate. Rivets along the edge had to be drilled out to lift the armor. Additional rivets would have to be drilled out to open the 50# STS second deck below. The opening in the 50# STS deck is much narrower than the opening in the armor plate.

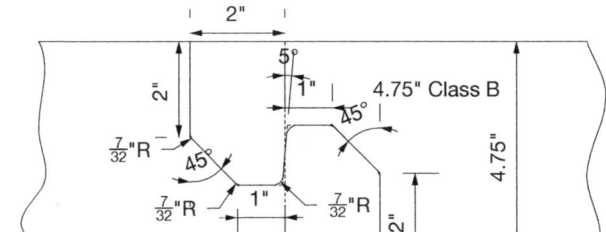

Figure 4.4: Removable Plate Joint (1:4). This is an enlargement of the scarf joint shown in Plan 4.4.

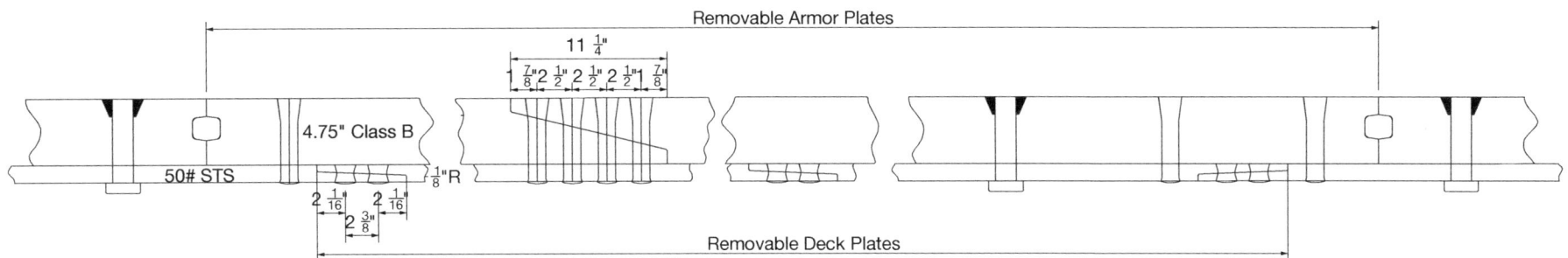

Figure 4.5: Transverse Section Through Removable Plates (1:12). The removable sections consisted of two or three plates. The plates were removable only because one longitudinal joint had a riveted scarf joint rather than being keyed.

Chapter 5: Third Deck

Rather than create a separate armored box for the steering gear as on the *Yamato*-class, the *Iowa*-class extends the citadel to encompass the steering gear. This has the advantages that the communications connections to the steering gear are within the citadel, creates more reserve buoyancy, and it protects the shafts but it has the disadvantage that non-vital storage areas end up being protected by armor. To reduce the amount of armor required, the top of the citadel drops from second deck to third deck aft of FR166. Third Deck is covered by tile in this area so the transition to armor is not visible from above.

A major difference between the third deck armor and the second deck armor is that the top of the former sits at the molded line of the deck. Where the second deck has a step up at the ends of the armor, the third deck armor is roughly even with the surrounding deck and, after the area was tiled, there is no visible indiction at third deck where the armor starts and ends.

The third deck armor has two different configurations: the extension from FR166 to FR189 and over the steering gear from FR189 to FR203.

5.18. Third Deck Extension Armor

The general layout of the armor third deck armor from FR166 to FR189 the armor is very similar to that at second deck. There is 30# STS deck plating that supports 5.6 inches of Class B ar

The armor plates were ordered from the mill oversized. The plates were all quadrilateral or ir-regular pentagons in shape. The yard had to mill the plates to the correct size to fit. The joints between plates were machined in the same way as those at second deck. The yard machined keyways along the longitudinal plate joints and fitted keys into them for alignment. Along transverse joints, the mill beveled the plates so that they could be welded to their neighbor and to third deck. There are no removable plates that had to be machined differently in this area. For additional strength, the joints on the armor plates are not aligned with the joints in the deck plates.

The armor plates were secured to third deck plates using quilting pins in the same manner they were used at second deck. The yard drilled holes through the armor and deck. The pins have a head and a shaft. The shafts were pushed through the holes from below and were welded to the top face of the armor plates. The pins were spaced 12 inches along plate joints and structural bulkheads and 24 inches elsewhere.

The third deck armor was attached directly to the belt armor (unlike the second deck armor) using a keyed hook joint. The yard machined the deck side of the joint to match the joint the mill machined in the belt plates.

After the deck armor was in place, a key was hammered into the joint then welded in place. The yard drilled and tapped holes about $8\frac{1}{8}$ inches for 2-inch tap bolts along the tongue the belt plates (avoiding plugs for lifting bolt holes). Bolts were screwed in and chipped flush. The connection of the third deck armor to the FR166 transverse armor are described in the latter's section.

5.19. Third Deck Steering Gear Armor

The third deck armor over the steering gear from FR189 to FR203 consists of thirteen 6.2 inch plates. The mill delivered the plates as oversized quadrilaterals with no other machining. The plates in this area serve as both the armor and the structural deck. The yard machined scarf joints between the deck plates. The scarf joints were secured using rivets. The scarf joints are arranged such that the plates at FR199 and FR202 could be removed to gain access to the rudder post and machinery.

The yard machined a keyed hook joint into the edges to mate with the belt and transverse plates. The joint is similar to that between FR166 and FR189 except that it was cut deeper to accept the thicker plates over the steering gear. The cross section of the joint at the top of the transverse plate is angled slightly to match the slope of the molded line of the deck. After the deck plates were in place, the yard hammered keys into the joints and welded them in place. The yard then drilled and tapped holes through the deck plates into the tongue of the belt plates. The holes were about $7\frac{1}{2}$ inches apart and avoided the plugs for the lilting bolt holds in the belt plates.

The forwardmost steering gear deck plates extend two feet forward of FR189 to create a transition area between the two third deck armor configurations. These plates were machined down to 5.6" from four inches aft of FR189 to their forward edge.

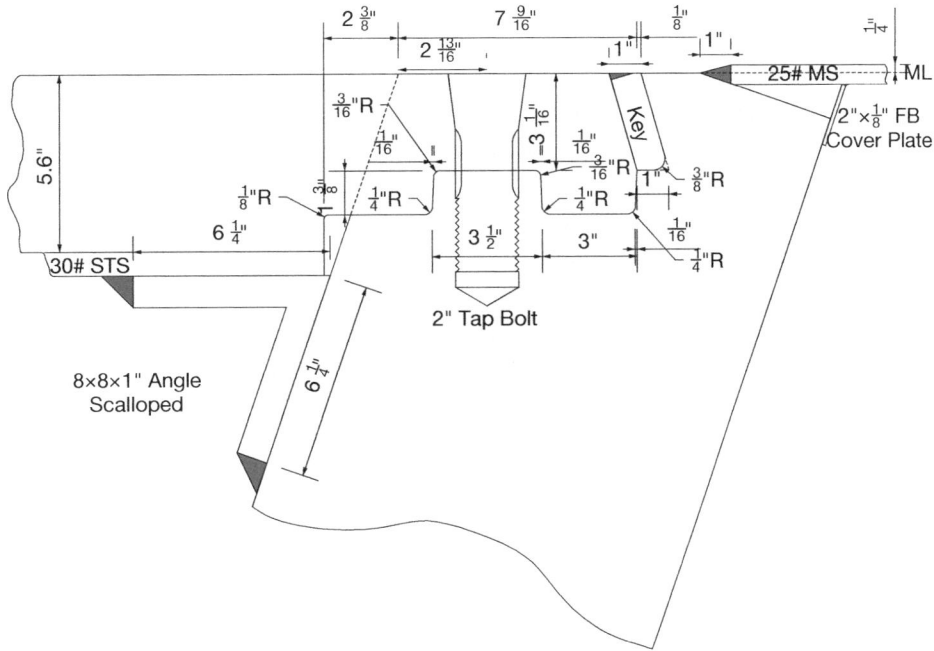

Figure 5.1: Connection of Deck Armor to Belt Between FR166 and FR189 (1:6). Inside corners on the deck and belt platers were rounded to $\frac{1}{4}$" for inside corners and $\frac{3}{16}$" for outside corners. Other than that, the joint profile is the same on both the deck and belt plates.

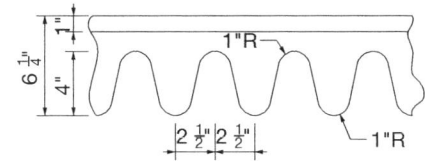

Figure 5.2: The scallop pattern used for the angle below Third Deck (1:12)

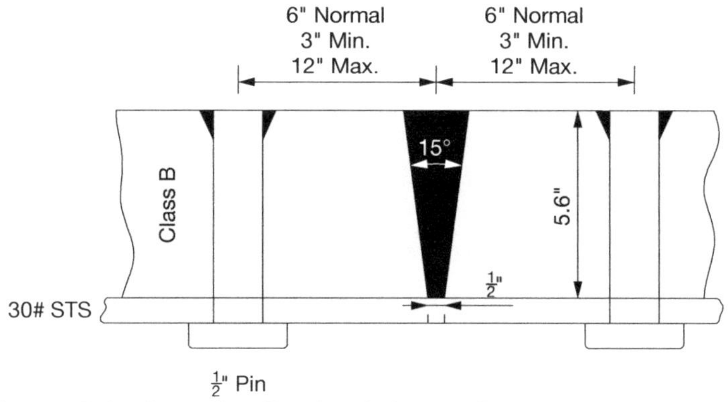

Figure 5.3: Longitudinal Joints Between Third Deck Plates Between FR166 and FR189 (1:6)

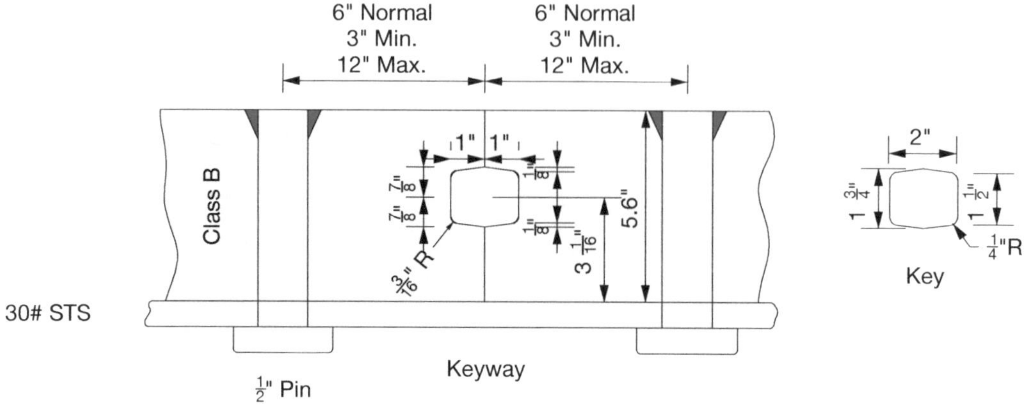

Figure 5.4: Transverse Joints between Third Deck Plates Between FR166 and FR189 (1:60).

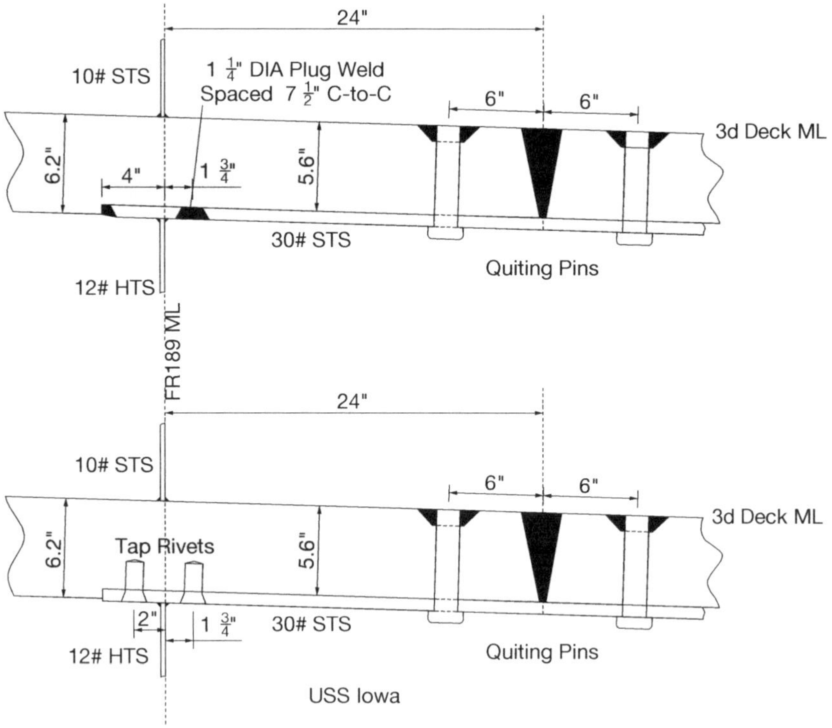

Figure 5.5: Third Deck Armor Transition at FR189 (1:12). The 6.2 inch steering gear deck plates were extended 24 inches forward of FR189 for strength. The plates had to be machined down to 5.6" thick forward of FR189 so that they would fit the shallower hook joint of the belt plates in this area.

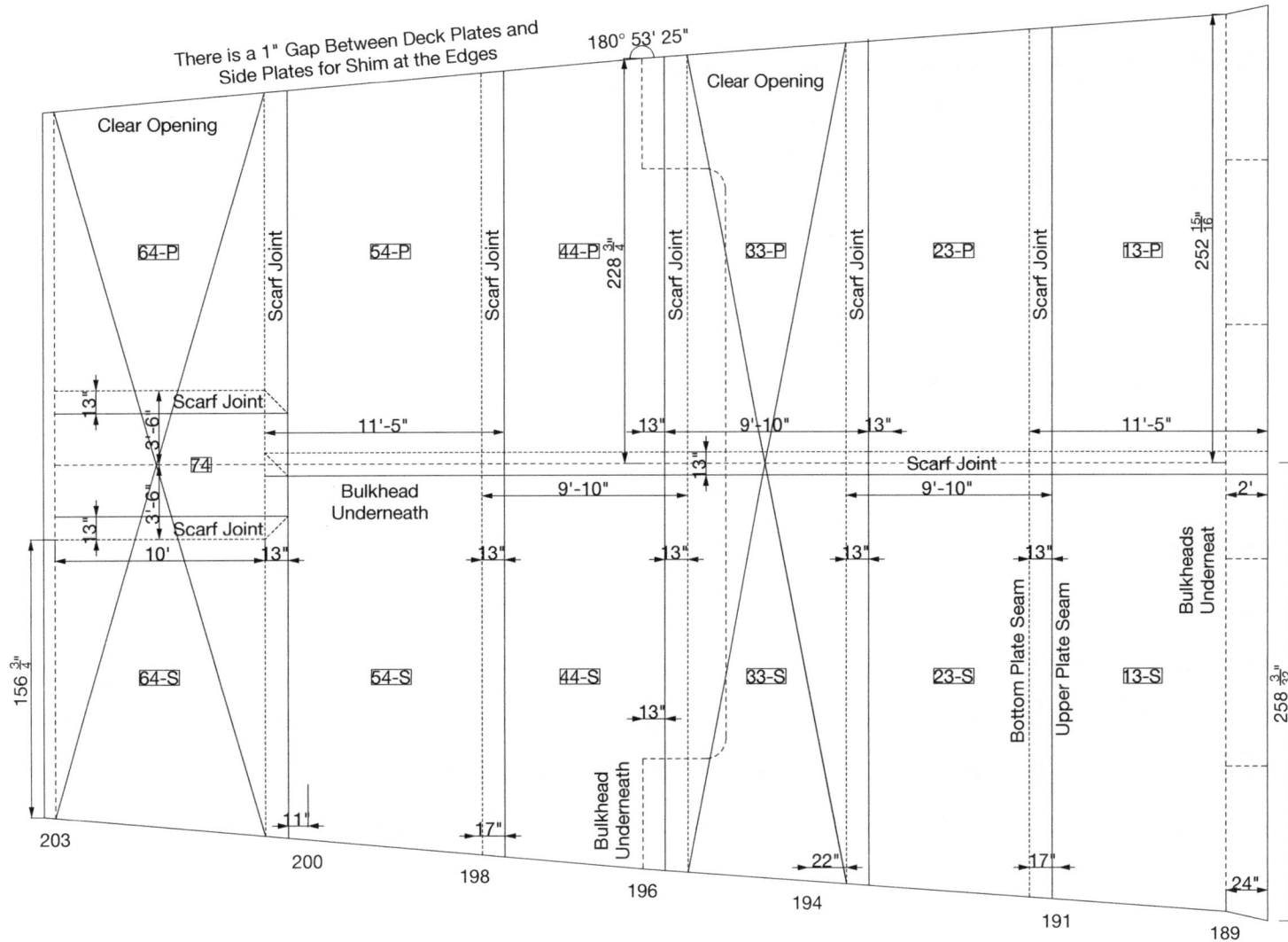

Figure 5.6: Third Deck Armored Plates FR188 $\frac{1}{2}$ to FR203. The plates in this range are 6.2-inches and serve as the structural deck. There are two places (indicated with X) where the plate joints are oriented such that they create removable areas.

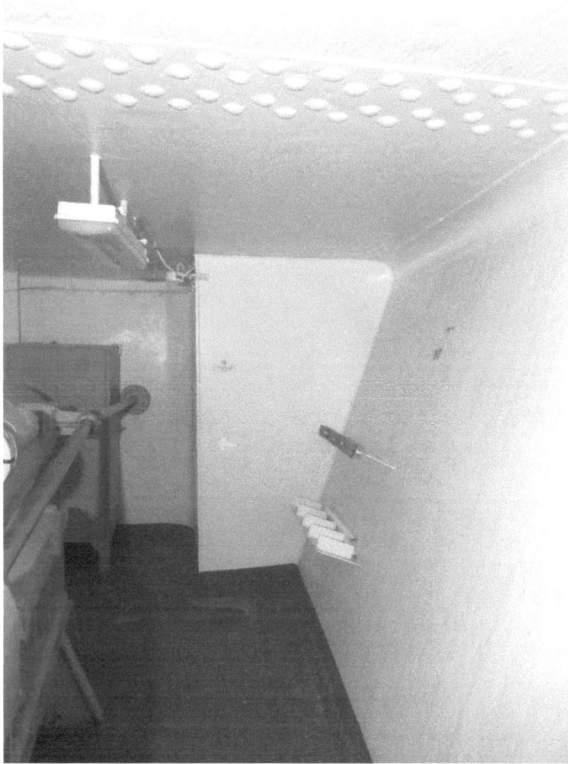

Photograph 5.1: This the starboard steering room of USS New Jersey looking forward. The belt armor is to the right and part of the rudder machinery is to the left. The bolts running through a scarf joint are at the top. The curve plate used to make the deck armor to belt armor joint watertight is visible at the upper right.

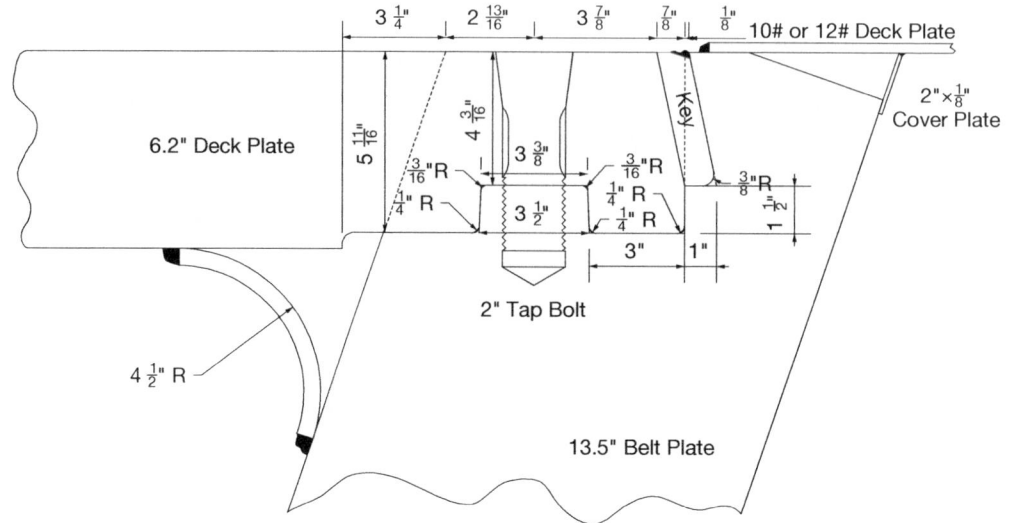

Connection of Third Deck Armor Plates to Belt
1:6

Figure 5.7: Connection of Third Deck Armor to Belt Between FR189 and FR203 (1:6). The joint in this range is basically the same as that used forward of FR189 except that it is cut deeper. The joint along the FR203 transverse armor is the same except it is sloped to match the molded line of the deck (see Figure 5.1).

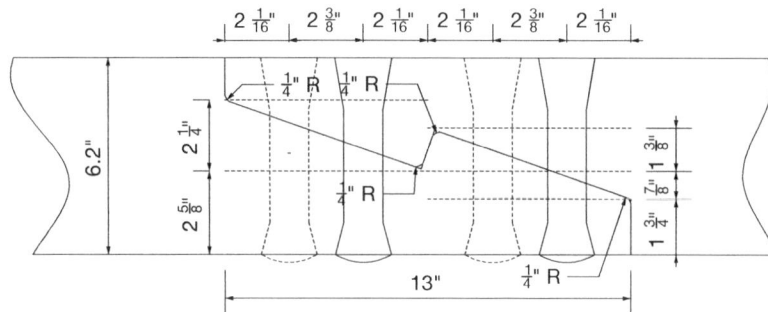

Figure 5.8: Scarf Joint Between FR189 to FR203 Third Deck Plates

Plate	Weight (lbs)
13-P	62,946
23-P	52,140
33-P	50,753
44-P	48,884
54-P	54,278
64-P	44,395
74	21,157
13-S	62,946
23-S	52,140
33-S	50,753
44-S	48,884
54-S	54,278
64-S	44,395
Total	647,949

Table 5.1: Weight of Third Deck Plates Between FR189 and FR203

Ship	Mill
USS *Iowa*	Bethlehem Steel
USS *New Jersey*	Bethlehem Steel
USS *Missouri*	The Midvale Co.
USS *Wisconsin*	Carnegie-Illinois Steel
USS *Illinois*	Bethlehem Steel
USS *Kentucky*	Carnegie-Illinois Steel

Table 5.2: Third Deck Plate Suppliers

Plate	Weight (lbs)
UD166-P	35,637
UC166-P	46,294
UB166-P	55,016
UA166-P	35,846
UA166-S	47,327
UB166-S	55,016
UC166-S	45,294
UD166-S	35,637
UD170-P	60,351
UC171-P	54,475
UB173-P	38,533
UA173	61,306
UB173-S	38,533
UC171-S	54,475
UD170-S	60,351
UD176-P	70,405
UC177-P	64,370
UA179-P	37,563
UA179-S	28,393
UC177-S	64,370
UD176-S	70,405
UD183-P	50,122
UC184-P	42,554
UC184	49,432
UC184-S	45,993
UB183-S	50,122
Total	1,297,820

Table 5.3: Weight of Third Deck Plates Between FR166 and FR189.

Photograph 5.2: Third Deck Connection to Belt Armor Aft of FR189. This is USS New Jersey looking upward. The belt armor is to the right and the girder that supports Third Deck outside the armor is to the left. The cover plate that closes off the gap between the top of the belt and third deck runs along the top of the belt.

Figure 5.9: Third Deck Armor Plates FR166 to FR189 (1:144). This diagram show the positions of the machined plates between FR166 and 24 inches forward of FR189. The plans do not completely specify the exact locations for the plates. The deck armor plates in this area are 5.6-inches and rest on a 30# STS structural deck.

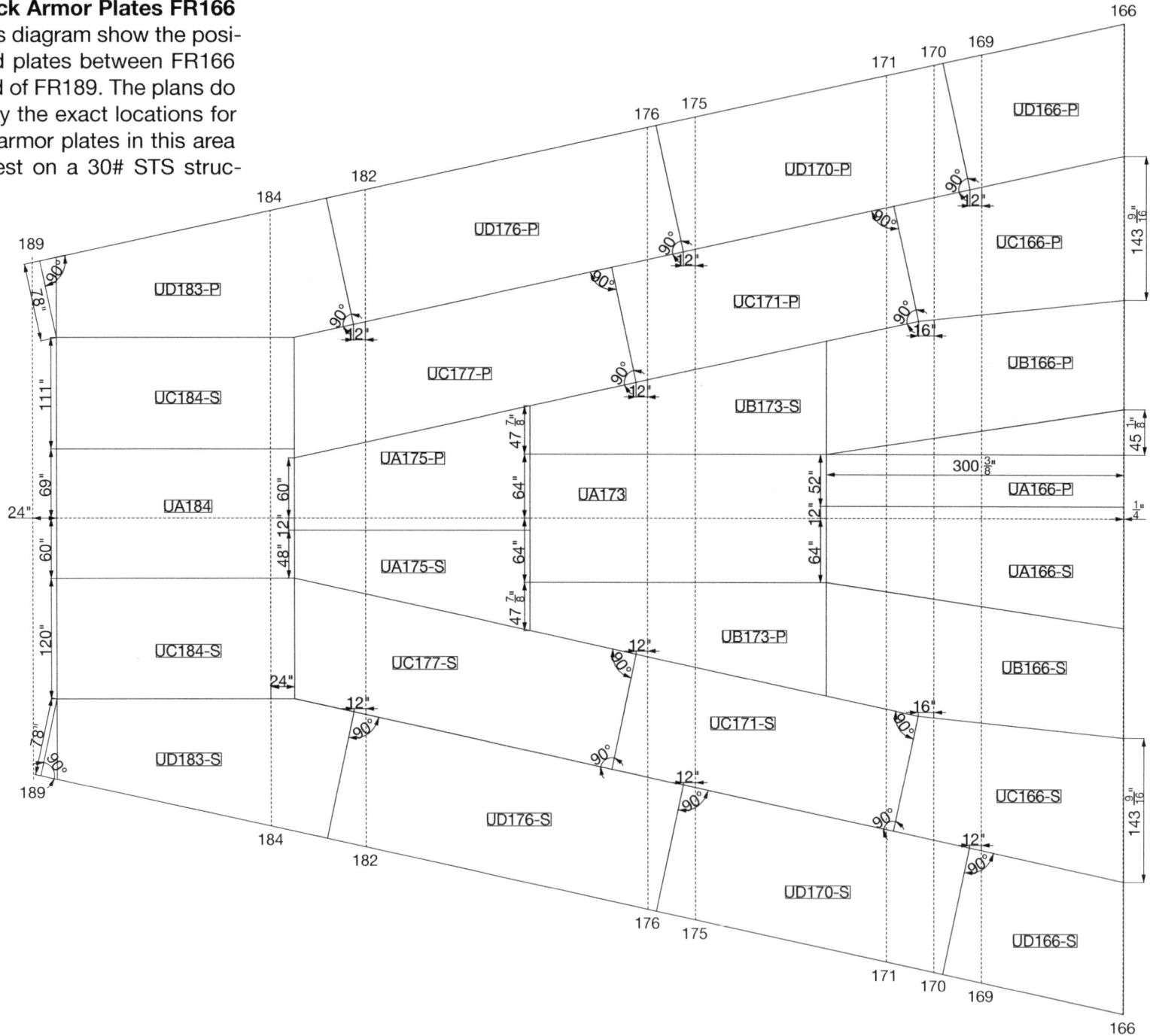

Chapter 6: Barbettes

The three turrets consist of a rotating turret that rests upon a fixed turret within the hull. The rotating turrets include the armored gunhouses visible above the deck and the pan level below that is obscured by the hull structure.

The fixed turret is cylindrical from second platform to third deck where it tapers to a truncated cone. A roller track runs along the top of the fixed turret and at the bottom of the pan level with bearing in between.

6.20. Upper Barbettes

The three cylindrical barbettes protect the turrets from the bottoms of the rotating gunhouses to the top of the citadel at second deck. The upper part of the barbettes are visible where they extend above main deck. While it appears the gunhouses rest on the barbettes, there is actually no structural connection between them, only a gas seal to keep out gases and water.

The interior faces of the barbettes have a radius of $223\frac{1}{2}$". To save weight the thickness of the barbettes varies from 11.6" at the ship's centerline and 17.3" at the sides. The three turrets sit at different heights above the baseline and second deck, so all three barbettes have different heights with the one for turret 2 being the tallest and the one for turret 1 being the shortest. Barbette 1 has 7 plates, barbette 2 has 12 plates, and barbette 3 has 11 plates. The arc coverage for the plates of each barbette have the same angle.

All four edges of each barbette had to be machined by the mill. At the top the mill drilled and tapped two pairs of bolts for attaching lifting pads. At the sides, the mill machined a welding groove and hourglass keyway into each edge. The mill also machined a slot for receiving a main deck support into one side of each joint. At the bottom the mill machined a rabbet along the inside bottom edge and drilled and tapped bolt holes around the rabbet. In effect, the rabbet along the inside created a tongue along the outside edge.

6.21. Lower Barbettes

A component of each barbette was a lower barbette. The lower barbettes are conical and extend from third deck to second deck. The weight of the barbette was transferred to the barbette support at second deck. Their combined weight was transferred to circular beam directly underneath below third deck. This beam is connected to the fixed turret. The lower barbettes are 120# STS. While officially they were not considered to be armor, lower barbettes provided additional protection for the turret below second deck.

The lower barbettes are composed of eight equal segments covering 45 degrees. The segments for lower barbette 2 are identical. the segments for lower barbette 3 are also identical. Second deck shear causes difference among the segments for lower barbette 1. The lower barbettes were milled by Carnegie-Illinois Steel. The mill machined a tapered keyway into the edge of each plate for alignment.

6.22. Assembly

The lower barbettes had to be assembled first. The yard inserted keys into the keyways for alignment then welded eight segments together on the inside and outside faces using scalloped butt straps. The yard welded the lower barbette to third deck along scalloped angles. The outer edges of the lower barbettes were welded to the underside of second deck using scalloped angles. The inner edge was joined to the deck with a more complex angle structure that was scalloped on the vertical flange but had large washers welded to the horizontal flange that were positioned under the bolt holes in the upper barbette. On USS *New Jersey* the horizontal flange was welded between the washers. It is likely that the lower barbettes had to be shimmed because their ordered height is $\frac{1}{8}$" short of ideal. Such shimming was probably done to the upper edges as second deck was fitted.

For each upper barbette, the yard had to machine a groove into the second deck armor around the turret opening to accept the tongue machined into the bottom of the barbette plates. Two approaches were used to machining this groove that constantly changed in width. The Brooklyn Navy Yard made the groove as narrow as possible for USS *Iowa* and USS *Missouri* and shimmed any gaps after the armor was installed. The yards for the other ships made the groove a constant width then back-filled the gap after the barbette was in place. Holes were drilled through the second deck armor at the bolt hole locations in the upper barbettes.

Brackets for deck supports were tap riveted into plates with slots before the plates were moved into position. After the plates were craned into position, hourglass keys were hammered into their joints and the plates were welded together along their inside face. Bolts with washers were screwed through the large washers that are part of the lower belt assembly and through second deck into the upper barbette plates. The bolts and washers were welded in place. Finally an angle bar with one scalloped flange was welded to the top of second deck against the barbette.

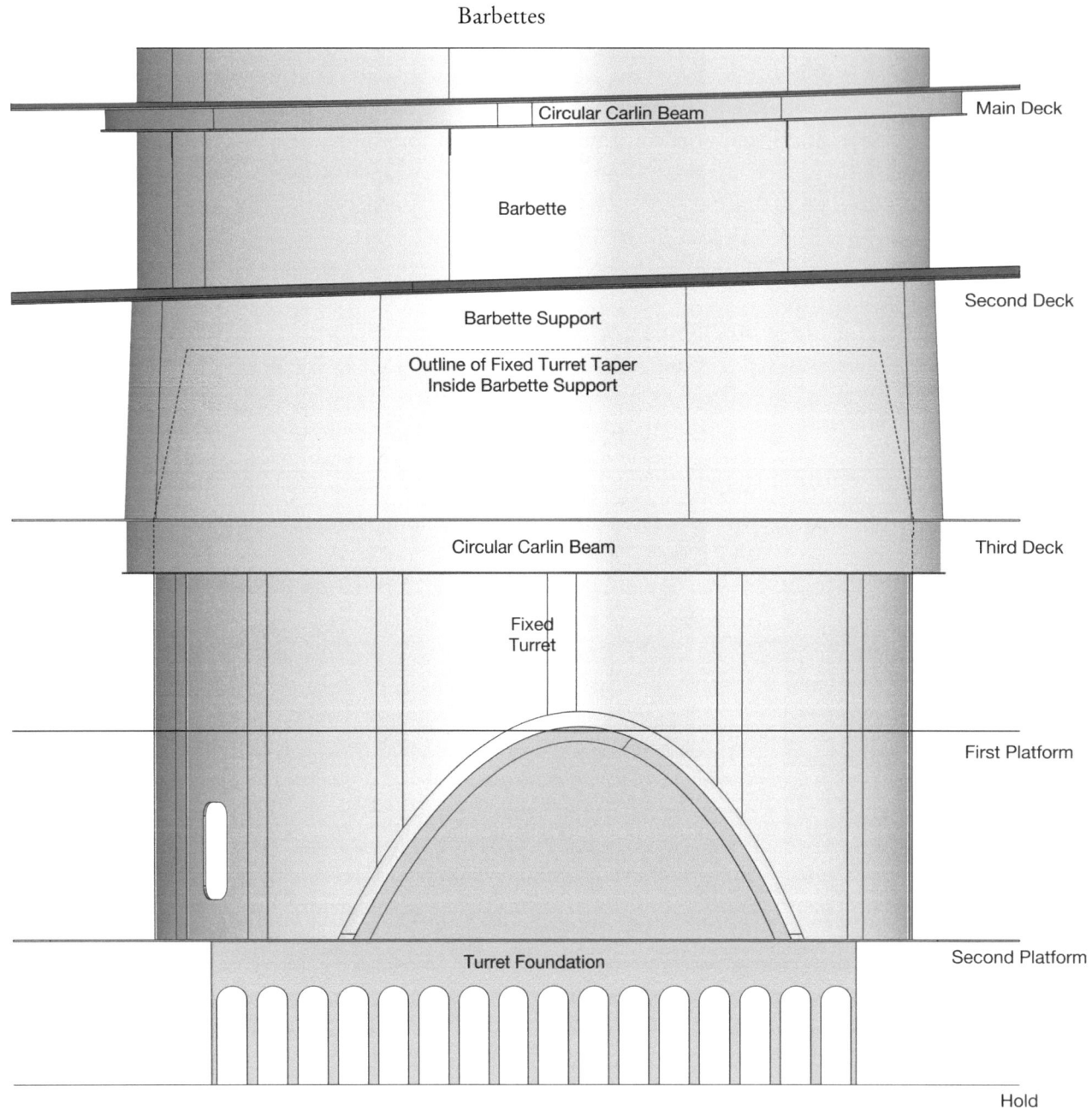

Figure 6.1: This rendering shows the structural components that support the barbettes. The fixed turret supports the entire structure. The barbette support rests on a circular Carlin beam attached to the fixed turret. The barbette sits on the barbette support. Turret 1 is shown. The other fixed turrets are not truncated at the side and lack the comb-shaped supports. The fixed turrets on USS Iowa and USS New Jersey are riveted as shown here. Those on USS Missouri and USS Wisconsin are welded.

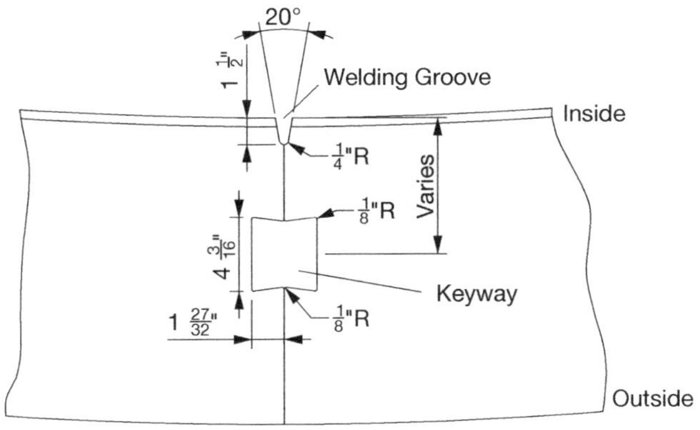

Figure 6.2: Plan of Joints Between Barbette Plates (1:12)

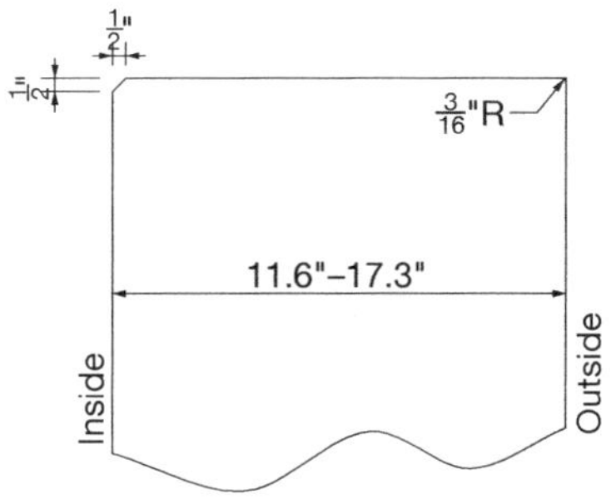

Figure 6.3: Section Through Upper Edge of Barbettes

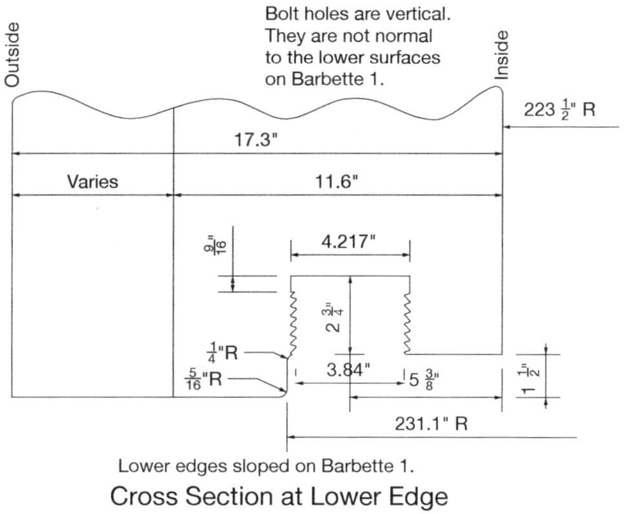

Bolt holes are vertical. They are not normal to the lower surfaces on Barbette 1.

Lower edges sloped on Barbette 1.

Cross Section at Lower Edge

Figure 6.4: Section Through Lower Edge of Barbettes (1:6)

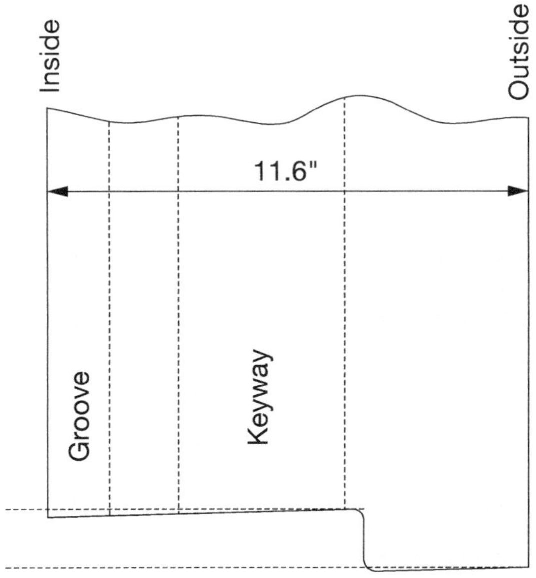

Figure 6.5: Section Through Lower Edge of Barbette 1 (1:6). Second deck slope causes edge of the rabbet of Barbette 1 to be sloped downward along the aft half and upward along the forward half. The face slopes fore–aft $\frac{5}{16}$" per foot.

Photograph 6.1: Lowering Turret No. 2 into the Barbette. This is upper half of the rotating turret on USS New Jersey. The lower half of the turret had already been craned in position. The upper turret lacks armor and armament. The small holes at the bottom are for attaching clips that keep the turret from pulling out. The larger openings are for inspecting and replacing roller bearings. The upper roller is visible above. This photograph illustrates how deep the turret support is within the barbette. (Nat'l Archives Phila.)

Photograph 6.2: Inside Face of Barbette 1 on USS New Jersey. The fixed turret is below the barbette. The turret stops at the very bottom are attached to the fixed turret. Above that is the rack gear for rotating the turret. Some of the roller bearings are visible under the canvas. (Nat'l Archives Phila.)

Figure 6.6: Barbette 1 Plan (1:96)

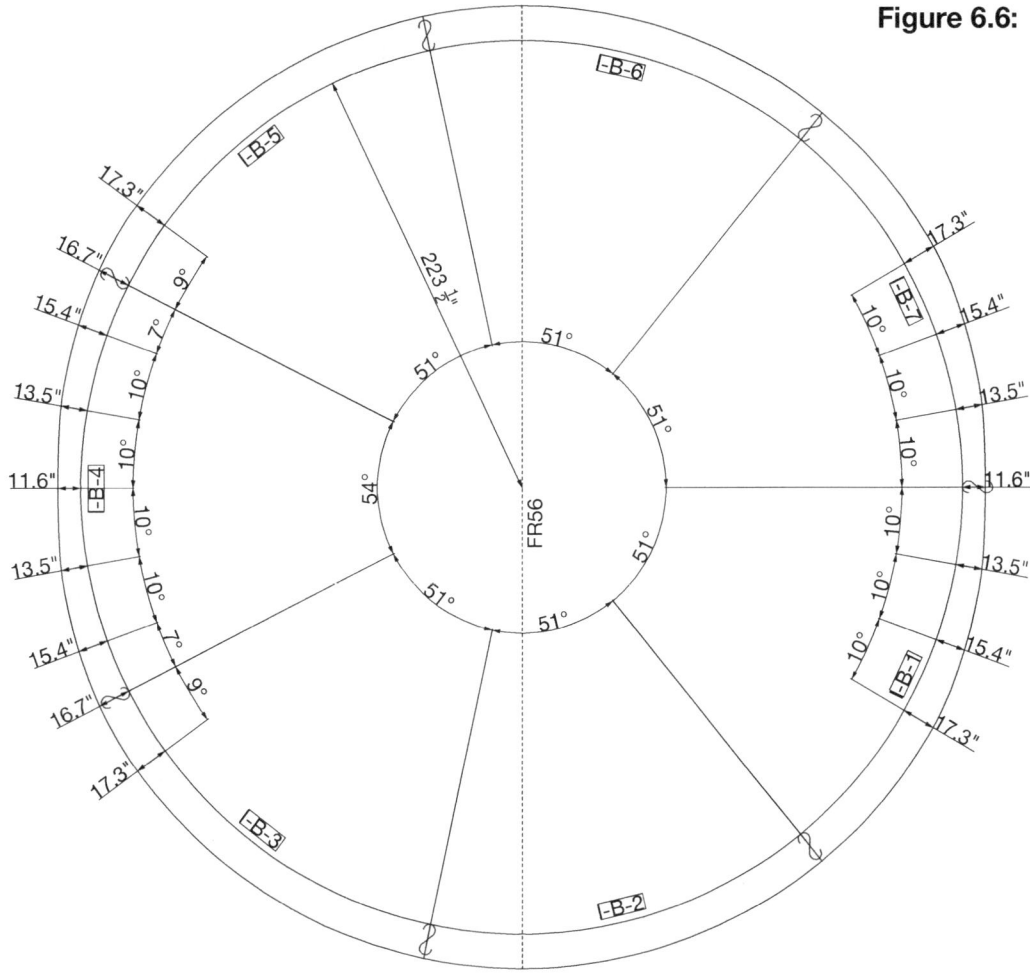

Figure 6.7: Barbette 1 Elevation (1:96). The upper smooth area is a faying surface with main deck. The lower smooth area is a faying surface with the angle bracing the barbette at second deck.

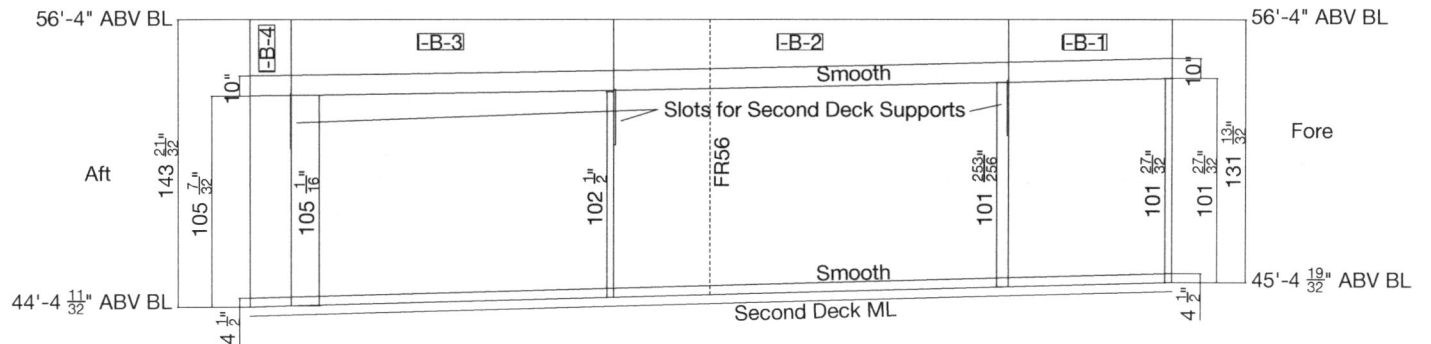

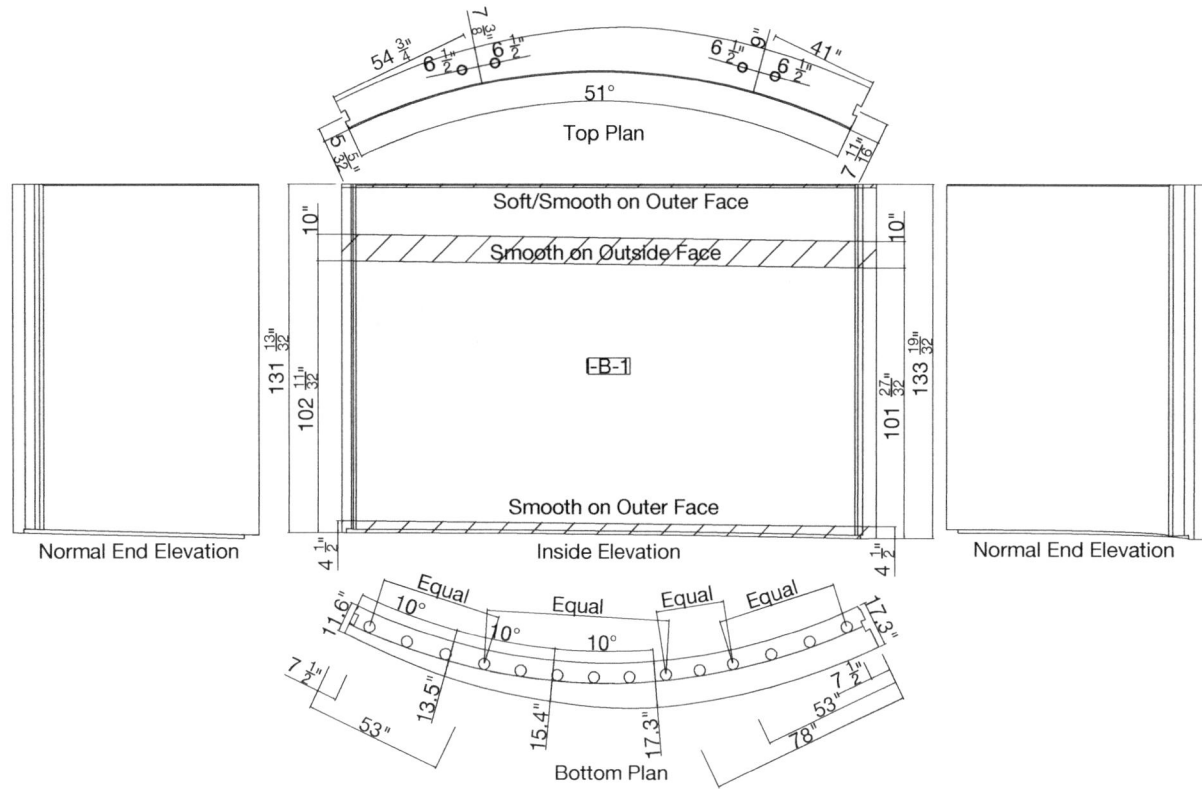

Figure 6.8: Plate I-B-1 (1:72)

Top Plan

Soft/Smooth on Outer Face

Smooth on Outer Face

Soft on Outside
Face Next to Slots
to 2 $\frac{7}{8}$" from
Edge

I-B-2

Smooth on Outer Face

Inside Elevation

Normal End Elevation

Slot $\frac{4}{8}$" Deep

Normal End Elevation

Slot $\frac{7}{8}$" Deep

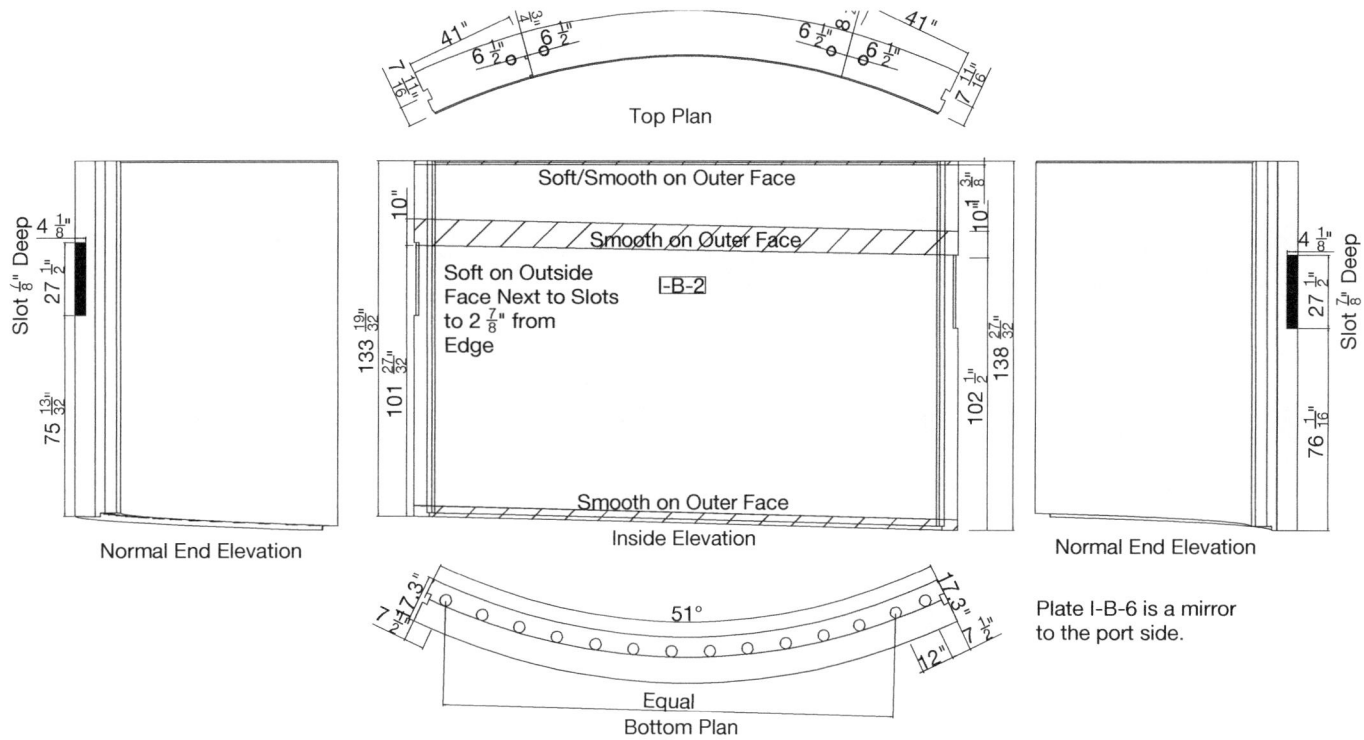

Plate I-B-6 is a mirror
to the port side.

51°

Equal
Bottom Plan

Figure 6.9: Plate I-B-2 (1:72). Each joint has a
slot for a second deck support bracket machined on
one side of the joint.

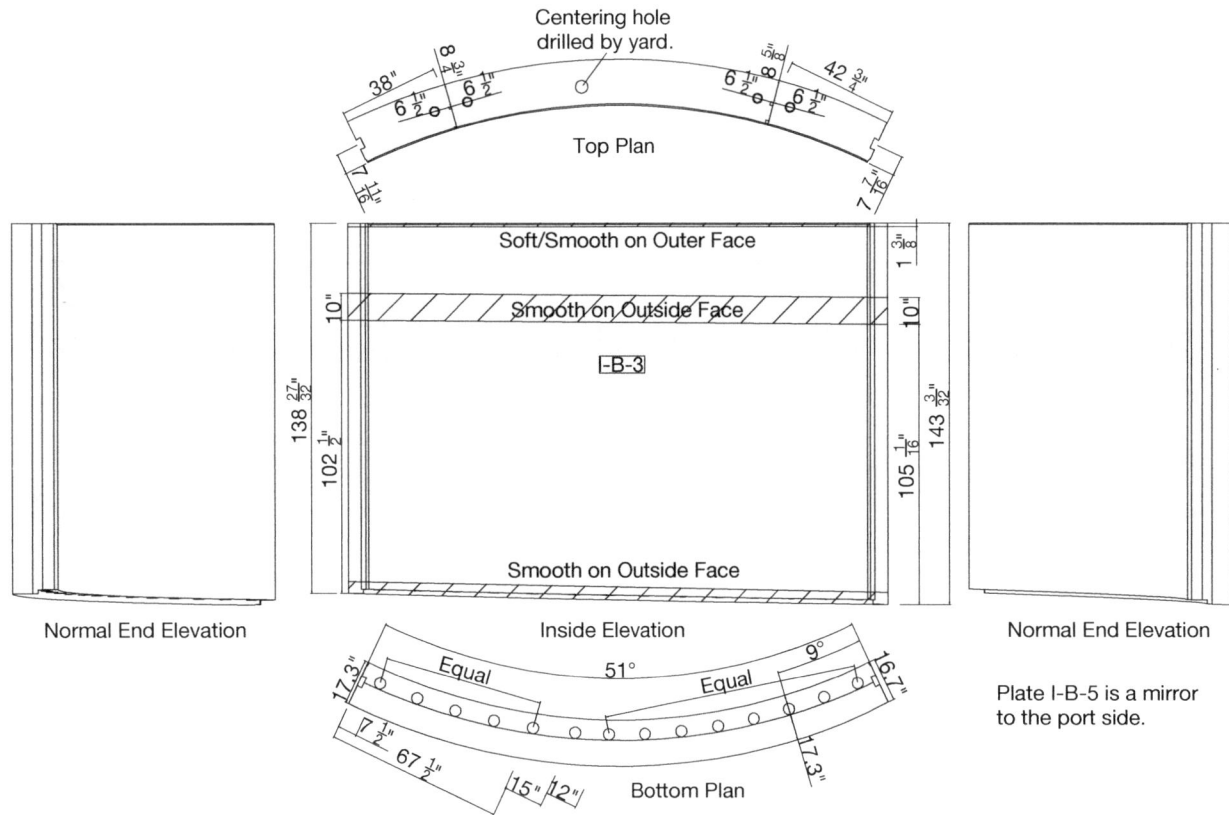

Centering hole
drilled by yard.

Top Plan

Soft/Smooth on Outer Face

Smooth on Outside Face

I-B-3

Smooth on Outside Face

Normal End Elevation

Inside Elevation

Normal End Elevation

Plate I-B-5 is a mirror
to the port side.

Equal   51°   Equal

Bottom Plan

Figure 6.10: Plate I-B-3 (1:72)

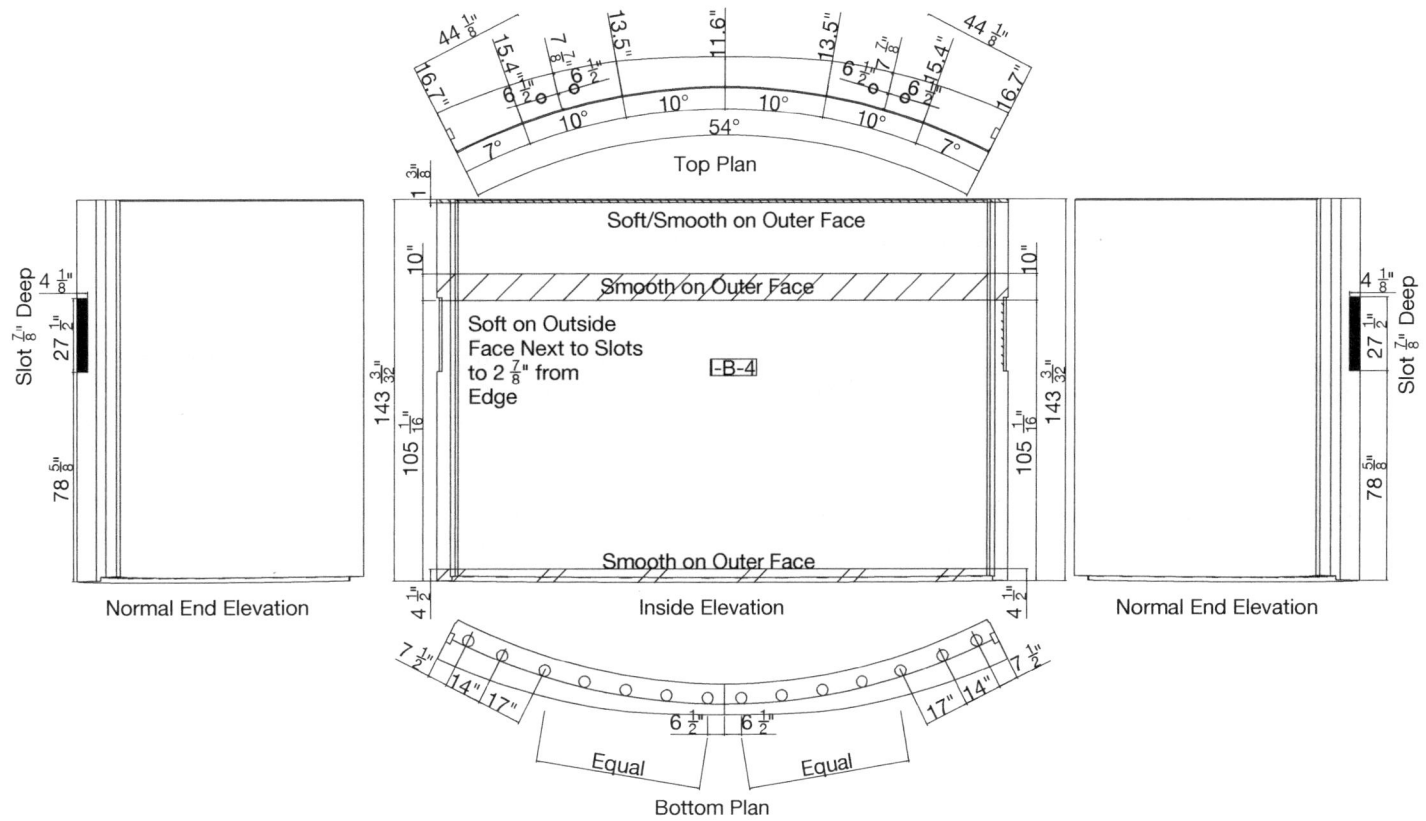

Figure 6.11: Plate I-B-4 (1:72)

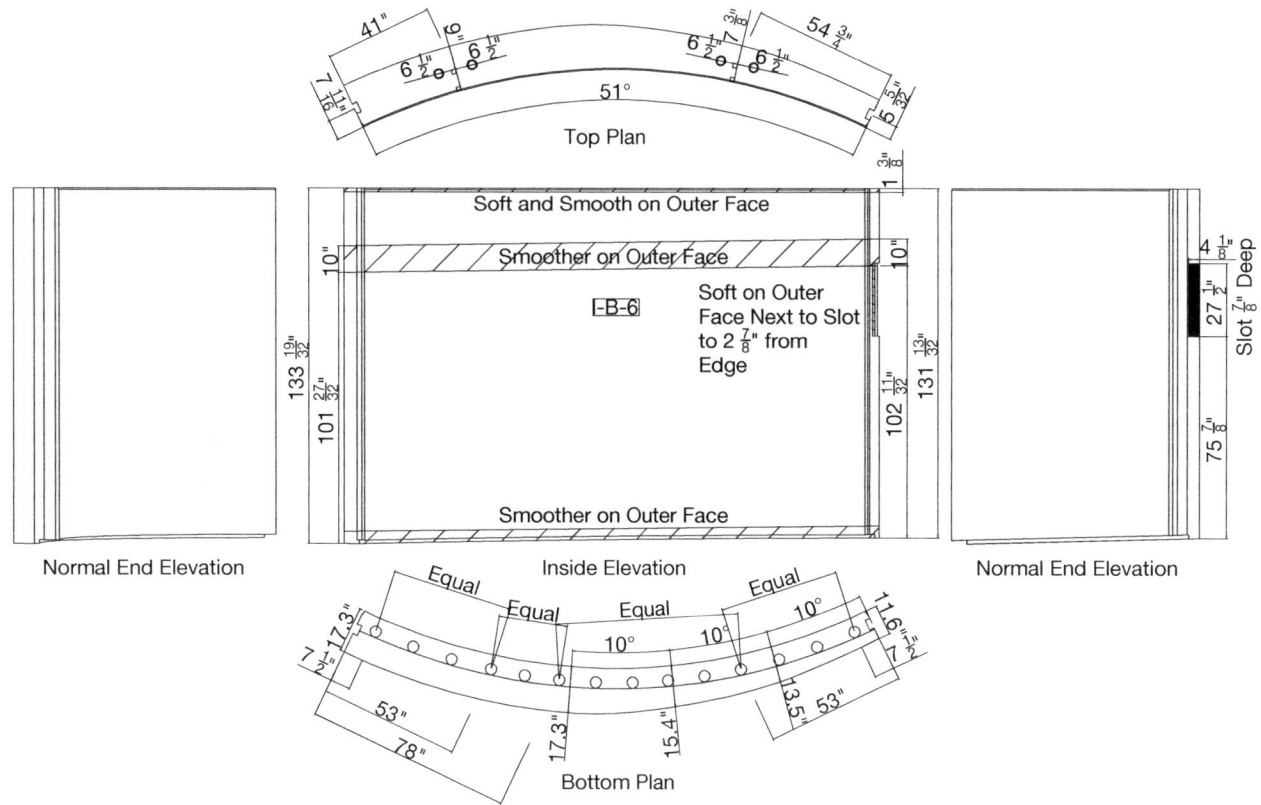

Figure 6.12: Plate I-B-7 (1:72)

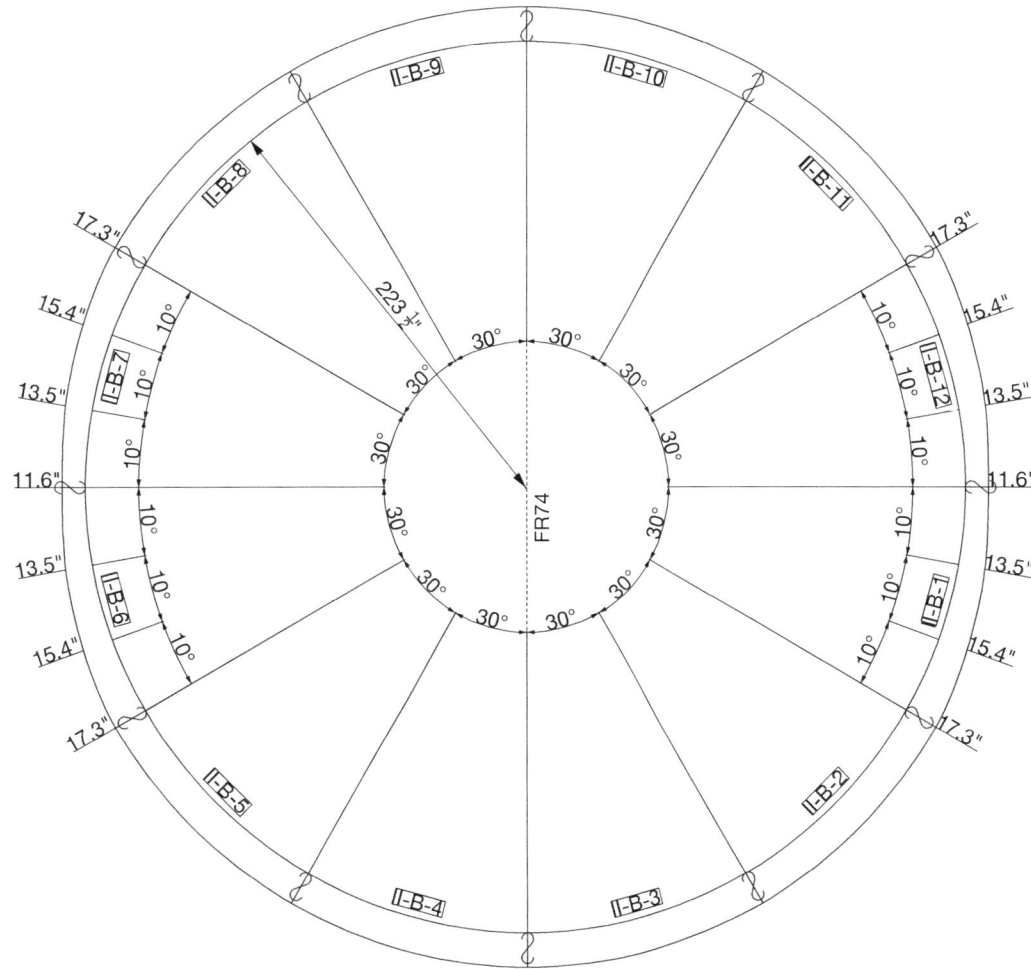

Figure 6.13: Barbette 2 Plan (1:96)

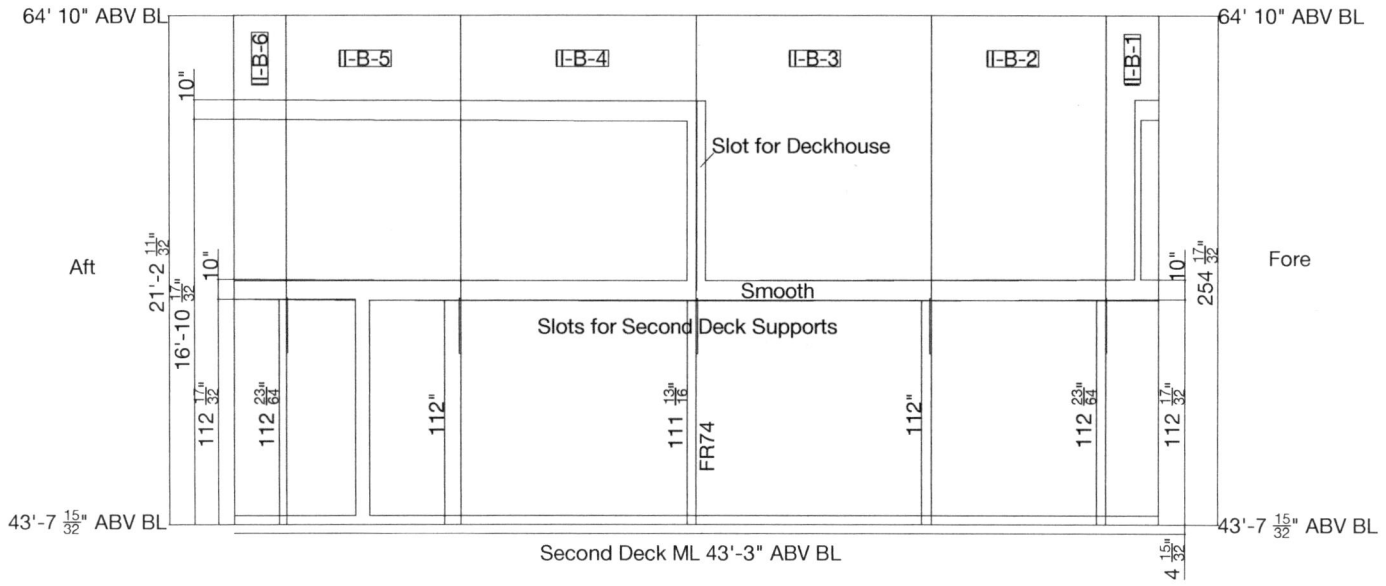

Figure 6.14: Barbette 2 Elevation (1:96).
Barbette 2 has smooth faying surfaces for the 01 level, deck houses, and bulkheads at second deck. A circular beam supports the 01 level at the barbette so there are no slots for brackets to support this level.

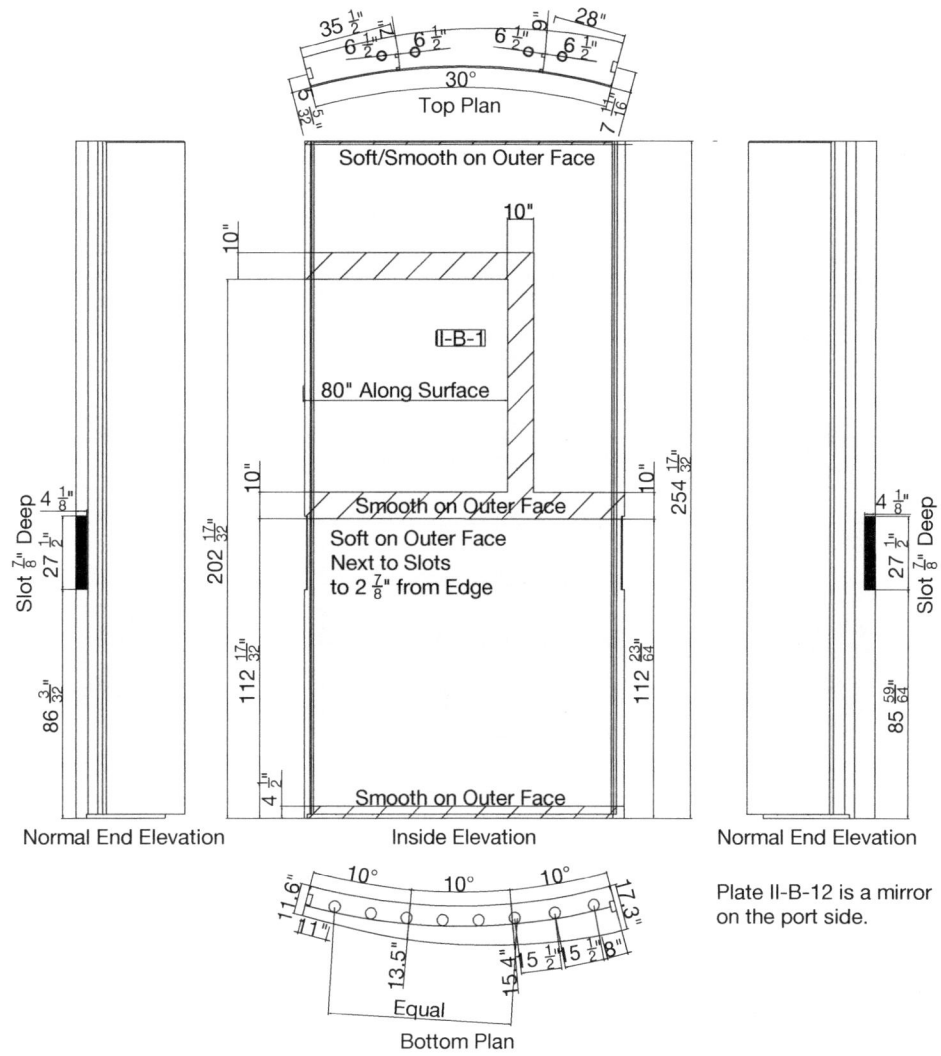

Figure 6.15: Plate II-B-1 (1:72)

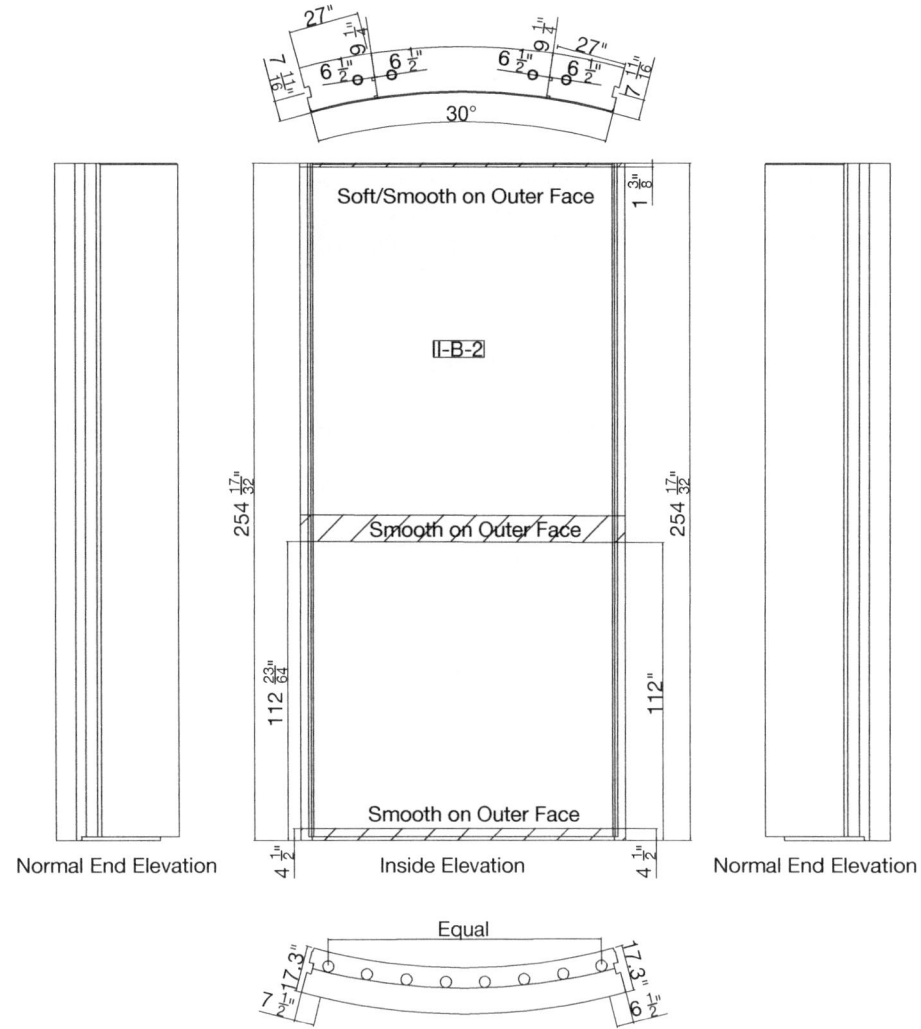

Figure 6.16: Plate II-B-2 (1:72)

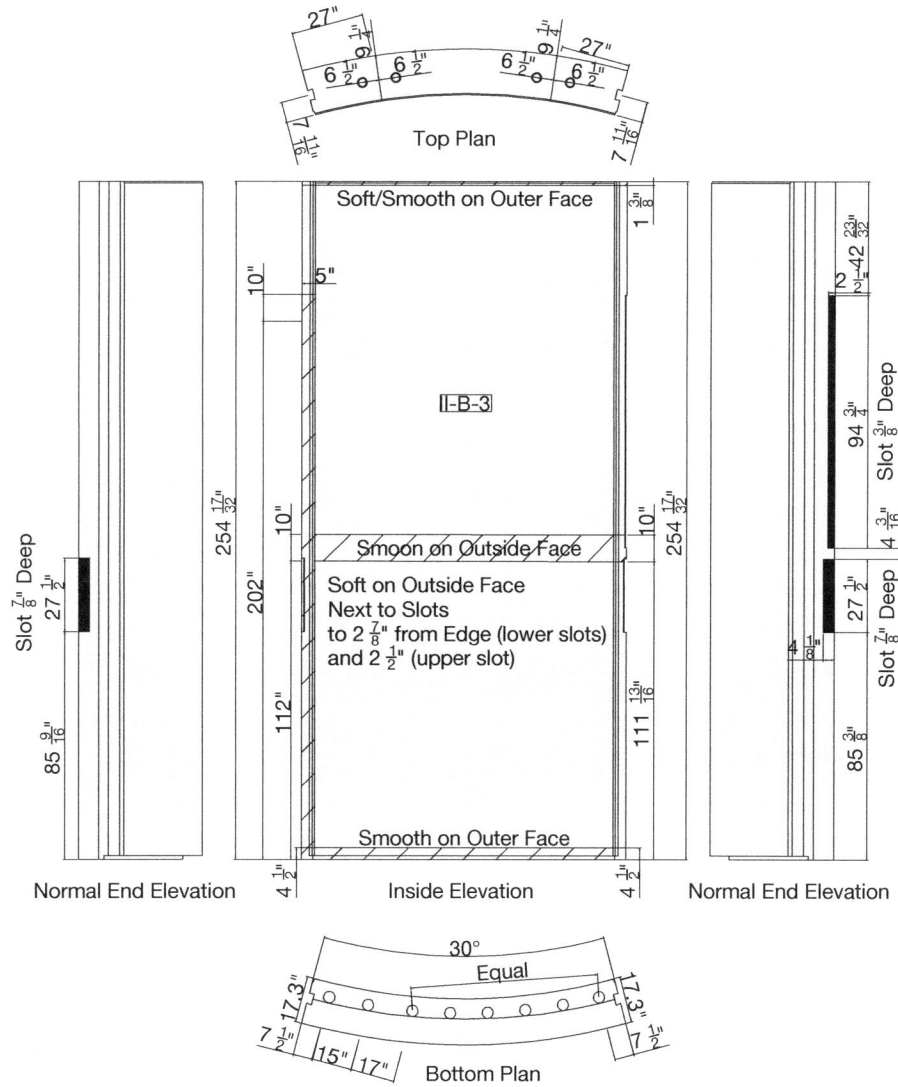

Figure 6.17: Plate II-B-3 (1:72). The Long slots above main deck are for attaching the deckhouse bulkhead.

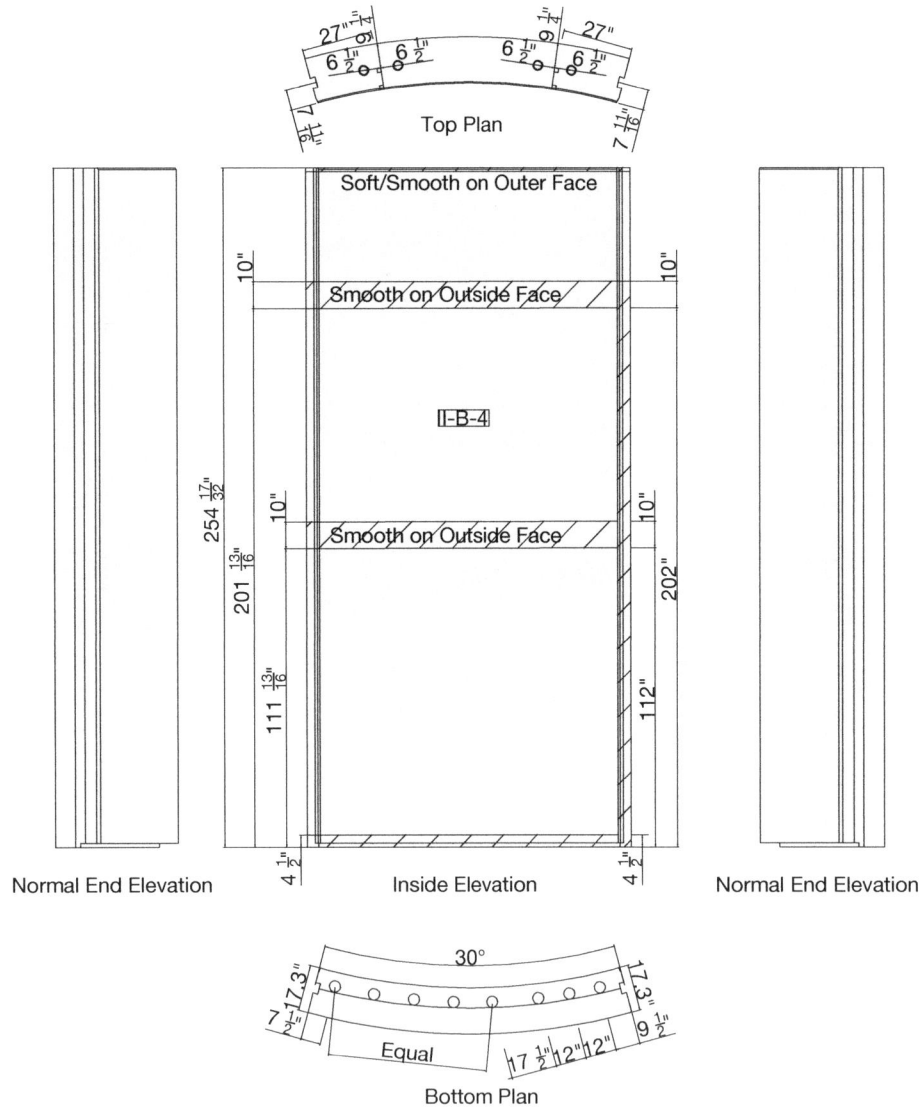

Figure 6.18: Plate II-B-4 (1:72)

Centering Pin Hole
Drilled By Yard

30"   6 1/2"   6 1/2"   6 1/2"   6 1/2"   30"

Top Plan

Soft/Smooth on Outer Face

1 3/8"

10"   Smooth on Outer Face   10"

II-B-5

Soft on Outer Face
Next to Slots
to 2 7/8" from Edge

Smooth on Outer Face

10"   10"

254 17/32"

202 11/32"

112"

10" Centered

112 11/32"

202 11/32"

254 17/32"

Slot 7/8" Deep   4 1/8"   27 1/2"

85 9/16"

Normal End Elevation

4 1/2"   Inside Elevation   4 1/2"

4 1/8"   Slot 7/8" Deep   27 1/2"

85 29/32"

Normal End Elevation

7 1/2"   17.3"   30°   17.3"   9"

Equal

Bottom Plan

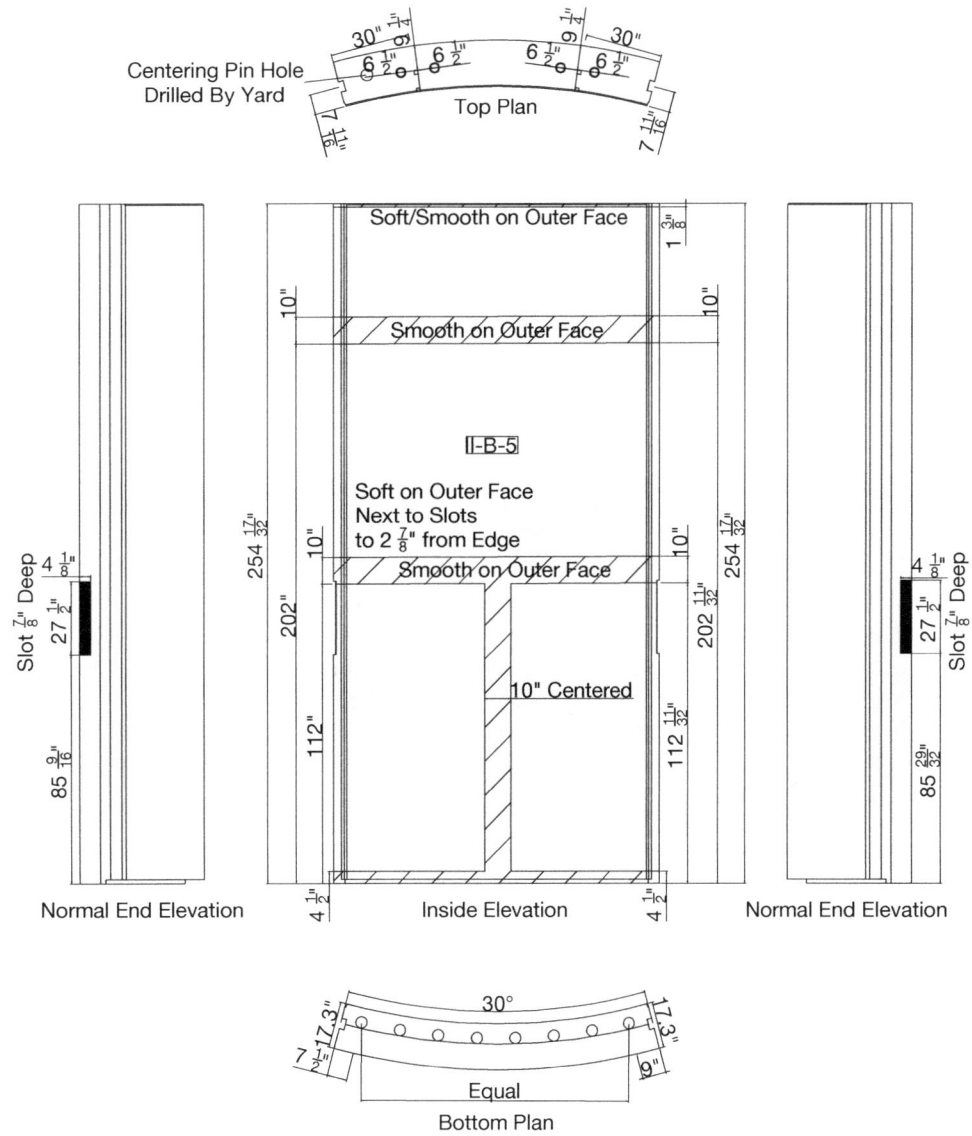

Figure 6.19: Plate II-B-5 (1:72)

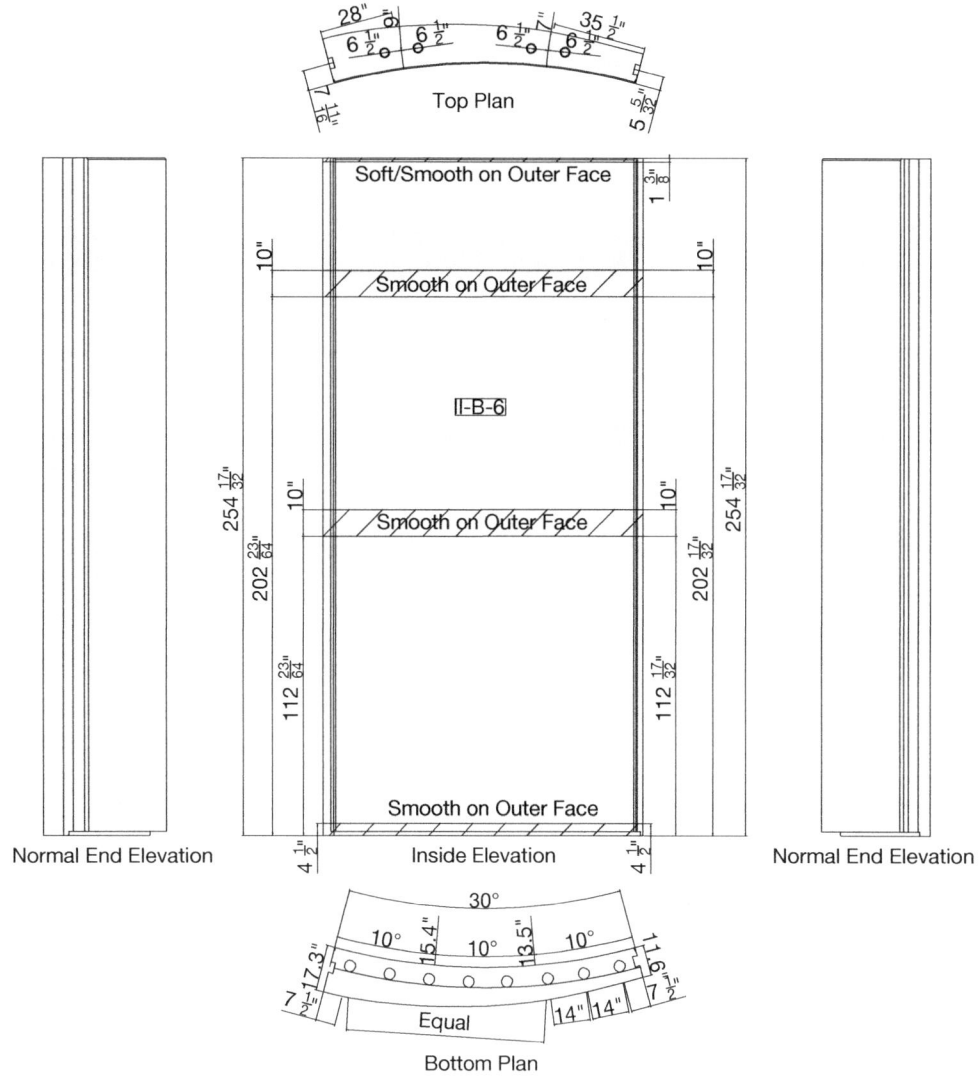

Figure 6.20: Plate II-B-6 (1:72)

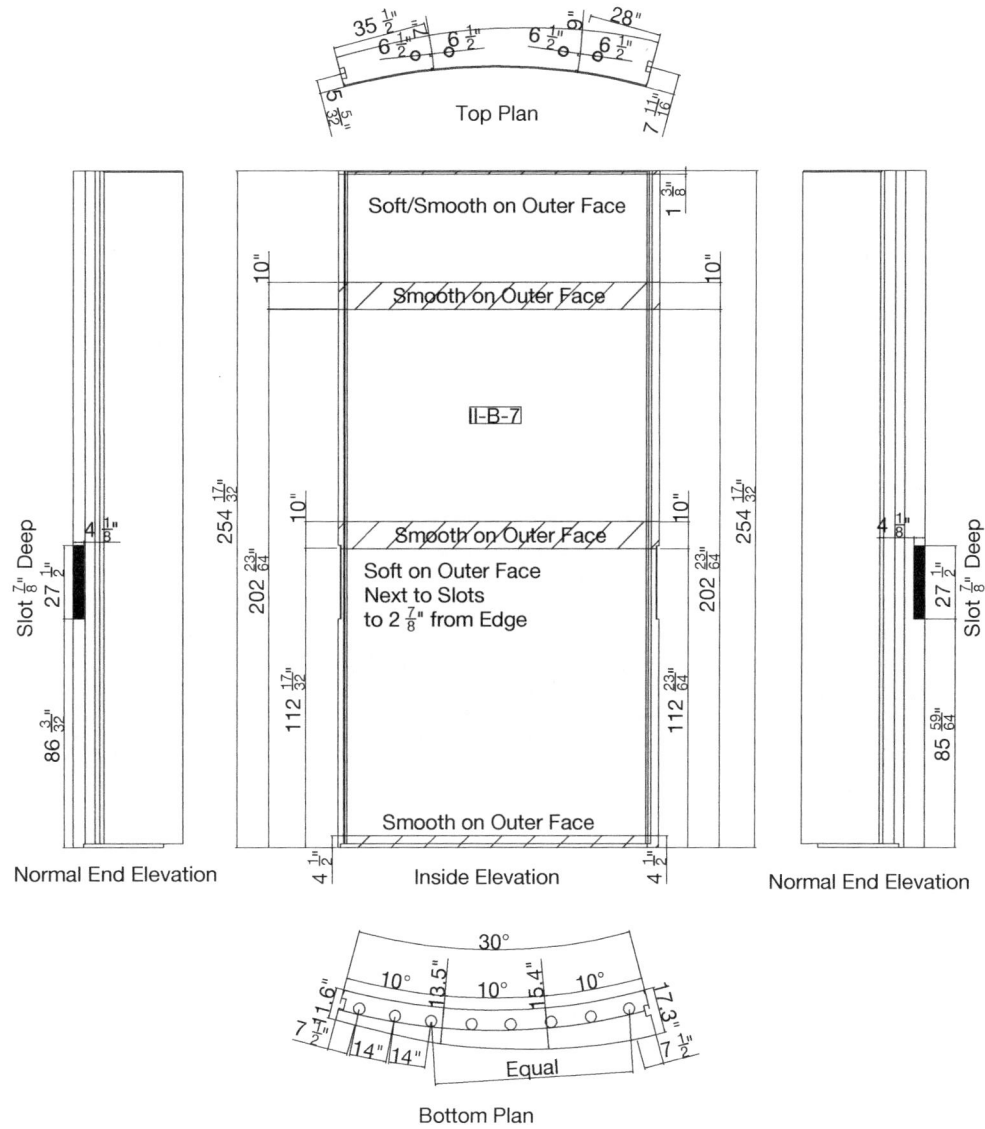

Figure 6.21: Plate II-B-7 (1:72)

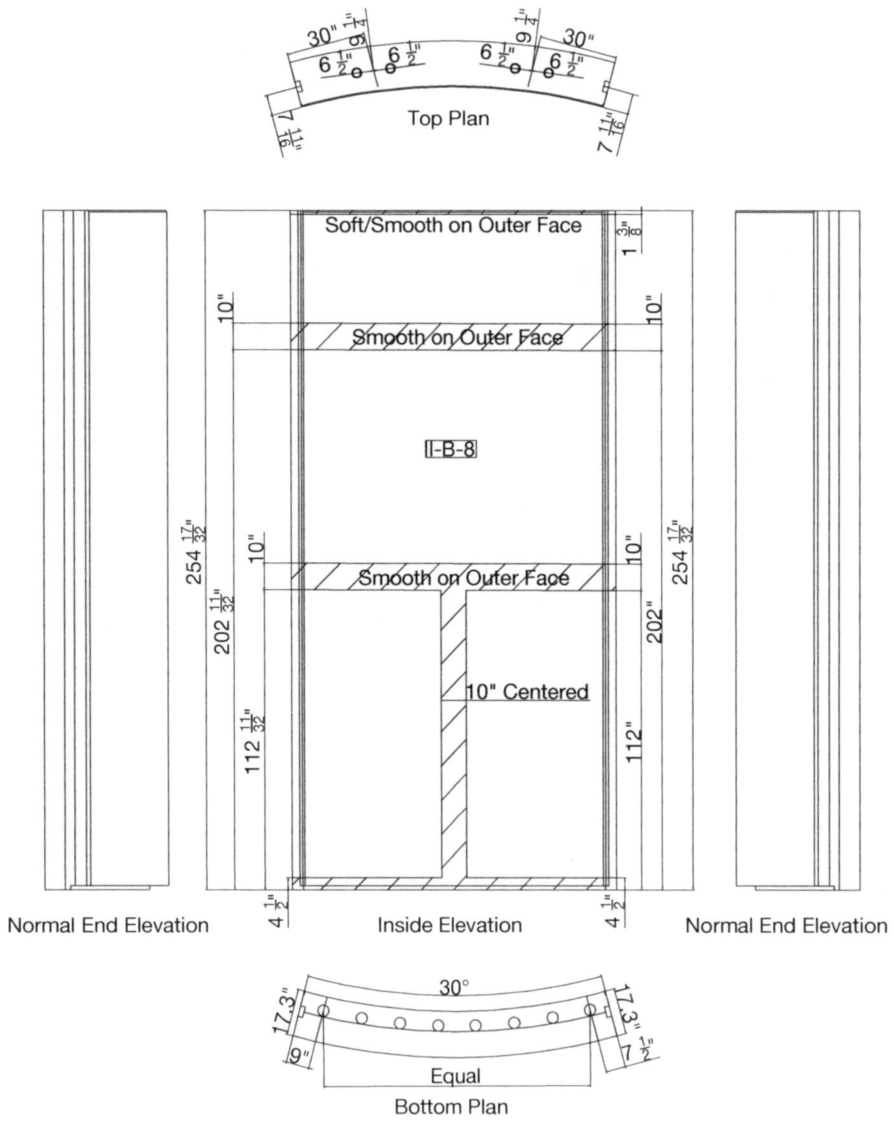

Figure 6.22: Plate II-B-8 (1:72)

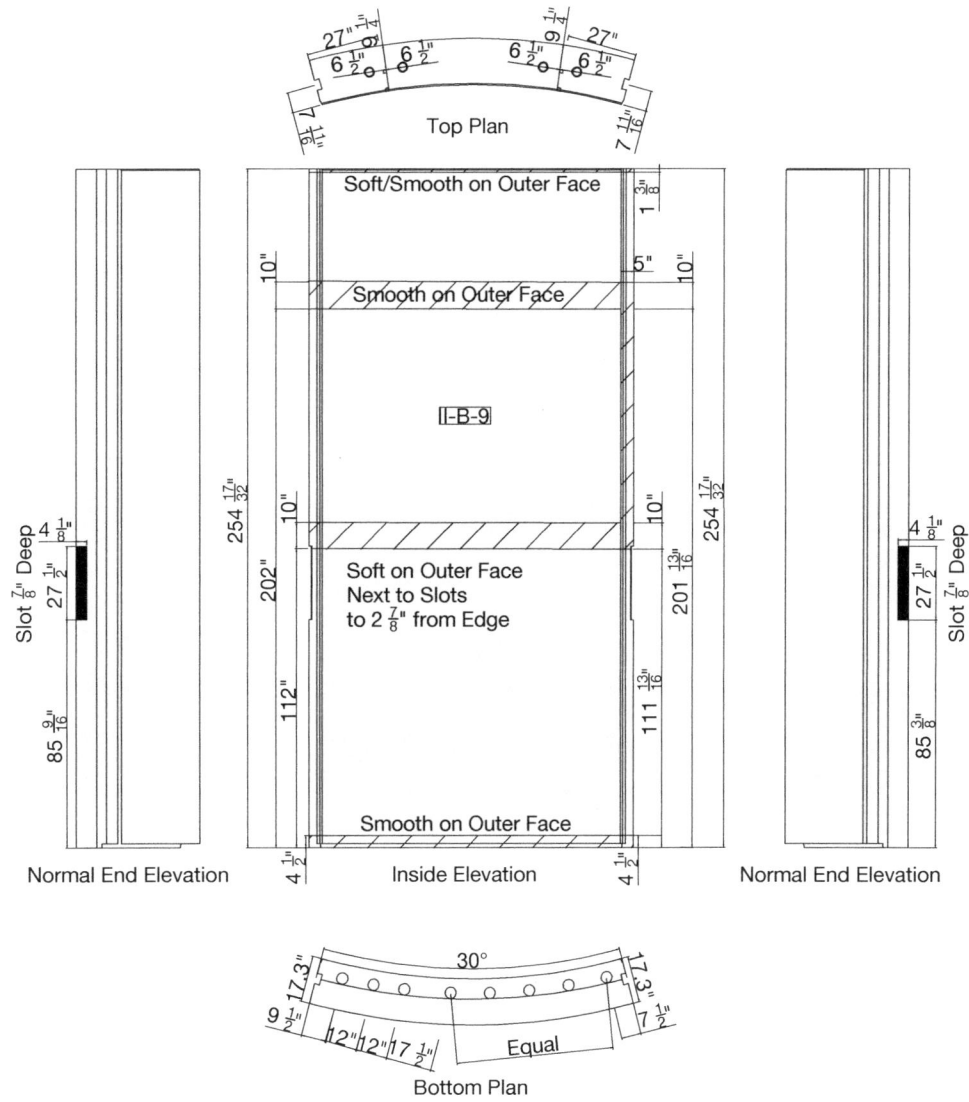

Figure 6.23: Plate II-B-9 (1:72)

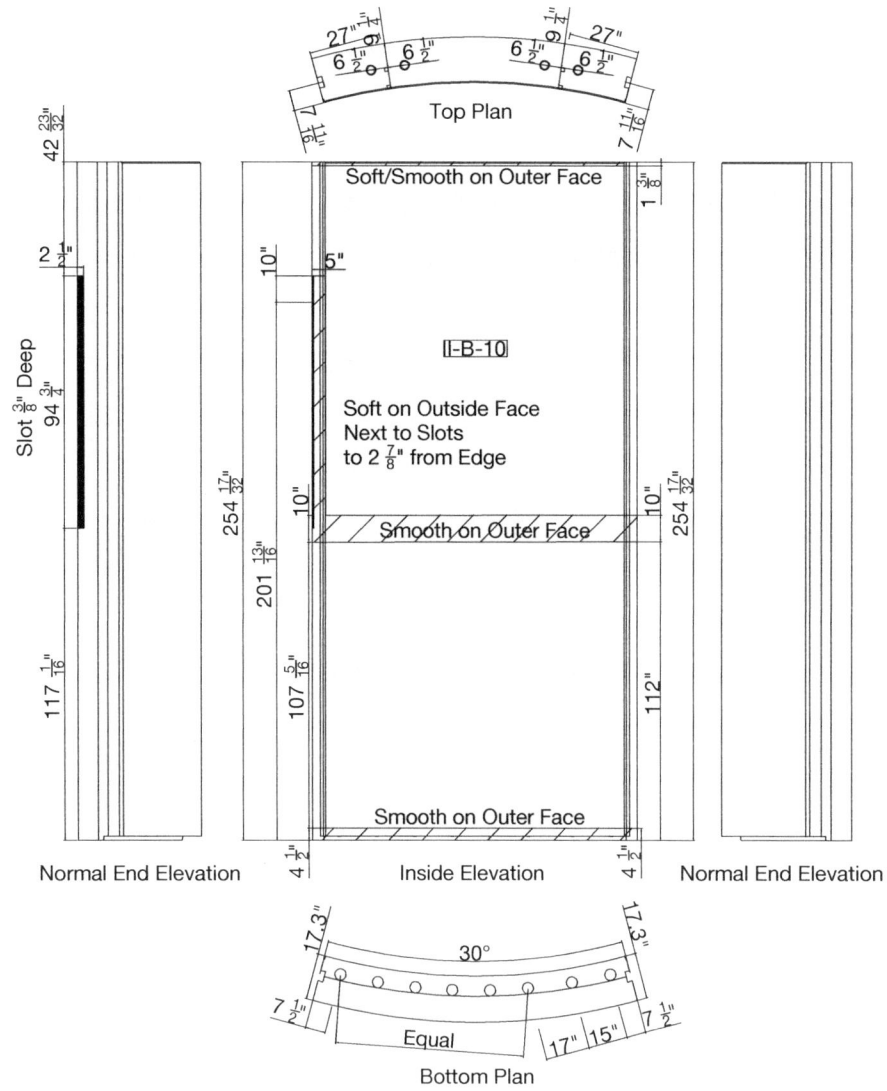

Figure 6.24: Plate II-B-10 (1:72)

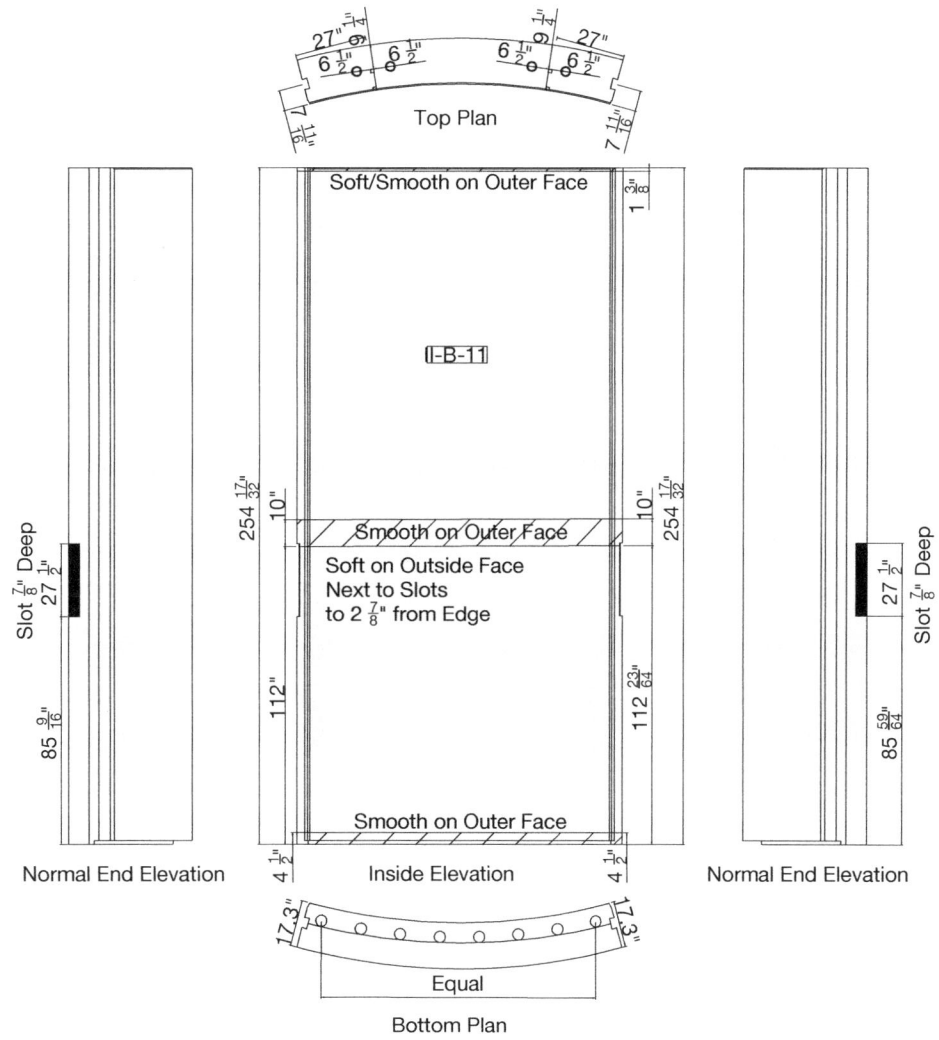

Figure 6.25: Plate II-B-11 (1:72)

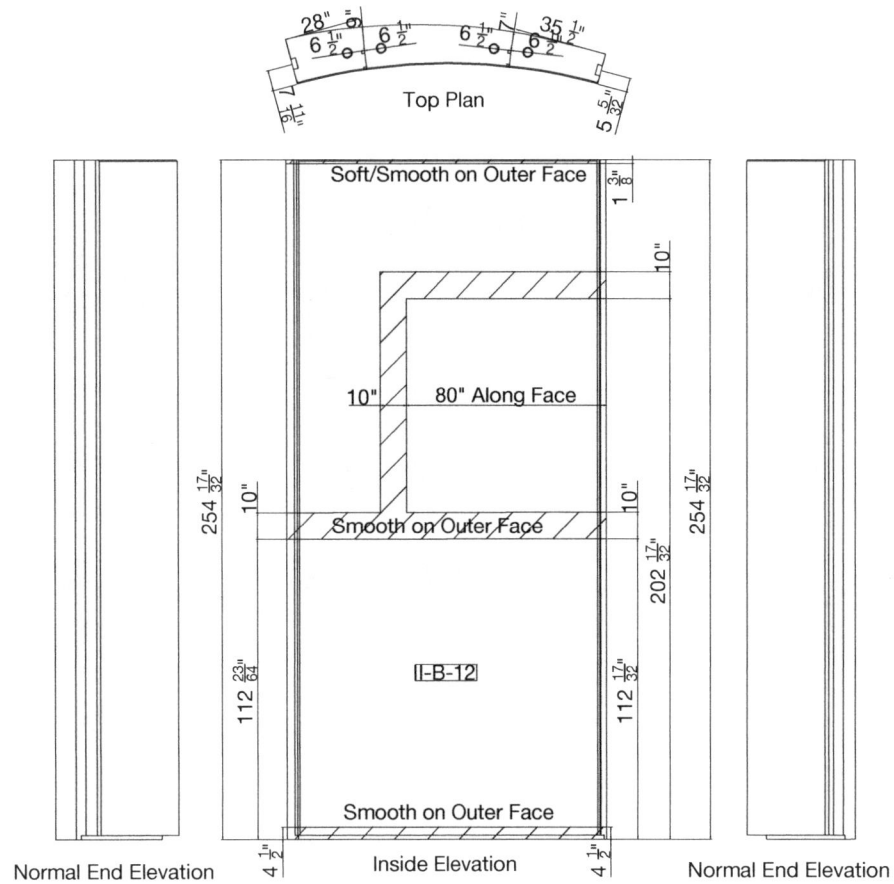

Top Plan

Soft/Smooth on Outer Face

10" 80" Along Face

Smooth on Outer Face

II-B-12

Smooth on Outer Face

Normal End Elevation   Inside Elevation   Normal End Elevation

The lower bolt openings are not mirrored from plate II-B-1.

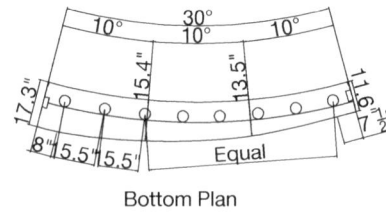

Bottom Plan

Figure 6.26: Plate II-B-12 (1:72)

Figure 6.27: Barbette 3 Plan (1:96)

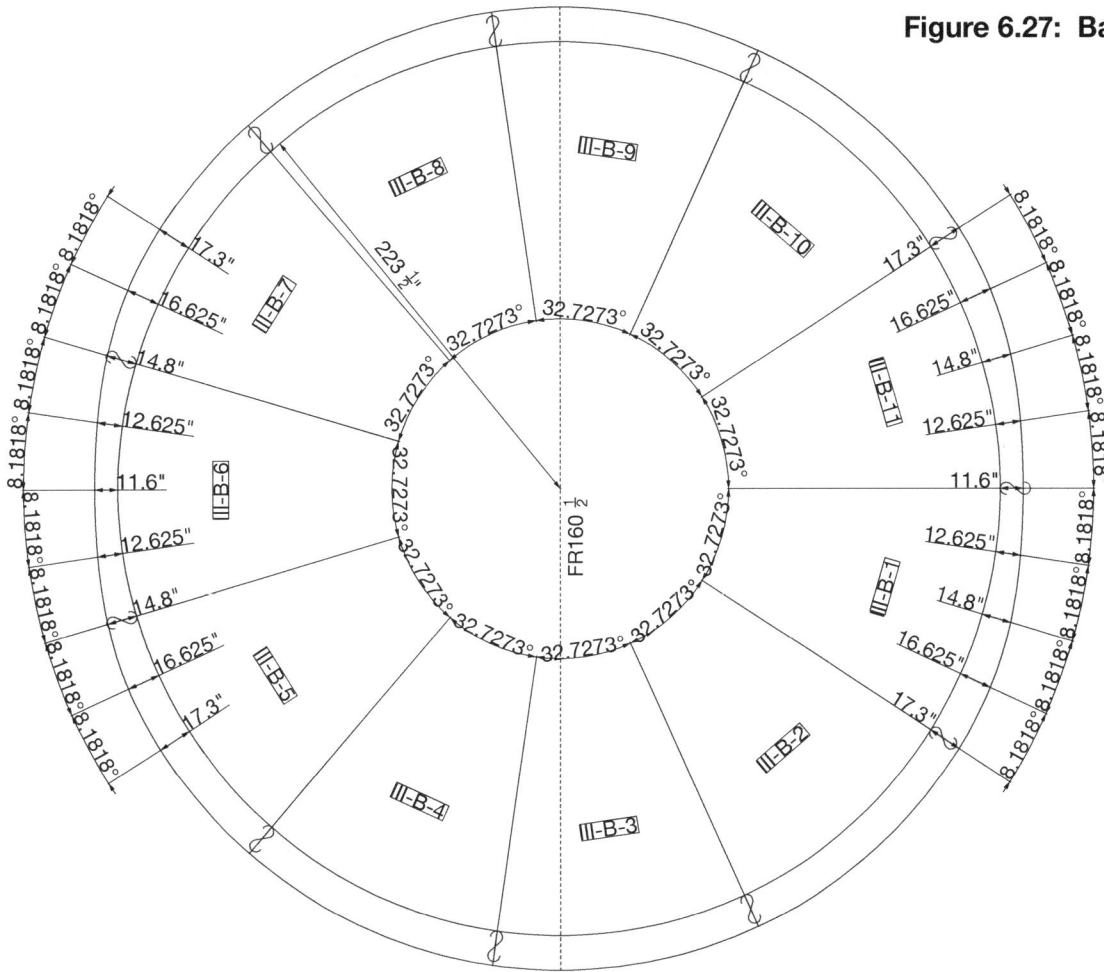

Figure 6.28: Barbette 3 Elevation (1:96). The smooth area at the top is the main deck faying surface. The smooth area at the bottom is the faying surface for the angle that is welded to second deck.

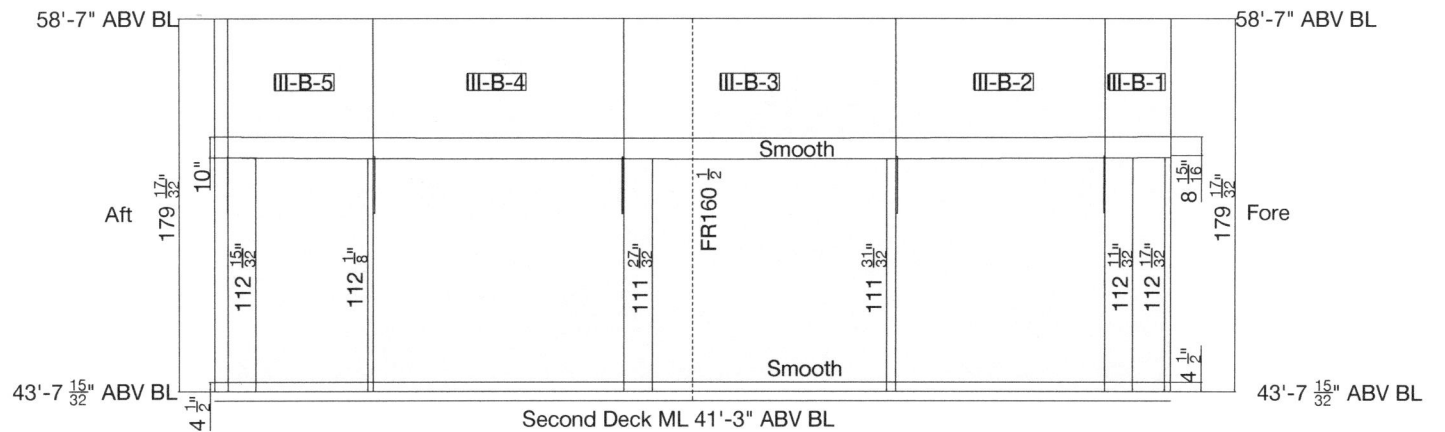

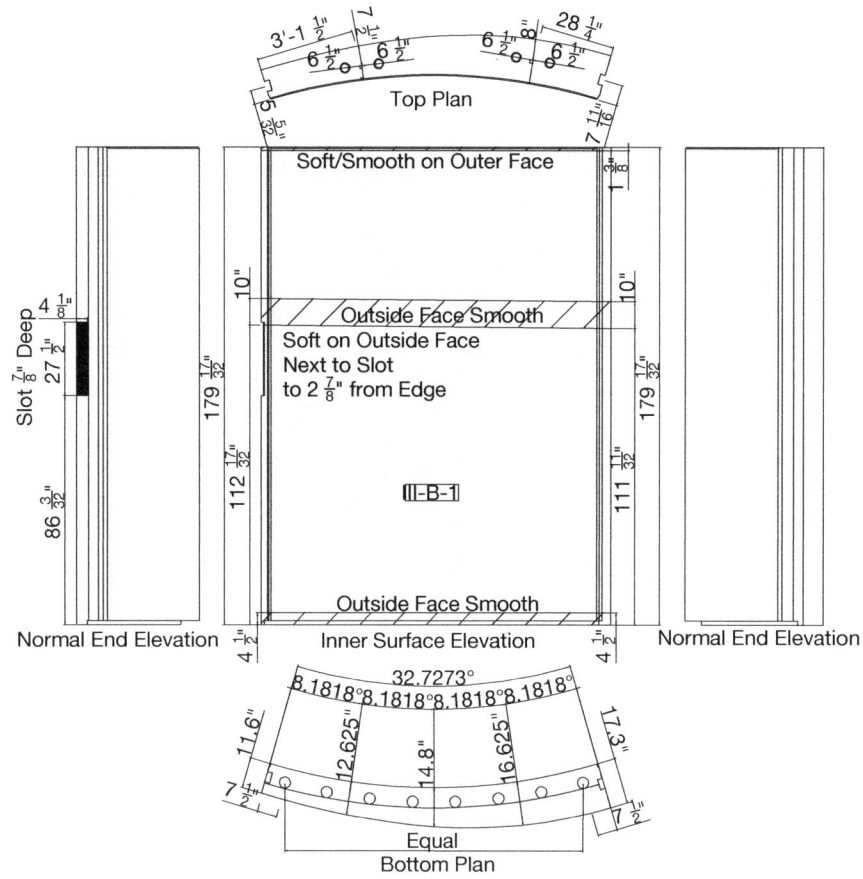

Figure 6.29: Plate III-B-1 (1:72)

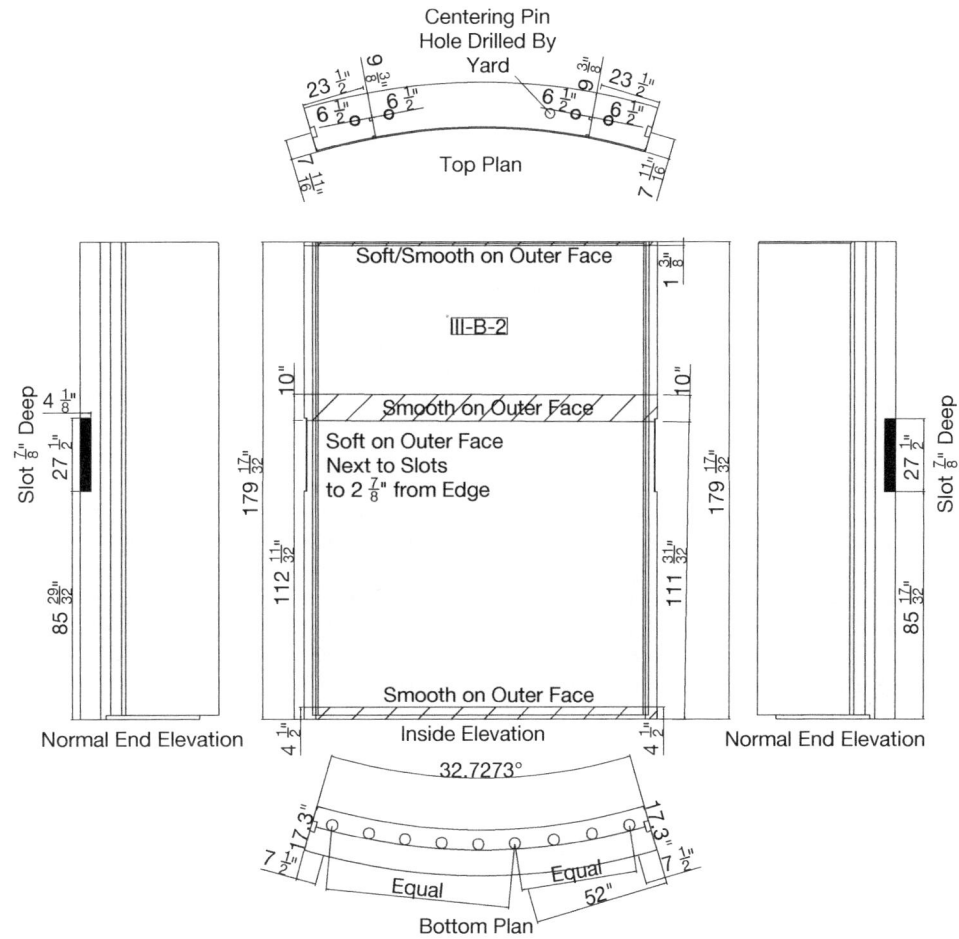

Figure 6.30: Plate III-B-2 (1:72)

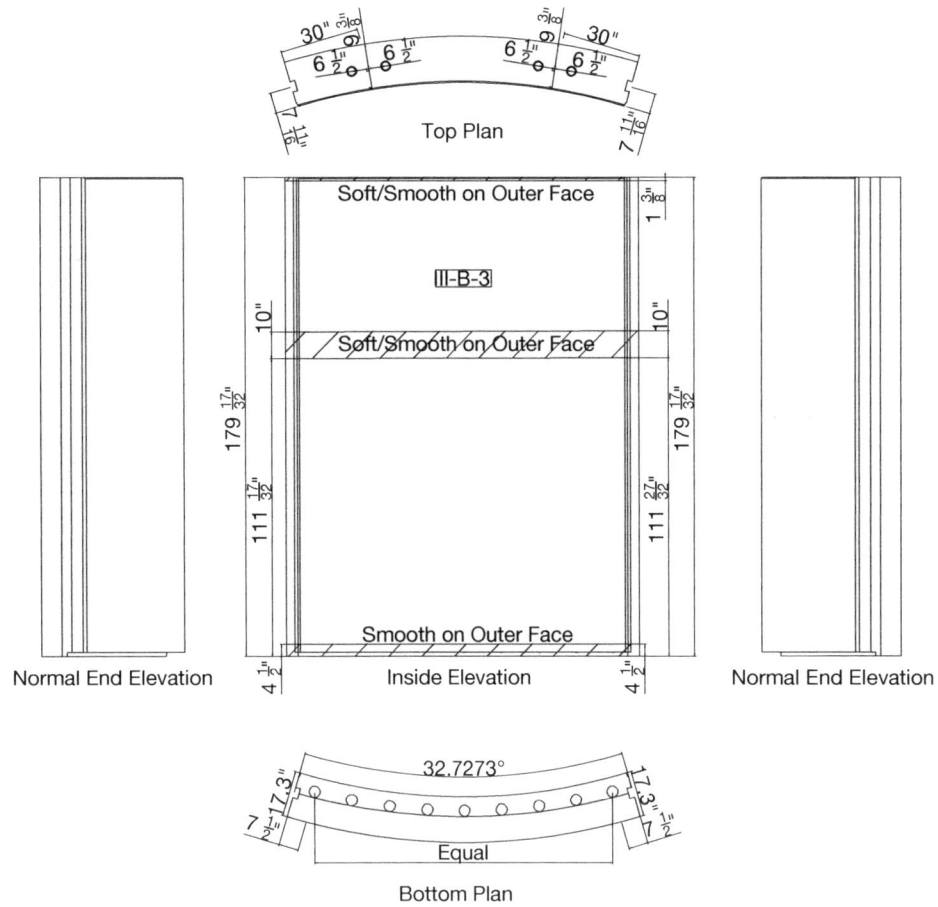

Figure 6.31: Plate III-B-3 (1:72)

Top Plan

Soft/Smooth on Outer Face

III-B-4

Smooth on Outer Face

Soft on Outer Face
Next to Slots
to 2 $\frac{7}{8}$" from Edge

Slot $\frac{7}{8}$" Deep

27 $\frac{1}{2}$"

85 $\frac{13}{32}$"

179 $\frac{17}{32}$"

107 $\frac{11}{32}$"

179 $\frac{17}{32}$"

112 $\frac{1}{8}$"

10"

10"

Smooth on Outer Face

4 $\frac{1}{2}$"

4 $\frac{1}{2}$"

Normal End Elevation

Inside Elevation

Slot $\frac{7}{8}$" Deep

27 $\frac{1}{2}$"

85 $\frac{11}{16}$"

Normal End Elevation

30"

6 $\frac{1}{2}$"

6 $\frac{1}{2}$"

6 $\frac{1}{2}$"

6 $\frac{1}{2}$"

30"

32.7273°

17.3°

17.3°

Equal

Equal

7"

9"

42"

17"

Bottom Plan

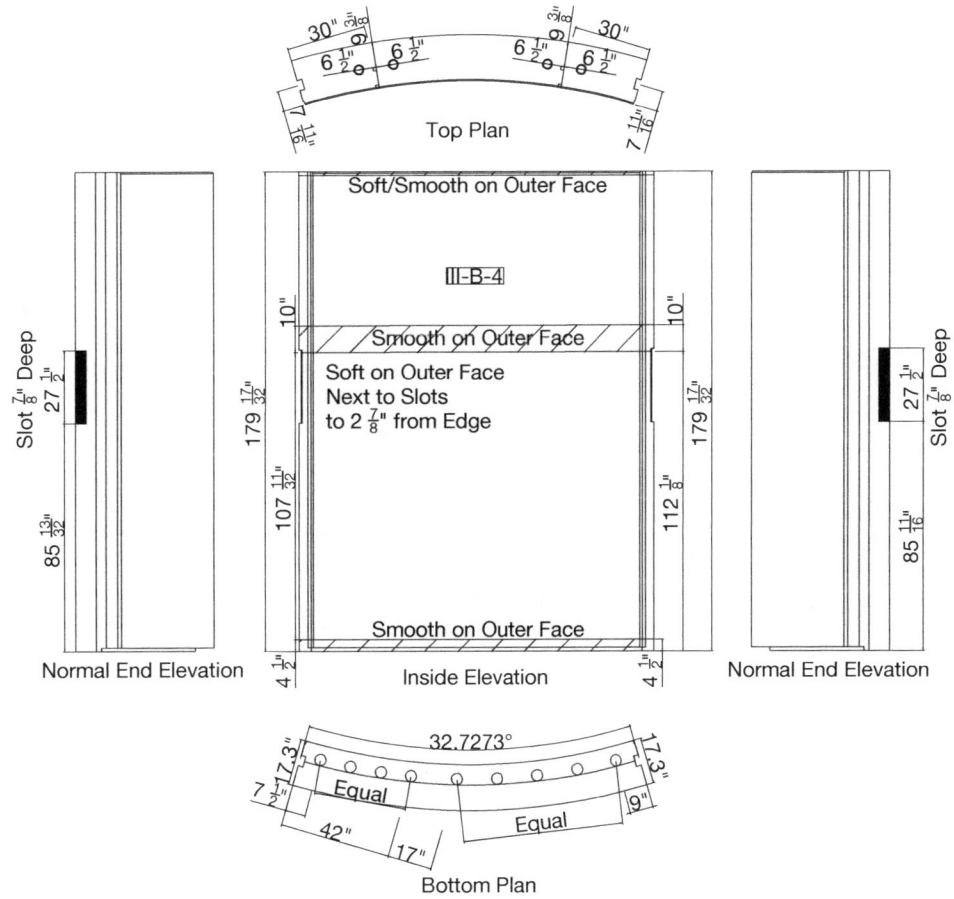

Figure 6.32: Plate III-B-4 (1:72)

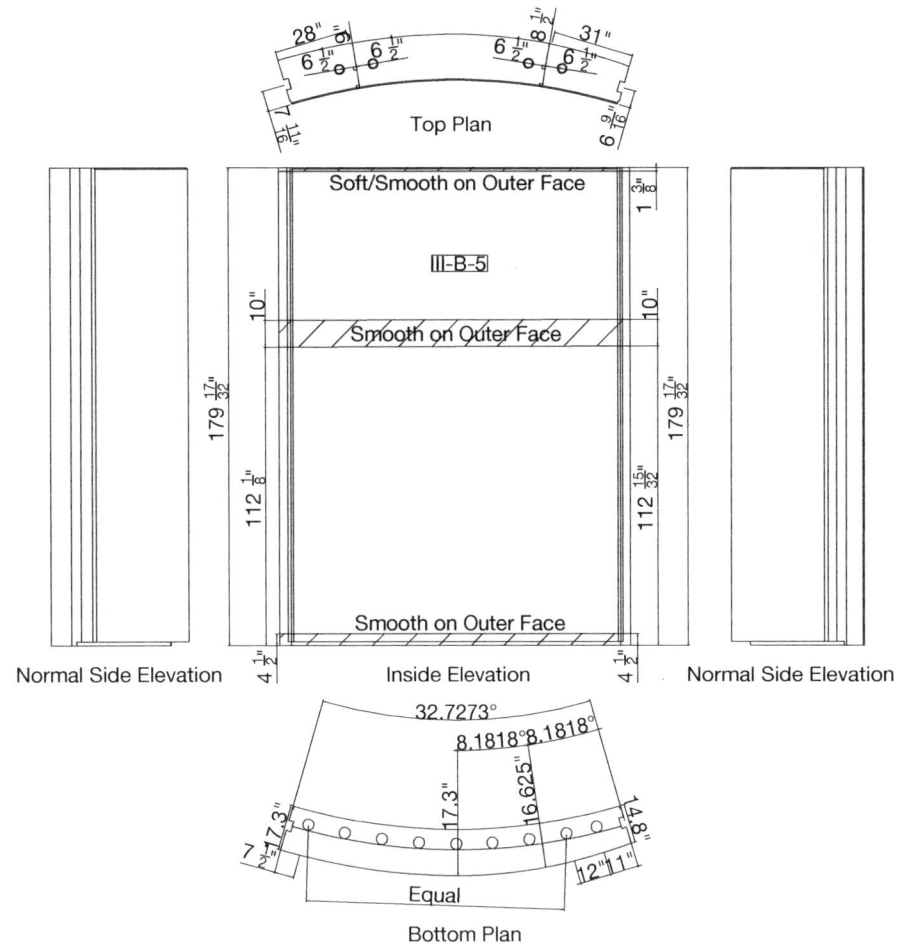

Figure 6.33: Plate III-B-5 (1:72)

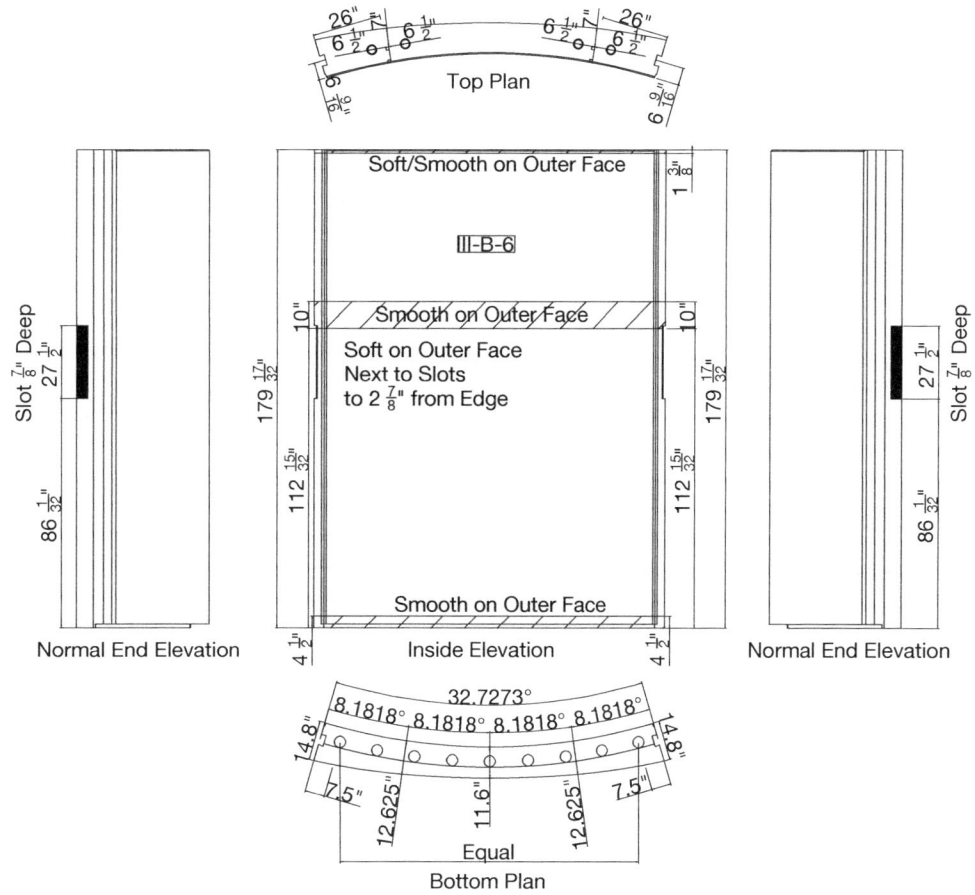

Figure 6.34: Plate III-B-6 (1:72)

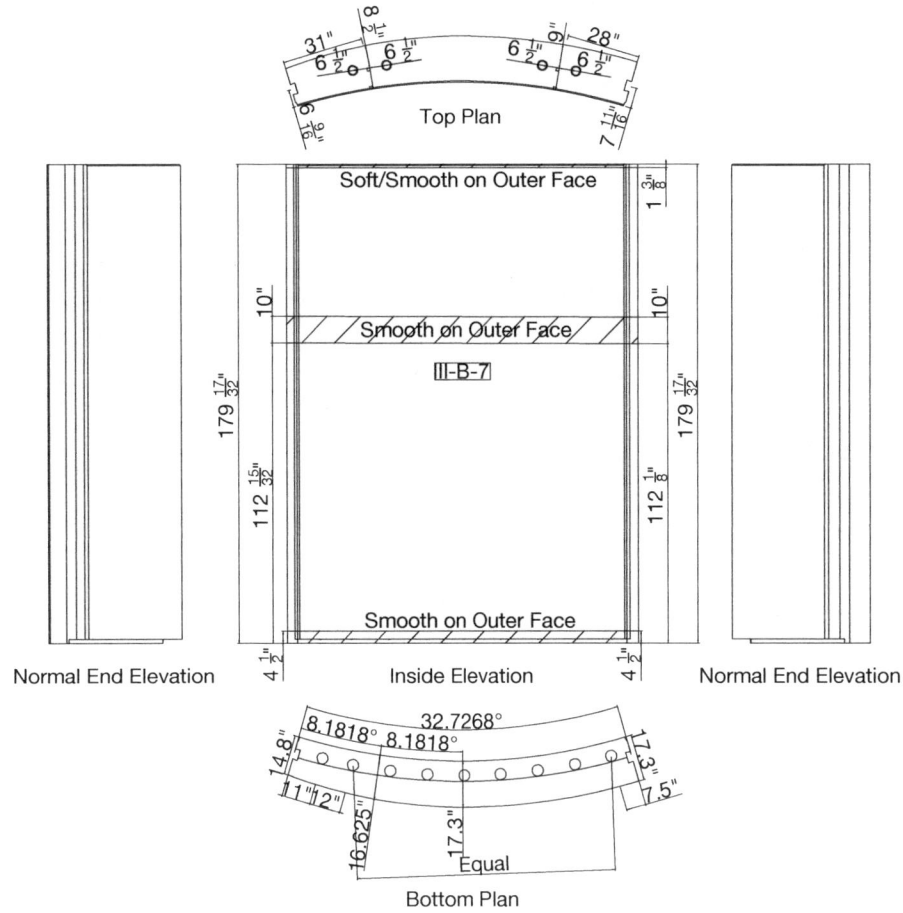

Figure 6.35: Plate III-B-7 (1:72)

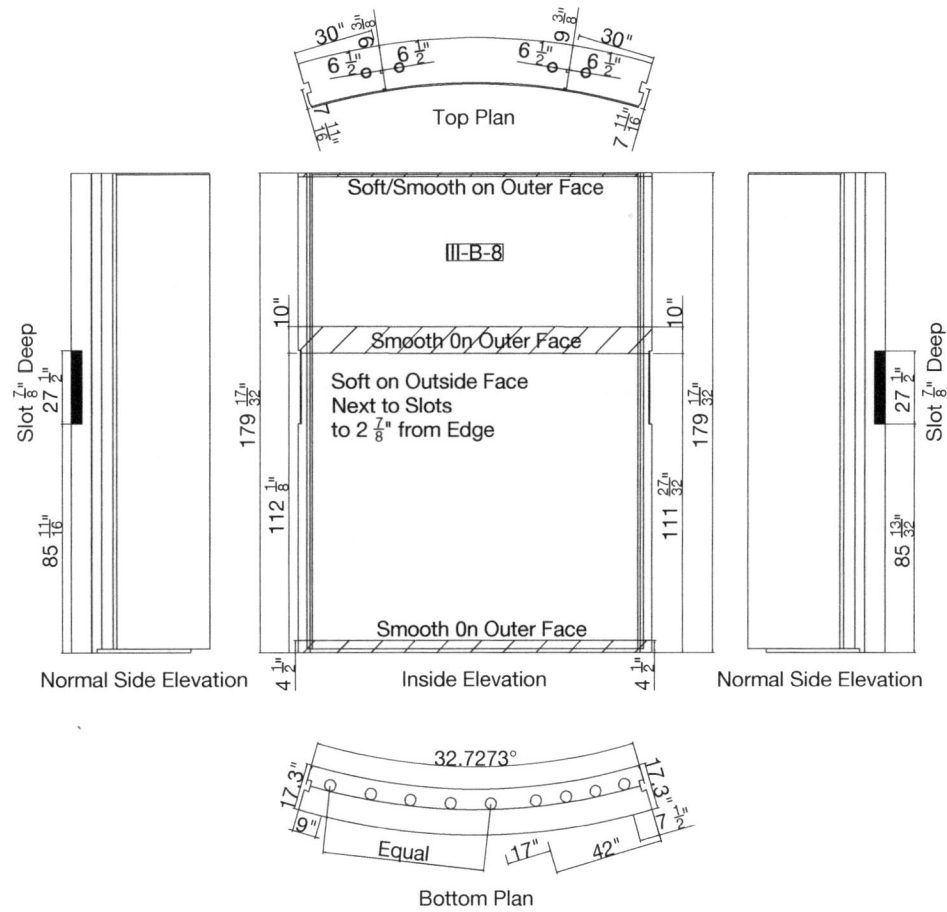

Figure 6.36: Plate III-B-8 (1:72)

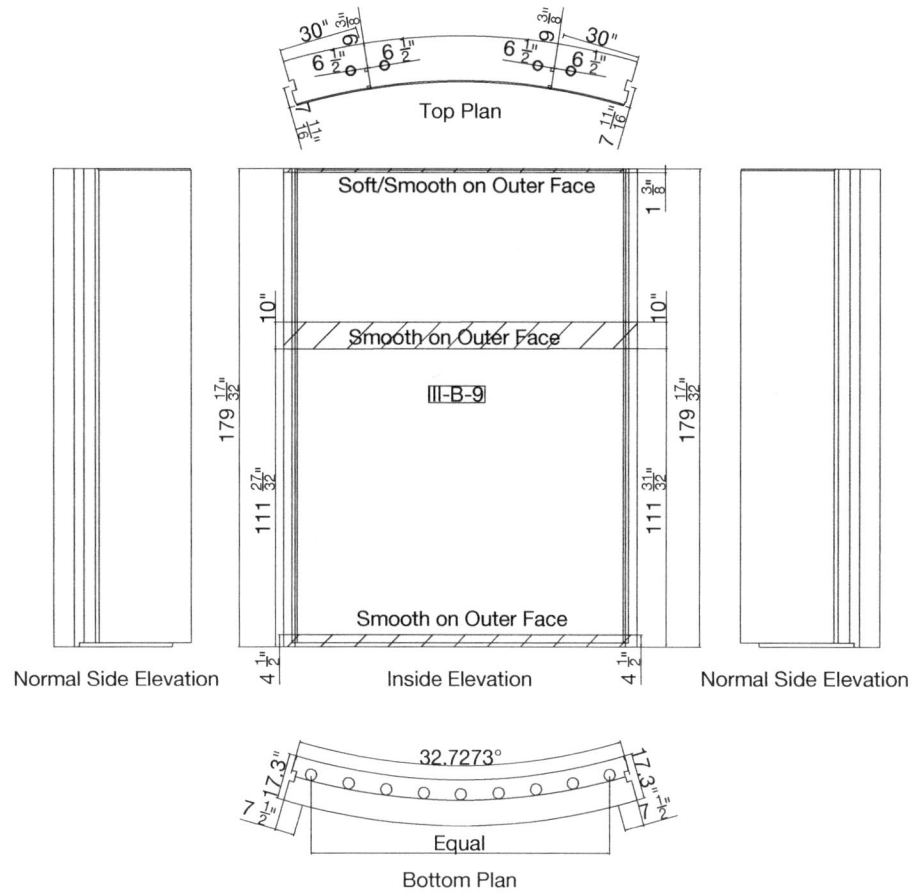

Figure 6.37: Plate III-B-9 (1:72)

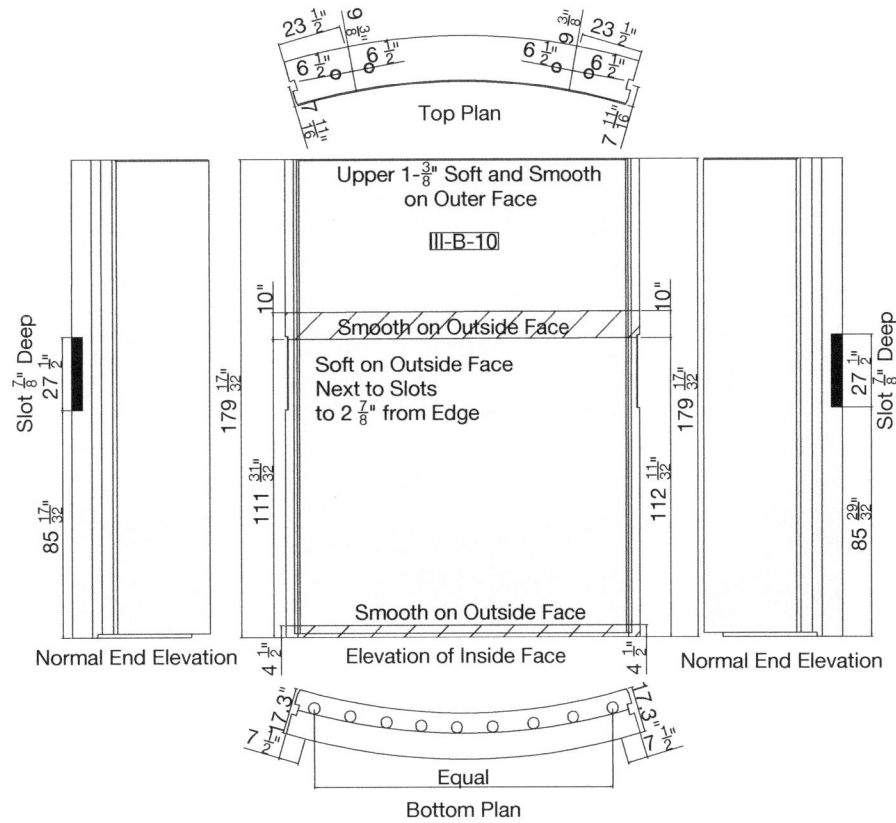

Figure 6.38: Plate III-B-10 (1:72)

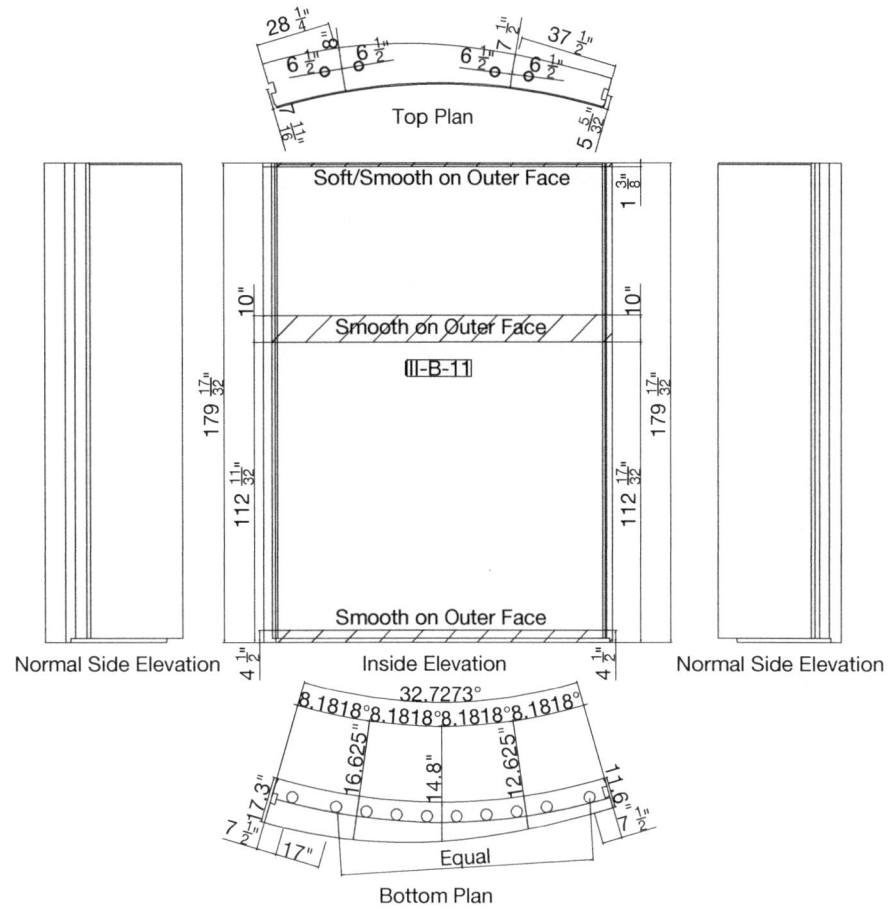

Figure 6.39: Plate III-B-11 (1:72)

BOTTOM EDGE OF TURRET ARMOR FRONT P.T
SOFT SPOT ALONG THIS EDGE

4" x ⅜" F.B.M.S. (1)

⅝" DIA. HEX.HD. TAP BOLTS 11NC-3
SPACED 5" CENTERS, SEE VIEW "3 A-3A"

DRILL AND TAP FOR
⅝ DIA. 11NC-3 BOLTS

(29)

(5)(6)(7)
WEB 12" M.S.

(2)

15" P.T. M.S. CUT TO SUIT WORK

DESIGNED CLEARANCE

LOCK WASHER (33)

4" x ¼" F.B.M.S. (12)

TO BE TRUE PLANE

20'-4¾" RADIUS TO TOE OF ANGLE
HORIZONTAL FLANGE TO BE CUT OR WIDENED TO SUIT WORK

2"

MINUS CURVATURE AND THICKNESS TOLERANCE

PLUS CURVATURE AND
THICKNESS TOLERANCE

11.6
MINIMUM THICKNESS OF BARBETTE

17.3
MAXIMUM THICKNESS OF BARBETTE

2¼"

20'-5⅝" RADIUS TO CENTER OF TUBE

20'-9⅝" RADIUS

18'-3'-½
NORMAL INNER RADIUS
OF BARBETTE

Plan 6.1: (7203ZA) The gas seal between the turret and the barbette. The bulb at the lower right is a grease trough. The angled bar shown in the bulb was attached to the turret and was clamped between the bulb and a flat ring attached to the barbette. The grease in the bulb would prevent water or air from passing from outside into the turret.

Photograph 6.2: The Gas Seal on Turret 1 on USS New Jersey. The grease trough was removed during the mothballing process.

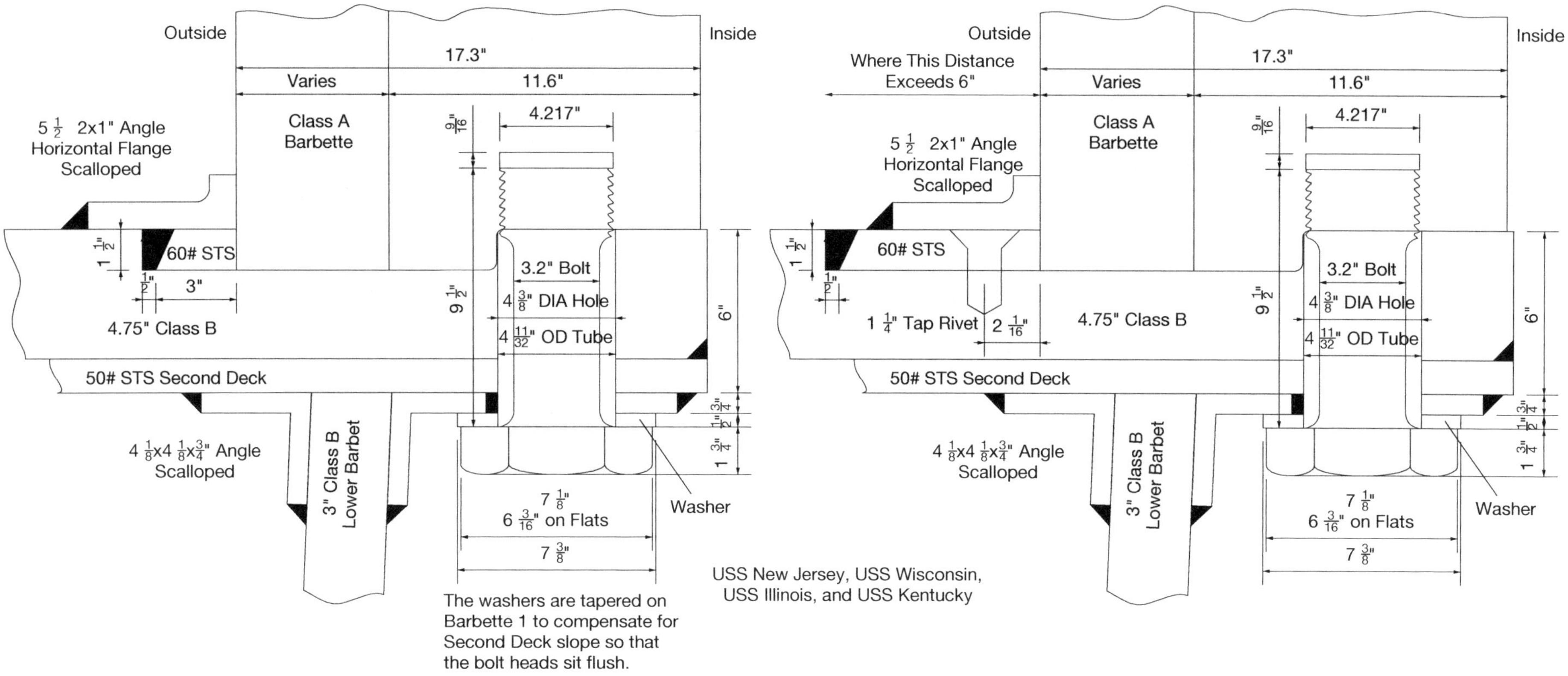

Figure 6.40: Connection of Barbettes to Second Deck (1:6).
The barbettes rest on top of the second deck armor and sit in
a groove cut in the deck plates. A generously wide groove was
used to accommodate variations in barbette plate thickness. The
excess space was filled as shown. See also Figure 6.53.

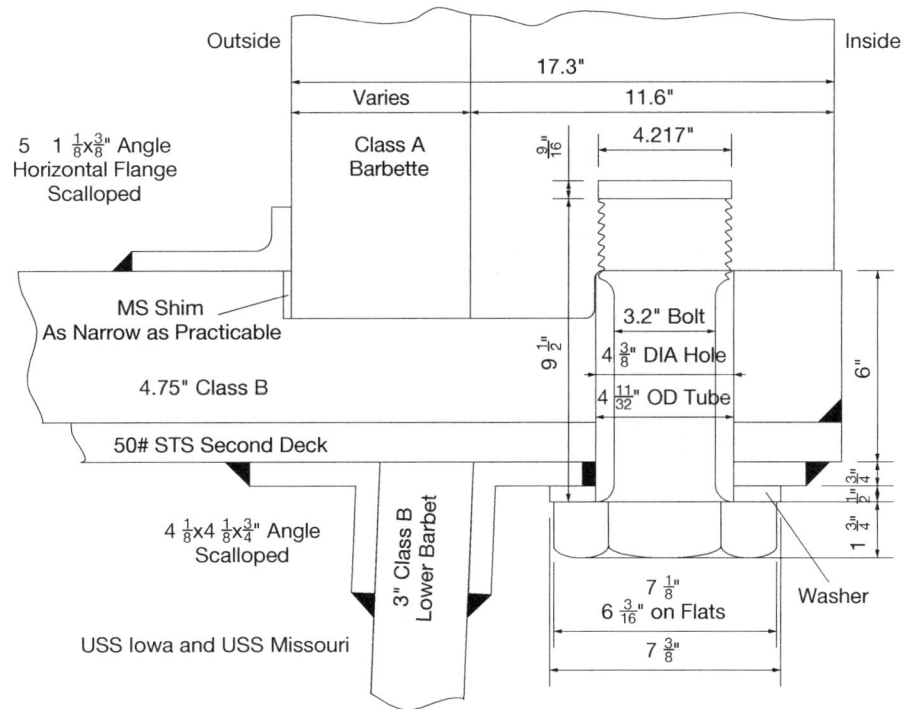

Figure 6.41: Connection of Barbettes to Second Deck (1:6). For USS Iowa and USS Missouri the Brooklyn Navy Yard cut the groove for the barbettes in second deck armor as close as possible to the plates.

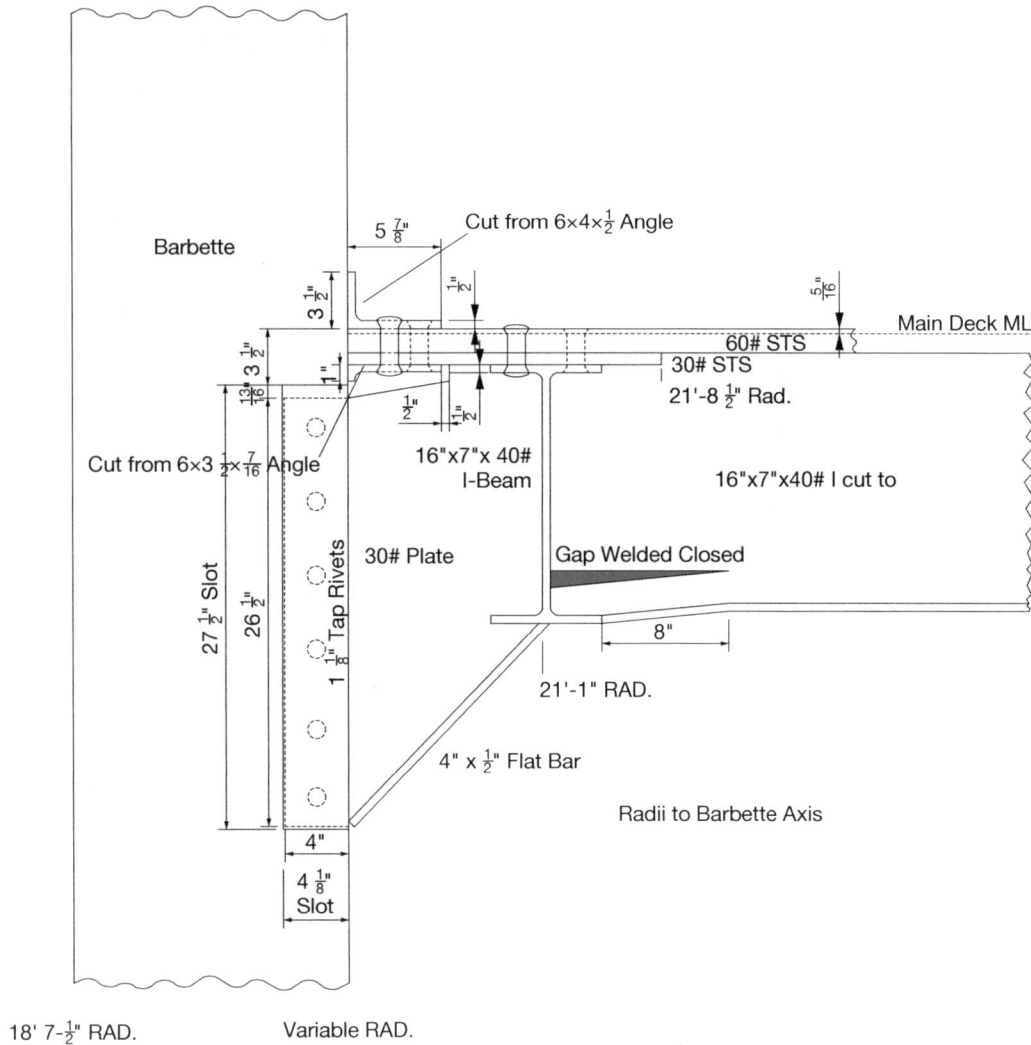

Figure 6.42: Main Deck Attachment to Barbettes (1:12). Each barbette joint has a slot for a bracket to support main deck. A circular Carlin beam wraps around the barbette to connect the brackets. Some of the brackets connect to beams or girders as shown here.

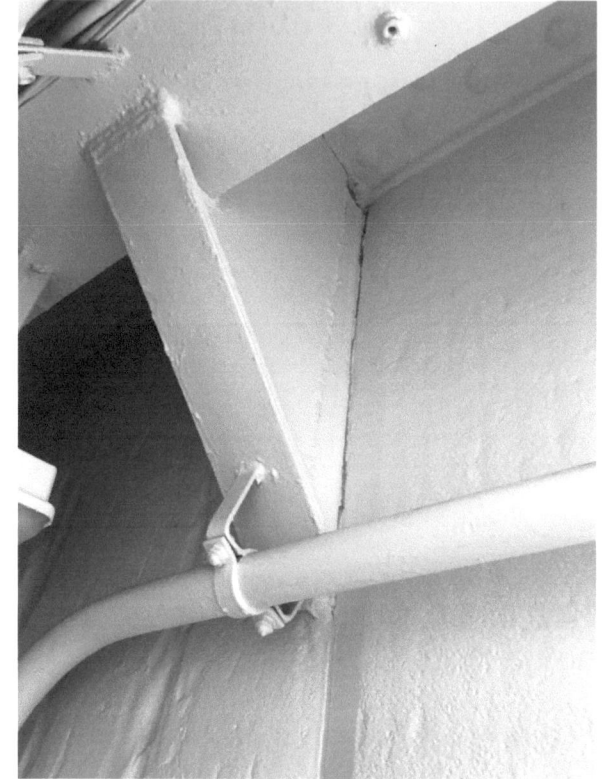

Photograph 6.3: A bracket in barbette 2 on USS New Jersey. This bracket only connects to a circular Carlin beam. Other brackets attach to beams or girders. The joint between plates is visible below the bracket. The slot for this bracket was machined in the plate to the left.

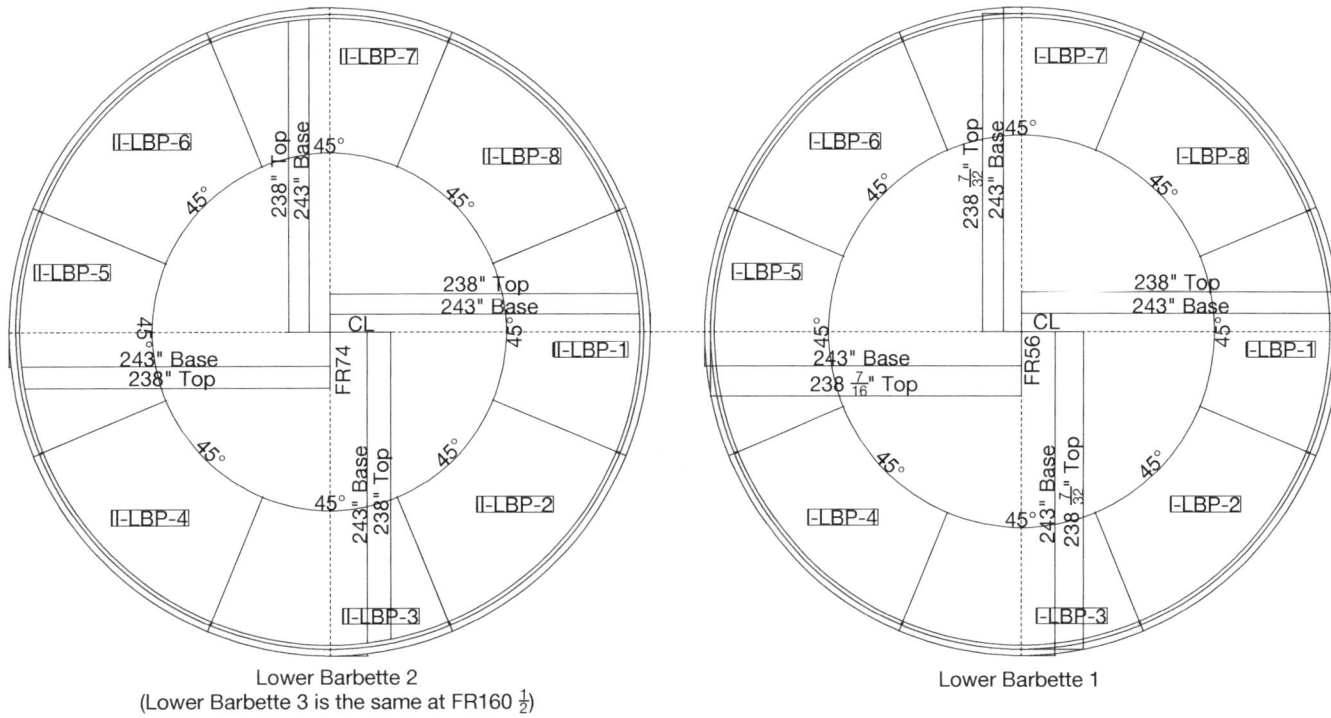

Figure 6.43: Lower Barbette Plans and Elevations (1:144)

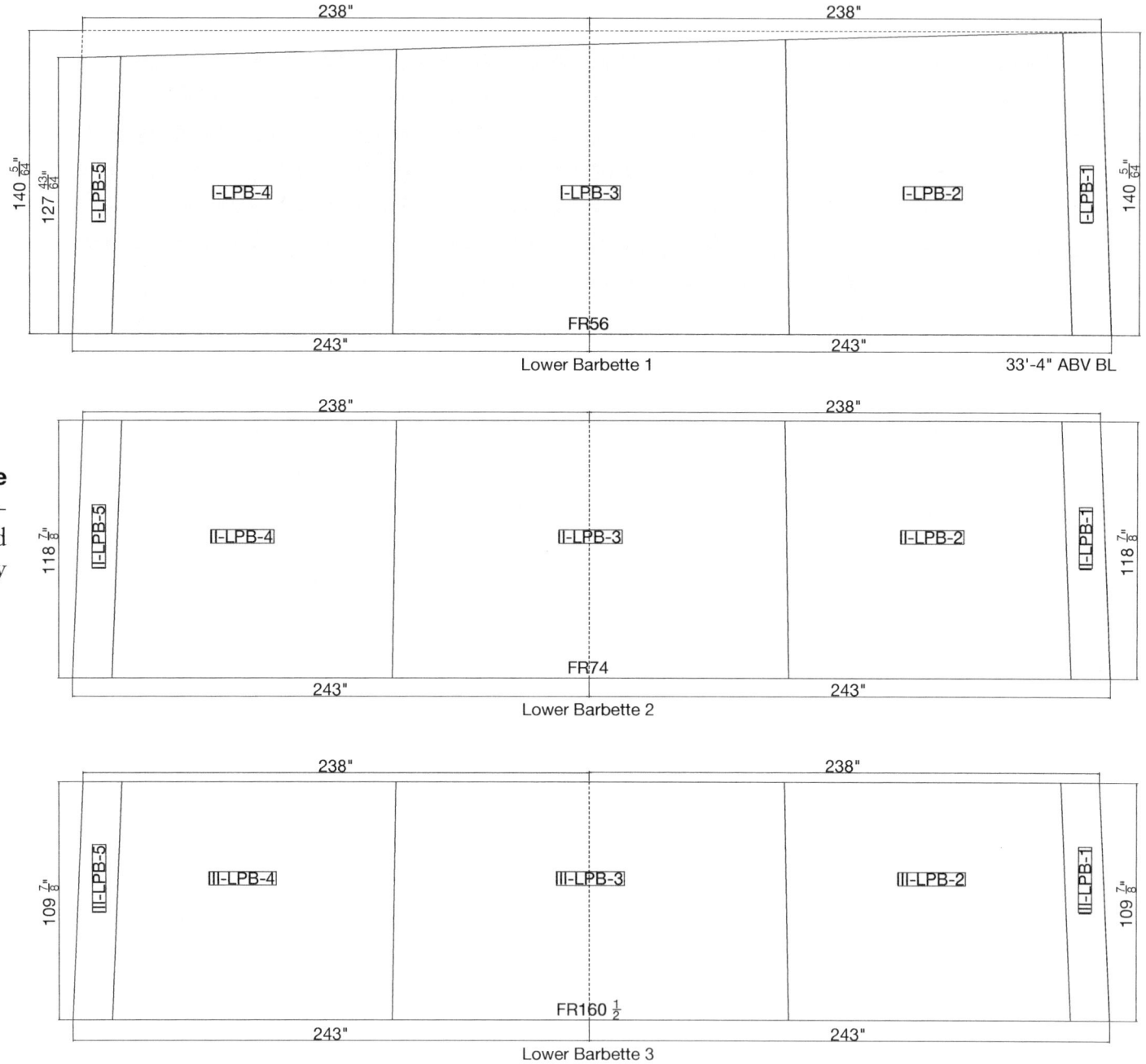

Figure 6.44: Lower Barbette Elevations (1:72). The lower barbettes are $\frac{1}{8}$" short from the designed molded lines. They were likely shimmed at the top.

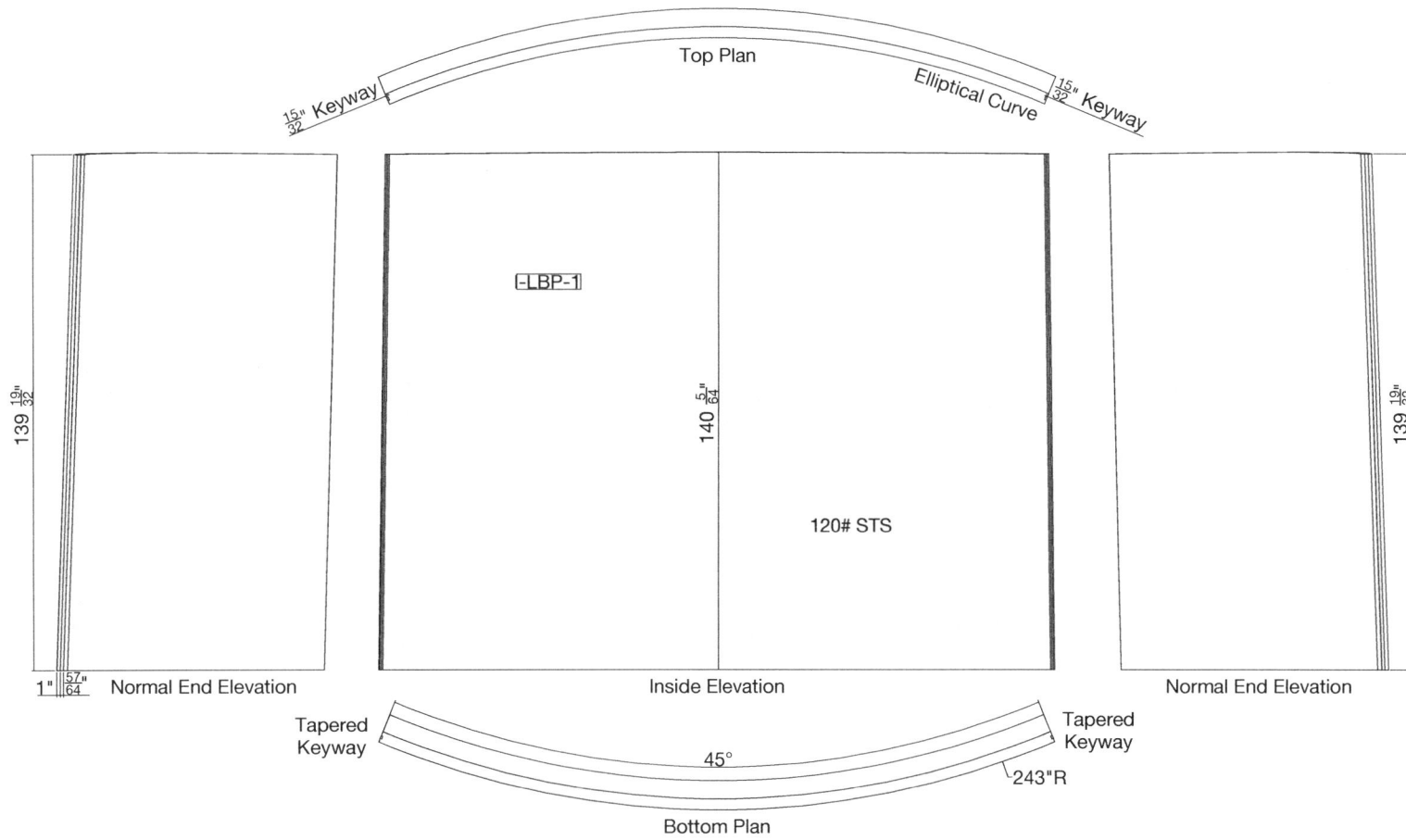

Figure 6.45: Lower Barbette Plate I-LBP-1 (1:48)

243"R

Top Plan

Elliptical Curve

$139 \frac{19}{32}$"

I-LBP-2

$138 \frac{1}{4}$"

120# STS

$136 \frac{15}{64}$"

Normal End Elevation

Inside Elevation

Normal End Elevation

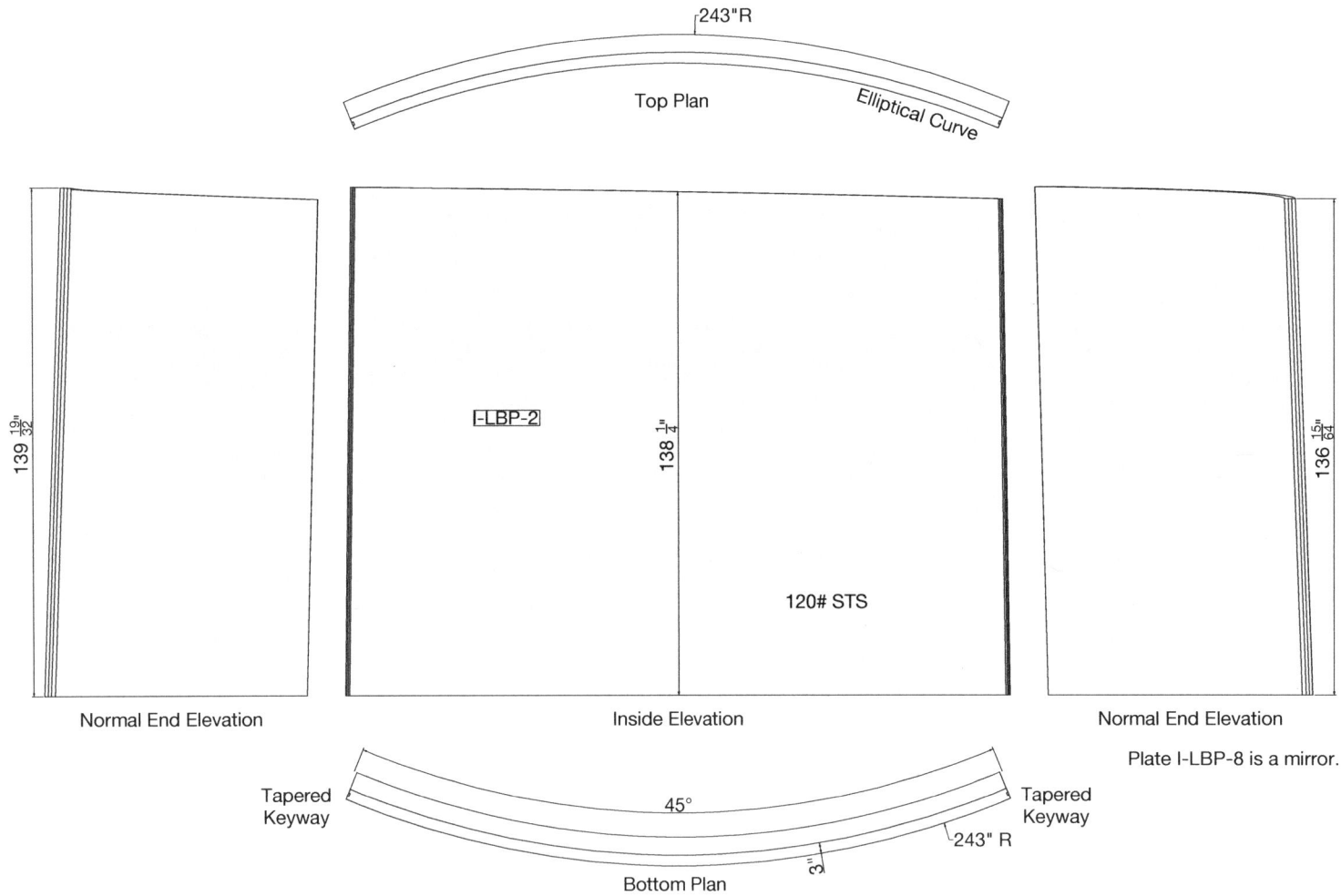

Plate I-LBP-8 is a mirror.

Tapered Keyway

45°

Tapered Keyway

243" R

3"

Bottom Plan

Figure 6.46: Lower Barbette Plate I-LBP-2 (1:48)

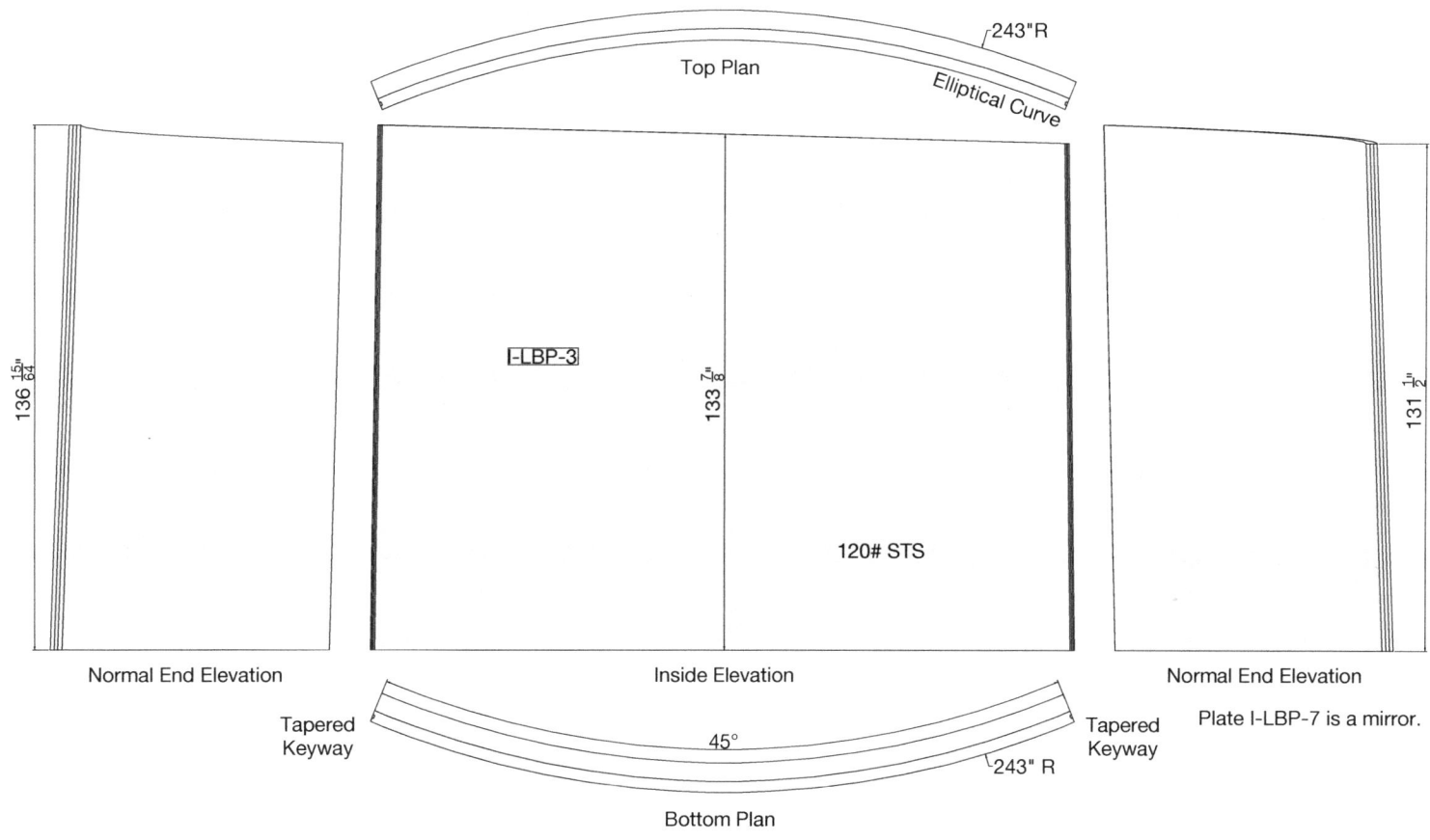

243"R

Top Plan

Elliptical Curve

136 15/64"

I-LBP-3

133 7/8"

131 1/2"

120# STS

Normal End Elevation

Inside Elevation

Normal End Elevation

Tapered Keyway

45°

Tapered Keyway

Plate I-LBP-7 is a mirror.

243" R

Bottom Plan

Figure 6.47: Lower Barbette Plate I-LBP-3 (1:48)

243"R

Top Plan

Elliptical Curve

I-LBP-4

129 $\frac{31}{64}$"

120# STS

131 $\frac{1}{2}$"

128 $\frac{9}{64}$"

Normal End Elevation

Inside Elevation

Normal End Elevation

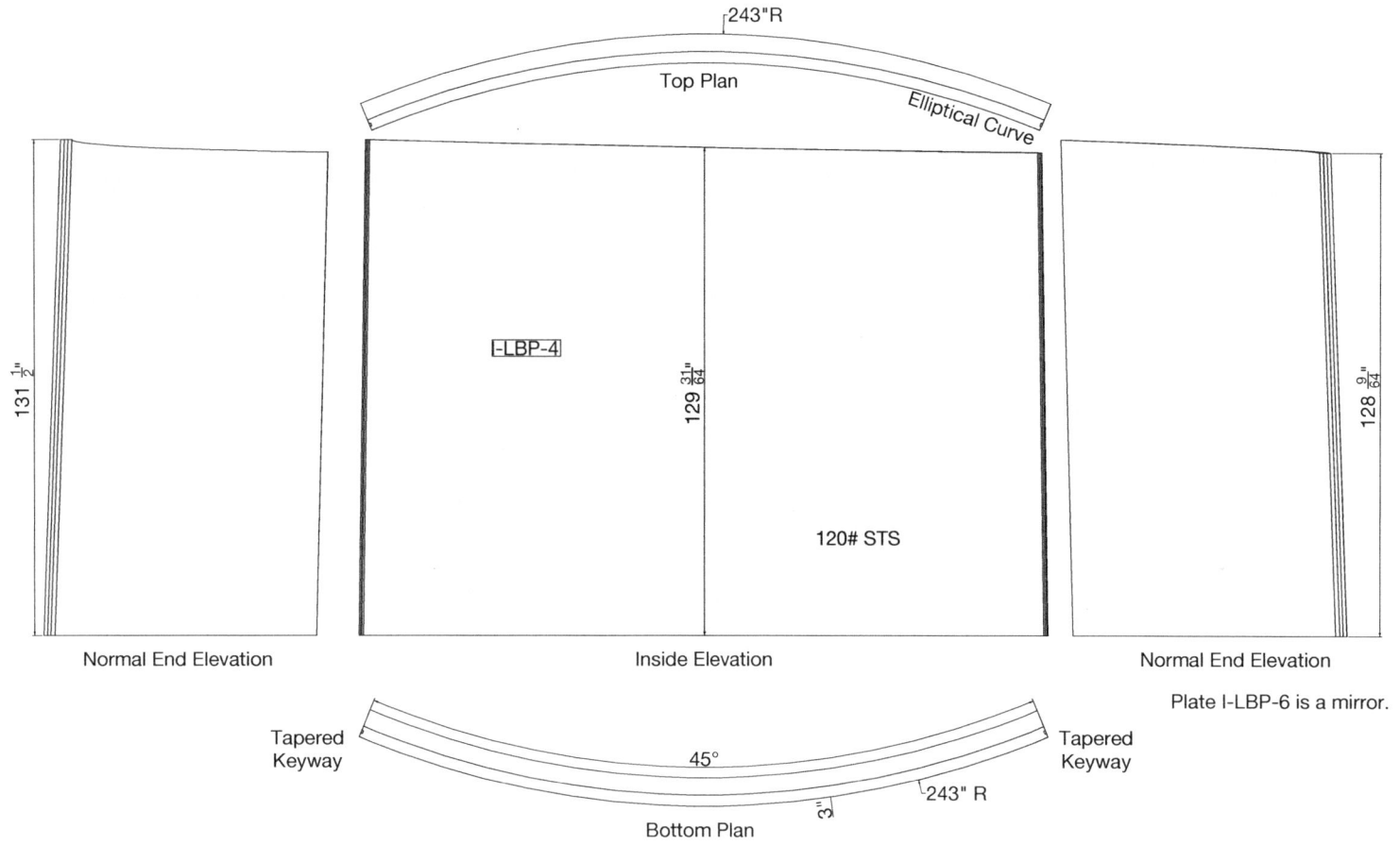

Plate I-LBP-6 is a mirror.

Tapered Keyway

45°

243" R

3"

Tapered Keyway

Bottom Plan

Figure 6.48: Lower Barbette Plate I-LBP-4 (1:48)

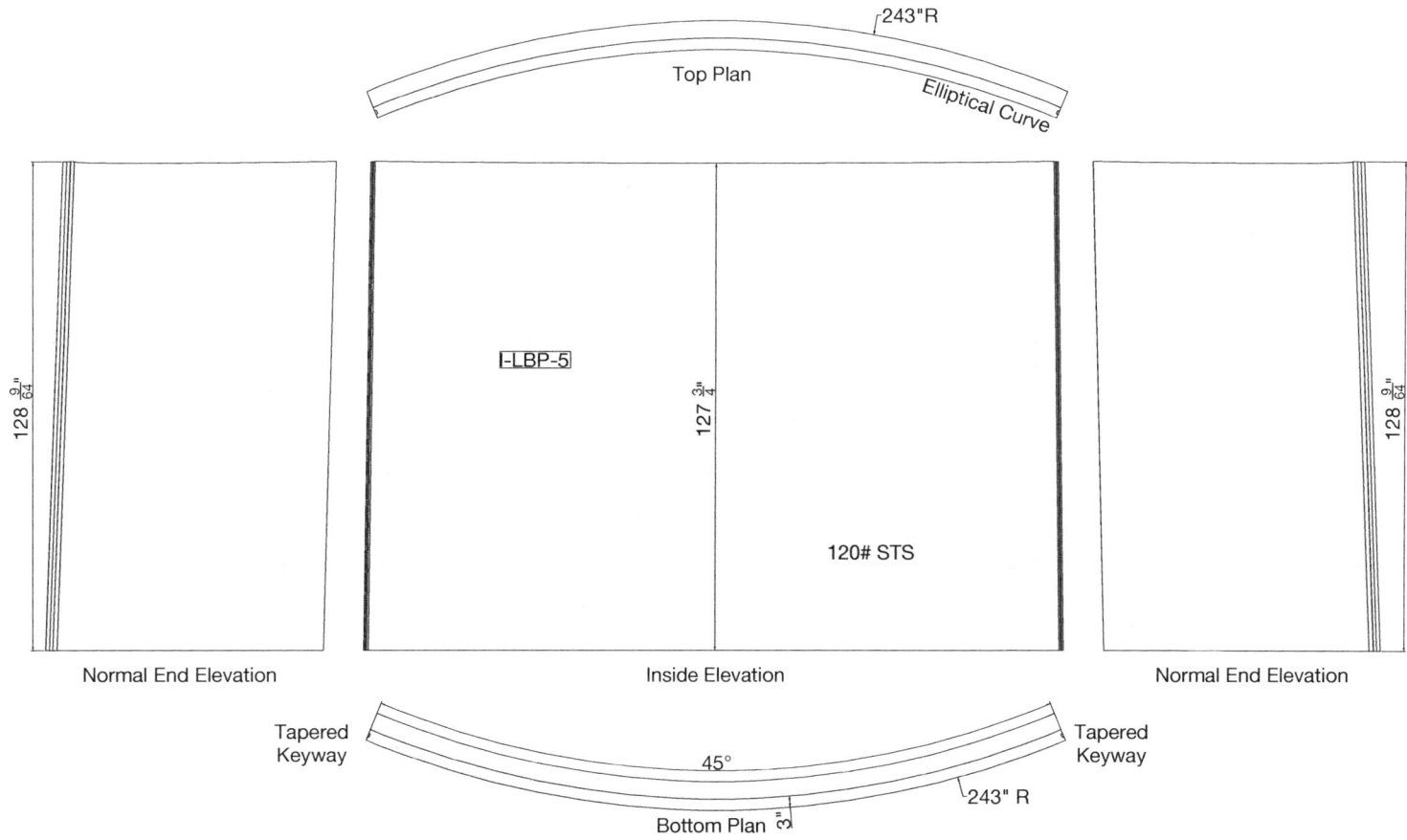

Figure 6.49: Lower Barbette Plate I-LBP-5 (1:48)

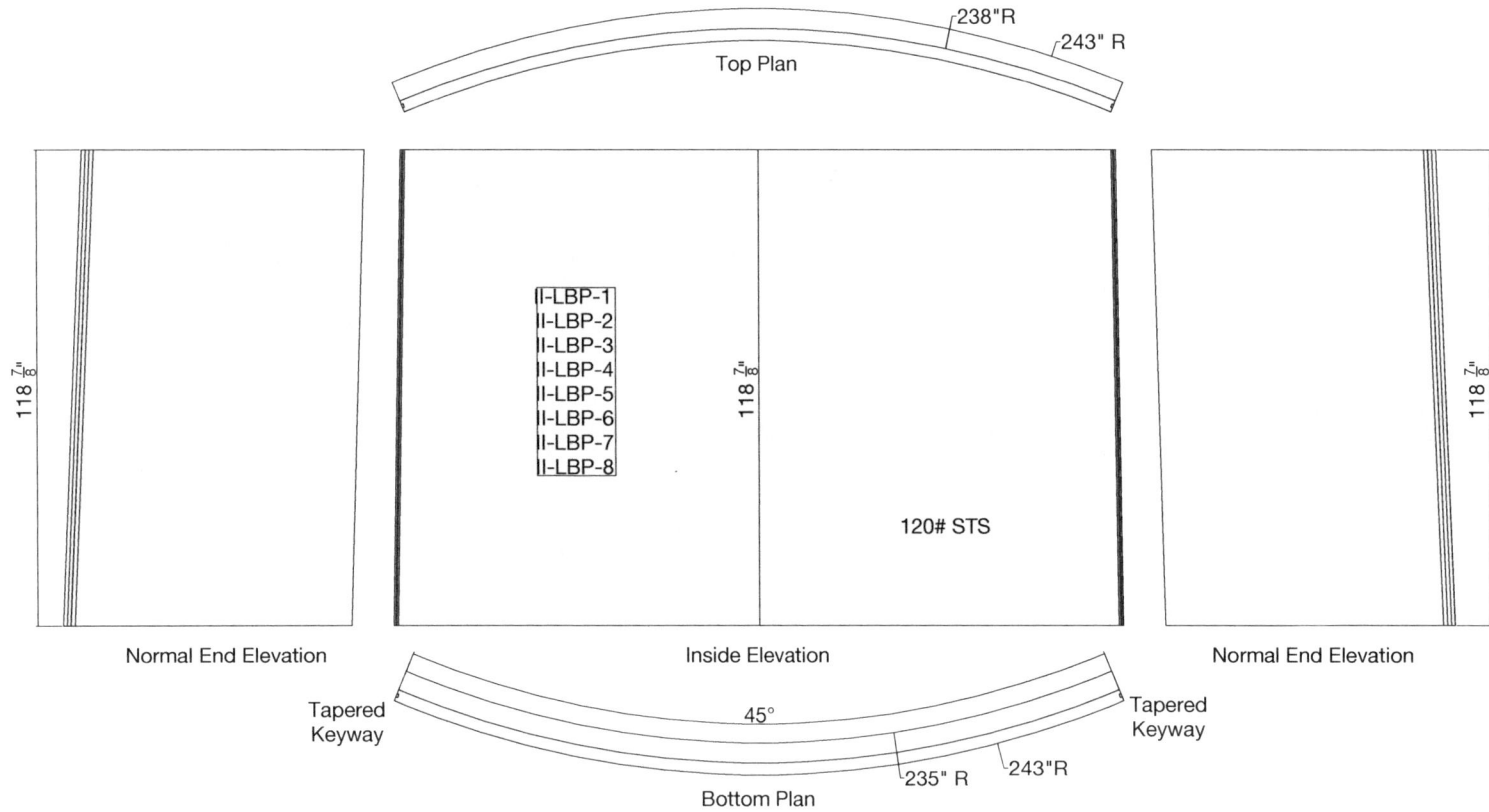

238"R
243" R

Top Plan

118 $\frac{7}{8}$"

II-LBP-1
II-LBP-2
II-LBP-3
II-LBP-4
II-LBP-5
II-LBP-6
II-LBP-7
II-LBP-8

118 $\frac{7}{8}$"

120# STS

118 $\frac{7}{8}$"

Normal End Elevation

Inside Elevation

Normal End Elevation

Tapered
Keyway

45°

Tapered
Keyway

235" R   243"R

Bottom Plan

Figure 6.50: Lower Barbette 2 Plates (1:48). The plates for lower barbette 2 are identical because there is no shear at second deck in this location.

238"R

243" R

Top Plan

Normal End Elevation

III-LBP-1
III-LBP-2
III-LBP-3
III-LBP-4
III-LBP-5
III-LBP-6
III-LBP-7
III-LBP-8

109 $\frac{7}{8}$"

109 $\frac{7}{8}$"

120# STS

Inside Elevation

109 $\frac{7}{8}$"

Normal End Elevation

Tapered
Keyway

45°

Tapered
Keyway

235" R   243"R

Bottom Plan

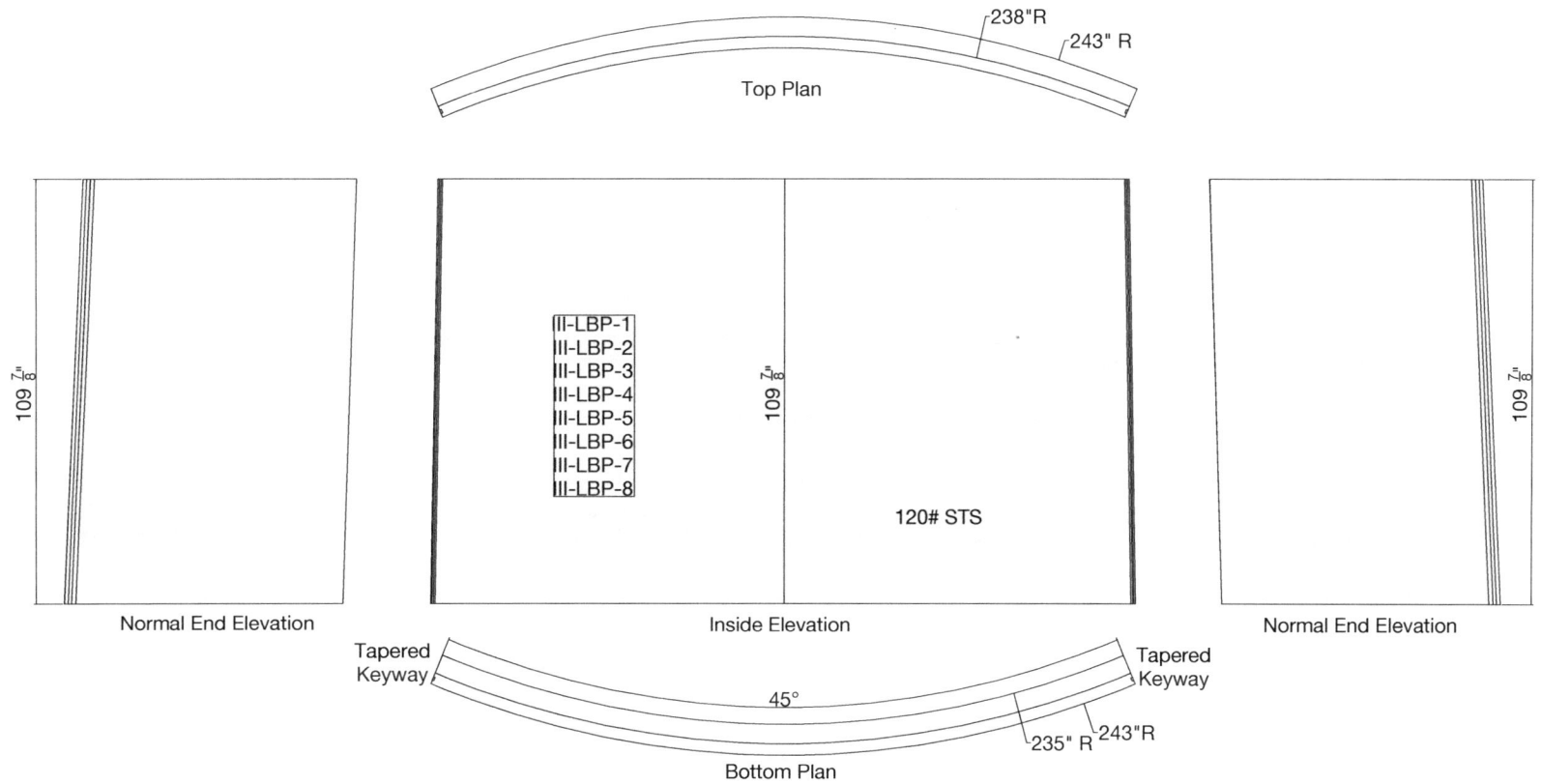

Figure 6.51: Lower Barbette 3 Plates (1:48). The plates for lower barbette 3 are identical because there is no shear at second deck in this location.

Side Elevation **Front Elevation**

$\frac{15}{16}$" $\frac{29}{32}$"

144" (Projected)

Varies

$\frac{1}{4}$"R $\frac{7}{8}$" $\frac{1}{4}$"R $\frac{1}{4}$"R $\frac{29}{32}$" $\frac{1}{4}$"R

Plan

$\frac{3}{16}$"R

Key

Plan

Inside

$\frac{1}{8}$"R

Keyway

Top $\frac{15}{32}$"

Plate Division

Face

Bottom
Outside

Section

1" $\frac{61}{64}$"

Projected

144"

Varies

Keyway

90°

1" $\frac{57}{64}$"

Keyway

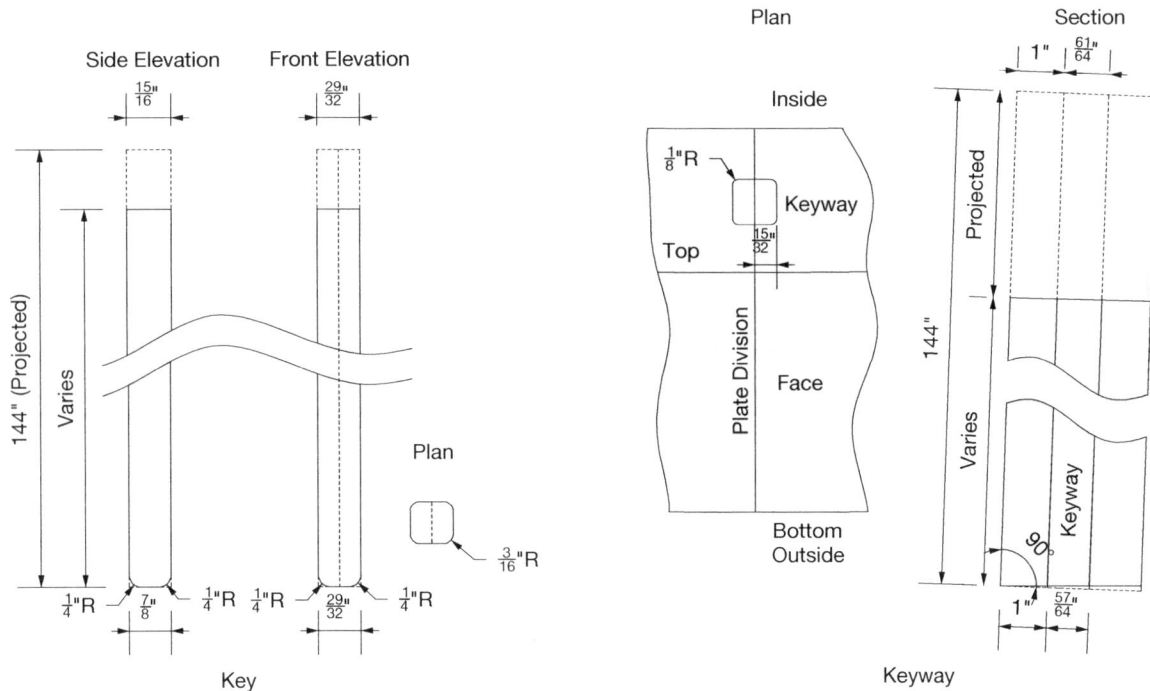

Figure 6.52: Lower Barbette Key and Keyway (1:4). The keys and keyways for the lower barbettes taper. The upper measurement point for the keyway is 144" above the bottom. This is a theoretical location above the top of the lower barbettes. The actual keys and keyways are shorter.

Photograph 6.4: USS Kentucky Under Construction. The large circular structure is the fixed Turret 2. The rim at the top is ready to accept the roller track that the turret rotates upon. The lower barbette and the outer scalloped angle for attaching second deck is visible around the fixed turret. (Naval History and Herritage Command)

4.75" Class B

50# STS Second Deck

Angle   Angle

120# STS Lower Barbette

Butt strap   Butt strap

Cross Section Through
Butt straps

2       $\frac{3}{4}$"R   4" $\frac{3}{4}$"R

$4\frac{1}{8}$"   3"

120# STS Lower Barbette

$4\frac{1}{8}$"   3"

$\frac{3}{4}$"R   4" $\frac{3}{4}$"R

Plan of Angles at Third Deck

Scallops Cut in
Both Flanges from
5 5 $\frac{3}{4}$ Angle

Bottom of Second Deck

Cut from 5 5 $\frac{3}{4}$" Angle
Both Flanges Scalloped

$3\frac{1}{16}$"

$8\frac{3}{8}$"

$\frac{13}{16}$"R

$\frac{13}{16}$"R

4"

$\frac{3}{4}$"R   4"   3"

30#
STS

For USS Iowa and USS
New Jersey, individual butt
straps were ordered from
the mill and scalloped at the
yard. For the other ships the
yards cut multiple butt
straps simultaneously from
a sheet.

Cut from 5 5 $\frac{3}{4}$" Angle
Both Flanges Scalloped

Top of Third Deck

Elevation of Inside and Outside Scallops
Connecting Lower Barbette Segments
and Lower Barbettes to Decks

$\frac{3}{4}$"R   4" $\frac{1}{4}$" $\frac{3}{4}$"R

$1\frac{1}{8}$" $2\frac{3}{4}$"

$2\frac{1}{8}$"   2" $6\frac{1}{8}$"   4"   Lower
Barbette

$\frac{7}{8}$"   3"

$\frac{5}{8}$"R

$2\frac{3}{16}$"R $4\frac{1}{2}$"R

Scallops vary due
to differences in
bolt spacing.

Plan of Angles at Second Deck
USS New Jersey Only

$\frac{3}{4}$"R   4" $\frac{1}{4}$" $\frac{3}{4}$"R

$1\frac{1}{8}$" $2\frac{3}{4}$"

$2\frac{1}{8}$" $6\frac{1}{8}$"   Lower
Barbette

$\frac{7}{8}$" $3\frac{1}{4}$"

$\frac{5}{8}$"R

$1\frac{1}{2}$"

$2\frac{3}{16}$"R $4\frac{1}{2}$"R

Plan of Angles at Second Deck

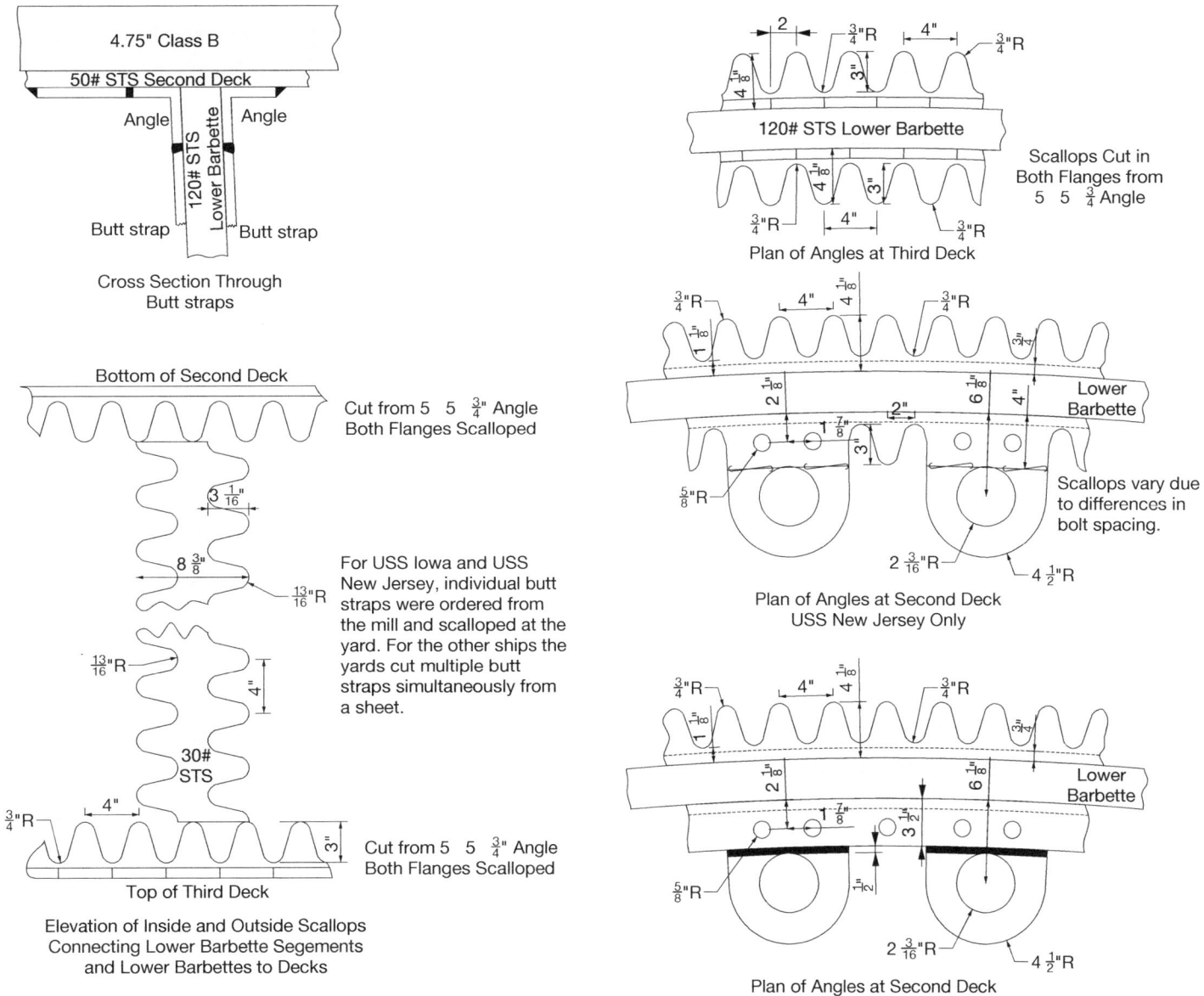

Figure 6.53: Lower Barbette Connections (1:12). The lower barbettes were welded together using scalloped buttstraps. They are welded to second and third deck using scalloped angles. The upper inside angles had extensions for attaching the barbettes with washers and bolts. See also Figure 6.40 and Figure 6.41.

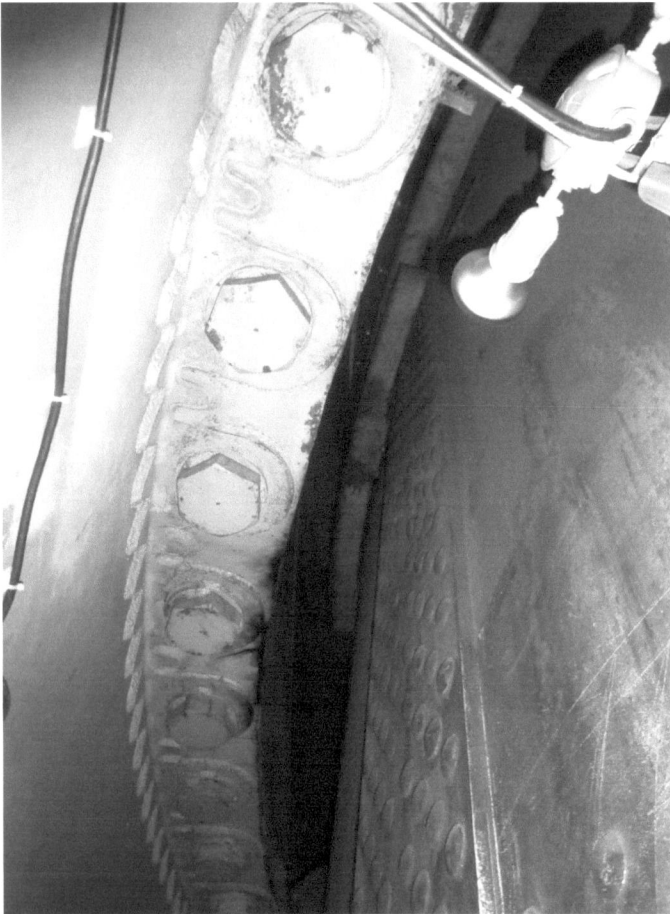

Photograph 6.5: Second Deck Viewed From Below. This photograph shows the connections of the barbette and barbette support for Turret 2 of USS New Jersey. The barbette support is to the left and the fixed turret is to the right. The angle above is welded to the barbette support and second deck. The bolts shown pass through the angle and second deck then into the bottom of the armor. The bolts are welded to the angle.

Photograph 6.6: The Barbette Support for Turret 2 on USS New Jersey. The museum cut the opening shown at third deck to give visitors access to the turret. The opening shows the size of the gap between the barbette and the turret. A scalloped seamstrap linking segments of the barbette support is visible to the right of the opening. The vertical flange of the scalloped angle at third deck is visible as well. The horizontal flange is covered by the tiling.

Photograph 6.7: The Connection of the Barbette Support (Right) for Turret 2 on USS New Jersey to Splinter Deck (Top). A scalloped angle runs along the joint. A bracket supporting splinter deck and second deck is at the center. The bracket is welded to a scalloped seamstrap connecting barbette support segments.

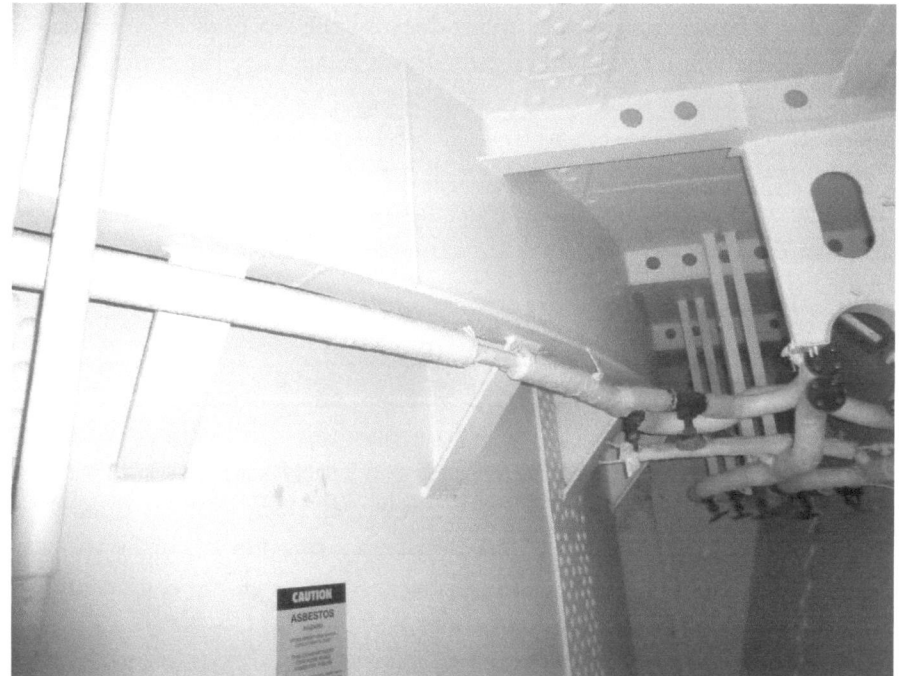

Photograph 6.8: Fixed Turret 1 on USS New Jersey at First Platform. The circular beam shown below third deck is directly underneath the barbette support. The brackets shown transfer the weight of the barbette and barbette support to the fixed turret.

Chapter 7: Conning Tower

The conning tower is the most heavily armored location on the *Iowa*-class battleships. The 17.3" sides of the conning tower are equaled only at the thickest areas of the barbettes. The conning tower sits on a rectangular-shaped pyramid foundation that extends from main deck to the 03 level. The conning tower has three levels: Flag (entrance at 03), Navigation (entrances at 04), and Fire Control (entrances at 05). USS *Iowa* (like USS *South Dakota*) was intended to be a fleet flagship and the conning tower is armored at all three levels. The levels within the conning tower are not aligned with the superstructure levels so this chapter references them by name. The flag levels of the conning tower ob the other three *Iowa*-class battleships is 25# STS and the space is largely unused except for door opening machinery and wiring trunks. An armored tube connects the conning tower to the citadel.

The plan of the conning tower consists of four tangent circular arcs that combined give the appearance of being elliptical. Features along the conning tower armor are located using angles from one of the four arc centers. The transverse axis of the conning tower is 12-inches aft of FR86.

The structure of the conning tower is different for each of the four completed ships. As mentioned, USS *Iowa* is unique in having the conning tower armored with vision peepholes at the 03 level. The conning tower on USS *New Jersey* is structurally the same as that of USS *Iowa* except that the bottom is raised to the

04 navigation level. The conning towers of the later ships were simplified to make construction simpler. The specific differences are discussed below. Complete plans have only appeared for USS *New Jersey*. These plans contain details for the other ships and have been used to create drawings marked as "inferred" for the other ship. Such inferred plans only include features shown in existing plans so they lack details that are likely the same as for USS *New Jersey*. When a plan only applies to specific ships, those ships are listed on the plan. If a plan applies to the entire *Iowa*-class, no ships are listed on it.

7.23. Armor Plates

A total of nine Class B armor plates make up the conning tower. There are five 17.3" plates at the sides (C-T-1 through C-T-5). The floor plate (C-T-6) is 4.0-inches, the roof plate (C-T-7) is 7.25-inches, and the two segments of the connecting tube (C-T-8 and C-T-9) are 16-inches thick. In addition, there are three doors (C-T-10, C-T-11, and C-T-12) with USS *Iowa* having one additional door (C-T-13). The five side plates and doors have vertical faces. All of the side plates, except for the forward centerline C-T-1, cross the arc segment boundaries of the conning tower plans so they have two radii. The other plates have horizontal faces. The side plates on USS Iowa are taller than those of the other ships to enclose the Flag Level and the upper connecting tube (C-T-8) is shorter.

7.24. Features Machined at the Mill

The main entrances to the conning tower are doorways cut into the side. There are two doors at the navigation level that open inward to the

Plate	Location	Weight (lbs.)
C-T-1	Side	111,411
C-T-2	Side	82,965
C-T-3	Side	82,965
C-T-4	Side	82,623
C-T-5	Side	83,303
C-T-6	Bottom	22,074
C-T-7	Top	45,904
C-T-8	Upper Tube	185,391
C-T-9	Lower Tube	154,084
C-T-10	Fire Control Door	5,085
C-T-11	STBD Navigation Door	5,354
C-T-12	Port Navigation Door	5,354
Total		866,513

Table 7.1: Conning Tower Plate Weights for USS New Jersey

centerline. These door openings are split at the center by side plate boundaries. There is one door at the Fire Control level on the starboard side that opens outward from the centerline. On USS *Iowa* there is a door at the port side on the Flag Level that opens outward from the centerline. The other ships have an airtight door in the same location that gives access to the unarmored conning tower foundation. The doorways have a slot machined at one side that engages the door when closed. These slots would have kept the door from falling out if the hinges were destroyed in battle.

Peepholes provide direct vision from the conning tower. The peepholes are all two inches height. Each level has a 14-inch peephole forward on the centerline. This is flanked by 12-inch peepholes which are flanked by 9-inch peepholes. The orientation of the peepholes is the same on the Flag and Fire Control levels. The Navigation Level has 8-inch peepholes in each of its doors.

Ideally a peephole would only flare out on the inside. Such an arrangement would have been impracticable. The curve and thickness of the conning tower would have required peepholes to flare to nearly six feet wide. Instead, the peepholes flare on both sides with most of the flare on the inside. The inner flare of the door peepholes extend outward into the sides of the door openings. Holes were drilled and tapped for lifting pads at the top edge of the side plates. The mill drilled and tapped four holes for two type M pads at the top edge of each side plate for craning the side plates into position.

The mill drilled five holes in the top platfor periscopes and, in the bottom plate, drilled a hole for the connecting tube and escape hatch. The mill also drilled and tapped four holes at the top of both of these plates for type G lifting eyes. Near the upper end of the connecting tube there are holes at the rear for cabling from the fire control system to enter. The upper faces of the connecting tubes were both drilled and tapped for four type E lifting eyes.

7.25. Plate Installation

7.25.1. Connecting Tubes

The mill machined the lower edge of the bottom tube on USS Iowa, USS *New Jersey*, and USS *Wisconsin* to have a tongue. The mill drilled and tapped eight evenly spaced holes along the

tube. The yard machined a groove with matching holes in the second deck armor to accept the tube. After the tube was in place, bolts were screwed through the bottom of second deck to secure the tube to the deck armor. A scalloped angle was welded along the upper joint between the tube and the armor.

This connection was simplified for USS *Missouri*. The mill machined a rabbet into the outer bottom edge. This created a nipple that allowed the tube to fit into a circular hole cut into the armor. Once the tube was in place, the tube was welded to the top and bottom of the second deck armor. The bottom weld had to be grinded flush to allow an extension of the 50# STS second deck to be welded to the bottom of the tube.

The joint between the tube sections was the same on all ships. The mill machined a tongue (lower) and groove (upper) joint to align the tube sections. The mill also machined welding grooves along the inner and outer surfaces of the joints. Once the upper tube was lowered in place, the two sections were welded together at the inside and outside.

7.25.2. Bottom Plate

The process for connecting the bottom plate to the tube depended upon the ship. For USS *Iowa*, USS N*ew Jersey*, and USS *Wisconsin*, the upper tube has a nipple with a slot cut long its outer edge. The bottom plate has a circular opening that fits the nipple. The mill drilled and tapped eight evenly spaced holes in the upper shoulder of the tube and drilled matching holes in the bottom plate. To install the bottom plate, a ring-shaped liner was placed over a nipple machined at the end of the tube. The liner was oversized

in thickness and had to be grinded to allow the lower plate sit at the right height. The nipple on the tube protrudes through the bottom plate. A retaining ring divided in halves was driven into the slot in the nipple to keep the tube and bottom plate from pulling away. The retaining ring was ordered oversized so it had to be grinded to the correct thickness to fit. Once in place the halves of the retaining rings were welded together. The retaining ring has eight holes that match those in the bottom plate and tube. Bolts were screwed through the holes to link the bottom, retaining ring, and tube together.

This connection was simplified for USS *Missouri*. The mill machined a shoulder in the upper tube the same thickness as the bottom plate and the opening in the bottom plate was made larger than on the other ships. The mill also machined a welding groove at the shoulder. The tube was welded to both sides of the bottom plate.

7.25.3. Side Plates

For USS *Iowa* and USS *New Jersey* the mill machined a tongue along the lower edges of the side plates and a matching groove along the outer upper edge of the bottom plate. The mill also drilled and tapped holes at regular intervals into the tongues and drilled matching holes in the bottom plate. After the side plates were lowered in place, bolts were screwed through the bottom plate into the side plate. A scalloped angle was welded to inside edge of the joint.

This joint was simplified for USS *Missouri* and USS *Wisconsin*. The mill machined a rabbet into the lower inside edge of the side plates to accommodate the bottom plate. After the side plates were in place, the side plates were welded to the top and bottom of the bottom plate.

The connections of the side plates to each other was the same for all ships. The mill machined an hourglass keyway at each side edge. It also machined welding grooves at each corner. Once two adjoining side plates were in place, an hourglass key was hammered through the keyway to align the plates. Keys had to be cut away where they passed through door and peephole openings. The plates were the welded together along the inside and outside.

7.25.4. Top Plate

For USS *Iowa* and USS *New Jersey*, the mill machined a keyed hook joint between the side

Photograph 7.1: USS Iowa in 1943 with its original open bridge. The vision slits at the Flag Level are visible under the Navigation Bridge. *USS Iowa* was unique in having three 20mm mounts on the roof of turret 2 during World War II rather than a 40mm quad mount that would have blocked the view from the flag level. (Nat'l Archives)

plates and the top plate. The tongue of the joint is located on the side plates. The tongue was drilled and tapped at regular intervals. Matching holes were drilled in the top plate. Once the top plate was lowered in place, bolts were screwed through the top plate into the sides. Keys were hammered into the joint and welded in place. Finally a scalloped angle was welded along the inside joint.

The side–top joint was simplified for the later ships. A rabbet was cut into the upper inner edge of the side plates. The top plate rests in the rabbet and was welded at the top and bottom to the side plates. The bolts and scalloped angle were omitted from the new design.

7.26. Machining at the Yard

The yard had to perform the machining for the door mechanisms and the airtight peephole covers. Machining for the peephole covers was relatively simple. The yard simply had to drill a countersunk hole below each peephole for a geared pinion shaft and ratchet mechanism.

The doors were more complex. The yard had to machine the shaft openings for the hinges then weld the hinges to the door. The mating face for the hinges at conning tower had to be grinded to match the curve of the conning tower.. The conning tower had to be drilled and tapped for bolts to secure the hinges. The yard had to drill three drill holes for the door bolts in each door and matching holes in the conning tower. The yard also had to drill holes for the door bolt pinion shaft that extended and retracted the bolts.

Photograph 7.2: *USS Missouri in 1986*. This view clearly shows how the unarmored Flag Level of the conning tower has no vision slits. (Nat'l Archives)

Figure 7.1: A rendering of the armored conning tower of USS New Jersey. The conning tower consists of five 17.3-inch side plates, a 7.25" top plate, a 4-inch bottom plate, and a 16-inch thick tube in two section that links the conning tower to the citadel at second deck.

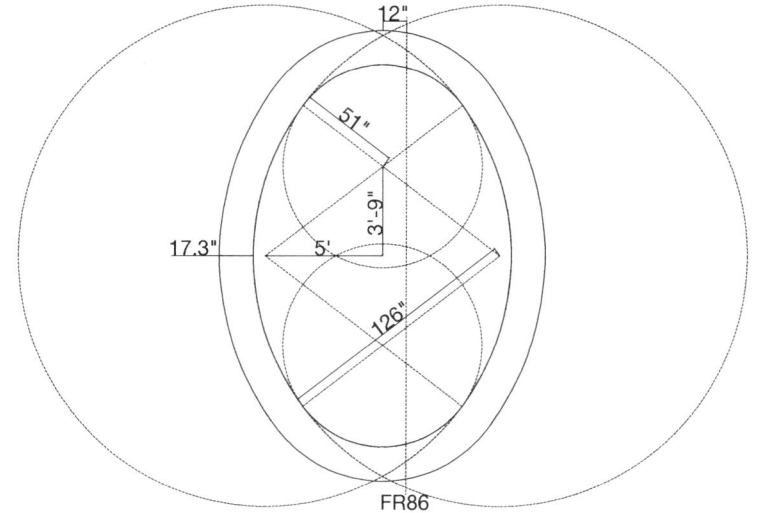

Figure 7.2: Conning Tower Layout (1:96). The conning tower plan consists of four circular arcs. The segments of the original curved bridges on USS Iowa and USS New Jersey share their centers with those the same arc centers at the conning tower armor.

Photograph 7.3: USS Iowa Conning Tower in 2022. The yard machined the edge of the bottom plate to be flush with the conning tower foundation and the side armor plates. (Jim Kurrasch)

Top of
Conning
Tower
96'-8" ABL

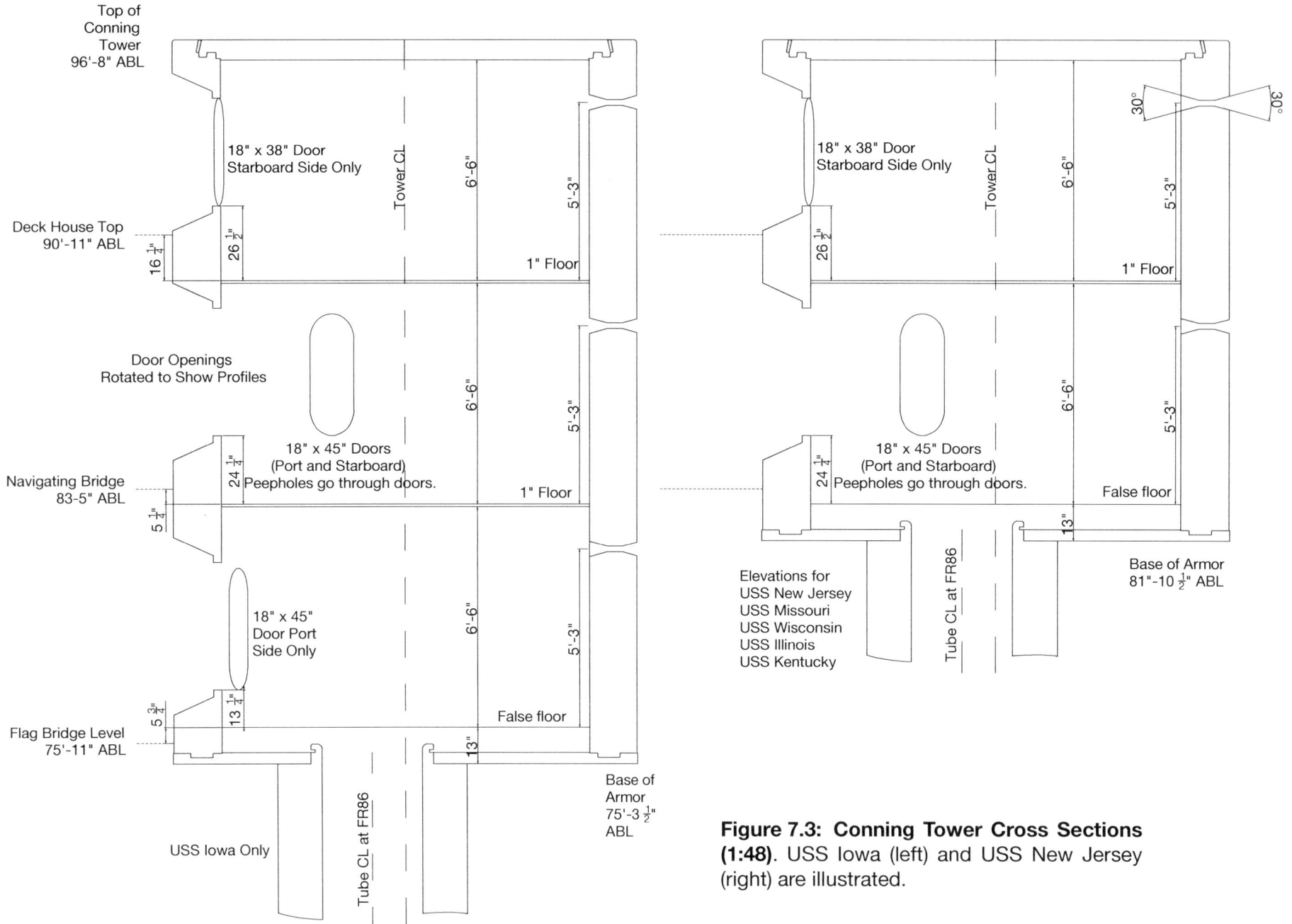

18" x 38" Door
Starboard Side Only

Tower CL

6'-6"

5'-3"

Deck House Top
90'-11" ABL

16 $\frac{1}{4}$"

26 $\frac{1}{2}$"

1" Floor

Door Openings
Rotated to Show Profiles

6'-6"

5'-3"

18" x 45" Doors
(Port and Starboard)
Peepholes go through doors.

1" Floor

Navigating Bridge
83-5" ABL

24 $\frac{1}{4}$"

5 $\frac{1}{4}$"

6'-6"

5'-3"

18" x 45"
Door Port
Side Only

5 $\frac{3}{4}$"

13 $\frac{1}{4}$"

False floor

Flag Bridge Level
75'-11" ABL

13"

Base of
Armor
75'-3 $\frac{1}{2}$"
ABL

USS Iowa Only

Tube CL at FR86

18" x 38" Door
Starboard Side Only

Tower CL

6'-6"

5'-3"

30°   30°

26 $\frac{1}{2}$"

1" Floor

6'-6"

5'-3"

18" x 45" Doors
(Port and Starboard)
Peepholes go through doors.

24 $\frac{1}{4}$"

False floor

13"

Base of Armor
81"-10 $\frac{1}{2}$" ABL

Elevations for
USS New Jersey
USS Missouri
USS Wisconsin
USS Illinois
USS Kentucky

Tube CL at FR86

**Figure 7.3: Conning Tower Cross Sections
(1:48).** USS Iowa (left) and USS New Jersey
(right) are illustrated.

Figure 7.4: Renderings of the conning tower and conning tower foundation for USS Iowa (left) and USS New Jersey (right) viewed looking forward. USS Missouri and USS Wisconsin are similar to the latter. The pyramidal foundation rests on second deck and holds the conning tower upright. It is made of 25# STS. The foundation passes upwards through main deck and emerges at the 01 level (see Photograph 7.2). The pyramid truncates at the 03 level. On USS Iowa, this is where the armored conning tower starts. On the other ships the foundation continues to the 03 level. The areas within the foundation at the 02 and 03 levels were largely empty except for the door opening machinery. The extensions sticking out the back are ends of armored trunks (100# STS) that connected to the forward directors. Cables from the directors ran through these trunks and then passed through holes in the conning tower tube (see Figure 7.55). The openings in the director tube are just below the bottom of the conning tower. On USS Iowa the end of the trunk curves downward to those openings. On the other ships the trunk curves upwards. During the 1980's an additional armored trunk was added to the inside of the aft face of the foundation.

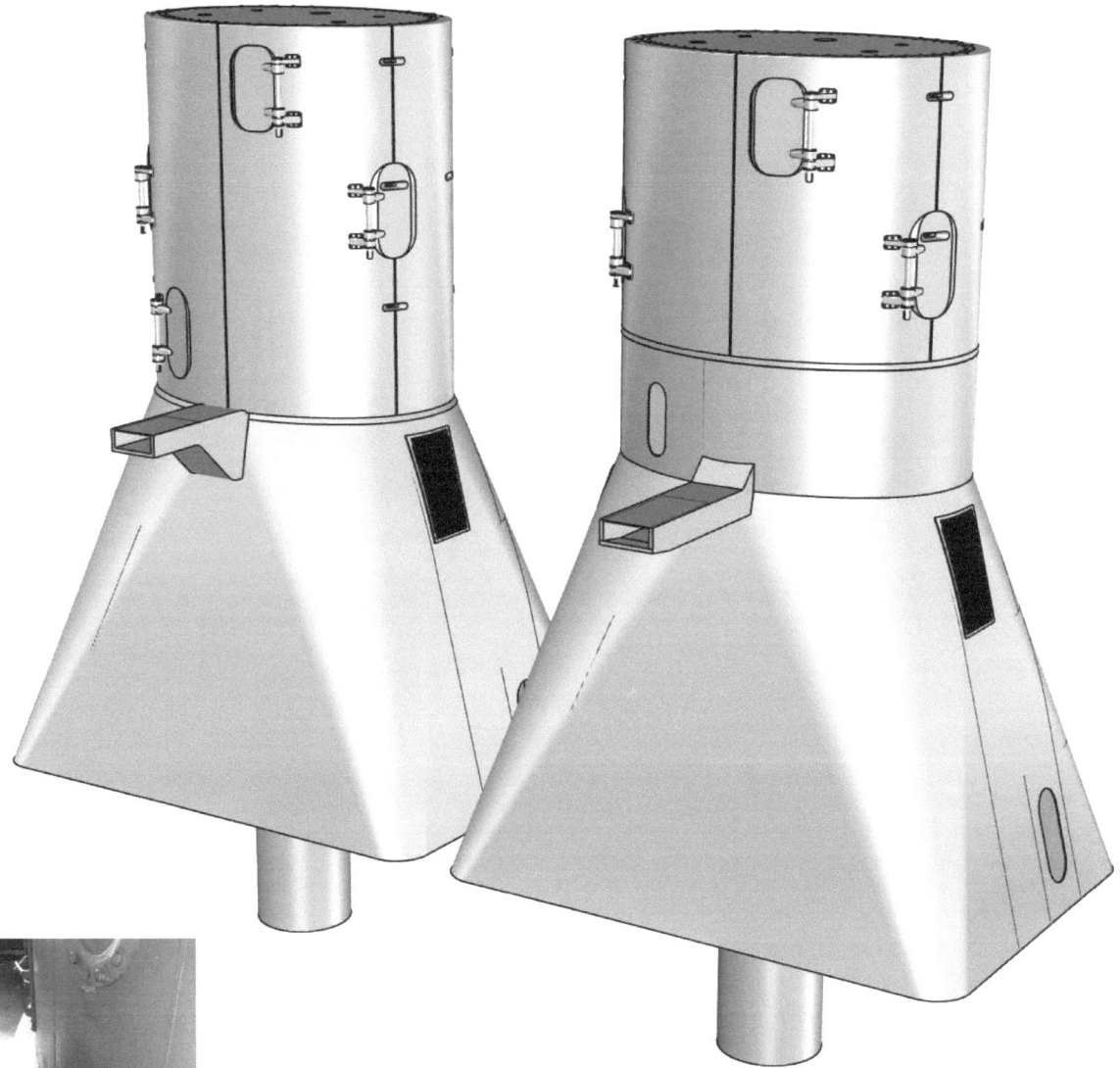

Photograph 7.4: Connection of Director Trunk to Conning Tower Foundation. This is the Flag Level of USS New Jersey. The armored trunk connecting to the conning tower slopes up to reduce cable chaffing.

Center of Door Openings: 93'-4 $\frac{1}{4}$" ABV BL
Center of Peepholes:94'-9 $\frac{3}{4}$" ABV BL

9" Peephole

C-T-2

12" Peephole

29°   44°

C-T-4

28° 45'

14" Peephole

11°

38" Door

C-T-1

C-T-5

28° 45'

29°   44°

12" Peephole

C-T-3

9" Peephole

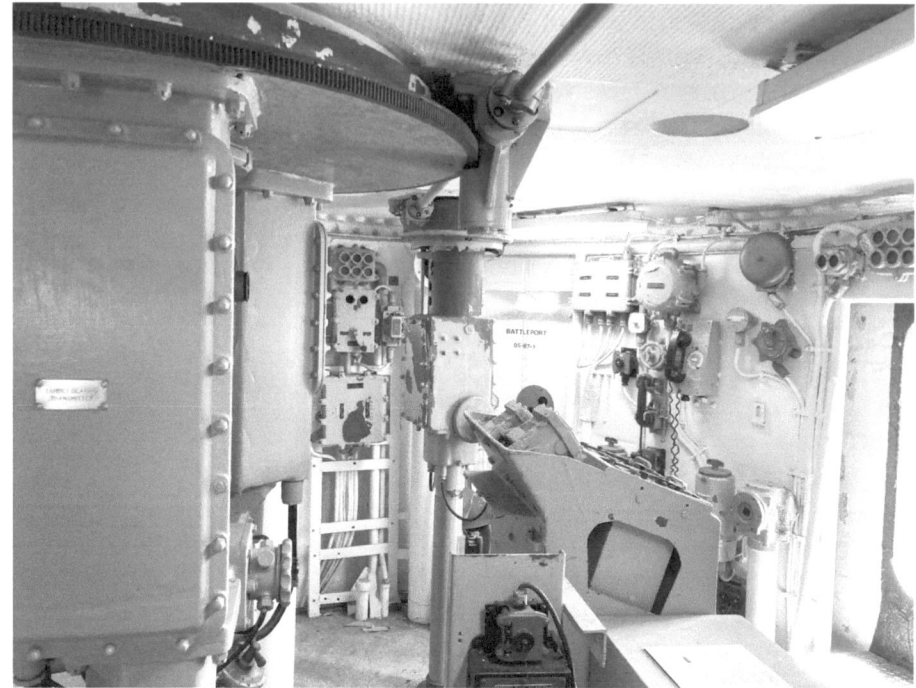

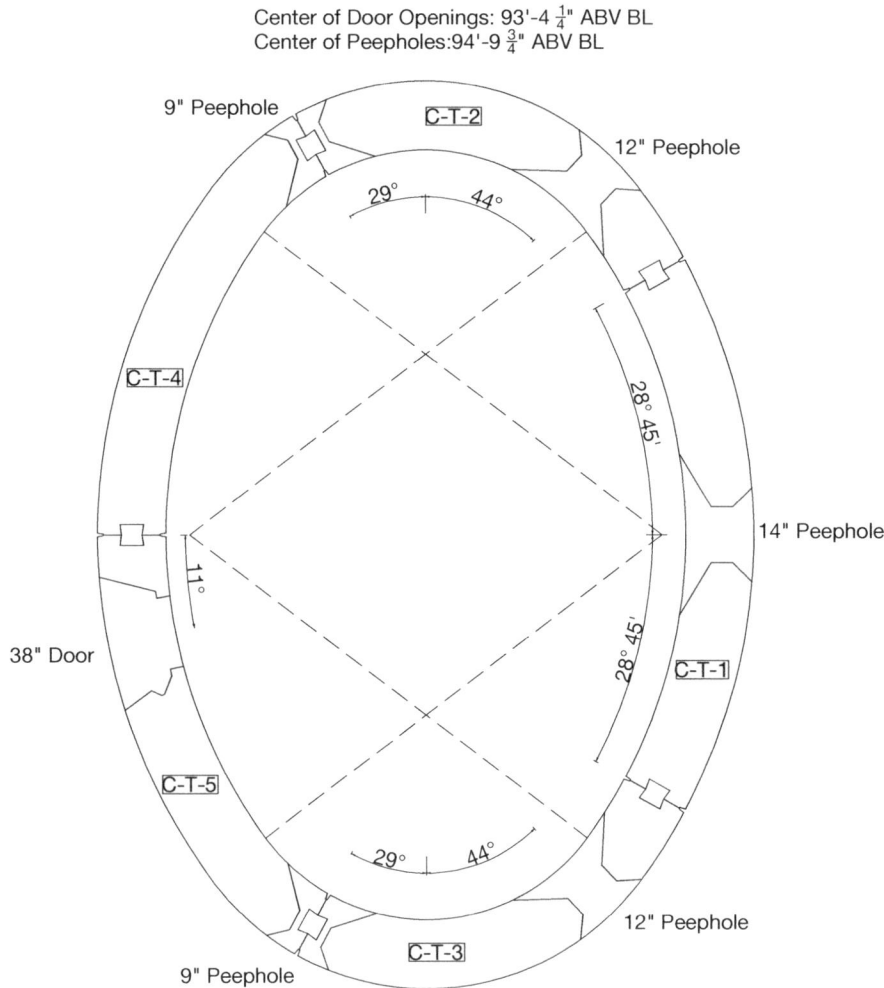

Figure 7.5: Layout of Fire Control Level (05) Openings (1:48). This figure shows the locations of doors, peepholes, and plate division.

Photograph 7.5: The Fire Control level of USS New Jersey looking to starboard. The fire control level houses the Mk.40 Director. The Mk.40 Director was intended for close in fighting if the less protected Mk.38 Directors had been put out of action. Unlike the Mk.38 Director, the Mk.40 did not have its own ranging equipment. It had to rely on the rangefinders in the turret or use its own radar repeater linked to an external radar. The Mk.40 could provide spotting, train, level, and cross level information to the fire control system. One of the two Mk.30 Periscopes is at the center background. The Mk.32 Periscope is at the left. A circular opening created for a World War II radar and later plated over is visible in the top plate.

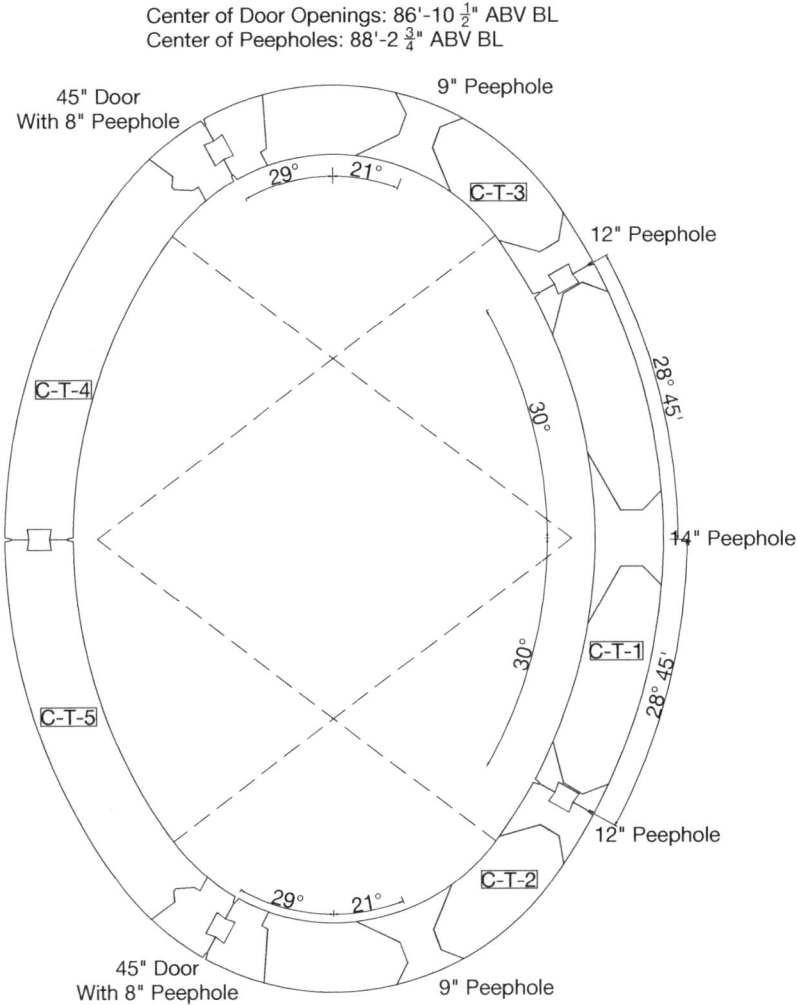

Center of Door Openings: 86'-10 $\frac{1}{2}$" ABV BL
Center of Peepholes: 88'-2 $\frac{3}{4}$" ABV BL

45" Door
With 8" Peephole

9" Peephole

29°   21°

C-T-3

12" Peephole

C-T-4

28° 45'

30°

14" Peephole

30°

C-T-1

28° 45'

C-T-5

12" Peephole

29°   21°

C-T-2

45" Door
With 8" Peephole

9" Peephole

**Figure 7.6: Layout of Navigation Level (04)
Openings (1:48).**

Photograph 7.6: Navigation level of USS New Jersey. The steering
stand is at the center. The door crank is attached to the manual door
opening/closing location.

**Photograph 7.7: Navigation Level of USS
Iowa**. Note the differences in how the deck
meets armor compared with USS New Jersey.
(Jim Kurrasch).

Photograph 7.8: Conning Tower Flag Level on USS Iowa Looking to Starboard. There are periscopes at both sides. The starboard periscope is next to the ladder. There is an communications area behind the partition at the right. Messages could be passed through rotating scuttles. (Jim Kurrasch)

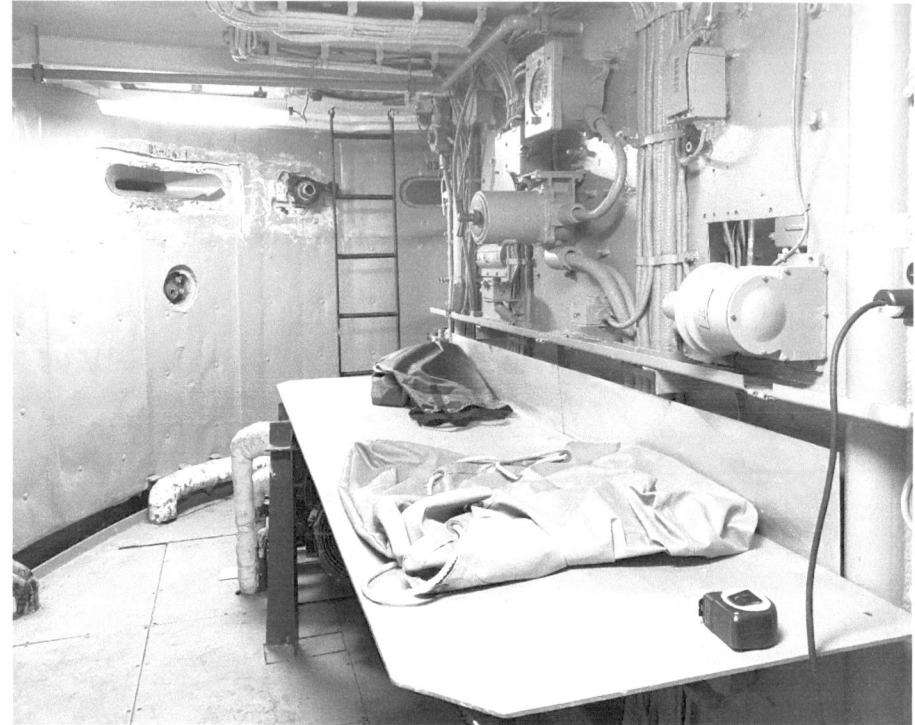

Center of Door Opening: 79'-4 $\frac{1}{2}$" ABL BL
Center of Peepholes: ABV BL 81'-7 $\frac{3}{4}$" ABV BL

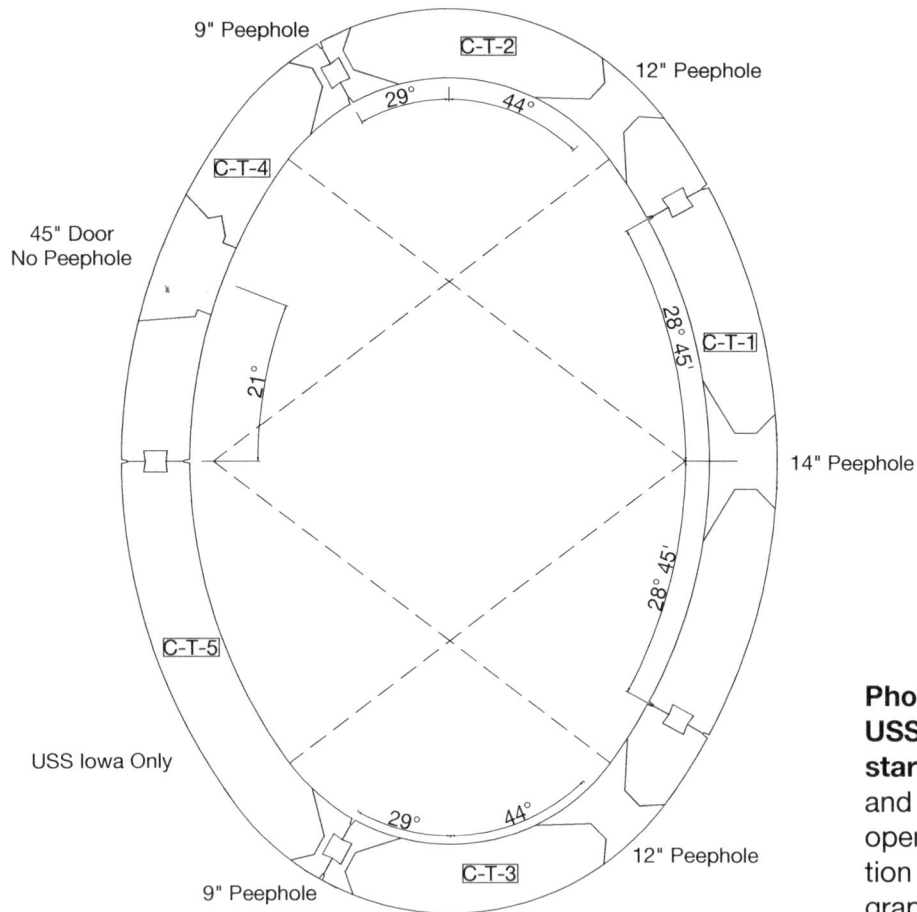

9" Peephole

C-T-2

12" Peephole

29°   44°

C-T-4

45" Door
No Peephole

21°

28° 45'

C-T-1

14" Peephole

28° 45'

C-T-5

USS Iowa Only

29°   44°

12" Peephole

C-T-3

9" Peephole

Figure 7.7: Layout of Flag Level (03) Openings (1:48).

Photograph 7.9: The Flag Level of USS New Jersey looking aft at the starboard side. It is largely empty and unarmored. The mechanism for opening the port door at the navigation level is at the top of the photograph. The airtight door to access this area from the 03 level is visible through the archway.

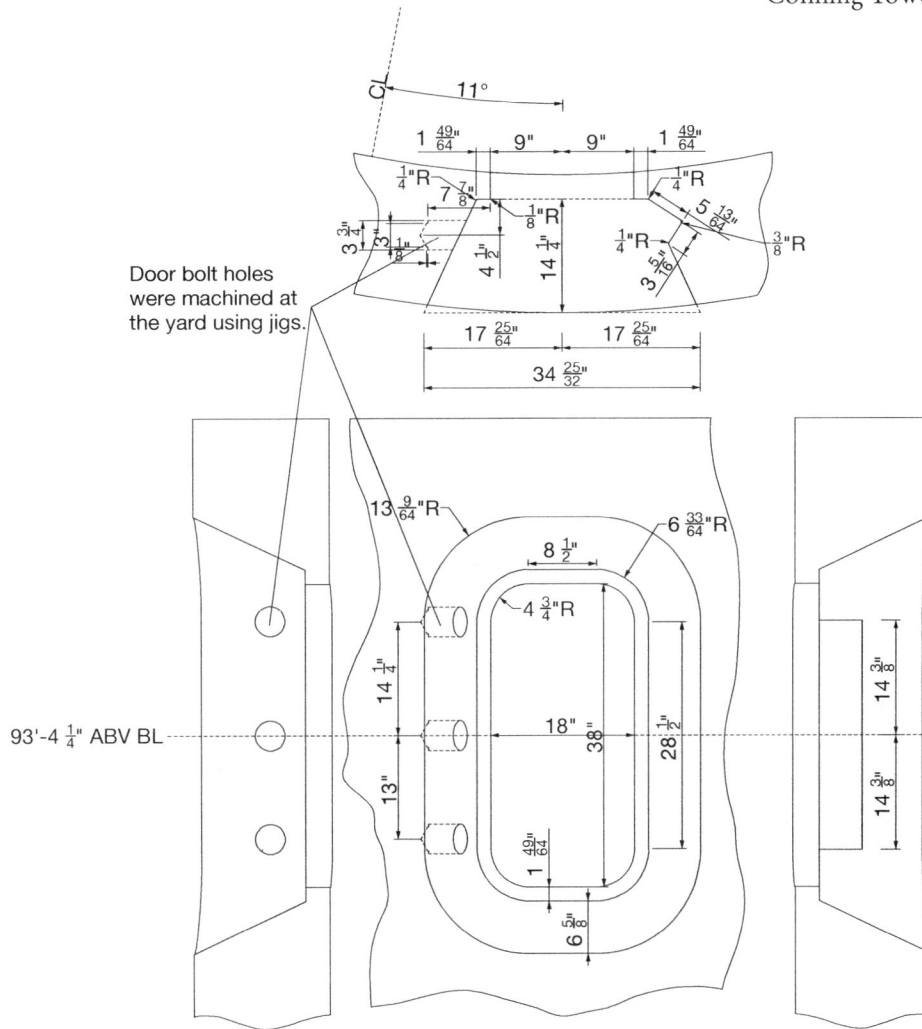

Figure 7.8: 38-Inch Door Opening at the Fire Control Level (1:24). The opening is slightly larger than the doors to allow room for door seals,

Figure 7.9: Fire Control Door. This is the door on USS New Jersey.

Photograph 7.10: Port Navigation Level Door on USS New Jersey. With the door closed and the bolts extended, the slot at the right would hold the door in place if the hinges were destroyed.

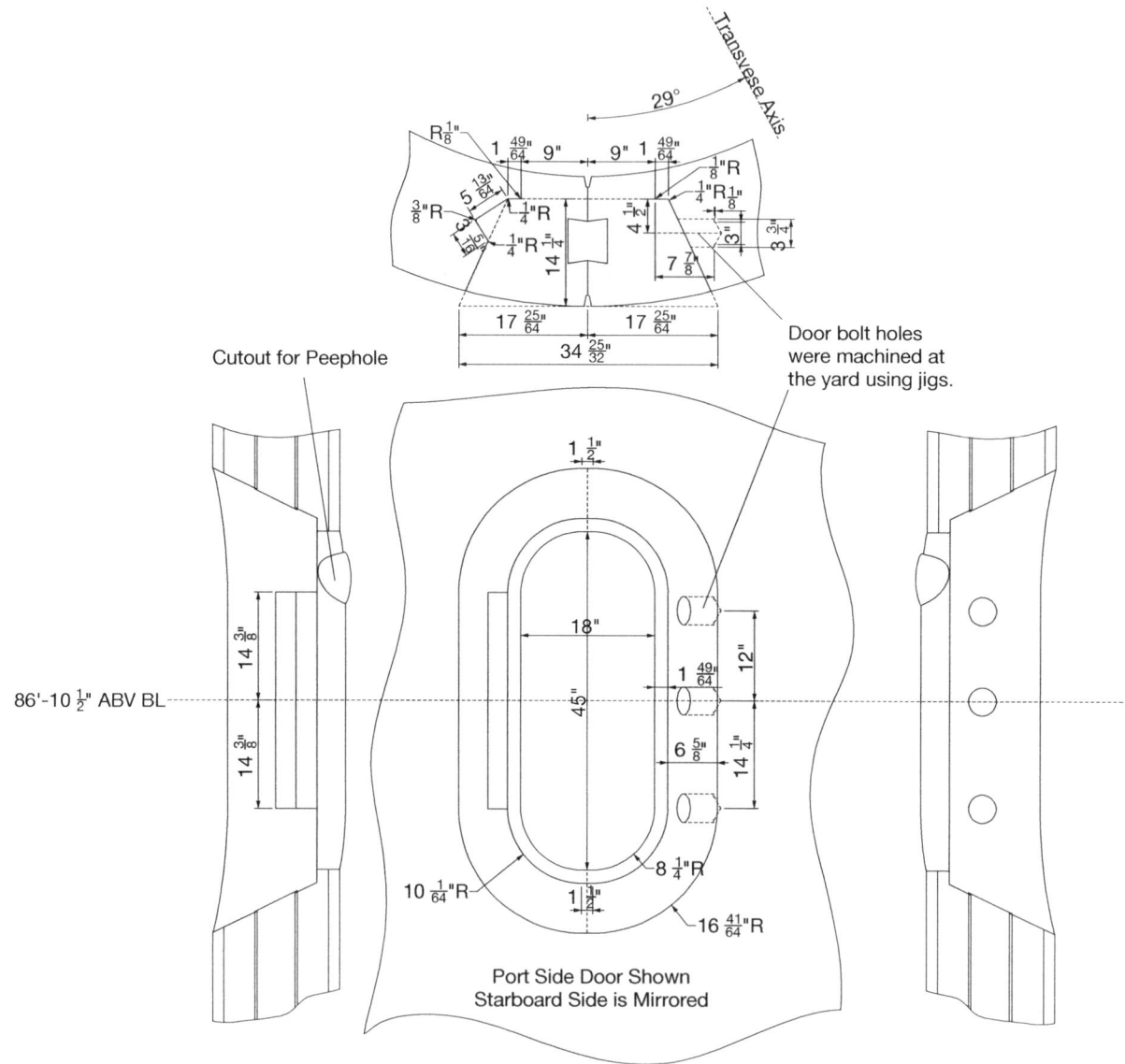

Figure 7.10: 45-Inch Door Openings at the Navigation Level (1:24). The starboard opening is shown. The port opening is a mirror. The doors at the Navigation Level have peepholes. The door peepholes extend from the door into the conning tower sides.

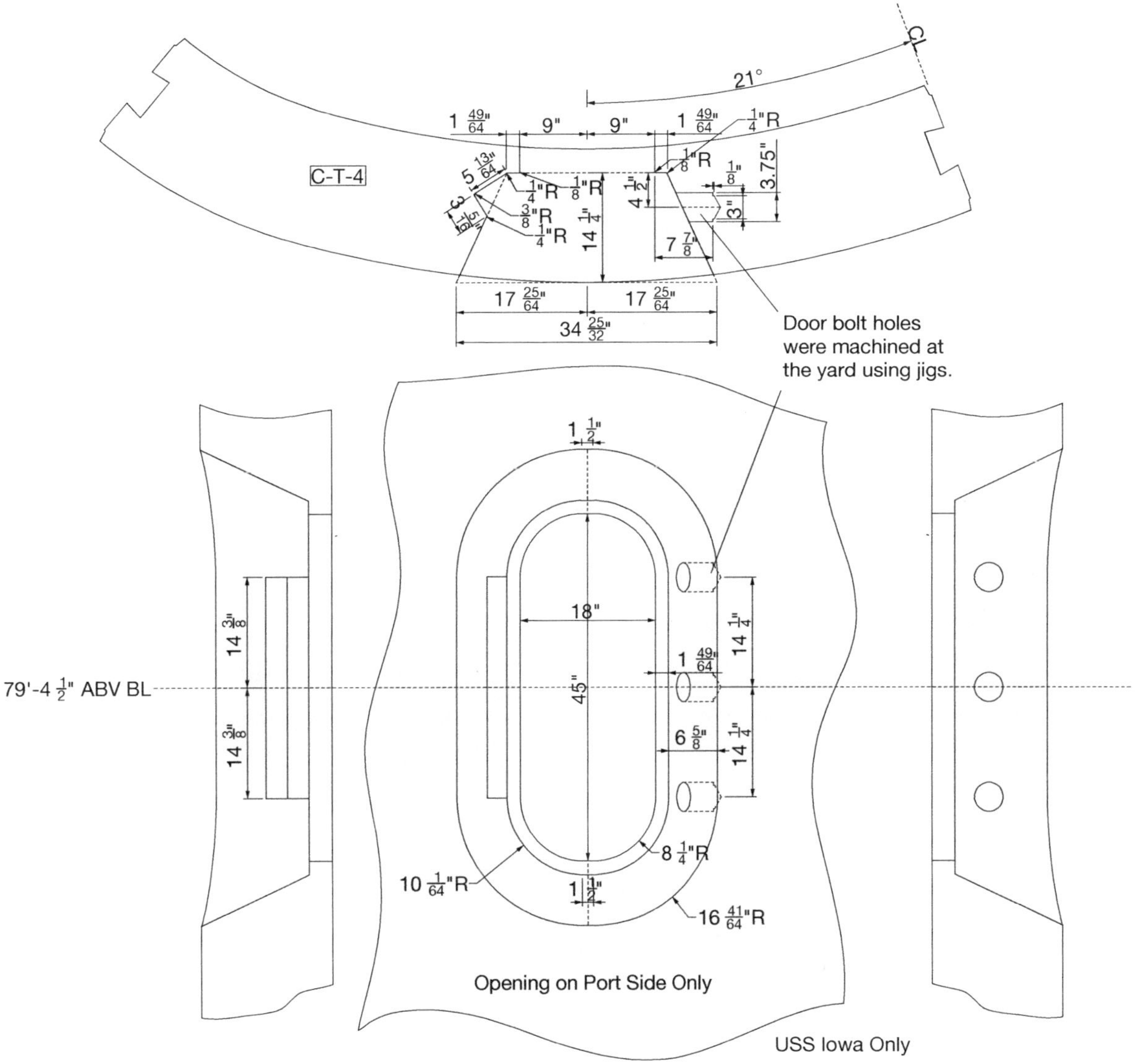

Figure 7.11: 45-Inch Door at Flag Level

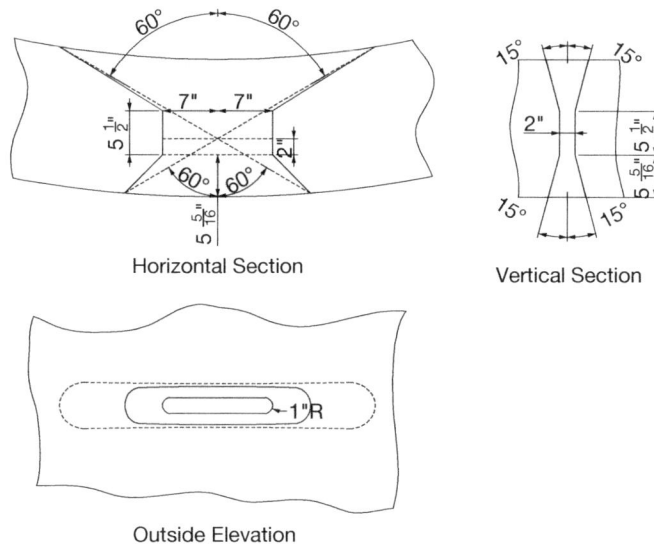

Horizontal Section

Vertical Section

Outside Elevation

Figure 7.12: 14-Inch Peephole (1:24). Each level has a 14-inch peephole facing forward on the centerline.

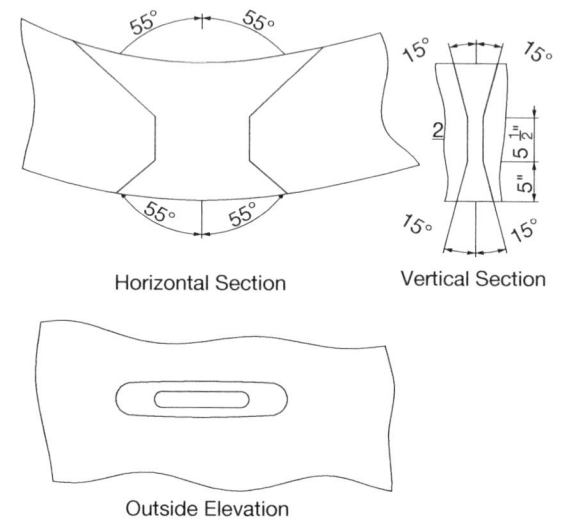

Horizontal Section

Vertical Section

Outside Elevation

Figure 7.13: 12-Inch Peephole (1:24). Each level has two 12-inch peepholes outboard of its 14-inch peephole.

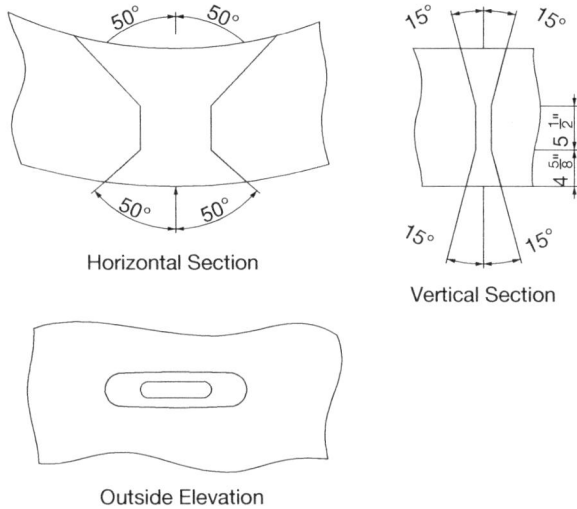

Horizontal Section

Vertical Section

Outside Elevation

Figure 7.14: 9-Inch Peephole (1:24). Each level has two 9-inch peepholes outboard of the 12-inch peepholes.

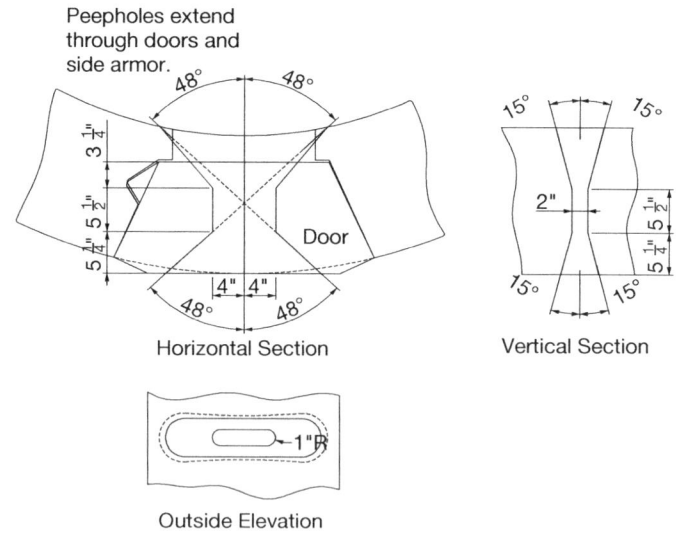

Peepholes extend through doors and side armor.

Door

Horizontal Section

Vertical Section

Outside Elevation

Figure 7.15: 8-Inch Peephole (1:24). These are only found in the doors of the navigation level. The peepholes extend into the side plates to allow a full range of view.

Figure 7.18: Rendering of the Conning Tower Side Plates. This shows the structure of the upper conning tower sides as on USS Iowa and USS New Jersey. The extension of the starboard navigation door peephole is visible. The other ships are similar except for the joint with the top plate.

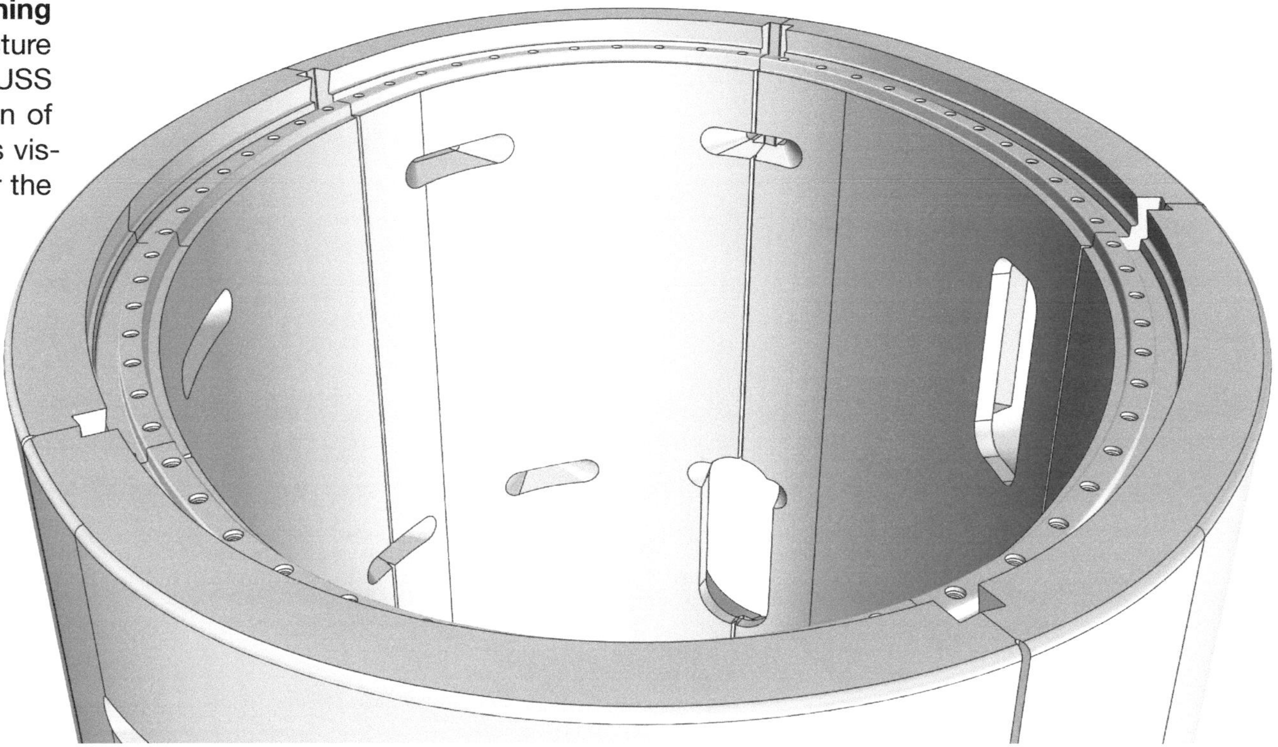

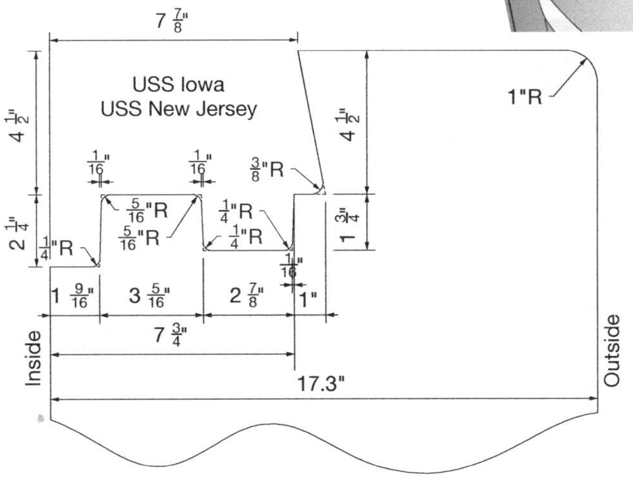

Figure 7.16: Upper Edge of Side Plates (1:6). This complex joint was used on USS Iowa and USS New Jersey to connect the side plates to the top plate.

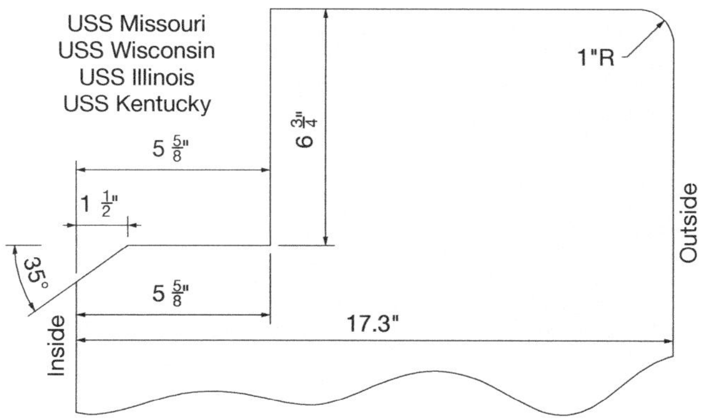

Figure 7.17: Upper Edge of Side Plates (1:6). The side to top plate joint was greatly simplified on the later ships.

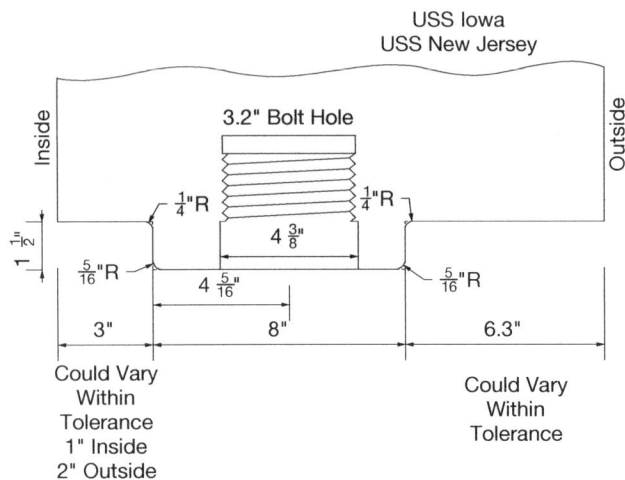

Figure 7.19: Lower Edge of Side Plates (1:6).The side plates connect to the bottom plate with a bolted tongue and grove joint on USS Iowa and USS New Jersey.

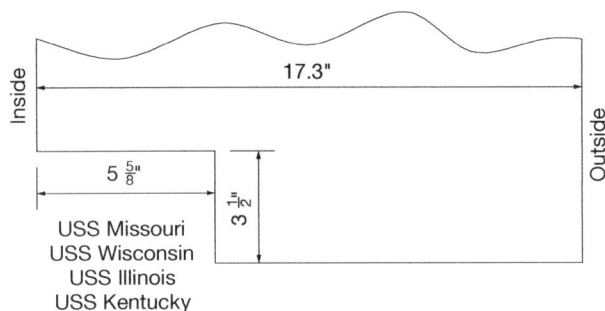

Figure 7.20: Lower Edge of Side Plates (1:6). This diagram shows how the bottom edges of the side plates were simplified to use a welded rabbet on the later ships.

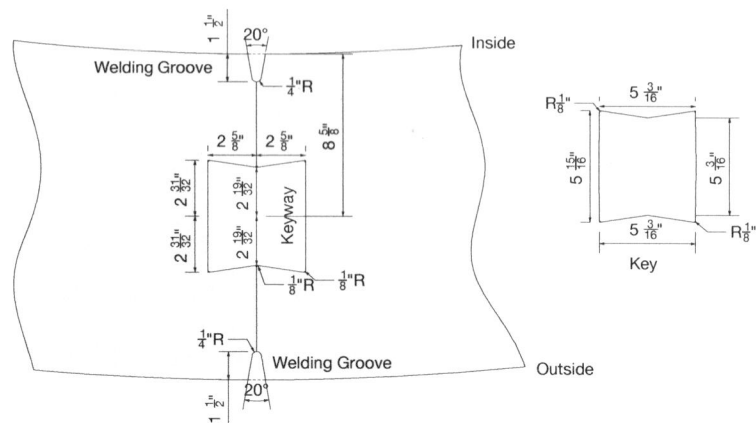

Photograph 7.11: Joint Between Side Plates (1:8). The edges of the side plates were machined with hourglass keyways and welding grooves. Adjoining plates were linked by hammering a key into the keyway and then welding along the grooves. In several places the keys pass through door openings or peepholes. In these areas the keys had to be cut away and grinded flush. The keys are faintly visible in the navigation level doorways.

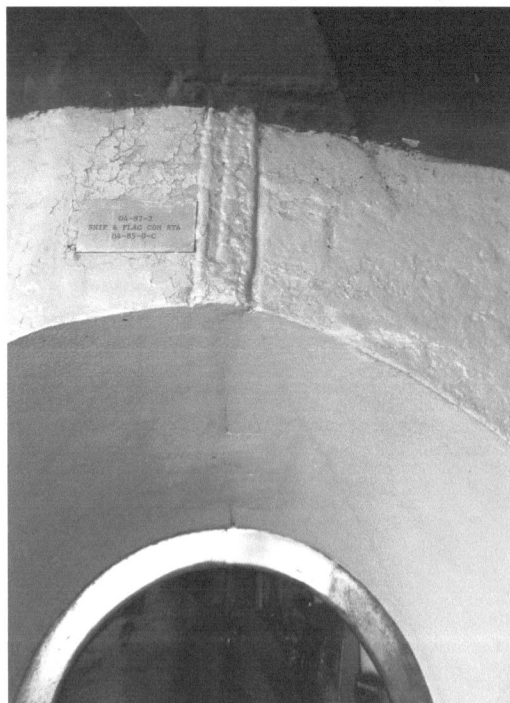

Photograph 7.12: Joint Between Plates. This is a close view of part of a Navigation Level door opening on USS New Jersey showing the joint. The outer weld bead is clearly visible. The hourglass key is faintly visible where it was cut away.

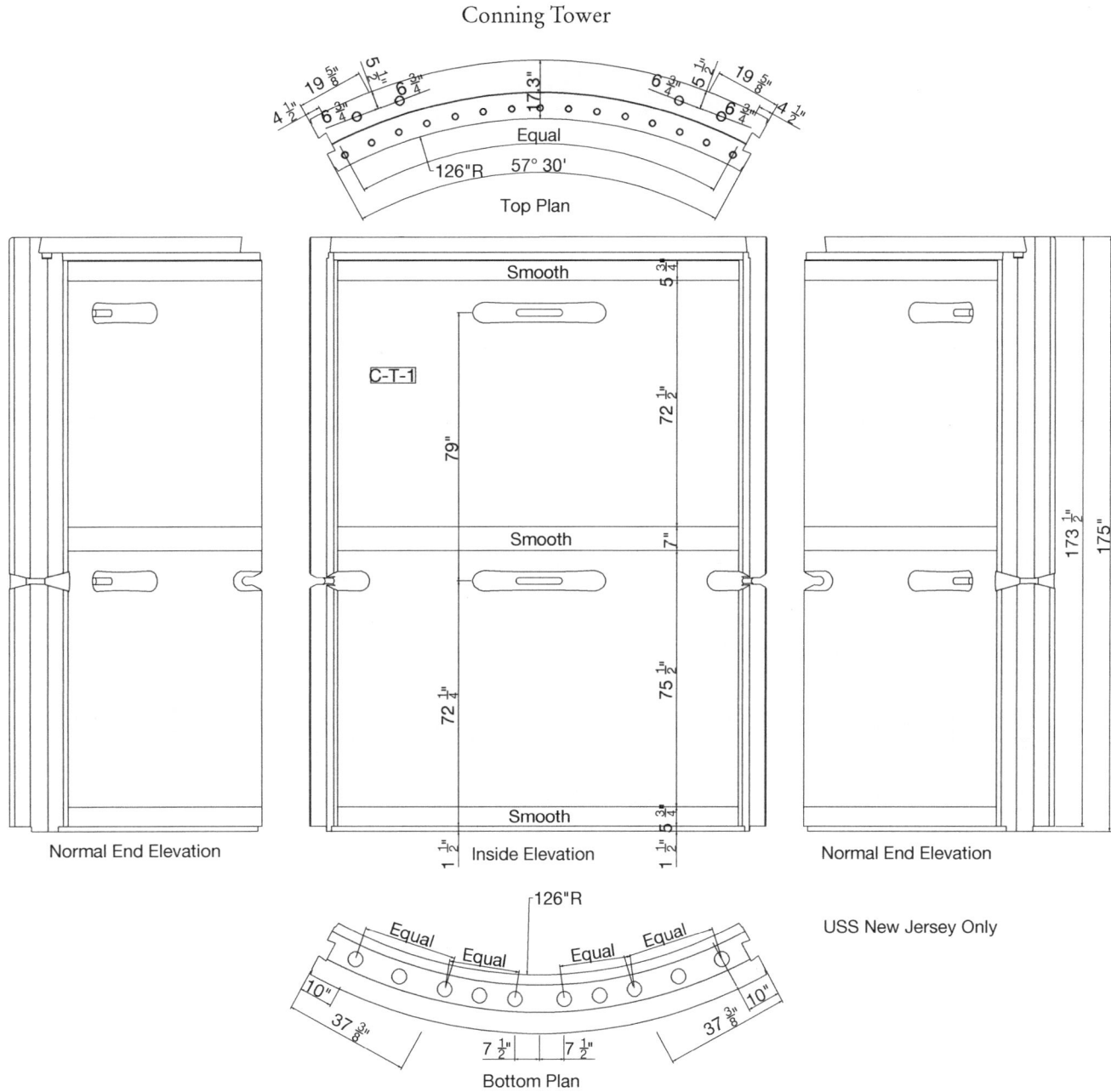

Figure 7.21: Plate C-T-1 on USS New Jersey (1:48). The holes along the bottom edge are tapped for bolts to attach the bottom plate. The openings in the outer lip are for lifting pads. Those on the inner tongue are for tap bolts that secure the top plate. The smooth areas at the top and bottom are for welding scallops to attach the top and bottom plates. The smooth area at the center is for the fire control level supports.

Figure 7.22: Plate C-T-2 on USS New Jersey (1:48). The areas marked smooth on the outside face correspond to the original enclosed bridge location. The bridge was changed on USS New Jersey shortly after commissioning. When a plate extends across two arc segments, the plans indicate the centers of both arcs.

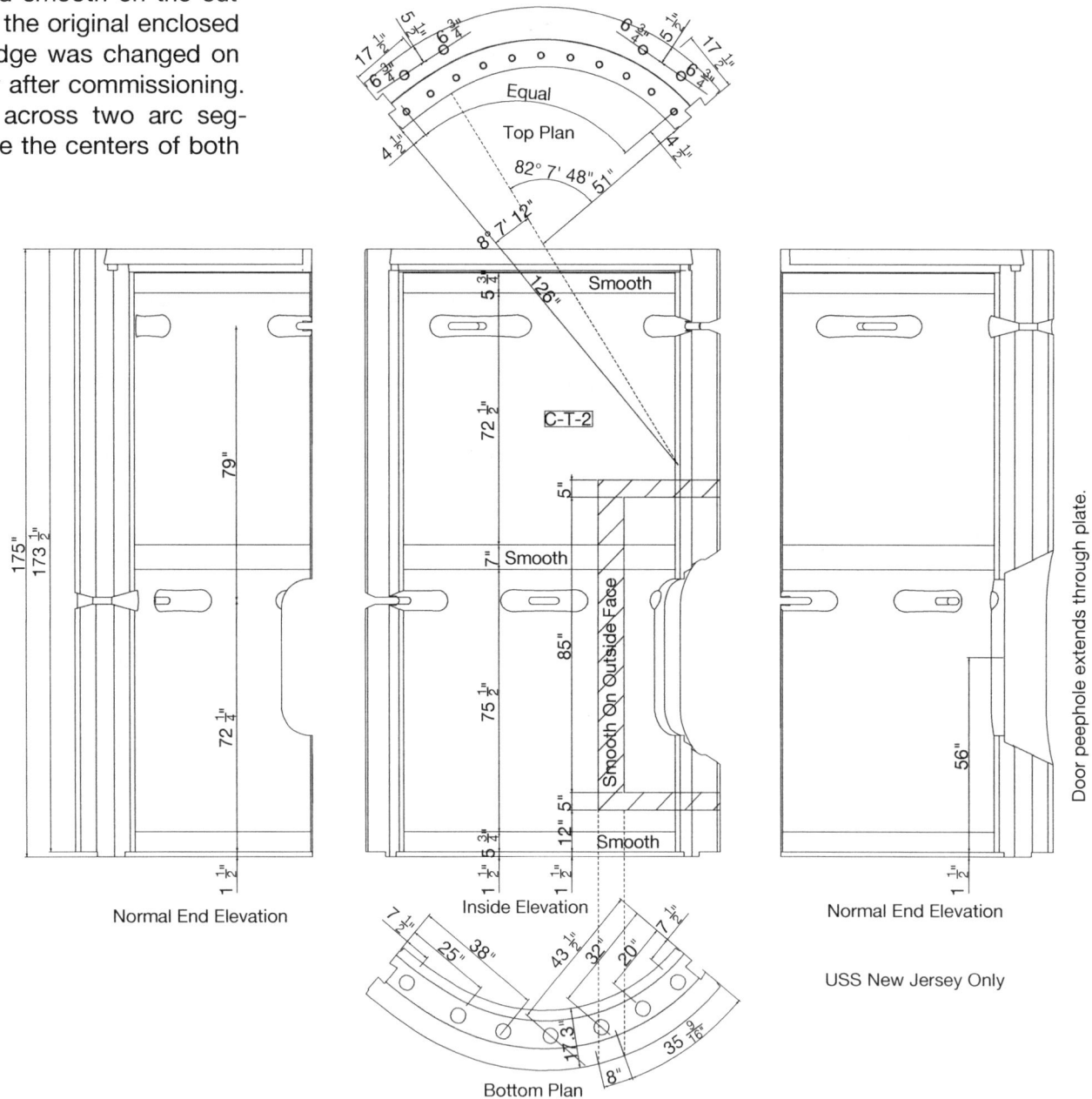

Top Plan

Normal End Elevation

Inside Elevation

Bottom Plan

Normal End Elevation

USS New Jersey Only

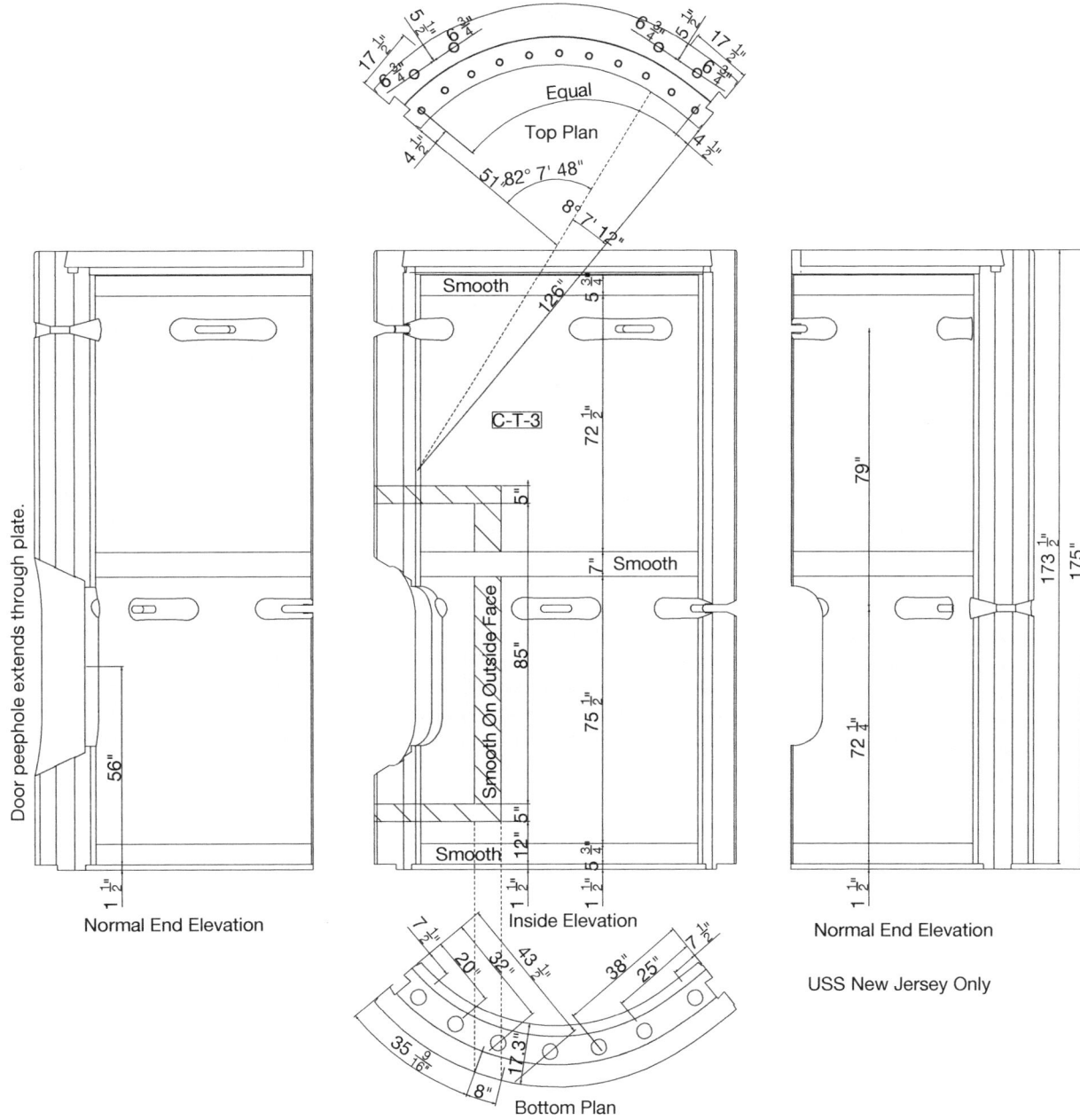

Figure 7.23: Plate C-T-3 on USS New Jersey (1:48)

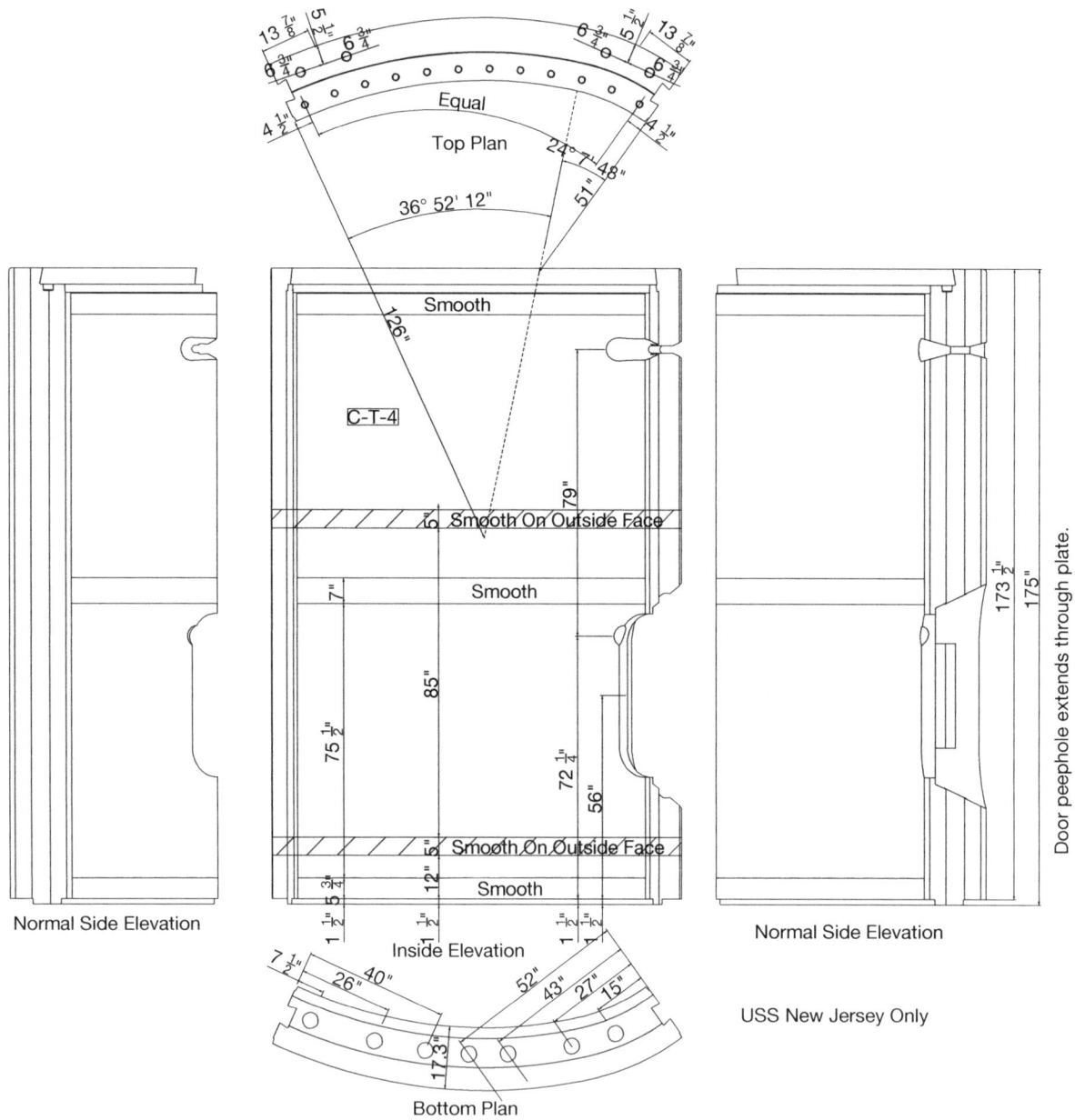

Figure 7.24: Plate C-T-4 on USS New Jersey (1:48)

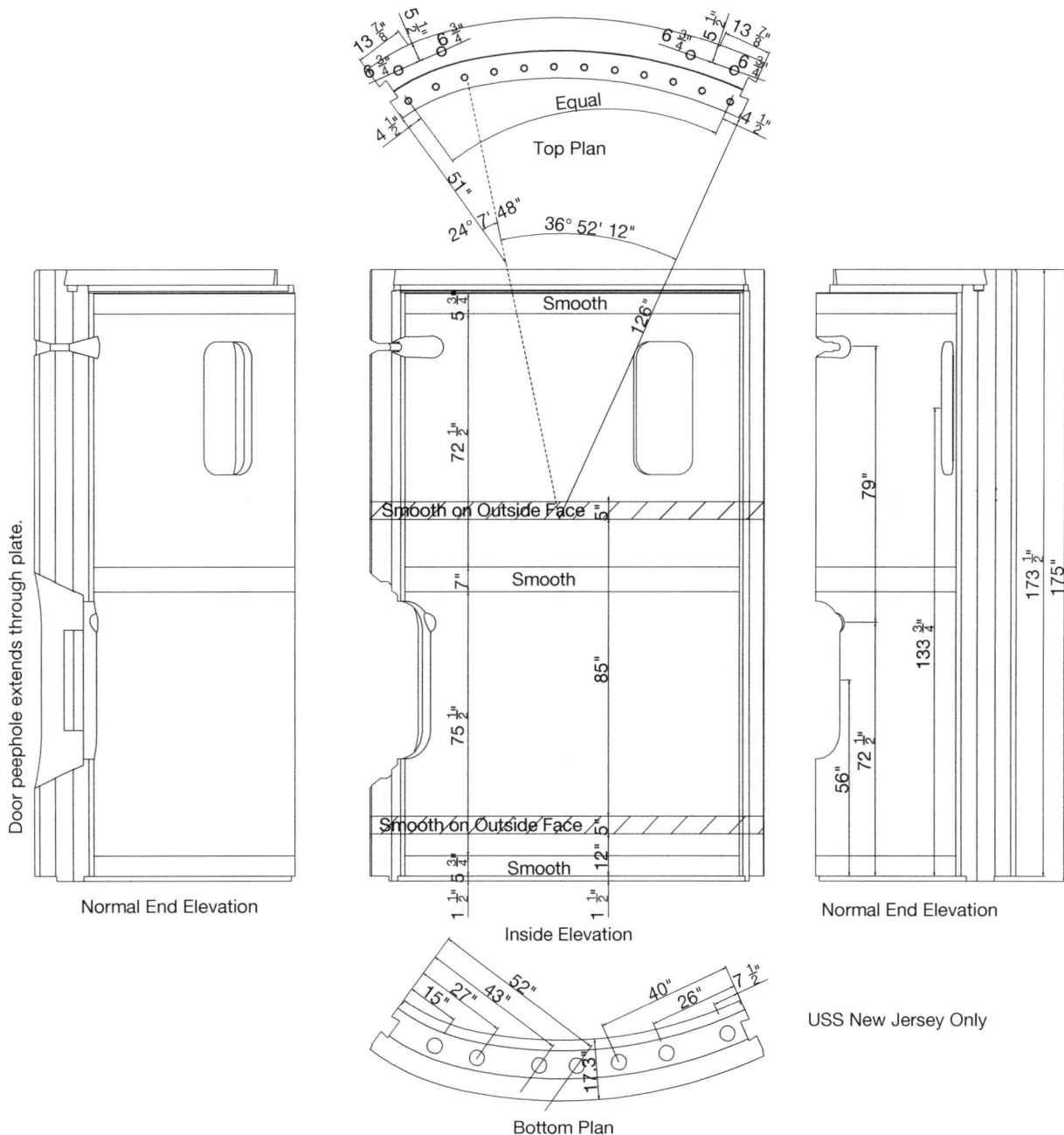

Figure 7.25: Plate C-T-5 on USS New Jersey (1:48)

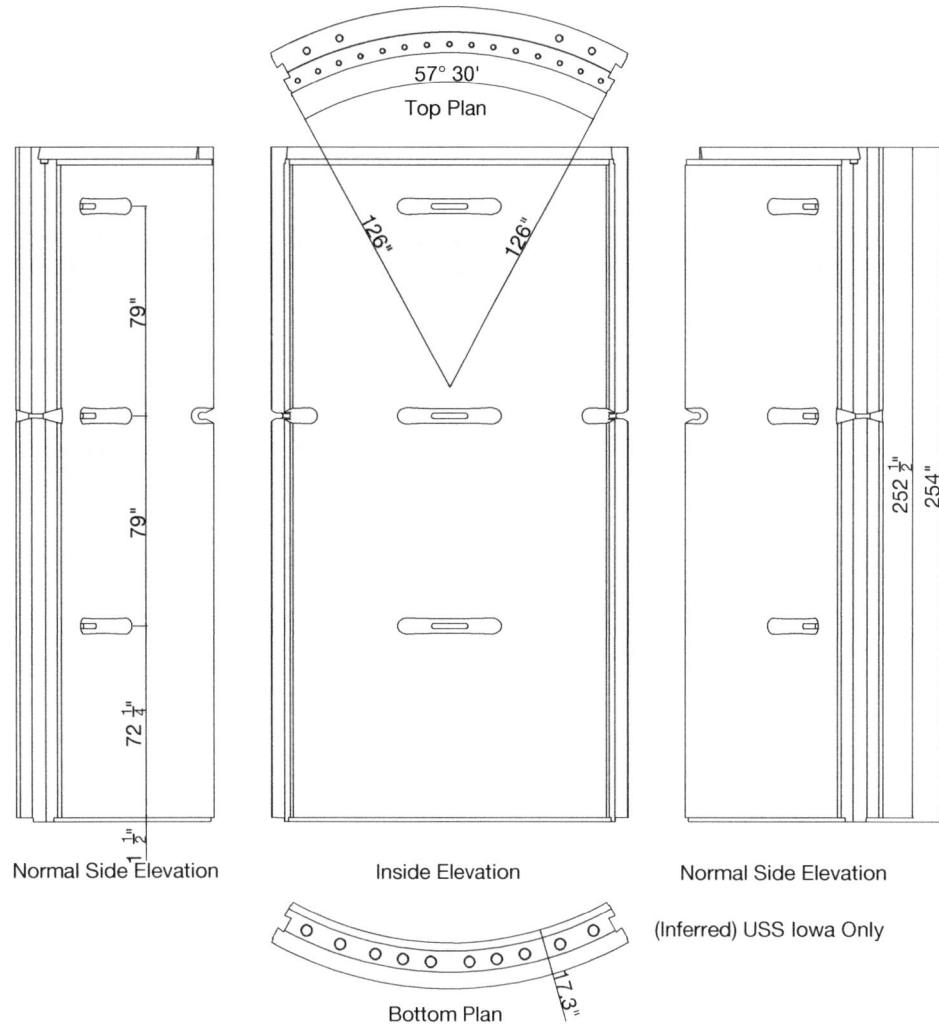

Figure 7.26: Plate C-T-1 on USS Iowa (1:72). These dimensions are inferred from various original plans.

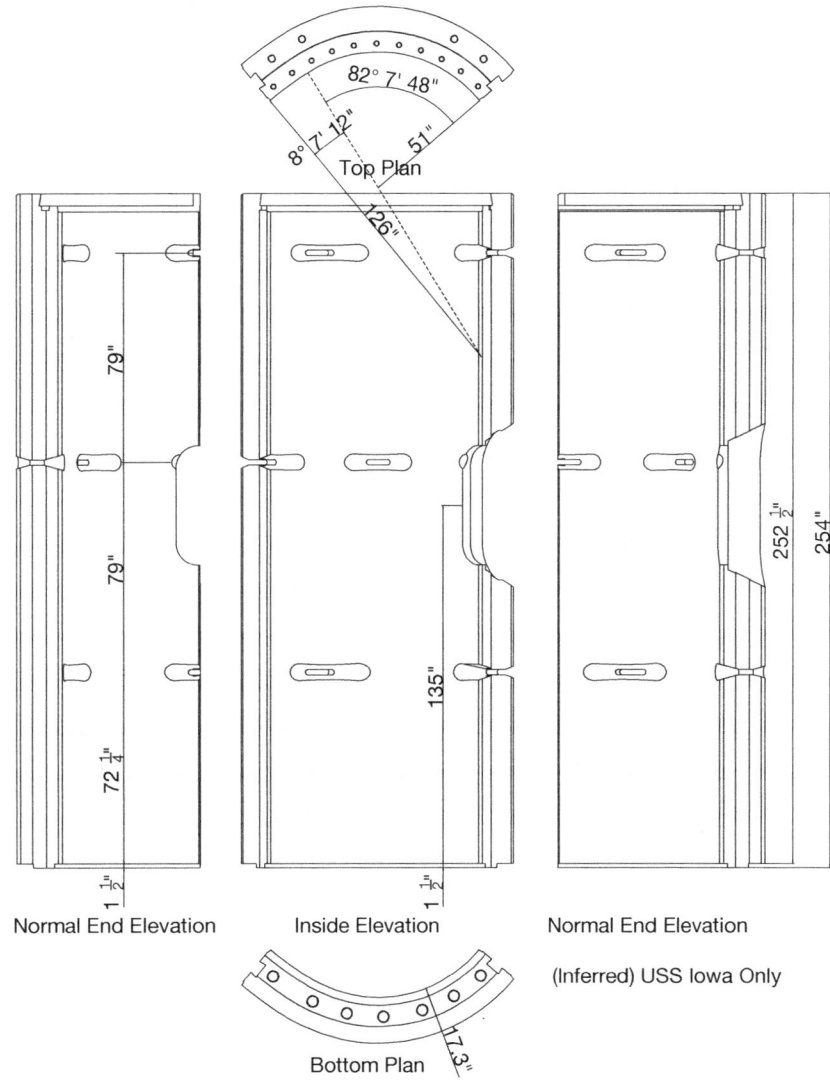

Figure 7.27: Plate C-T-2 on USS Iowa (1:72).

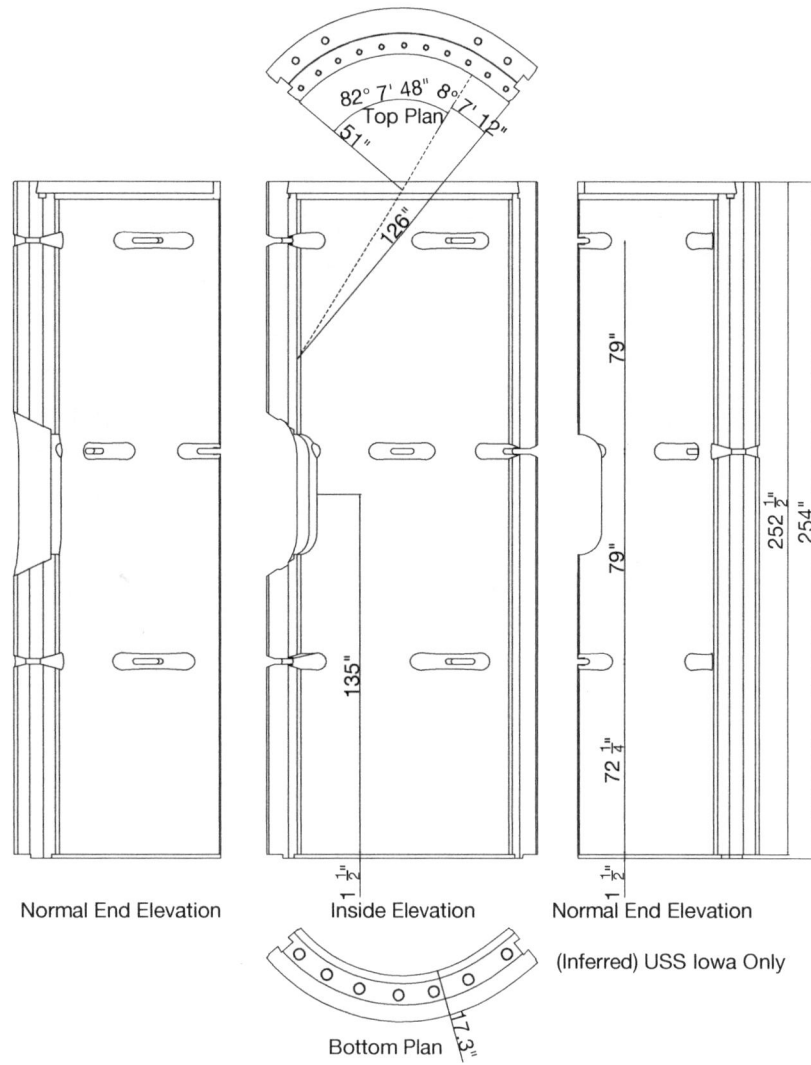

Figure 7.28: Plate C-T-3 on USS Iowa (1:72)

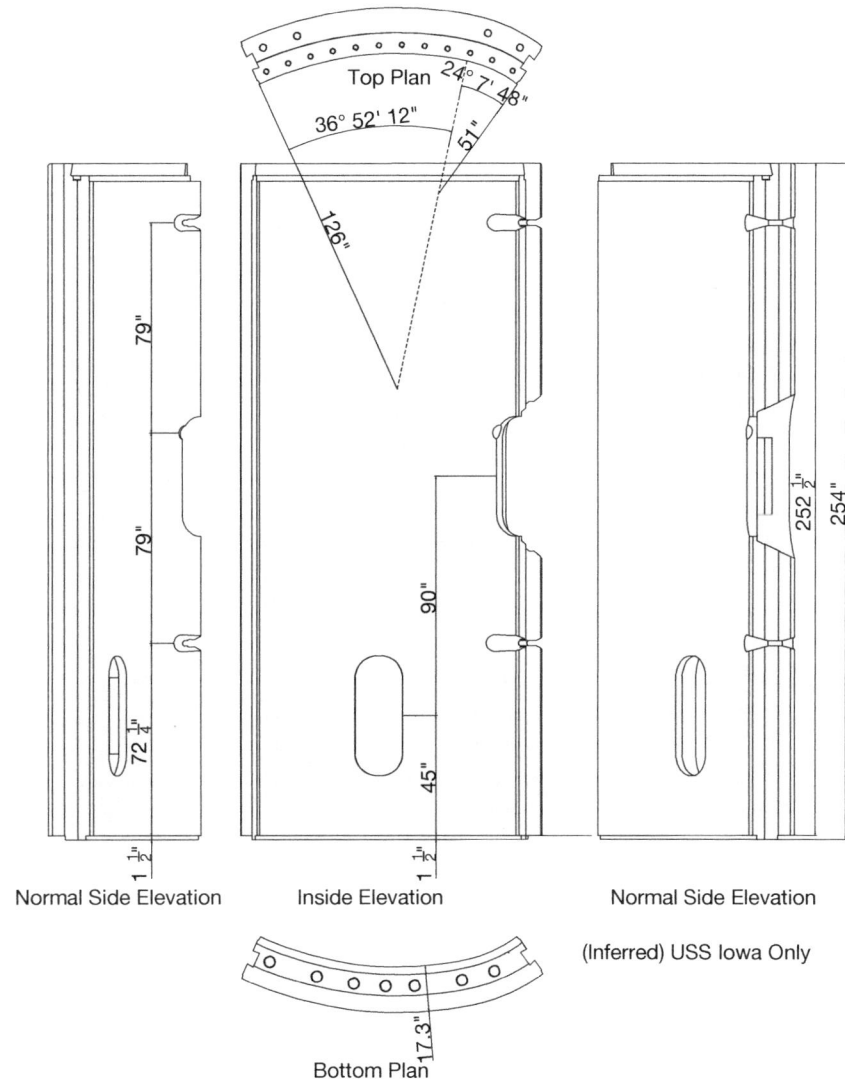

Figure 7.29: Plate C-T-4 on USS Iowa (1:72).

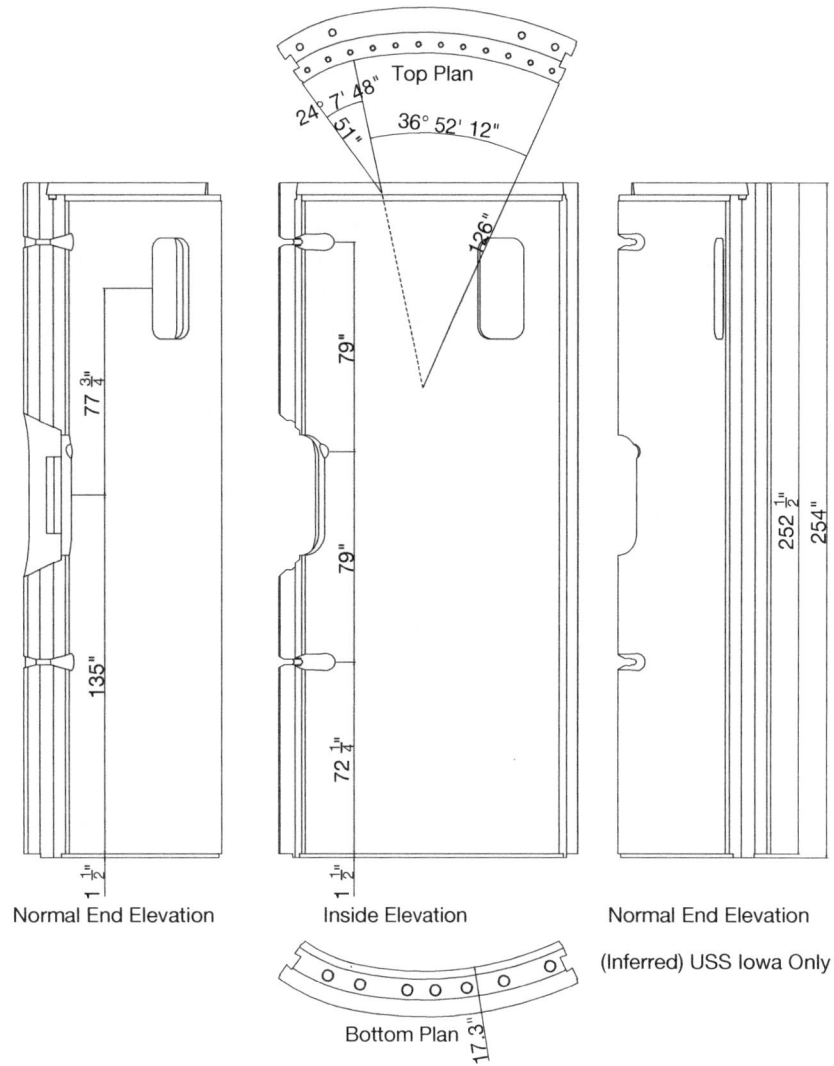

Figure 7.30: Plate C-T-5 on USS Iowa (1:72).

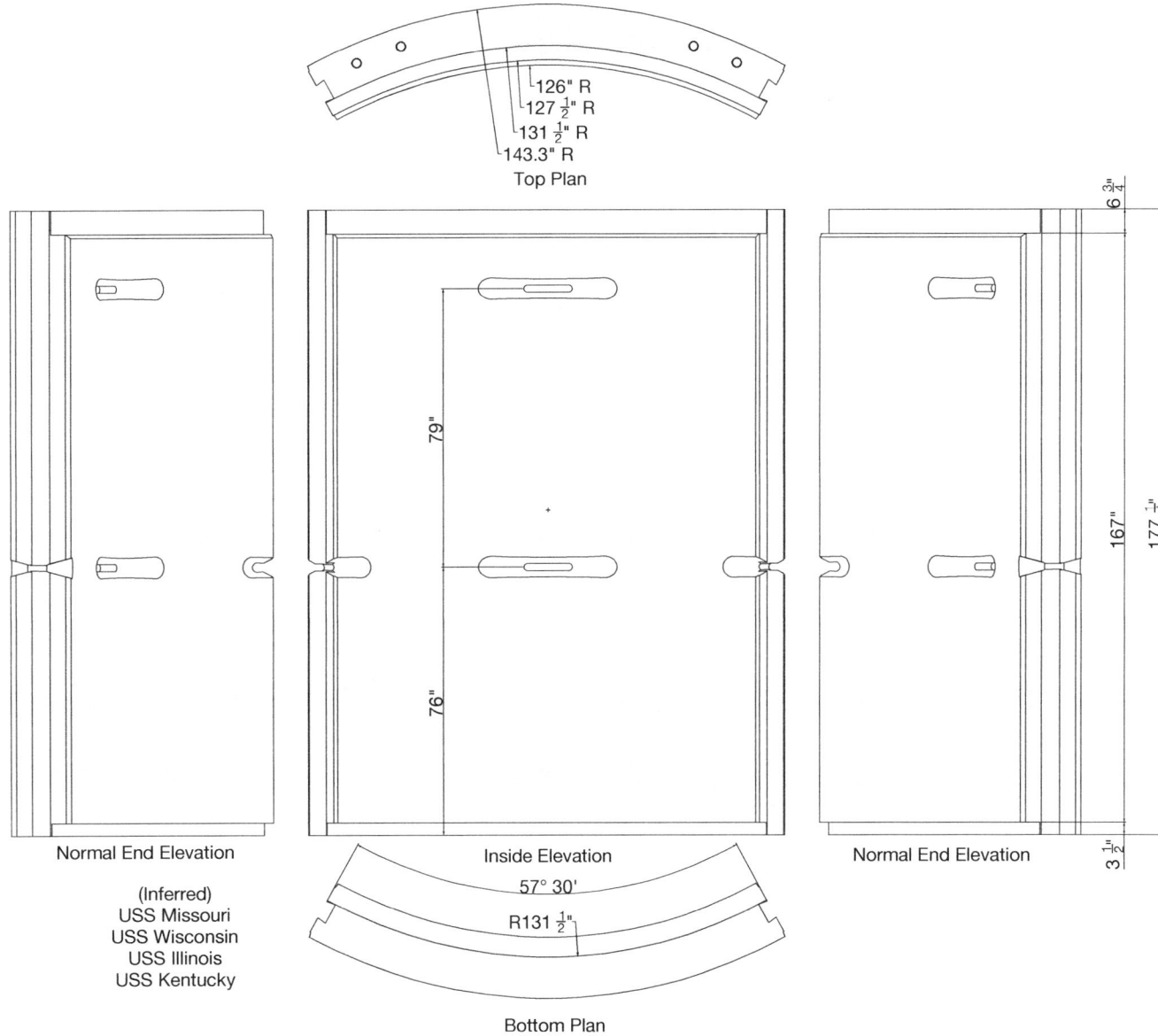

126" R
127 ½" R
131 ½" R
143.3" R

Top Plan

Normal End Elevation

(Inferred)
USS Missouri
USS Wisconsin
USS Illinois
USS Kentucky

79"

76"

Inside Elevation

57° 30'

R131 ½"

Bottom Plan

Normal End Elevation

6 ¾"

167"

177 ¼"

3 ½"

Figure 7.31: Plate C-T-1 on USS Missouri, USS Wisconsin, USS Illinois, and USS Kentucky. These dimensions are inferred from various original plans.

Top Plan

8° 7' 12"

82° 7' 48" 51"

126"

79"

76"

59 3/4"

Normal End Elevation

Inside Elevation

17.3"

Bottom Plan

6 3/4"

167"

177 1/4"

3 1/2"

Normal End Elevation

(Inferred)
USS Missouri
USS Wisconsin
USS Illinois
USS Kentucky

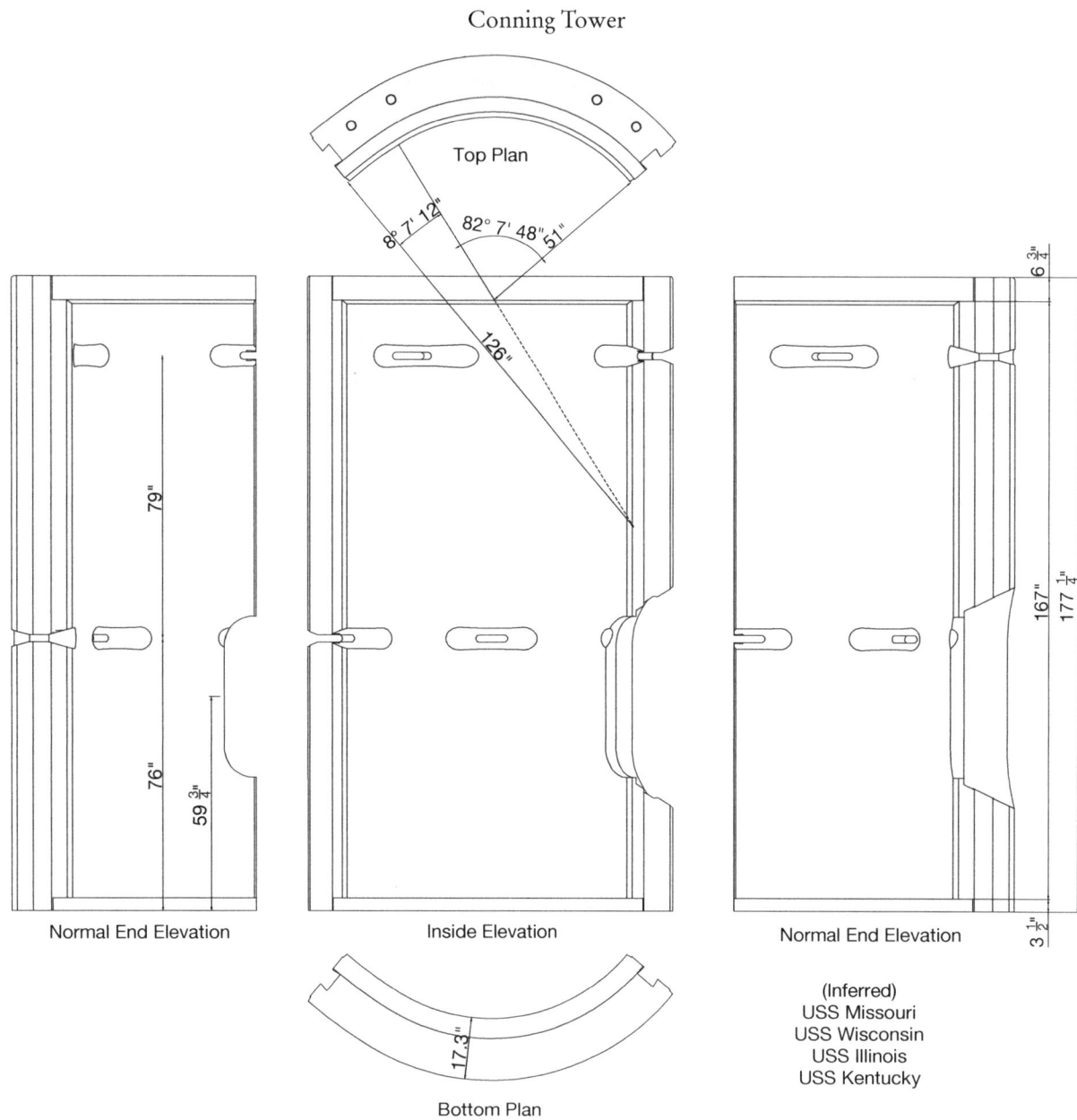

Figure 7.32: Plate C-T-2 on USS Missouri, USS Wisconsin, USS Illinois, and USS Kentucky

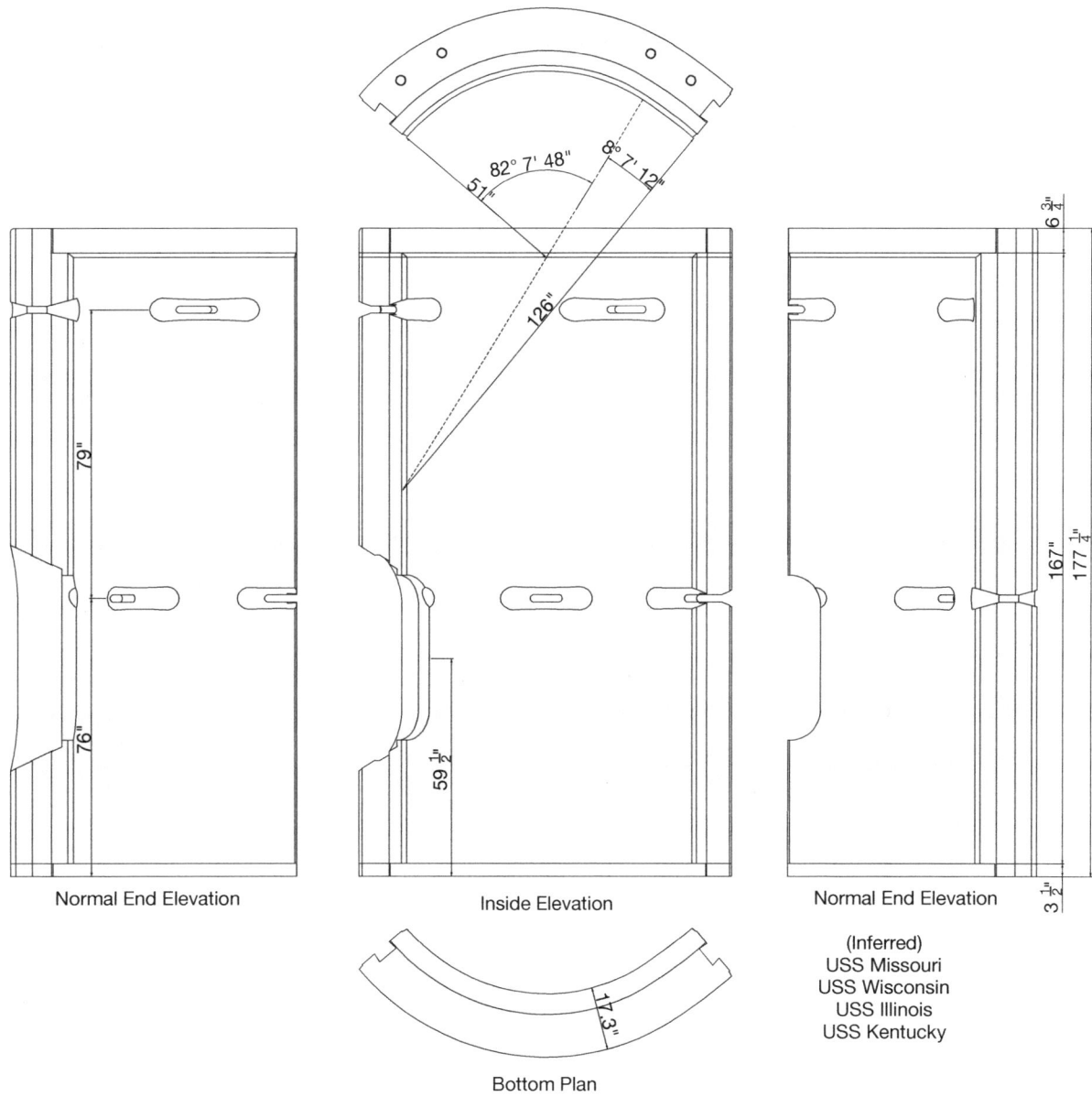

82° 7' 48"

51"

8° 7' 12"

126"

6 3/4"

79"

167"

177 1/4"

76"

59 1/2"

3 1/2"

Normal End Elevation

Inside Elevation

Normal End Elevation

(Inferred)
USS Missouri
USS Wisconsin
USS Illinois
USS Kentucky

17.3"

Bottom Plan

Figure 7.33: Plate C-T-3 on USS Missouri, USS Wisconsin, USS Illinois, and USS Kentucky

Top Plan

24° 7' 48"

36° 52' 12"

51"

126"

6 3/4"

79"

167"

76"

59 3/4"

3 1/2"

Normal End Elevation

Inside Elevation

Normal End Elevation

(Inferred)
USS Missouri
USS Wisconsin
USS Illinois
USS Kentucky

17.3"

Bottom Plan

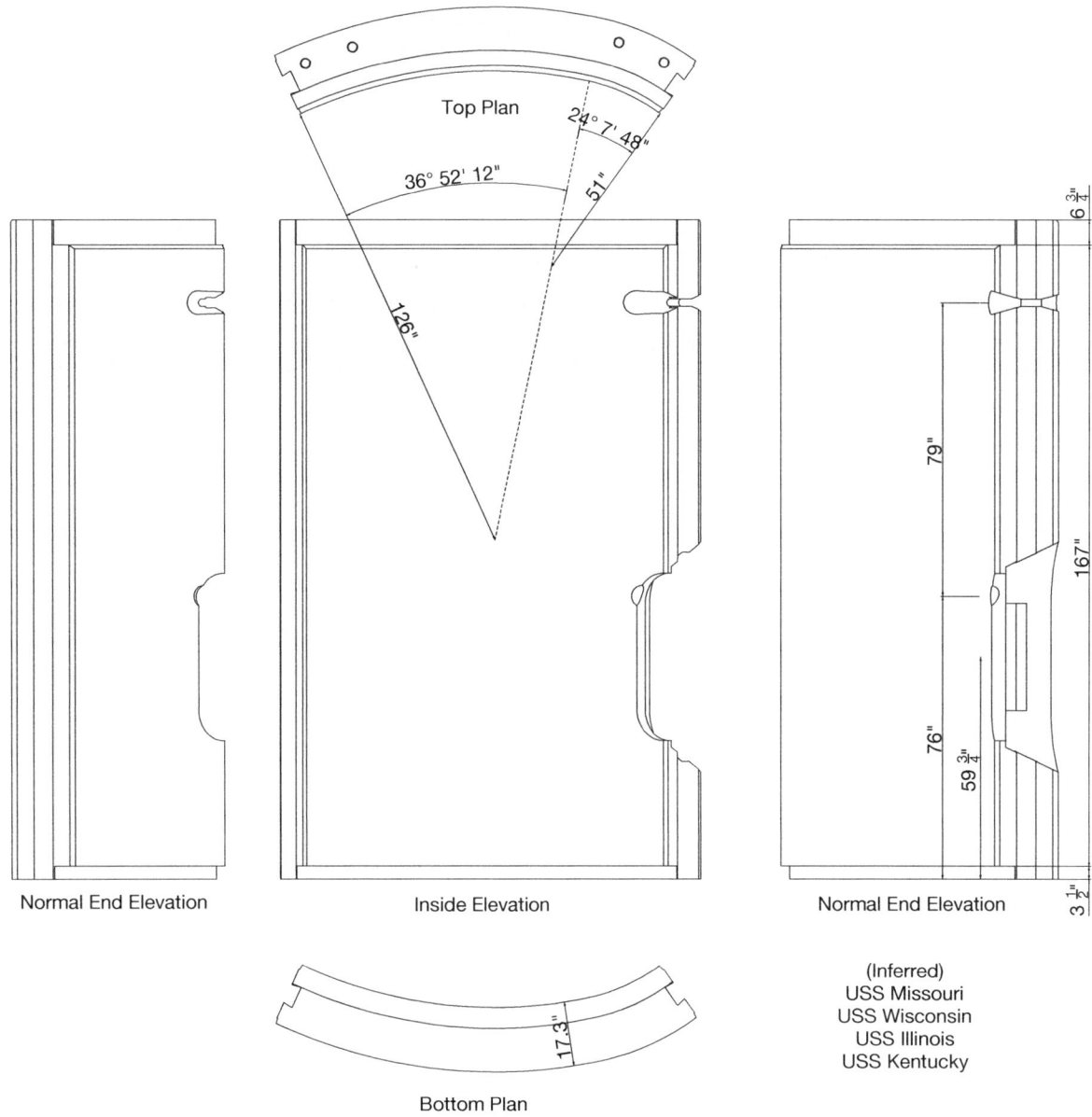

**Figure 7.34: Plate C-T-4 on USS Missouri, USS
Wisconsin, USS Illinois, and USS Kentucky**

Top Plan

24° 7' 48"

51"

36° 52' 12"

126"

79"

77 3/4"

137 1/2"

76"

59 3/4"

6 3/4"

167"

3 1/2"

Normal End Elevation

Inside Elevation

Normal End Elevation

(Inferred)
USS Missouri
USS Wisconsin
USS Illinois
USS Kentucky

17.3"

Bottom Plan

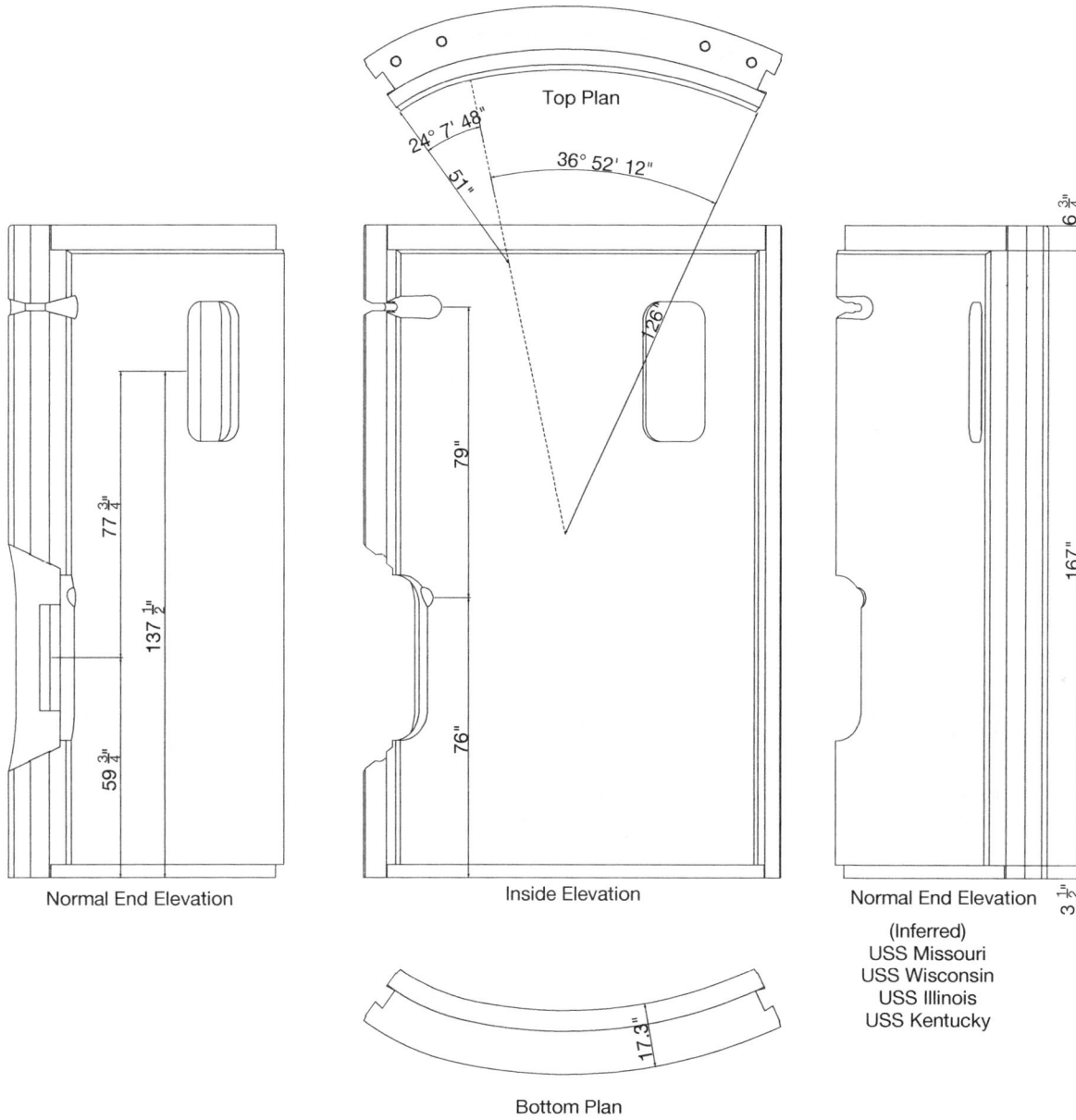

Figure 7.35: Plate C-T-5 on USS Missouri, USS Wisconsin, USS Illinois, and USS Kentucky

Figure 7.37: Edge of Top Plate (1:4). The profile of the top plate edge is the same as the side plate except for radii at the corners.

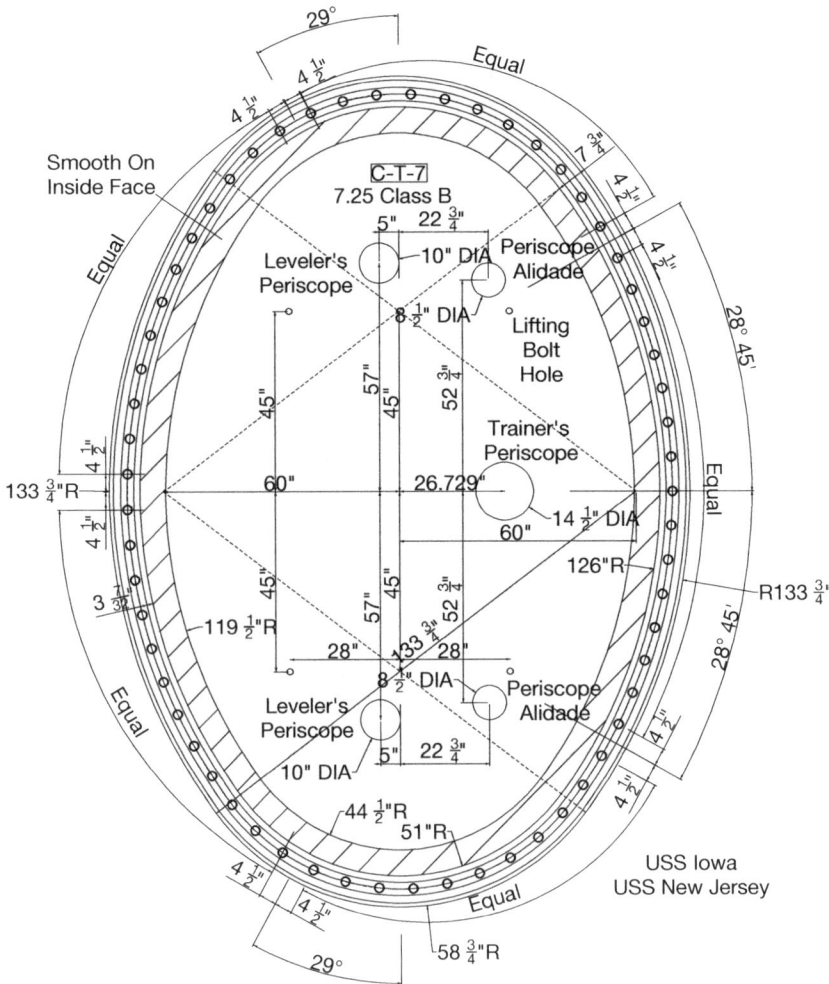

C-T-7

USS Iowa
USS New Jersey

C-T-7
7.25 Class B

Smooth On Inside Face

Leveler's Periscope

Periscope Alidade

Lifting Bolt Hole

Trainer's Periscope

Periscope Alidade

Leveler's Periscope

10" DIA
8 ½" DIA
14 ½" DIA

USS Iowa
USS New Jersey

Figure 7.36: Top Plate C-T-1 on USS Iowa and USS New Jersey (1:48)

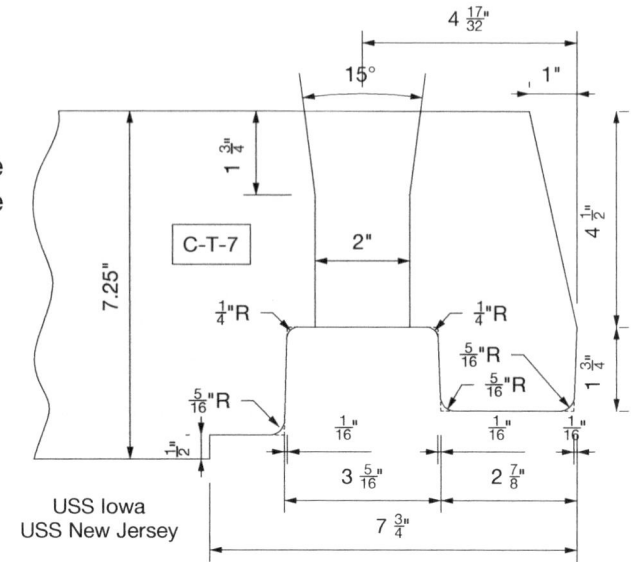

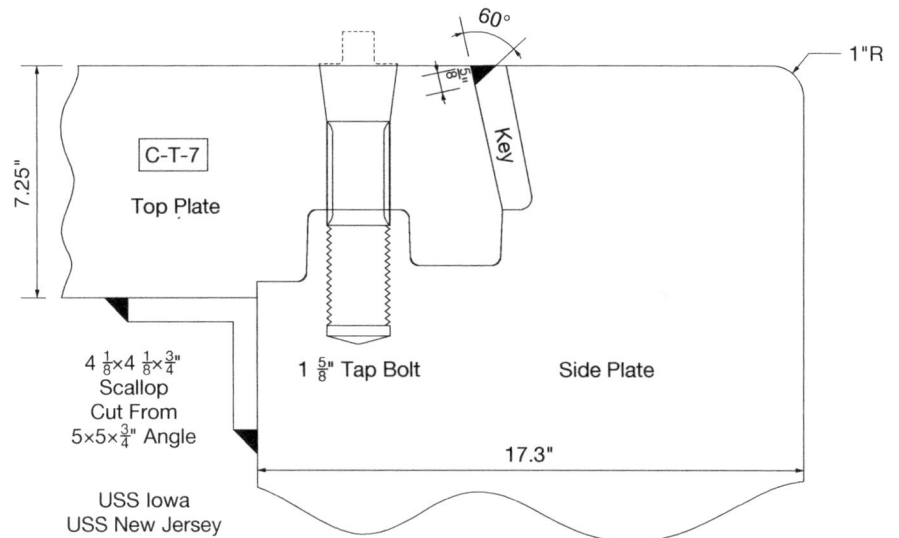

C-T-7
Top Plate

Side Plate

Key

$1\frac{5}{8}$ " Tap Bolt

$4\frac{1}{8}\times4\frac{1}{8}\times\frac{3}{4}$ " Scallop Cut From $5\times5\times\frac{3}{4}$ " Angle

17.3"

USS Iowa
USS New Jersey

Figure 7.38: Connection of the Top Plate to Side Plates (1:6). The top plate is secured at the top with tap bolts and welded key and at the top with a welded scallop at the bottom.

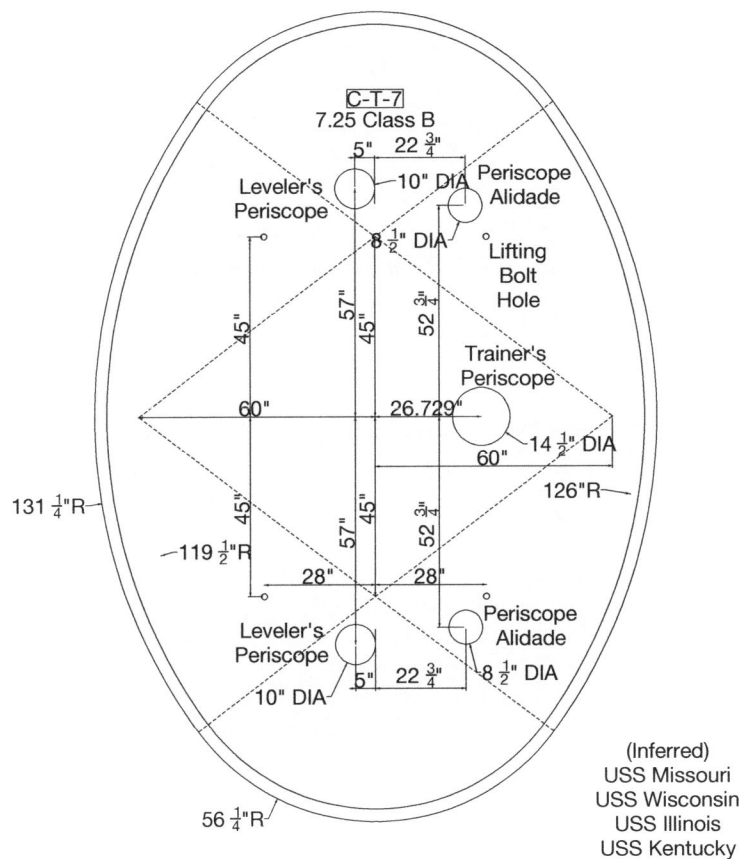

Figure 7.39: Plan and of the top plate C-T-7 on USS Missouri, USS Wisconsin, USS Illinois, and USS Kentucky (1:48). The dimensions are inferred from multiple original plans.

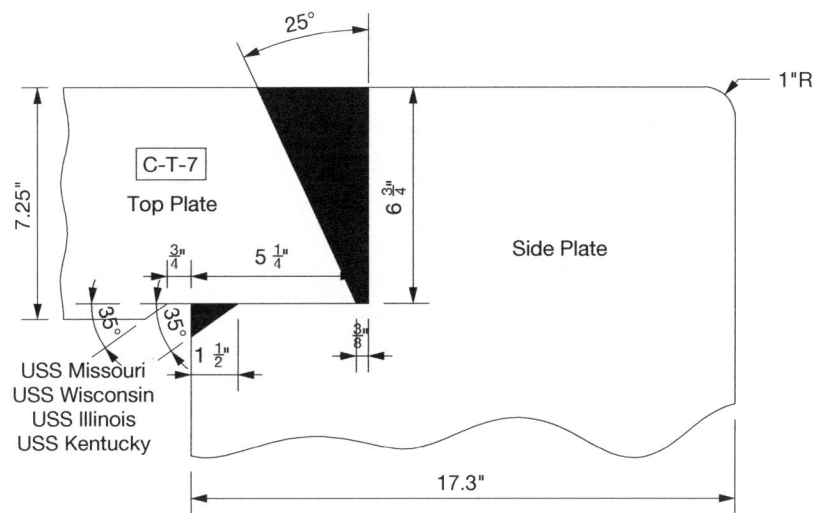

Figure 7.40: Section of the top plate C-T-7 on USS Missouri, USS Wisconsin, USS Illinois, and USS Kentucky (1:6). The side to top joint was greatly simplified on these later ships.

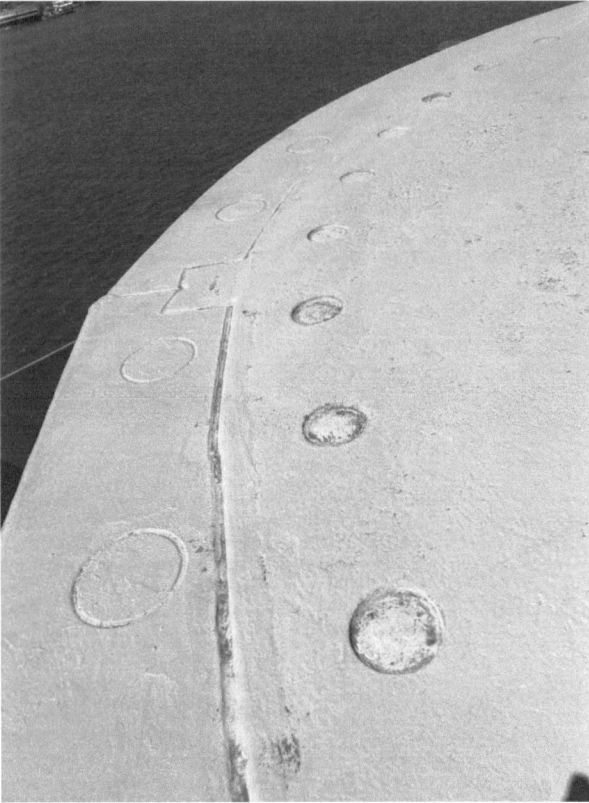

Photograph 7.13: The Conning Tower Top of USS New Jersey. The ring of bolts through the top plate is at the right. Plugged holes for lifting pads are at the right. An hourglass key is at the right center.

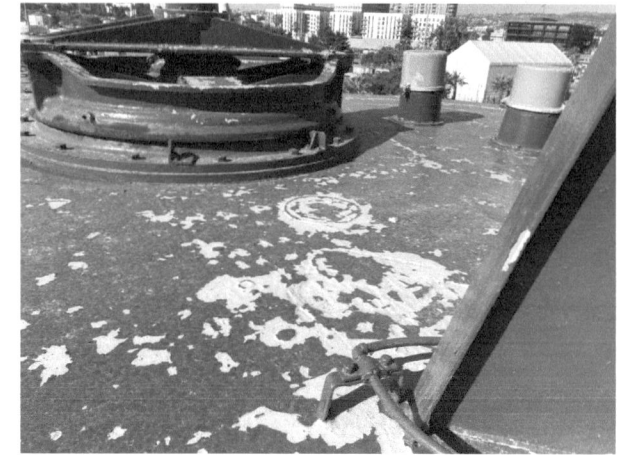

Photograph 7.14: Top Plate of USS Iowa. The foundation for the Mk.32 Periscope is at the upper left. A hole drilled for a World War II radar at the center of the photograph has been welded shut. (Jim Kurrasch)

Photograph 7.15: Conning tower Top Plate of USS Wisconsin. The extremely wide weld is at the right (see Figure 7.40). Am hourglass key between roof segments is at the top. Plugged holes for lifting pads are at the left. (Keith Nitka)

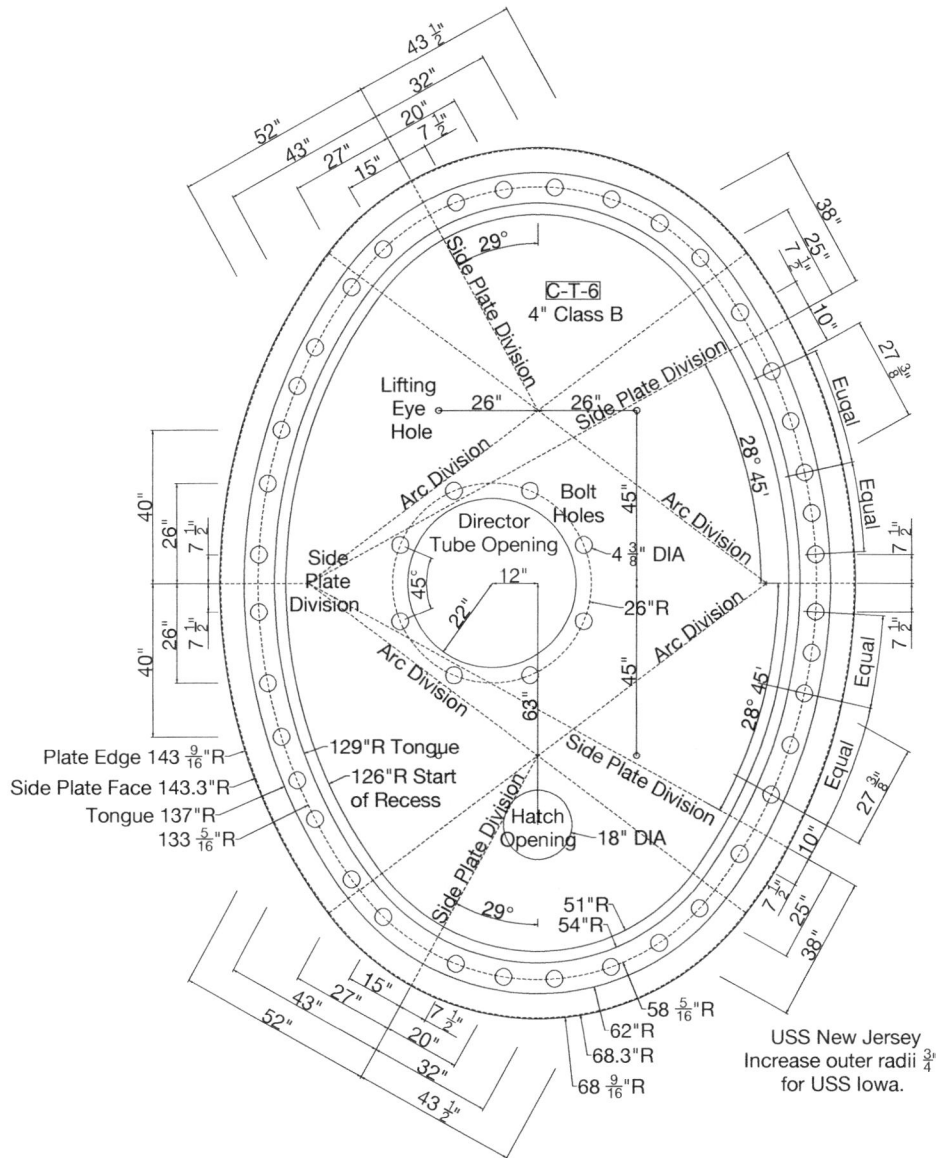

Figure 7.41: Plan of the bottom plate C-T-6 on USS Iowa and USS New Jersey (1:48). The exterior dimensions for USS Iowa would have been larger so that the edge could be grinded to match the sides of the conning tower foundation.

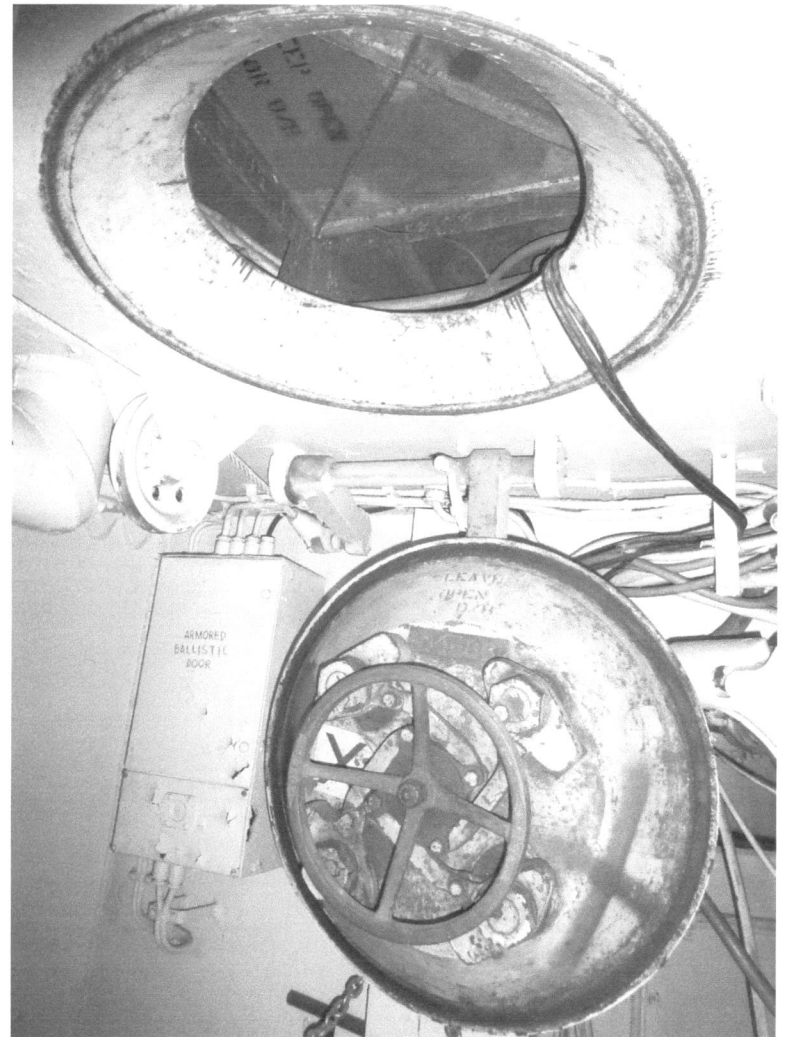

Photograph 7.16: Escape Hatch in the Bottom Plate on USS New Jersey. This hatch in the bottom plate provides a circituous escape route out of the conning tower.

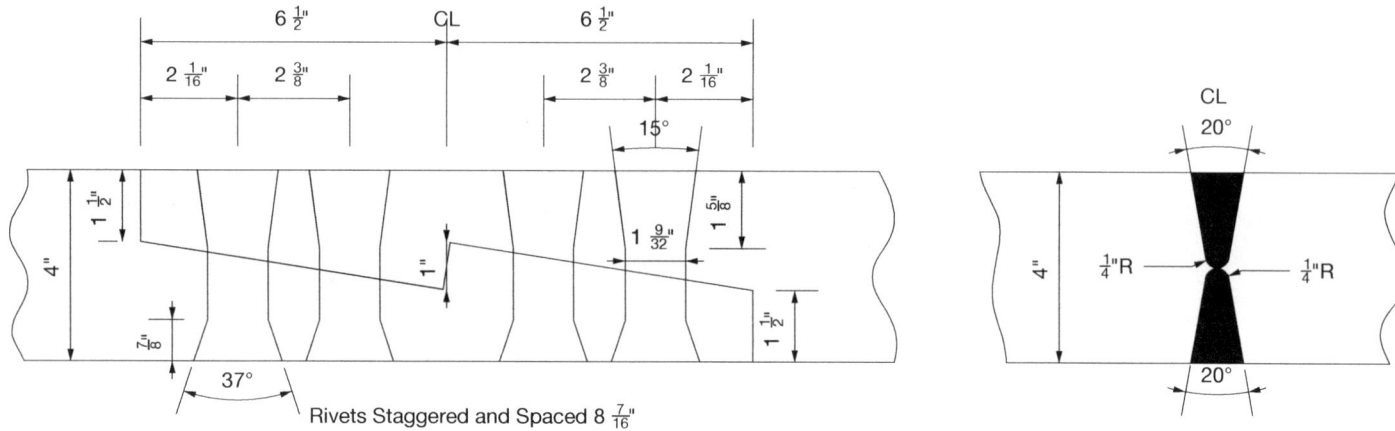

Figure 7.42: Alternatives for Jointing Bottom Plate Halves When Made in Two Pieces (1:4). The ship plans gave the armor maker the choice of making the bottom plate in one piece or dividing into two pieces along the center line and riveting or welding the halves together as shown here. This diagram is just a historical note because the plates were made in one piece.

Figure 7.43: Conning Tower Scallop Pattern (1:4). The same scallop pattern was used throughout the conning tower.

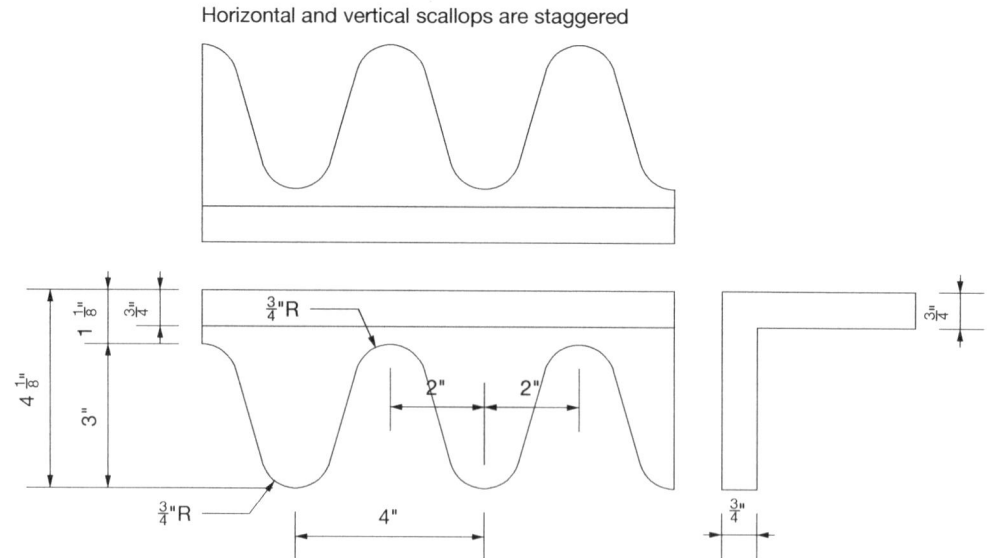

17.3"

3" | 8" | 6.3"

C-T-6

$\frac{5}{16}$"R $\frac{5}{16}$"R

$\frac{1}{4}$"R $\frac{1}{4}$"R

81'-10 $\frac{1}{2}$" ABV BL USS New Jersey
75'-3 $\frac{1}{2}$" ABV BL USS Iowa

Opening for 3.2" DIA Bolt

10 $\frac{1}{4}$

4 $\frac{3}{8}$"

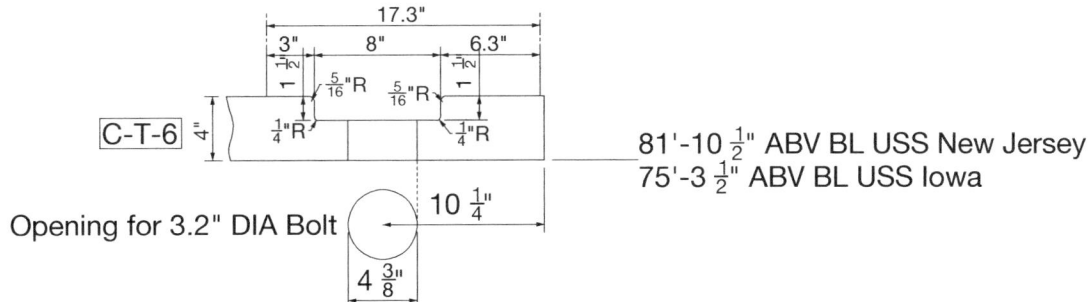

Figure 7.44: Outer Edge of Bottom Plate (1:12). The mill machined a groove around the outer edge of the bottom plate C-T-6 to match the tongue on the side plates of USS Iowa and USS New Jersey.

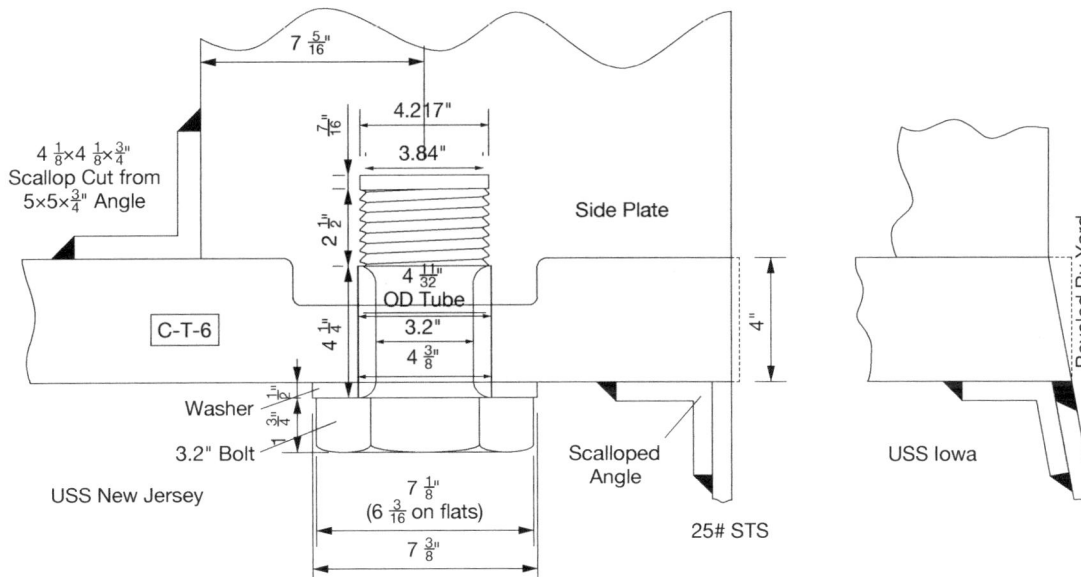

7 $\frac{5}{16}$"

4.217"

3.84"

4 $\frac{1}{8}$×4 $\frac{1}{8}$×$\frac{3}{4}$"
Scallop Cut from
5×5×$\frac{3}{4}$" Angle

Side Plate

2 $\frac{1}{2}$

4 $\frac{11}{32}$"
OD Tube

C-T-6

4 $\frac{1}{4}$

3.2"

4 $\frac{3}{8}$"

4"

Beveled By Yard

Washer

3.2" Bolt

USS New Jersey

Scalloped
Angle

USS Iowa

7 $\frac{1}{8}$"
(6 $\frac{3}{16}$ on flats)

25# STS

7 $\frac{3}{8}$"

Figure 7.45: Connection of Bottom Plate to Side Plates (1:6). The edges of the bottom plate were grinded to create a smooth transition between the conning tower sides and the conning tower foundation.

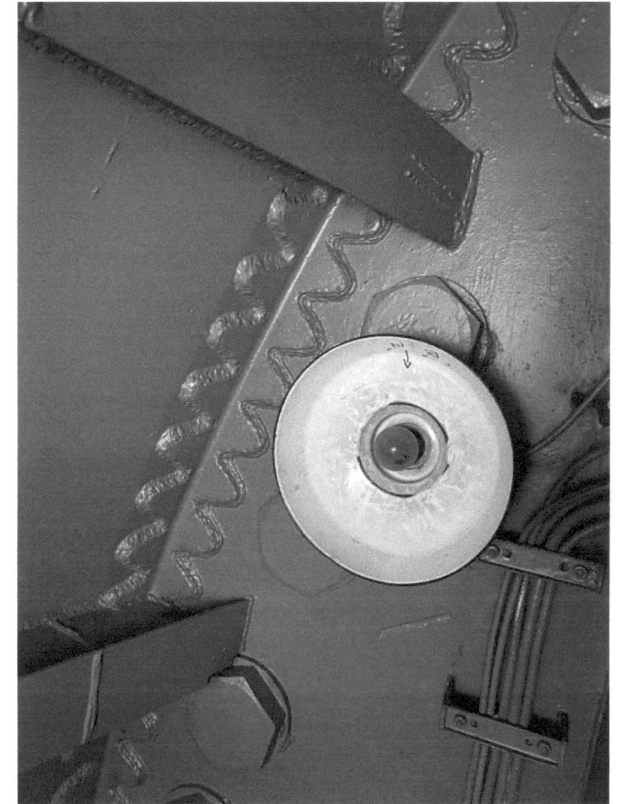

Photograph 7.17: The bottom plate of USS Iowa looking up. The bolts attach the side plates to the bottom plate. The welded scallops attach the conning tower to its foundation. There is an original World War II era light fixture. (Jim Kurrasch)

Figure 7.47: Plan of Bottom Plate (1:48). For the later ships the tongue and groove joint around the edge of the bottom plate was eliminated. USS Wisconsin retained the joint connecting the tube used in USS Iowa and USS New Jersey (see Figure 7.41). The other ships used a simplified joint with an enlarged opening. The dimensions here are inferred from multiple original plans.

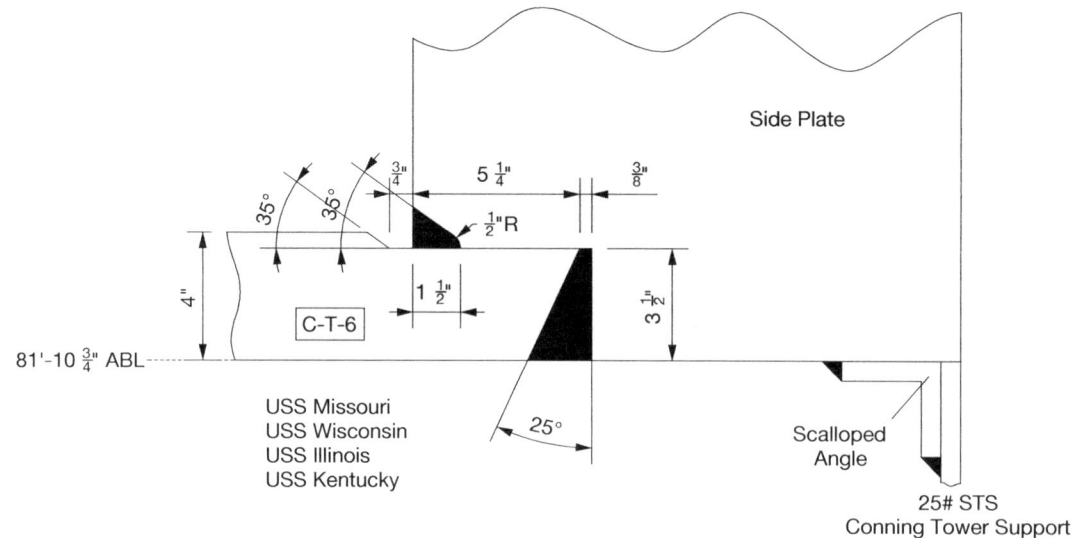

C-T-6
4" Class B

Opening For
USS Wisconsin

Opening For
USS Missouri
USS Illinois
USS Kentucky

$4\frac{3}{8}$" DIA

45°

12"

22"R

30"R

63"

131$\frac{1}{4}$"R

18" DIA

(Inferred)
USS Missouri
USS Wisconsin
USS Illinois
USS Kentucky

56$\frac{1}{4}$"R

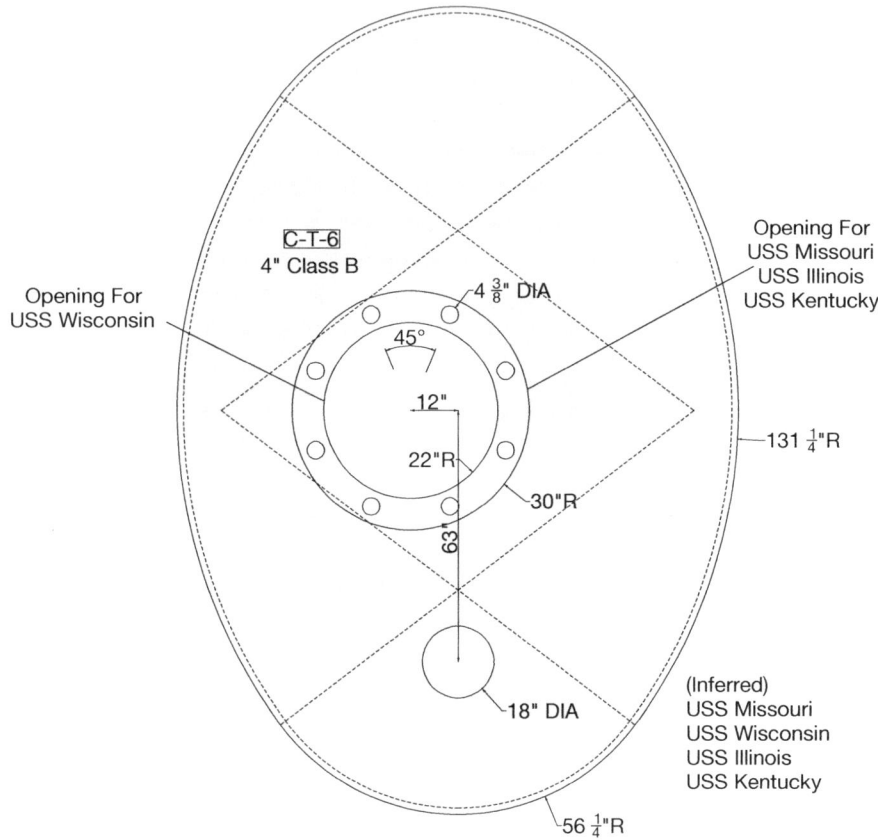

Figure 7.46: Connection of Bottom Plate to Side Plates (1:6). The side to bottom plate joint was simplified for USS Missouri and USS Wisconsin. Direct welding replaced the bolts and welding along scalloped angles. This figure is taken directly from original plans.

Side Plate

35°

35°

$\frac{3}{4}$"

5$\frac{1}{4}$"

$\frac{3}{8}$"

$\frac{1}{2}$"R

4"

C-T-6

1$\frac{1}{2}$"

3$\frac{1}{2}$"

81'-10$\frac{3}{4}$" ABL

USS Missouri
USS Wisconsin
USS Illinois
USS Kentucky

25°

Scalloped
Angle

25# STS
Conning Tower Support

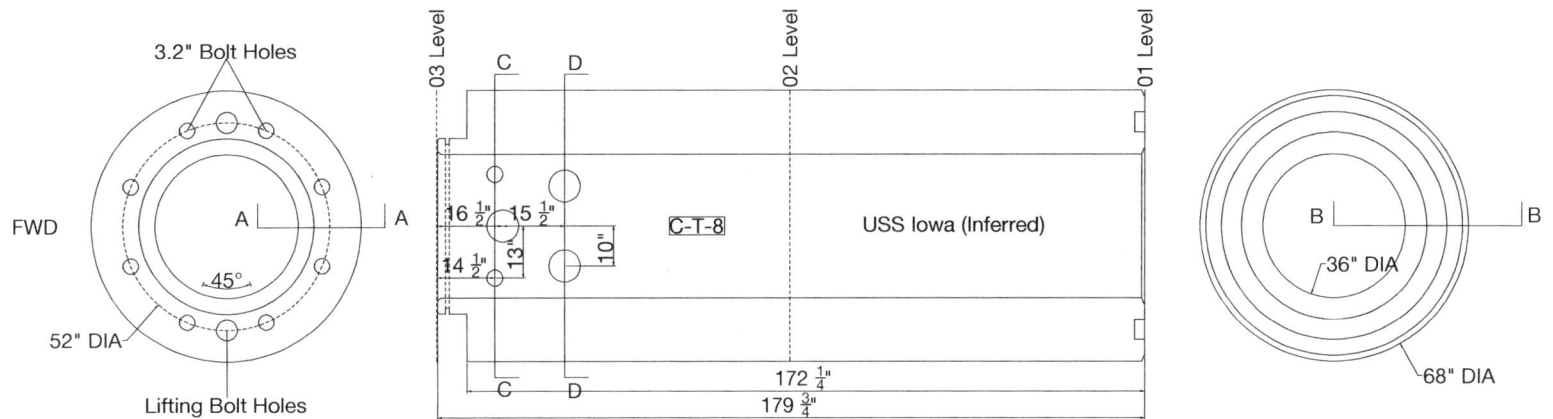

Figure 7.48: Upper Conning Tower Tube (1:48). The conning tower tube consists of two segments. On USS Iowa the upper conning tower tube is shorter than those of the other ships. The five openings in the side of the tube are for fire control cabling. A 100# STS wiring trunk connects the forward main battery director to those openings that runs under Third Deck. On USS Iowa the trunk goes down slightly as it connects to the tube (see Figure 7.185).

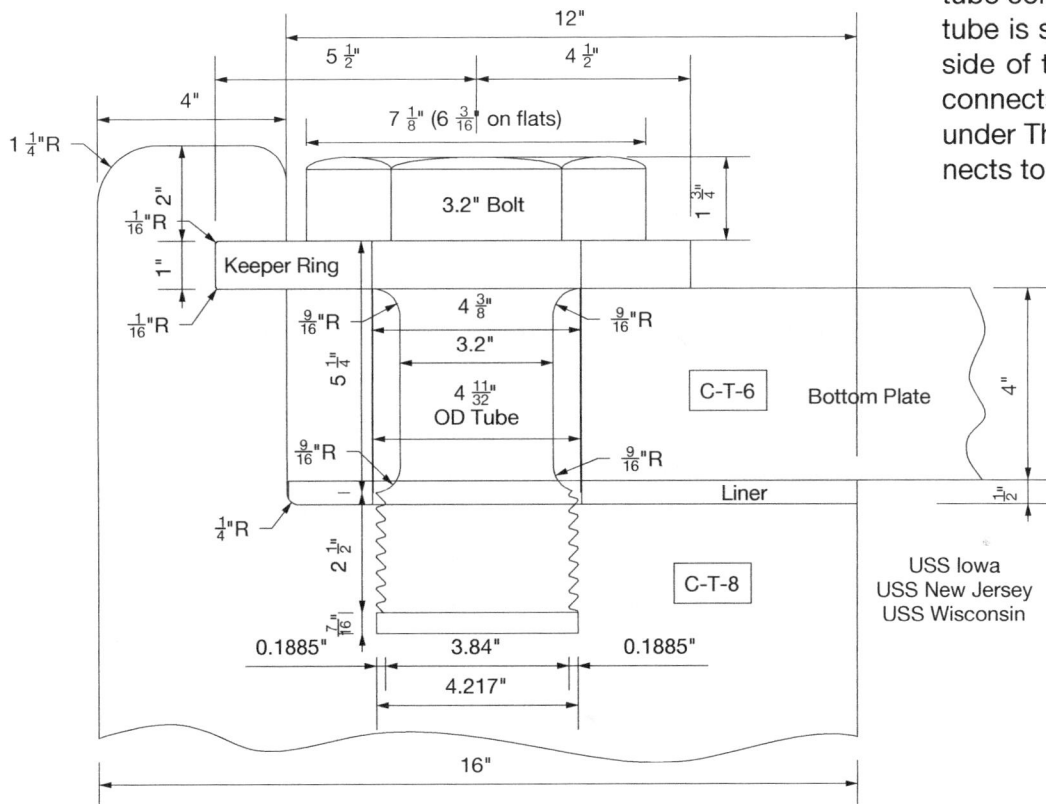

Figure 7.49: Section A–Upper Edge (1:4). On USS Iowa, USS New Jersey, and USS Wisconsin, the upper end of the tube has a slotted nipple. A Liner Ring was placed over the nipple to account for any error in spacing between the bottom plate and the tube. The liner was ordered oversized in thickness and had to be grinded to fit. The nipple on the tube passes through the bottom plate. The two halves of the Keeper Ring were inserted into the slot in the nipple to keep the tube from pulling out. The Keeper Ring was then welded into one piece and bolts were driven through the ring and bottom plate into the tube.

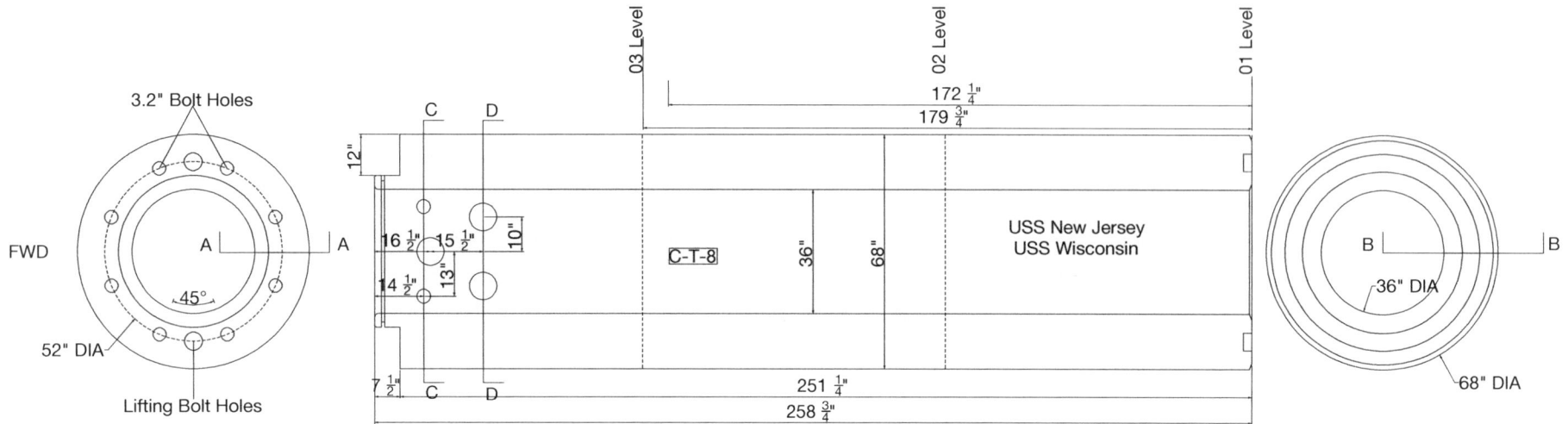

Figure 7.50: Upper Conning Tower Tube (1:48). On USS New Jersey and USS Wisconsin the conning tower tube had to be lengthened to reach the higher bottom plate. On the later ships the director cabling trunk goes upwards to reach the openings in the tube (see Figure 7.185).

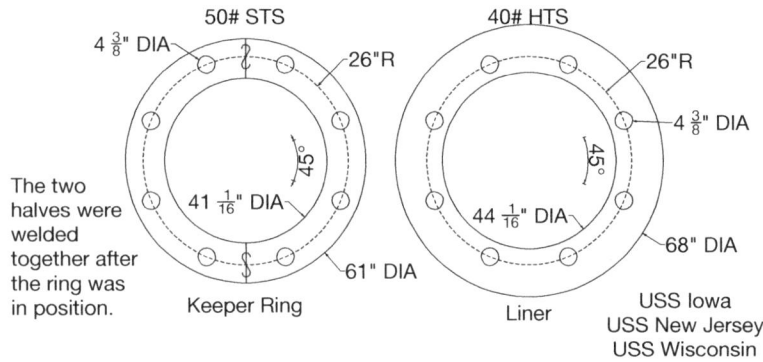

Figure 7.51: Tube Connecting Rings (1:48). These were ordered oversized in thickness and had to be grinded for fit. The locking ring is in halves that were welded together after it was in place. See Figure 7.49.

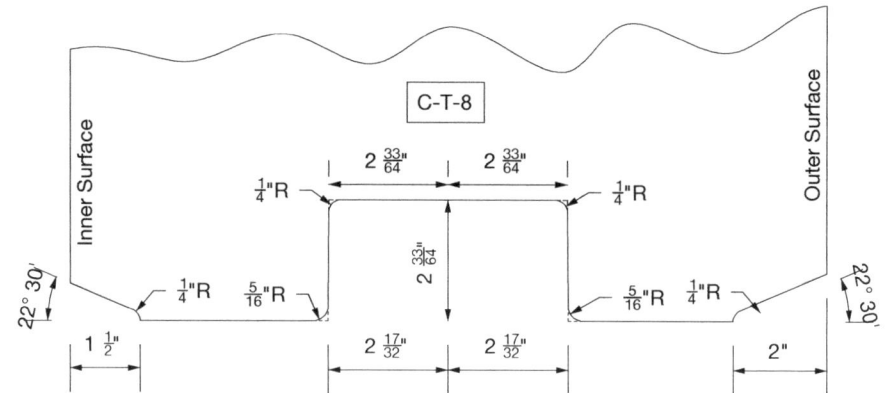

Figure 7.52: Section B–Lower Edge (1:4). The lower edge of the tube has a groove that matches a tongue in the lower conning tower tube. This joint was the same on all ships.

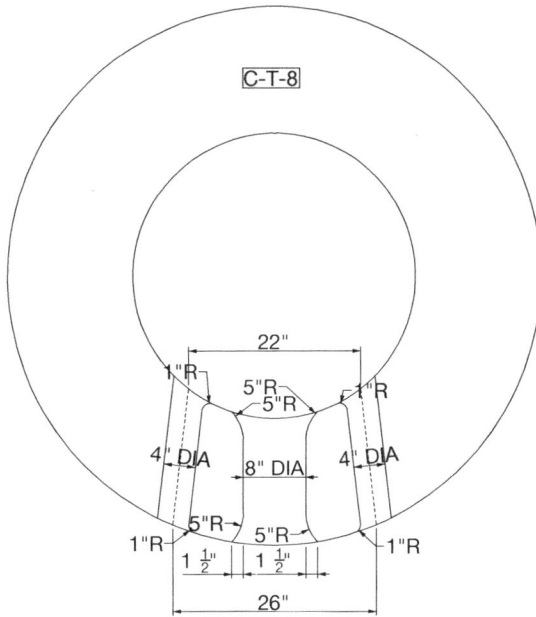

Figure 7.53: Section C–Through Upper Row of Tube Openings (1:24). This is the same for all ships.

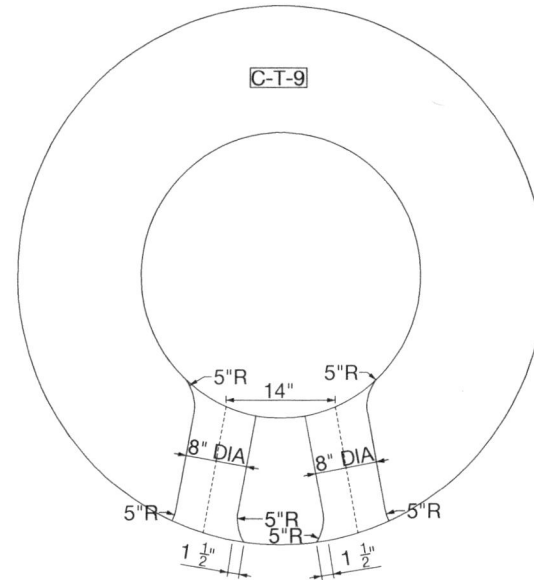

Figure 7.54: Section D–Through Lower Row of Tube Openings (1:24). This is the same for all ships.

Figure 7.55: A rendering of the upper end of the conning tower tube on USS New Jersey and USS Wisconsin. The tube on USS Iowa appears the same from this viewpoint. The holes at the side allow cables from directors to enter the tube.

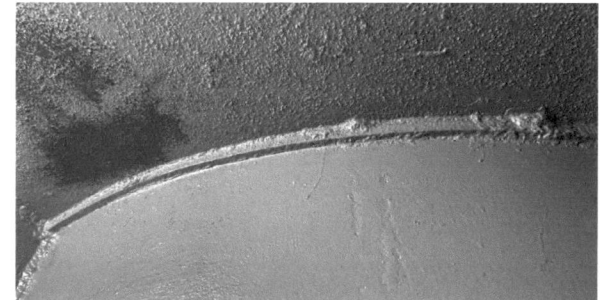

Photograph 7.18: The connection of the tube to the bottom plate of USS Iowa. The liner used to shim the gap between the tube and bottom plate is visible. The liner diameter is greater than specified in the plans. (Jim Kurrasch)

FWD

E — E

45°

52" DIA

Lifting Bolt Holes

C D   03 Level   02 Level   01 Level

13"

13 ½ 15 ½

13"

11 ½

10" 10"

C-T-8

36"

68"

USS Missouri
USS Illinois
USS Kentucky
(Inferred)

3 ½

252 ¼"

255 ¾"

C D

B — B

36" DIA

68" DIA

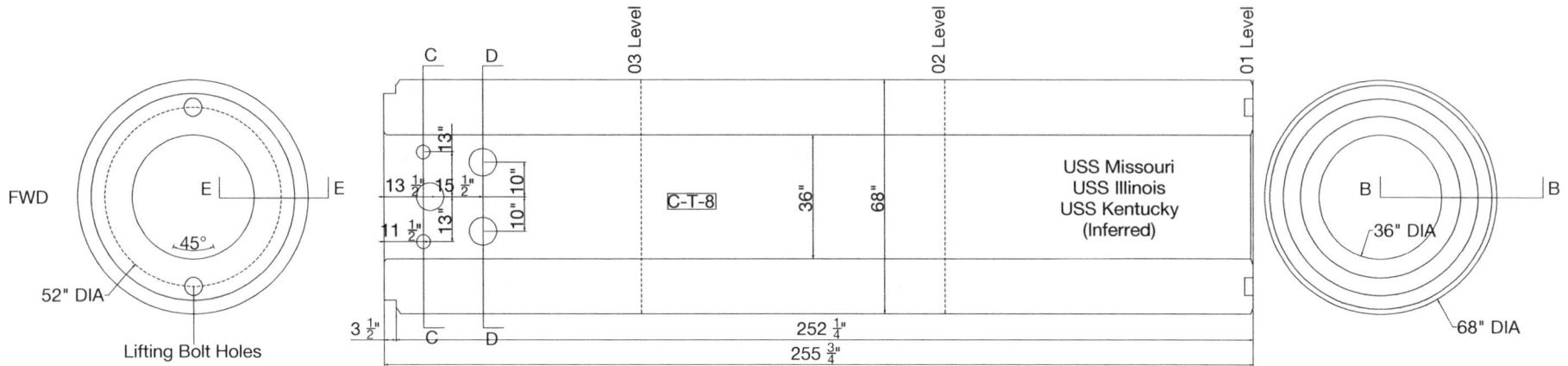

Figure 7.56: Upper Conning Tower Tube (1:48). On USS Missouri, USS Illinois, and USS Kentucky the USS Wisconsin connection between the tube and bottom plate was simplified.

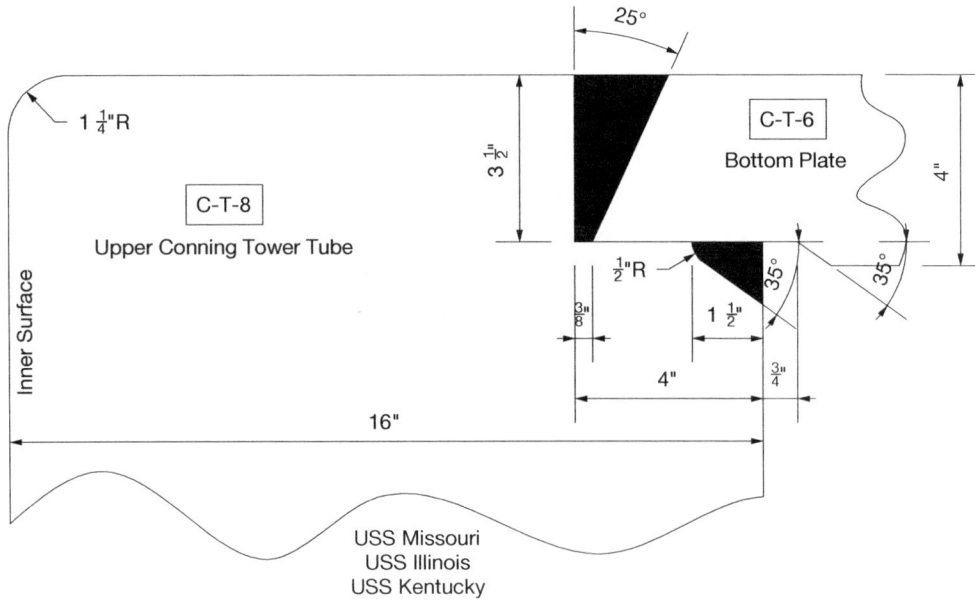

25°

1 ¼"R

3 ½"

C-T-6

Bottom Plate

4"

C-T-8

Upper Conning Tower Tube

½"R

35°

35°

⅜"

1 ½"

¾"

4"

Inner Surface

16"

USS Missouri
USS Illinois
USS Kentucky

Figure 7.57: Section E–Upper Edge (1:48). The simplified upper connection on USS Missouri, USS Illinois, and USS Kentucky the USS Wisconsin. The joint was entirely welded.

Photograph 7.19: This is the connection between the bottom plate and tube on USS Iowa. The false floor of the flag level is at the top and the cables run down into the tube. The bolt in the foreground sits on top of the Keeper Ring. (Jim Kurrasch)

Figure 7.58: This is a rendering of the connection of the tube to the bottom plate. The Keeper Ring is ghosted to show how it slips into the slot of the nipple on the tube.

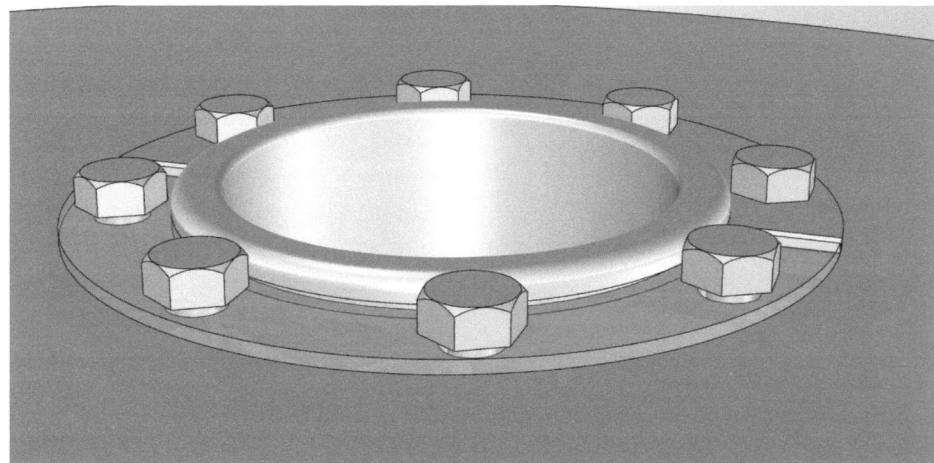

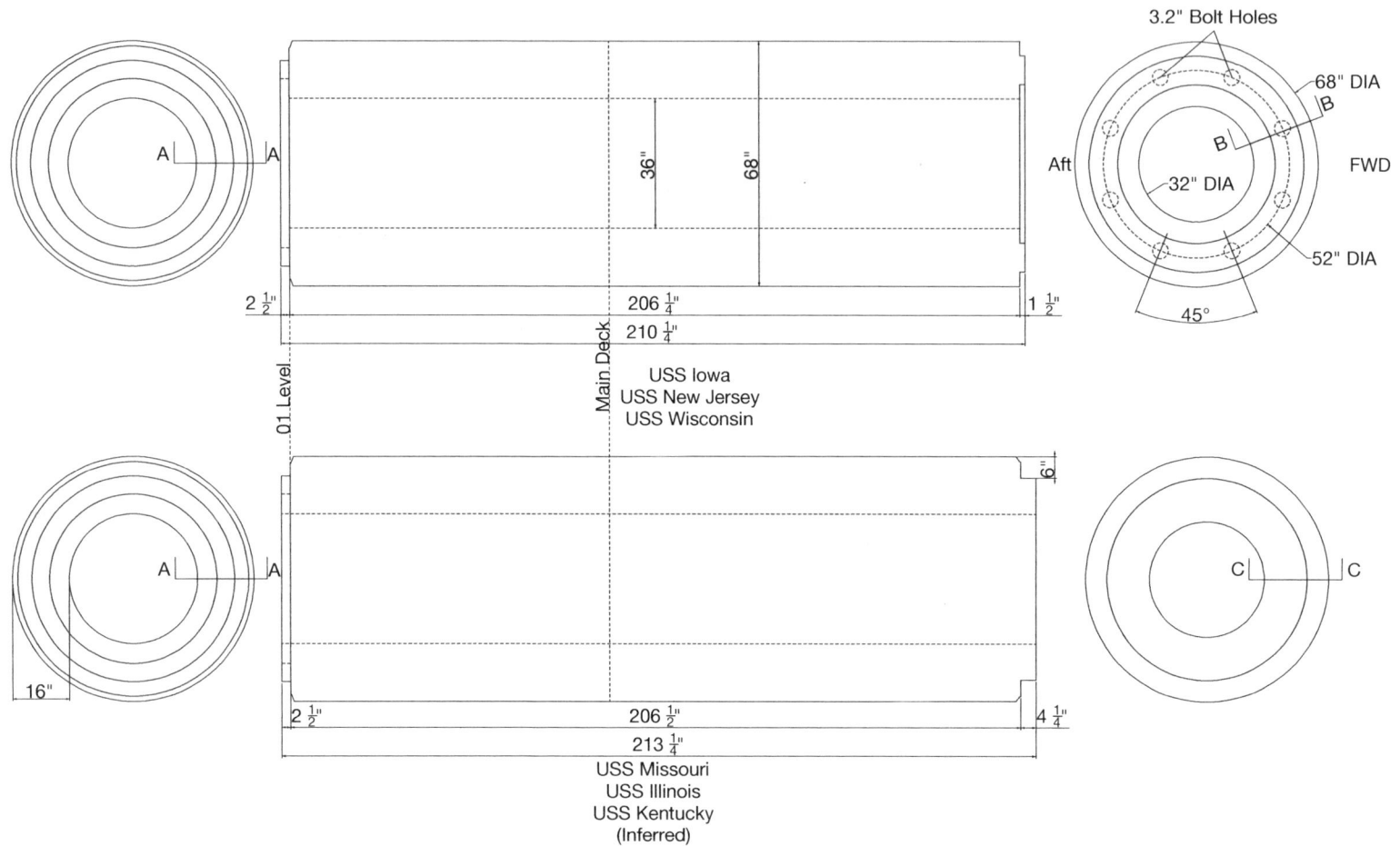

3.2" Bolt Holes

68" DIA

B

Aft   FWD

32" DIA

52" DIA

45°

36"

68"

A   A

2 ½"   206 ¼"   1 ½"

210 ¼"

Main Deck

01 Level

USS Iowa
USS New Jersey
USS Wisconsin

A   A

C   C

16"

6"

2 ½"   206 ½"   4 ¼"

213 ¼"

USS Missouri
USS Illinois
USS Kentucky
(Inferred)

Figure 7.59: C-T-9 Lower Conning Tower Tube (1:48). The lower tubes differ in their joint connecting to second deck. USS Iowa, USS New Jersey, and USS Wisconsin have a bolted tongue and groove joint. The later ships used a simplified welded rabbet joint.

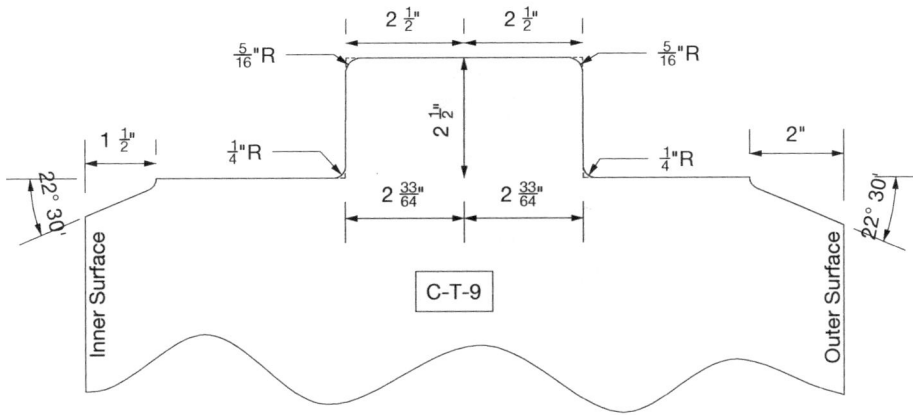

Figure 7.60: Section A–Upper Edge (1:4). The joint between the upper and lower tubes is the same on all ships. It is not a precision fit as in many other locations. The groove on the upper tube is slightly wider and deeper than the tongue on the lower tube.

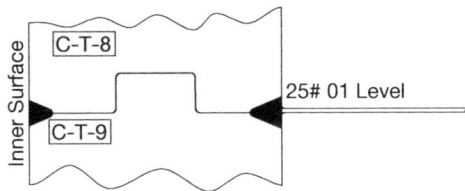

Figure 7.61: Connection Between Upper and Lower Tubes (1:12). The tubes were welded together along the inside and outside grooves.

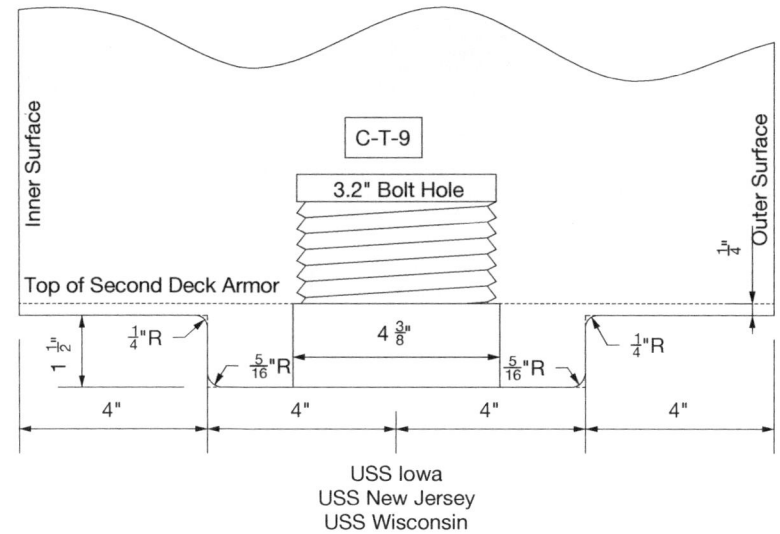

Figure 7.62: Section B–Lower Edge (1:4). On USS Iowa, USS New Jersey, and USS Wisconsin the mill machined the lower edge connection to second deck with a bolted tongue.

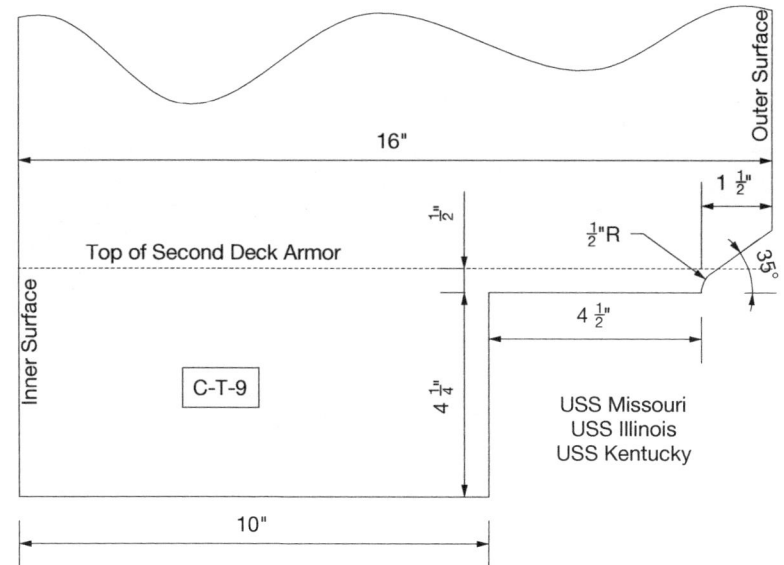

Figure 7.63: Section C-Lower Edge (1:4). The second deck connection was simplified on later ship to use a welded rabbet joint.

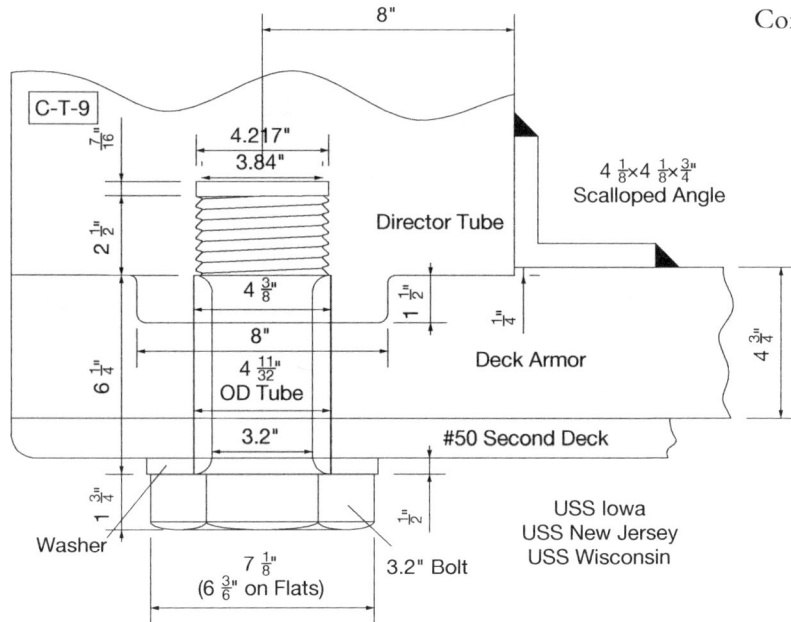

Figure 7.64: Connection of Conning Tower Tube to Second Deck (1:6). USS Iowa, USS New Jersey, and USS Wisconsin used a bolted tongue and groove joint with welding along scalloped angles at the outside. The other ships used a simpler rabbet joint that was welded.

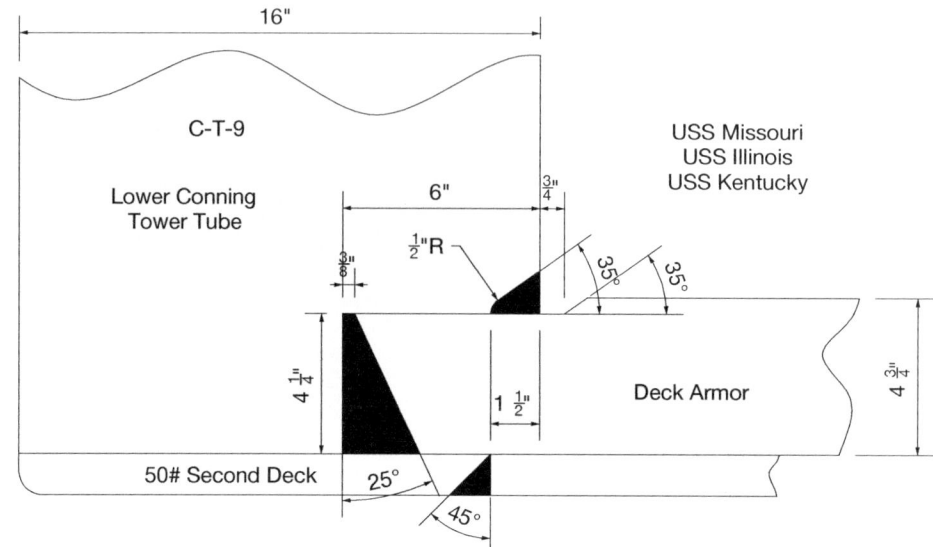

Figure 7.65: Connection of Conning Tower Tube to Second Deck (1:6). The simplified connection used on later ships. It lacked the bolts and the scalloped angle.

Photograph 7.1: Connection of the Lower Tube to Second Deck on USS New Jersey.

Photograph 7.2: Connection of the Lower Tube to Second Dec on USS Missouri. (Franklin Clay)

Photograph 7.3: The false floor of the Flag Level on USS Iowa. Part of the floor has been lifted to access the escape hatch in the bottom plate. The false floor is not structural and just rests on raised supports that allow wiring to pass. On the other ships the false floor and escape hatch are at the Navigation Level. (Jim Kurrasch)

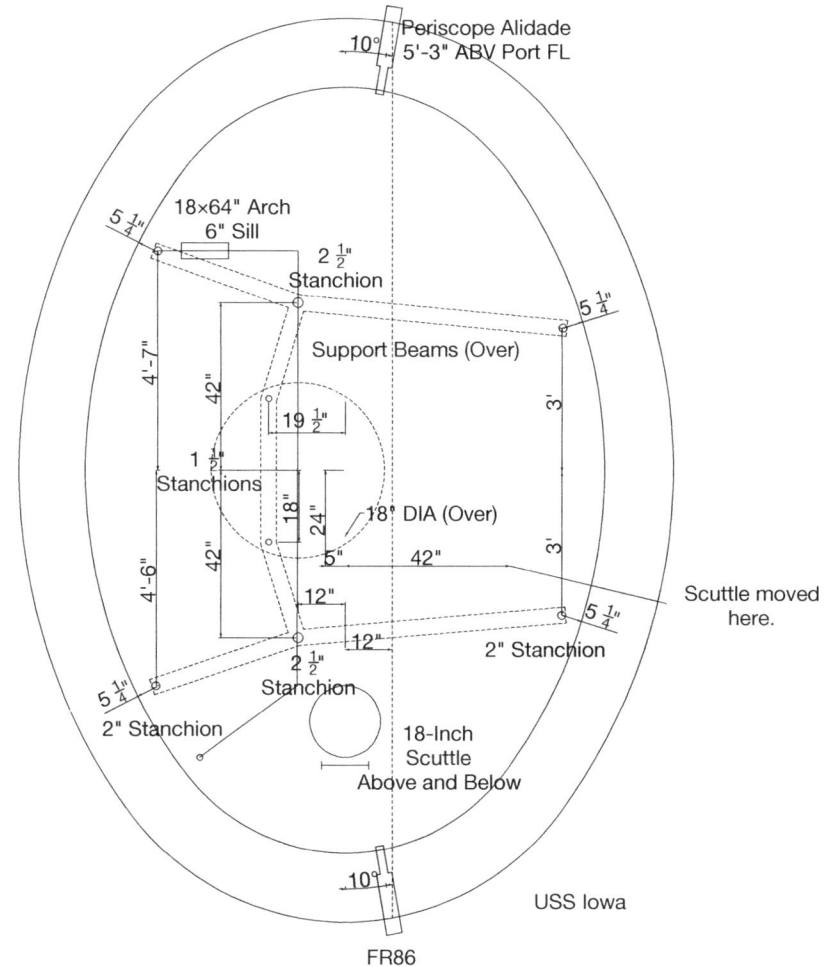

Figure 7.66: Plan of Flag Level (1:48). The structures over the removable floor at the Flag Level of USS Iowa. The partition shown creates a separate area for communications in the aft section of the level.

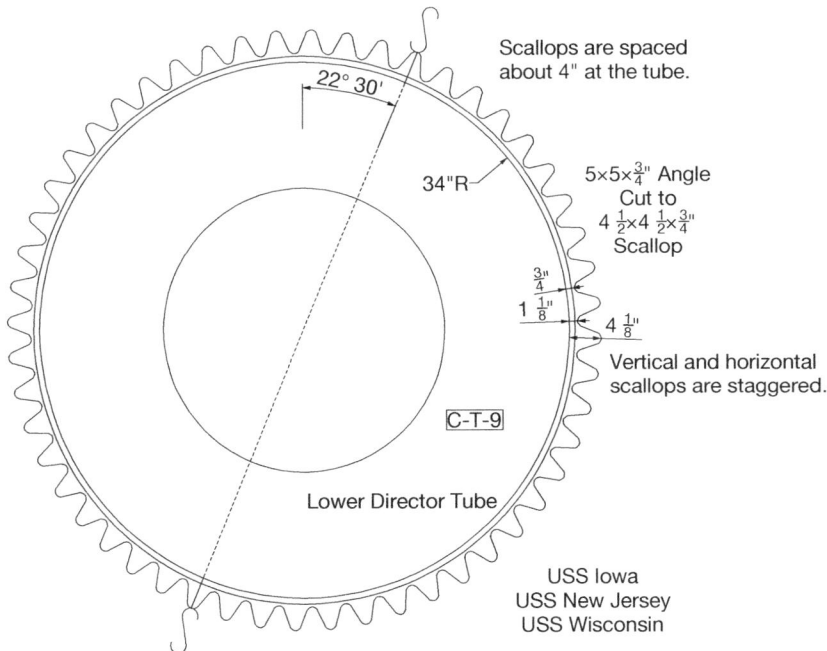

Figure 7.67: Scalloped Angle at Second Deck (1:24)

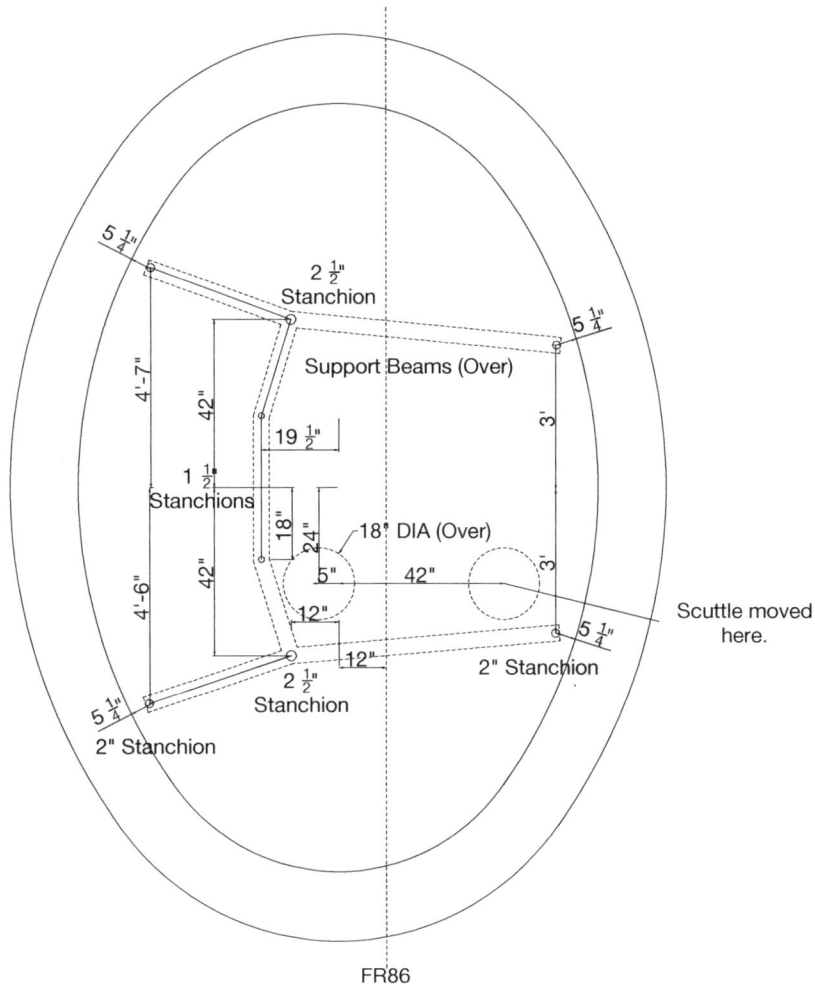

Figure 7.68: Plan of Navigation Level (1:48).
A partition that separates the radio area of the Navigation Level and the overhead beams that support the Fire Control level.

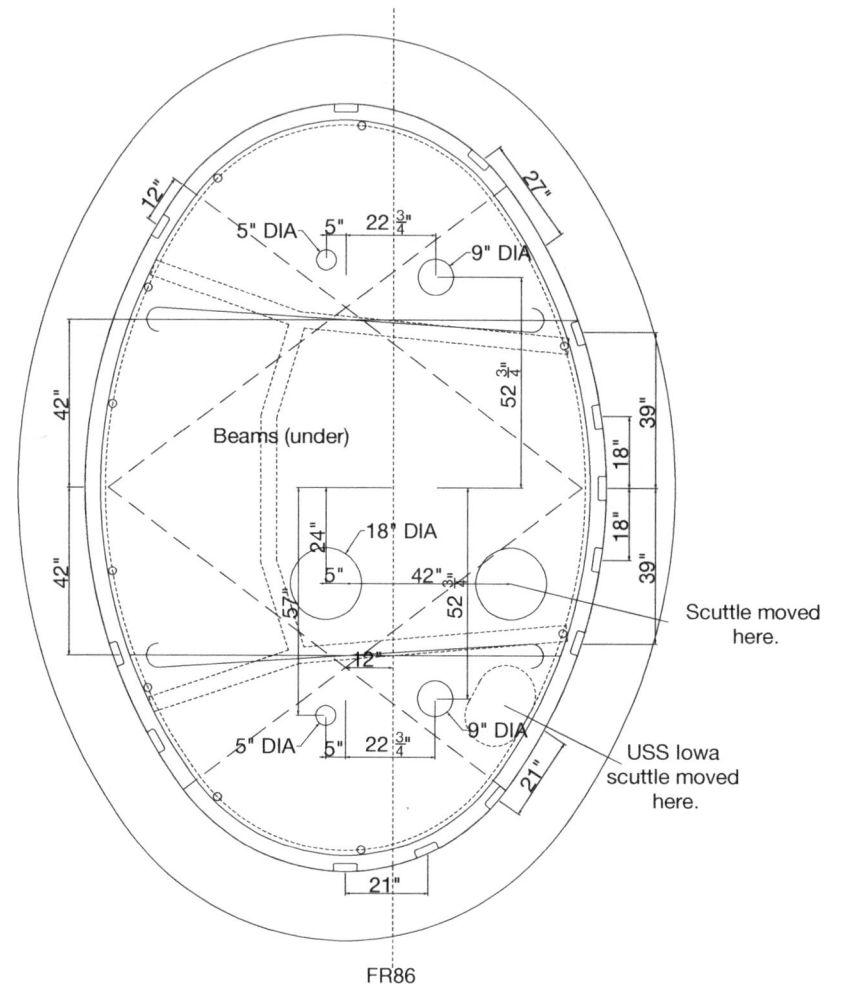

Figure 7.69: Plan of Fire Control Level (1:48)

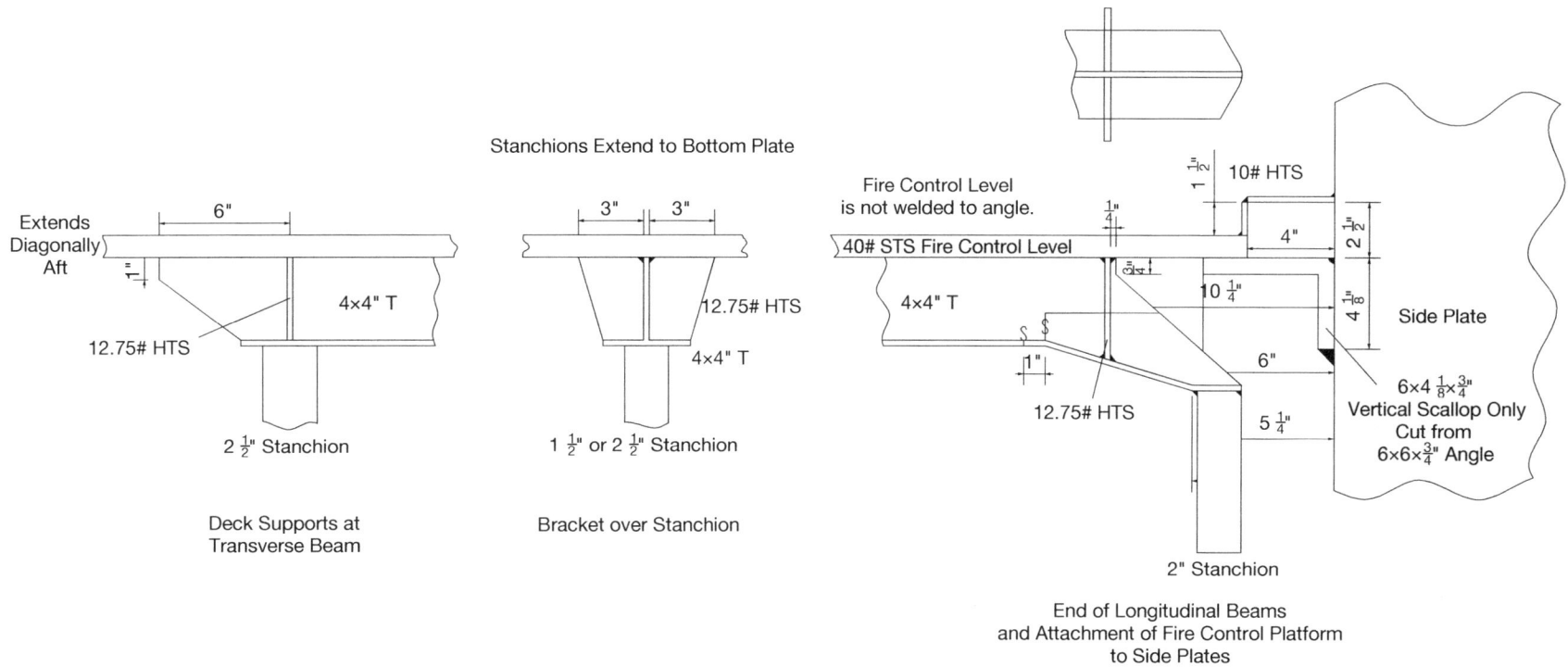

Extends Diagonally Aft

6"

1"

4×4" T

12.75# HTS

2 ½" Stanchion

Deck Supports at Transverse Beam

Stanchions Extend to Bottom Plate

3" 3"

12.75# HTS

4×4" T

1 ½" or 2 ½" Stanchion

Bracket over Stanchion

Fire Control Level is not welded to angle.

10# HTS

1 ½"

40# STS Fire Control Level

¼"

4"

2 ½"

4×4" T

¾"

10 ¼"

4 ⅛"

Side Plate

12.75# HTS

1"

6"

5 ¼"

6×4 ⅛×¾"
Vertical Scallop Only Cut from 6×6×¾" Angle

2" Stanchion

End of Longitudinal Beams and Attachment of Fire Control Platform to Side Plates

Figure 7.70: Fire Control Level Supports (1:8). The deck for the Fire Control level is supported in such a way that it is not rigidly connected to the armor. An impact that distorted the shape of the side armor less than four inches would not buckle the deck. On USS Iowa the Navigation Level is supported in the same way. The Flag level on USS Iowa and Navigation levels on the other ships are false floors elevated by supports resting on the bottom plate.

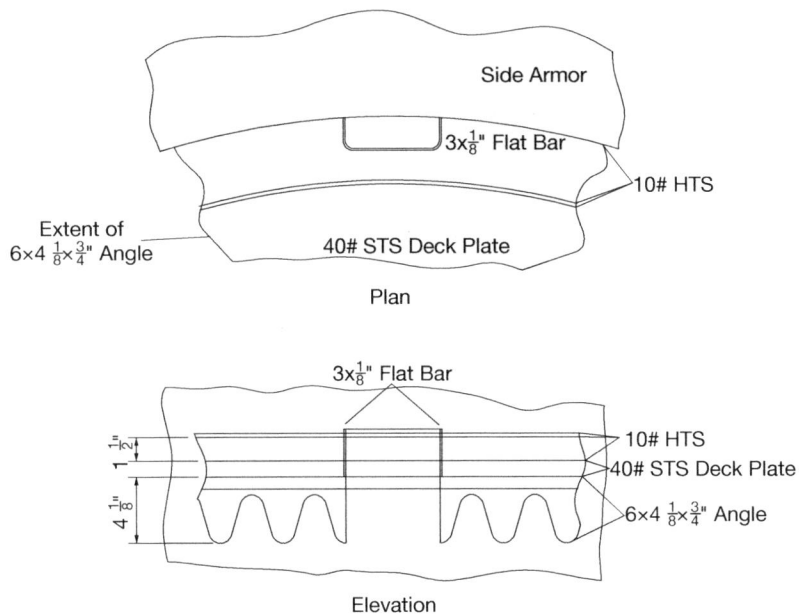

Figure 7.71: Vents in Fire Control Level Platform (1:12).
The navigation level on USS Iowa has similar vents.

Photograph 7.4: Outer Edge of the Fire Control Level on USS New Jersey. Note the vent holes.

Photograph 7.5: The starboard Navigation Level door on USS New Jersey. The crank shown is used to extend and retract the door bolts by rotating the three Pinion Shafts. The same crank can be repositioned to manually open the door. Above and to the left of each Pinion Shaft there is a grease fitting for lubricating its bolt. Above and to the right of each is a loop to attach a Bolt Toggle Pin to the door. The Bolt Toggle Pin fits into a hole in the Pinion Shaft to keep it from rotating. The pins and the chains are missing. Here the door is being held open by extending the center bolt into the bracket at the right.

Figure 7.72: Fire Control Level Door (1:24)
The extensions from the sides of the doors fits into recesses in the door openings. The extension is on the opposite side from the bolts. If the door hinge were destroyed, the bolts would hold the door in place on one side and the extension would hold it in place on the other side. The inside face of each door was drilled and tapped for lifting eye-bolts used for crane attachment points. The eye-bolts were removed after the door was installed and the holes were plugged.

End Elevation   Outer Elevation   End Elevation   Inside Elevation

Flag Level door C-T-13 on
USS Iowa is similar but omits
Peephole and the radius is
143.3".

143.3"R

25°

Top Plan

Starboard door C-T-12 is a mirror.

8" Peephole

C-T-11

Outside Elevation

16 ½"R

10"R

Lifting Eye-Bolts

10"R

10"R

16 ½"R

Inside Elevation

C-T-13

Outside Elevation
Flag Level Door
(USS Iowa Only)

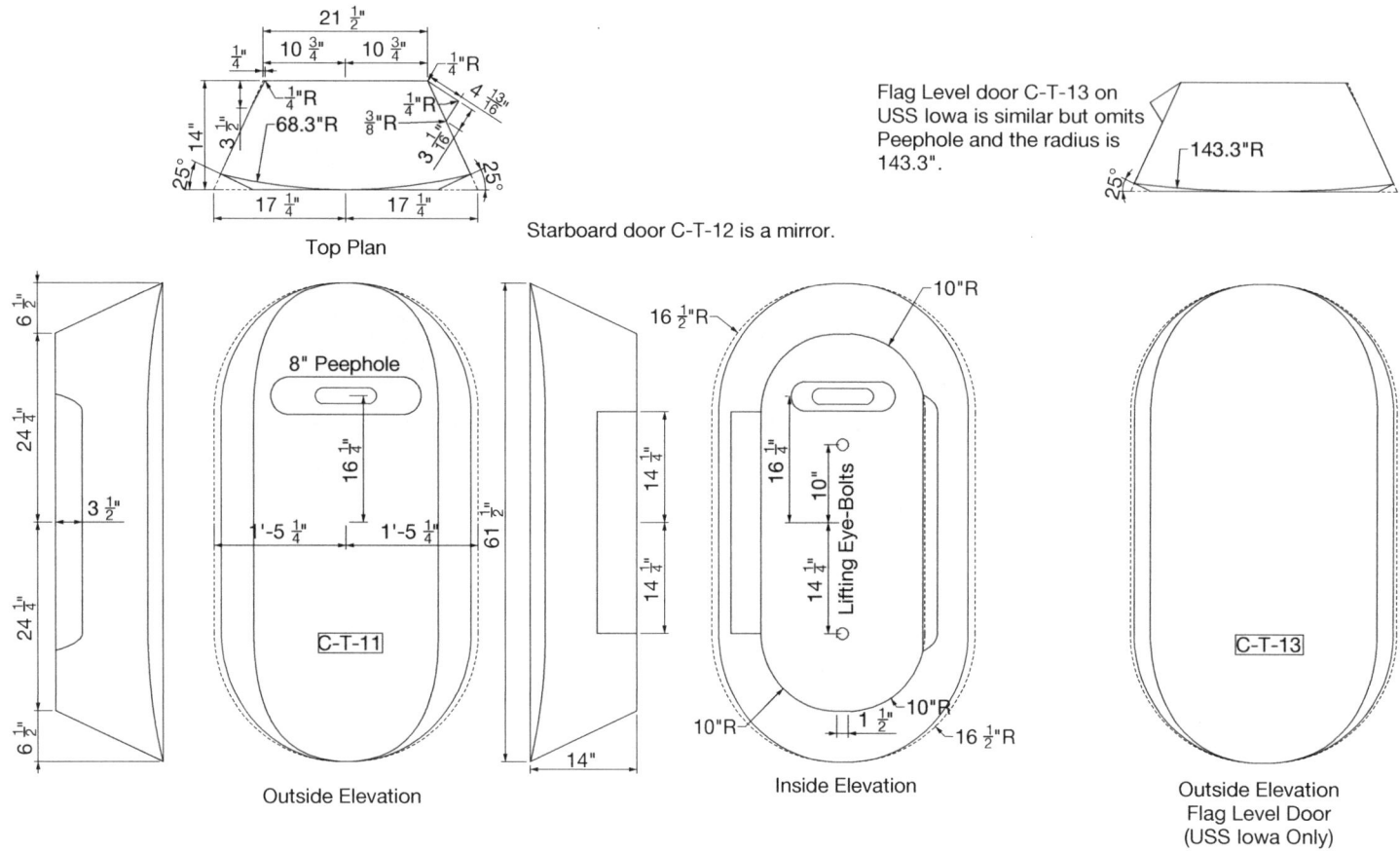

Figure 7.73: Port Navigation Door and Flag Level Door (1:24). The doors at the Navigation Level and fFag Level on USS Iowa have the same basic dimensions. The navigation doors sit in the smaller radius arc of the conning tower plan while the flag door is in the larger radius arc. The difference in radius causes the outer edges to be different as well. The conning tower doors are slightly smaller than their openings to allow room for seals. The port side navigation level door is shown. The starboard door is a mirror.

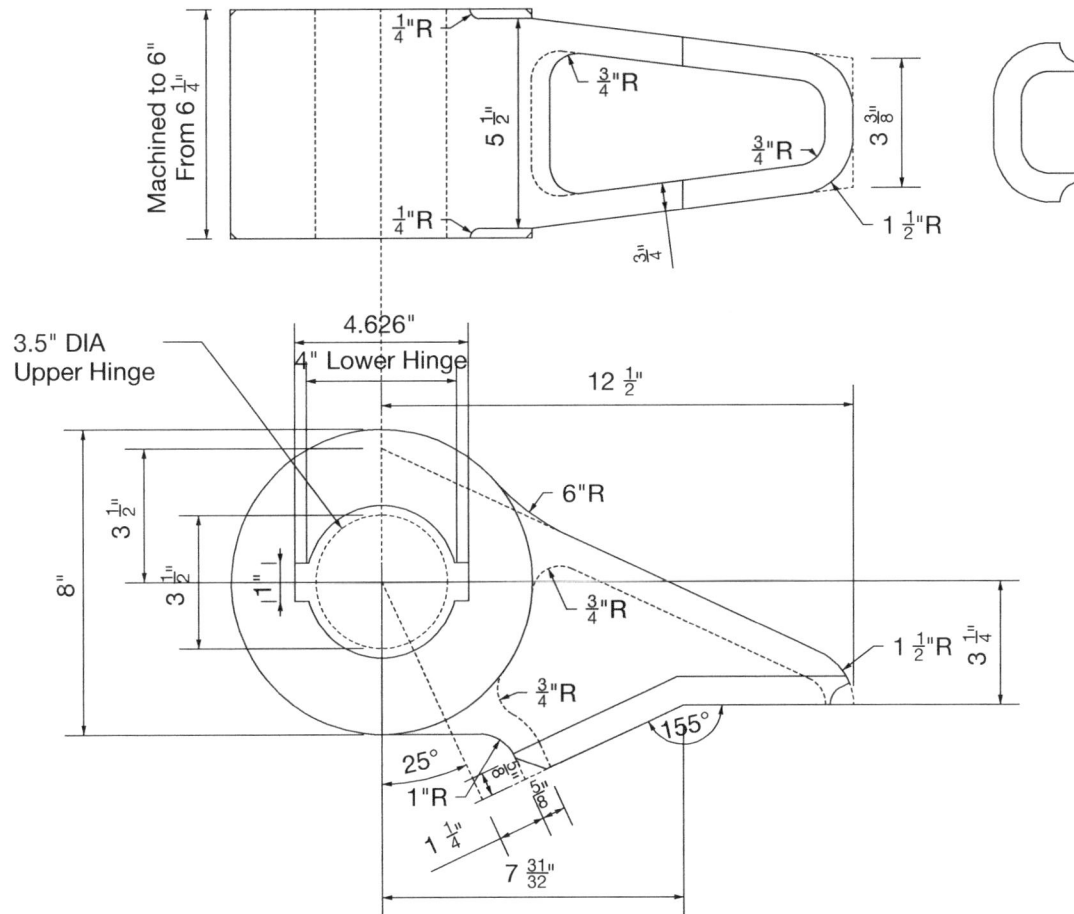

Figure 7.74: Hinge Leaf for Fire Control and Flag Level Doors (1:6). The doors at the Flag and Fire control Levels open away from the centerline. The upper and lower hinge leaves are the same except for the machining of the bore. The lower hinge bore is larger and has slots for keys to link the rotation of the door shaft to the door. The knuckle was cast over-sized so that the yard could grind the faces smooth. The leaves were welded to the door along weld grooves at the side.

Photograph 7.6: Flag Level Door on USS Iowa. The upper bolt toggle pin attachment is missing. As on *USS New Jersey*, the pins and attaching chains are gone. (Jim Kurrasch)

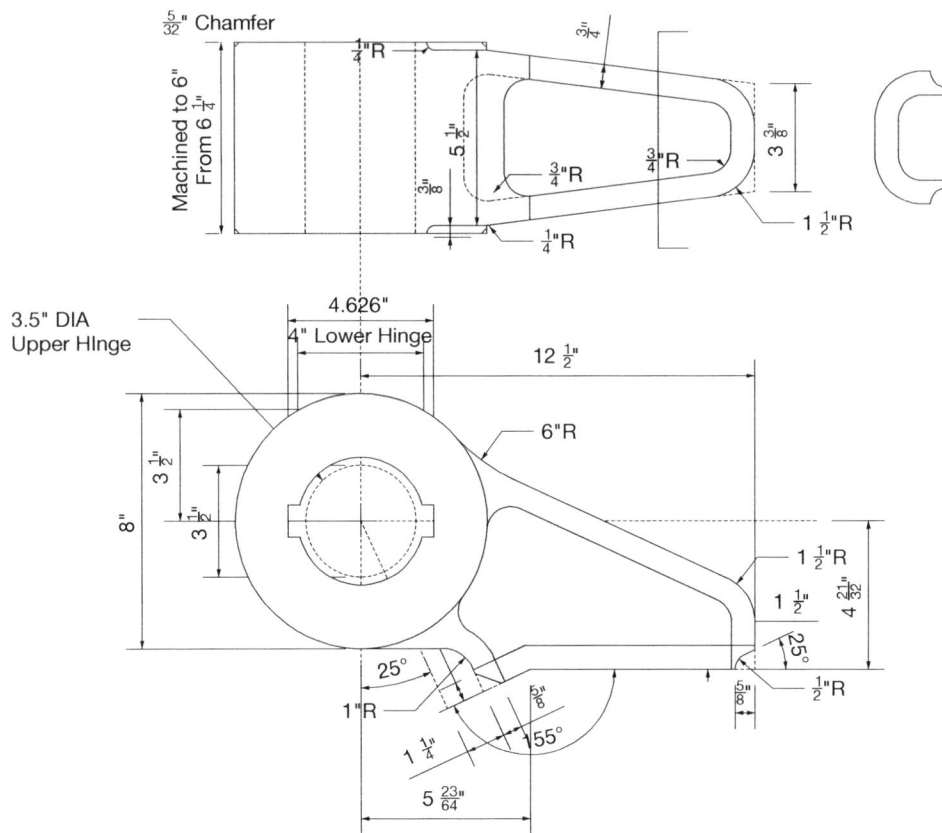

Figure 7.75: Hinge Leaf for the Navigation Level Doors (1:6). The doors at the 04 level open towards the centerline. The machining of the bores is the same as for the doors at the other levels. The door hinges for the navigation level door have a different shape because the navigation level doors are located within the smaller diameter arcs of the conning tower plan while the fire control and flag doors are located within the larger diameter arcs. The yard welded the hinges to the door face along the grooves at the side.

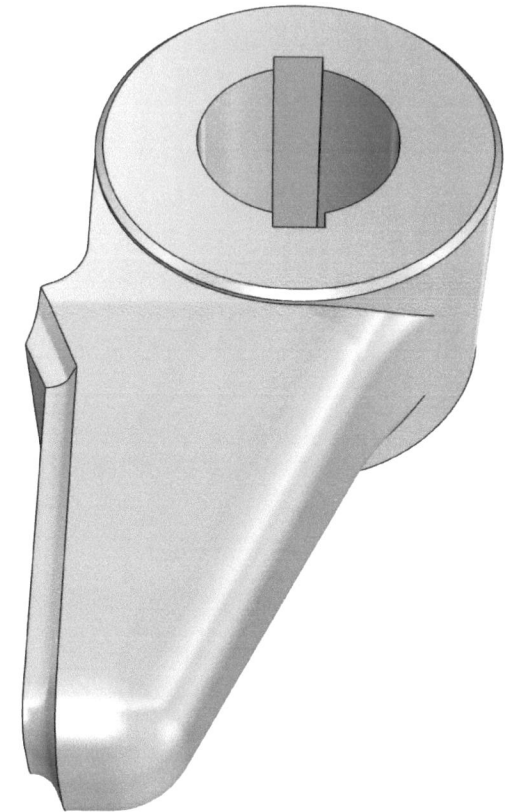

Figure 7.76: Rendering of an upper hinge leaf. The slot in the knuckle opening allows the door shaft to engage with the leaf when opening and closing the door. The recess around the edge at the left is for welding the leaf to the door. The upper leaf was cast the same as the lower leaf but the shaft opening was machined smaller and the slots were omitted.

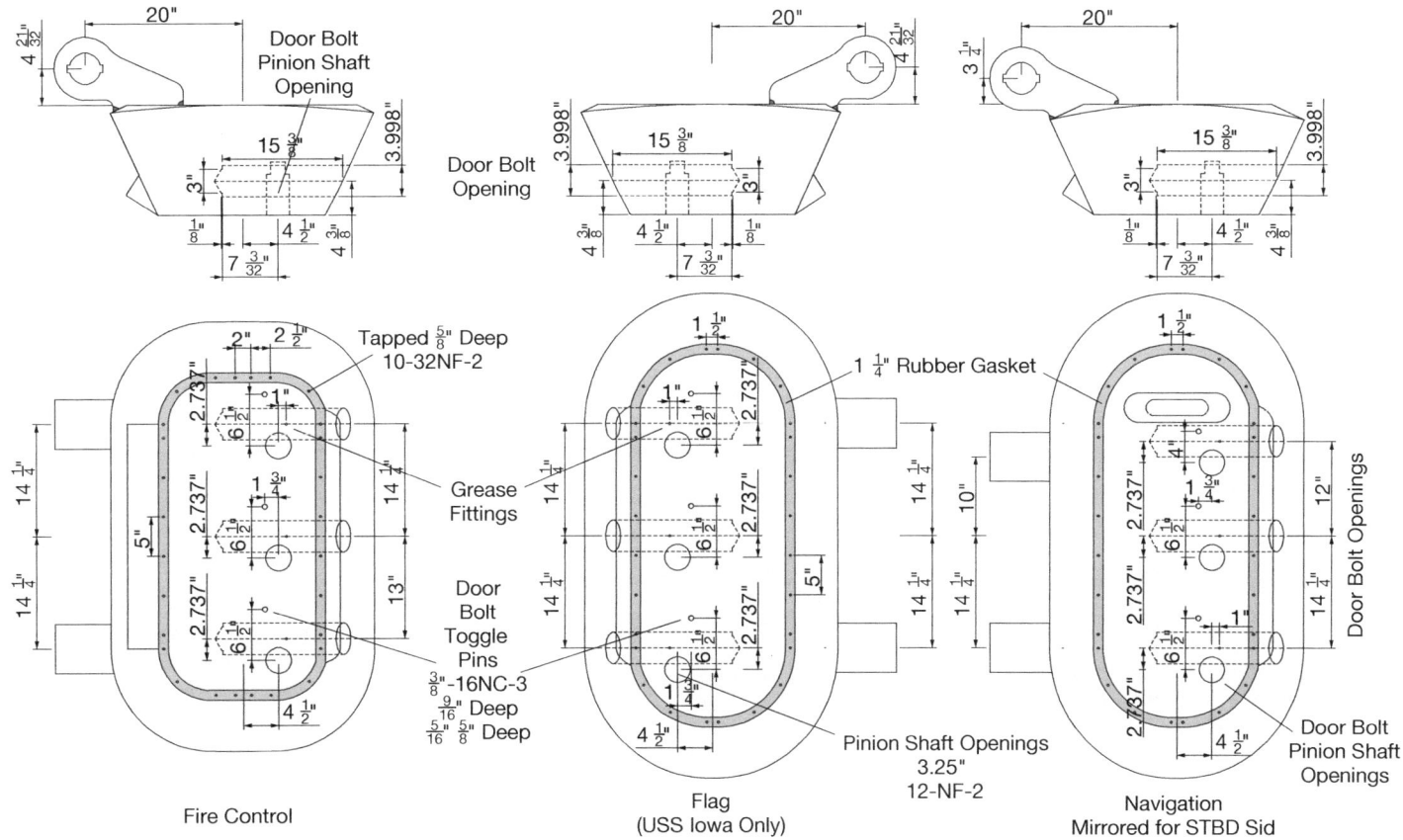

Figure 7.77: Drilling for Bolts, Attachment of Hinges and Gaskets (1:24). This figure shows the location of the hinges and openings for the gasket and door machinery. The yard performed these operations.

Figure 7.79: Lower Hinge Leaf (1:6). The lower leaf supports the weight of the door so it is larger than the upper hinge that holds the door vertically. The steps in the shaft opening through the knuckle align bearings and packing seals.

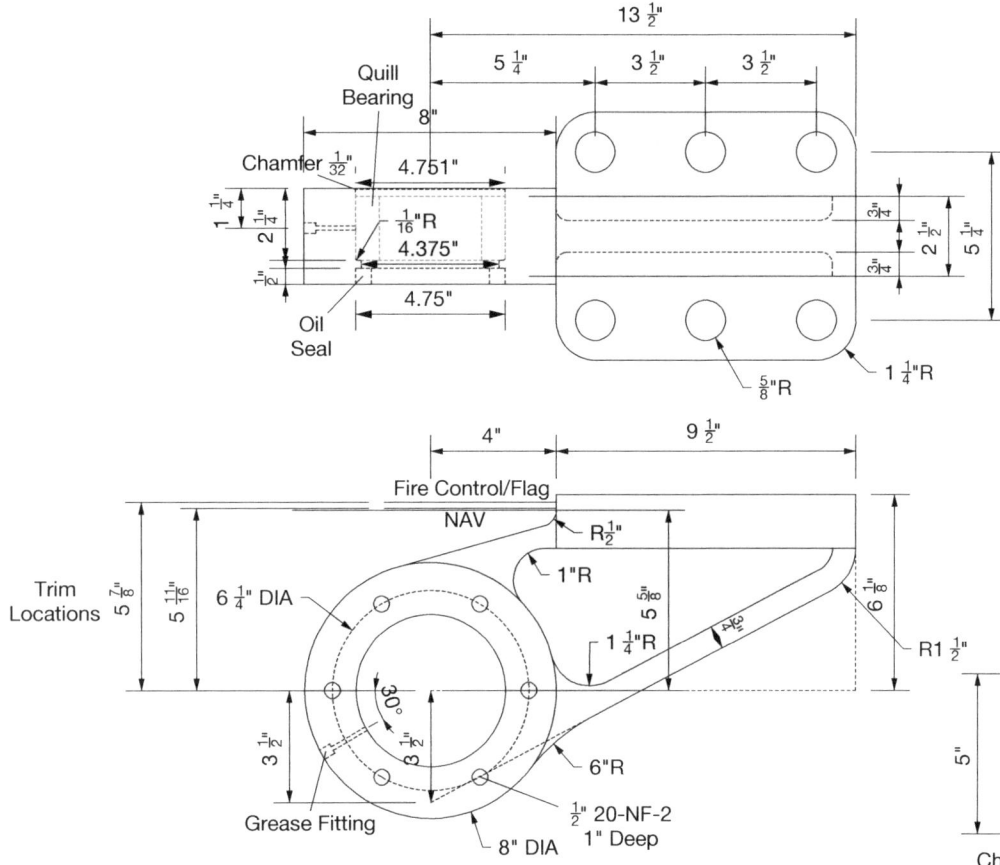

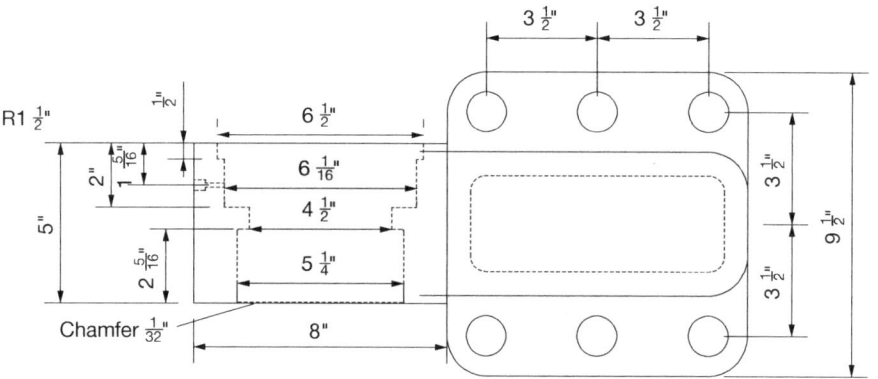

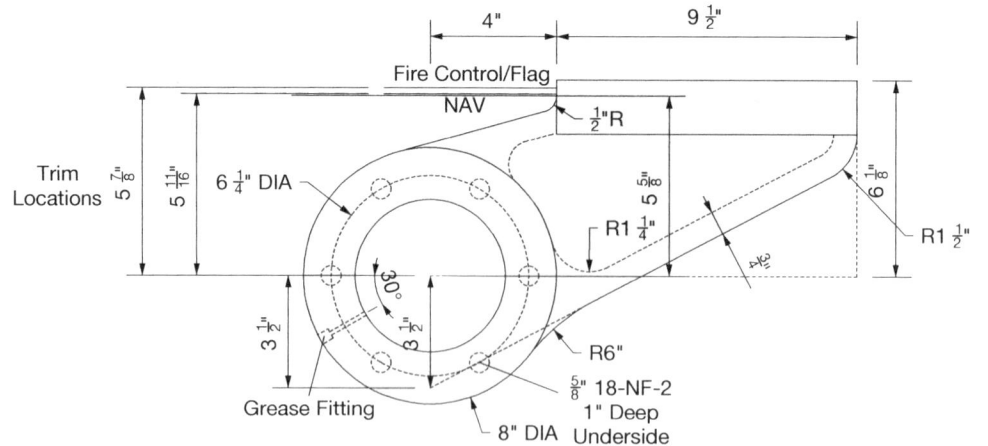

Figure 7.78: Upper Hinge Leaf (1:6). The upper conning tower hinge leaves were cast the same for both opening directions. The yard machined the openings and bolt holes so that left and right leaves were mirrors. The mounting plates were trimmed so that the face was at the location indicated for the particular door. The mounting face was then grinded to a curve to fit the curve of the conning tower. The hinges were bolted to the conning tower.

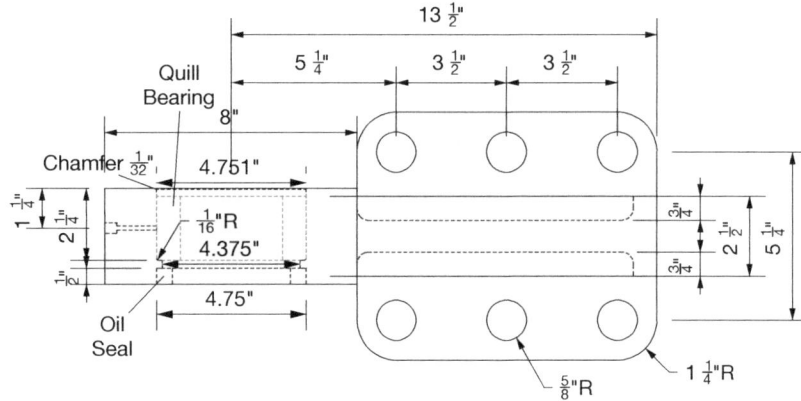

Fire Control and
Flag Levels

Navigation Level

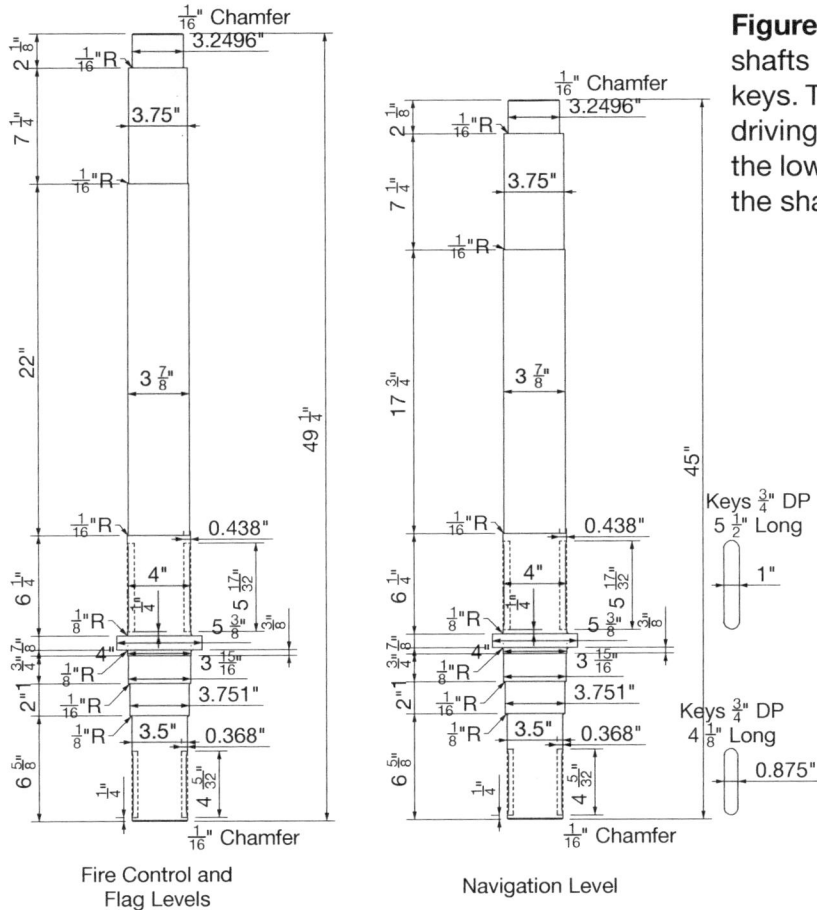

Figure 7.80: Door Shaft (1:12). The door shafts are slotted for two pairs of opposing keys. The pair at the lower end connects to the driving machinery. The pair above connects to the lower hinge of the door. The widest area of the shaft sits on a thrust bearing in the hinge.

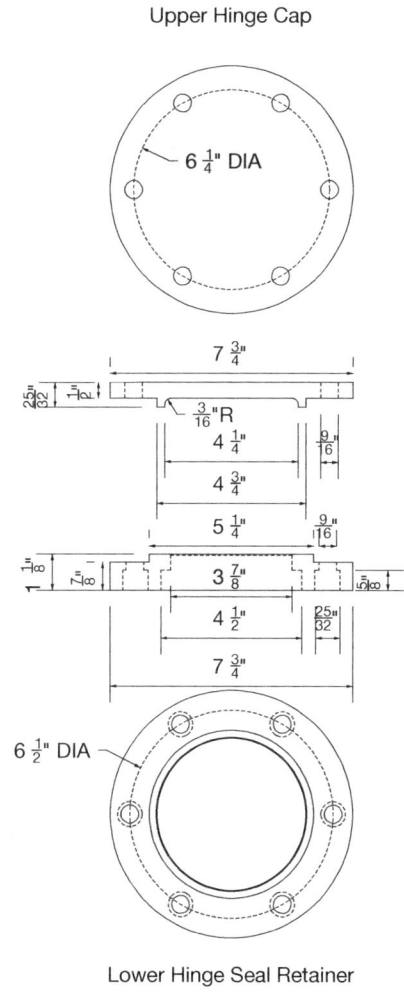

Upper Hinge Cap

Lower Hinge Seal Retainer

Figure 7.82: Rendering of a Lower Hinge Leaf. A thrust bearing rests on the wide step and an oil seal rests on the narrow step above. Another oil seal sits with the narrowest section of the opening and a quill bearing sits below. See Figure 7.84.

Figure 7.81: Upper hinge cap and Lower Seal Retainer (1:6). These hold the bearings and oil seals around the door shafts in place.

Fire Control Door
(Flag Door Mirrored on USS Iowa Only)

Upper Hinge Cap

Quill Bearing

Oil Seal

$\frac{1}{4}$"

Upper Hinge Leaf

Upper Door Hinge Leaf

Door Shaft

$14\frac{1}{4}$"

$14\frac{1}{4}$"

Keys

Lower Door Hinge Leaf

Thrust Bearing

$\frac{1}{4}$"

Oil Seals

Lower Hinge Leaf

Quill Bearing

Keyways for Attaching Drive Machinery

Lower Hinge Seal Retainer

Port Navigation Door
(Starboard Mirrored)

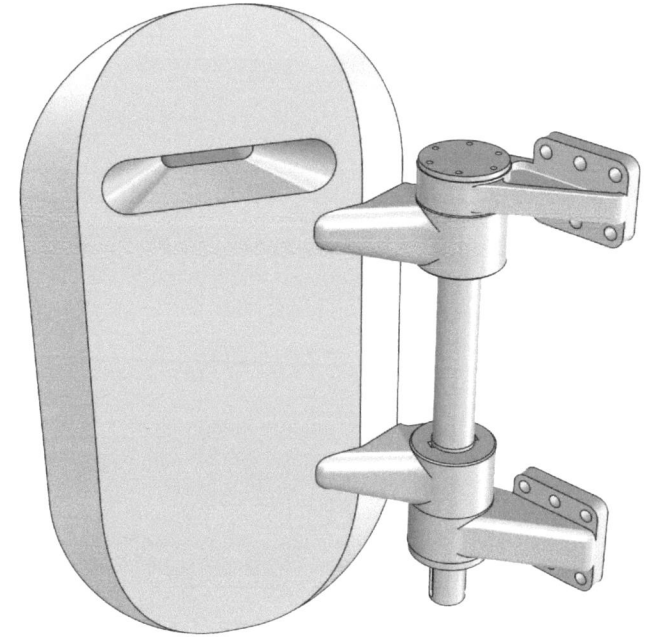

Upper Hinge Cap

Quill Bearing

Oil Seal

$\frac{1}{4}$"

Upper Hinge Leaf

Upper Door Hinge Leaf

10"

Door Shaft

Vertical Door Center

$14\frac{1}{4}$"

Keys

Lower Door Hinge Leaf

Thrust Bearing

$\frac{1}{4}$"

Oil Seals

Lower Hinge Leaf

Quill Bearing

Lower Hinge Seal Retainer

Keyways for Attaching Drive Machinery

Door Hinge Assembly

1:12

Figure 7.83: Rendering showing the door hinge components in place. This example is the port navigation door,

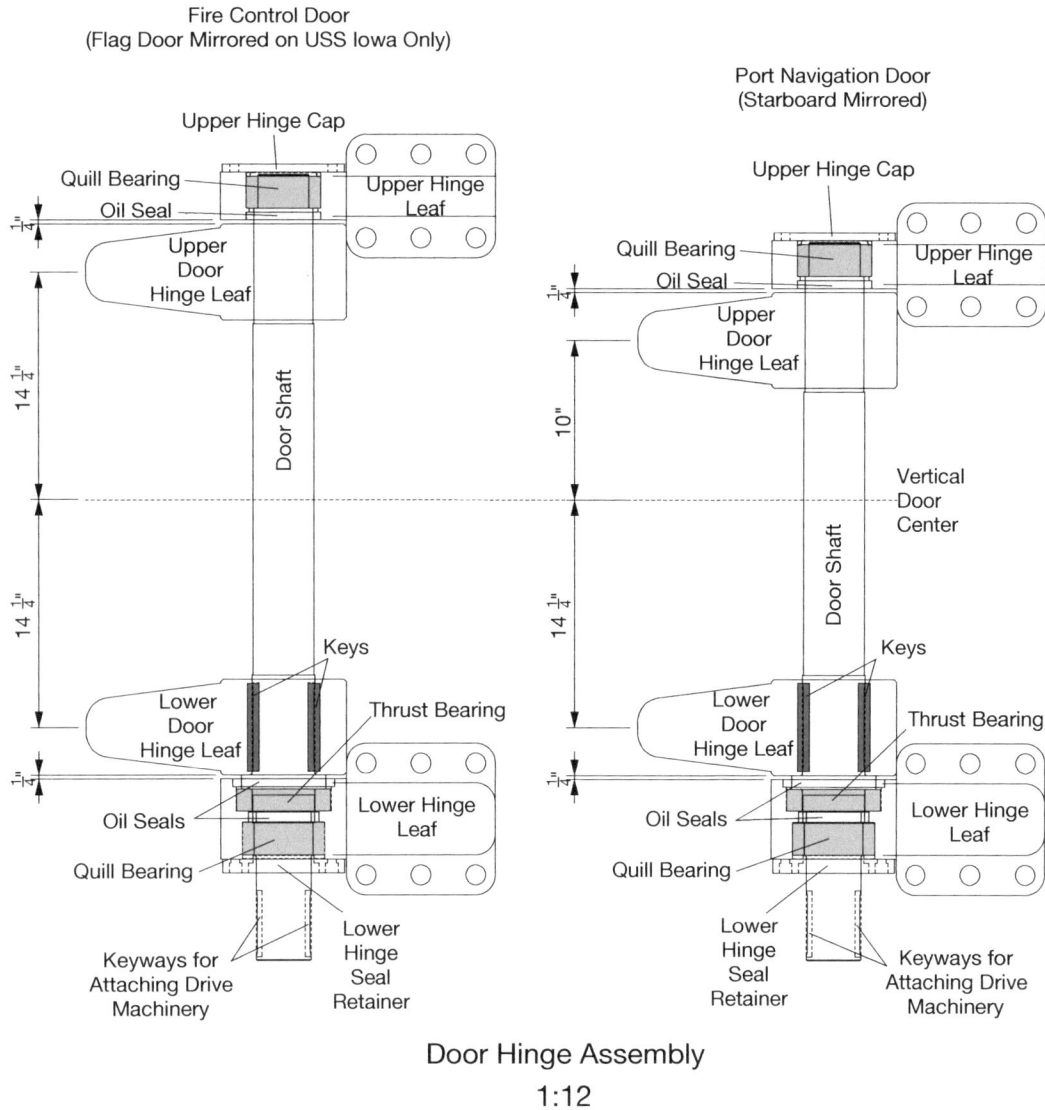

Figure 7.84: Door Hinge Assembly (1:12). This is how the shaft and hinge components are arranged. The widest part of the door shaft rests on a thrust bearing. Quill bearings are located at either end of the assembly. Oil seals fill the remaining gaps in the hinges.

Plan 7.1: (1305BN). The plans for the door mechanism are clear enough to reproduce here. The bolts are steel and the bushings are made of bronze. At the left is a view of a door bolt. There is a rack gear at the bottom for moving the bolt. A hole runs through the center to prevent compressing air as the bolt opens. At the lower left is the inner bushing installed at the bottom of the bolt hole that guides the bolt. At the lower right is the outer bushing that is installed just inside the bolt hole opening.

(5) DOOR BOLT BUSHING (INNER)

(6) DOOR BOLT BUSHING (OUTER)

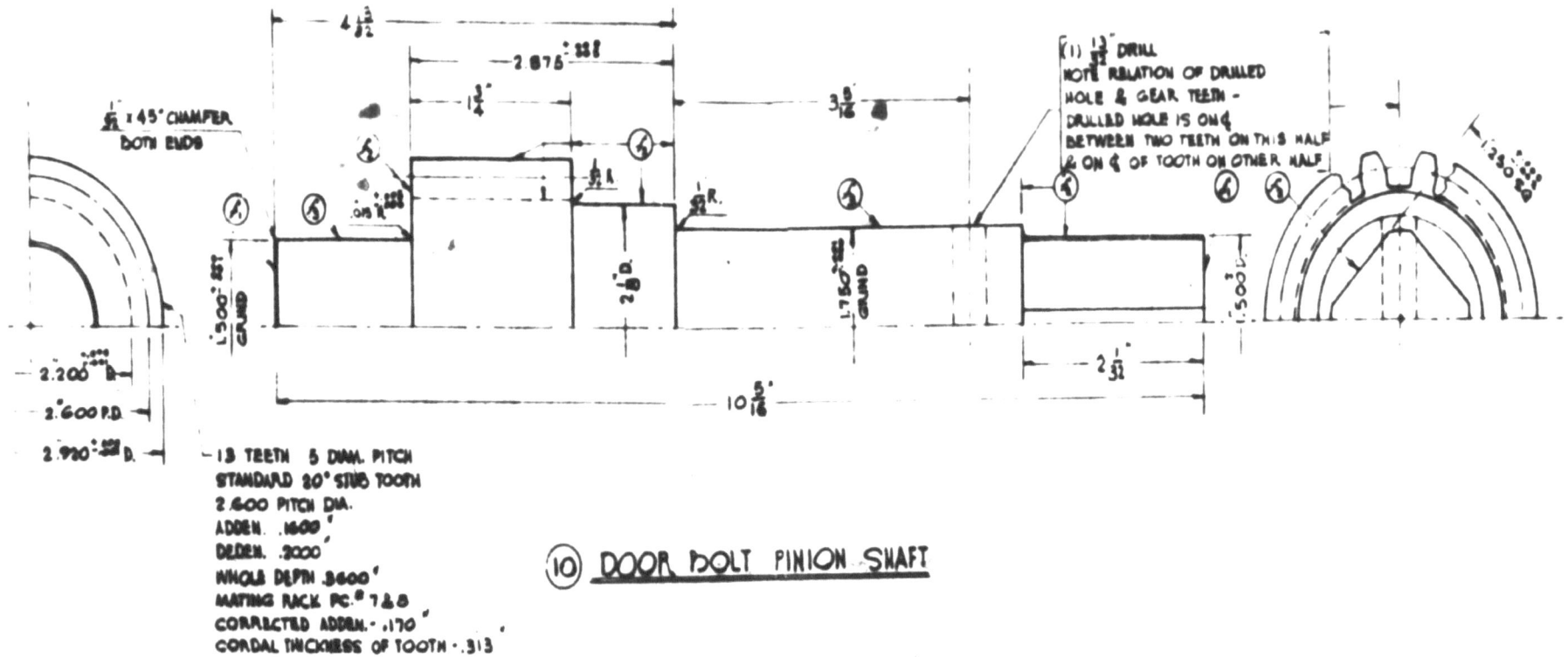

(10) DOOR BOLT PINION SHAFT

Plan 7.2: (1305BN). The door bolt pinion shaft is the heart of the door mechanism. The hole for the pinion was drilled countersunk at three diameters. The outer 2" were drilled and tapped for 5.25" 12-NF-2. The hole was then drilled at 3" to $5\frac{9}{32}$" deep. The full depth of $6\frac{25}{32}$" was drilled to 1.875" diameter. The bushing shown was inserted into the deepest part of the opening to guide the pinion. After the pinion was in place, the pinion shaft plug was screwed into the opening to hold the pinion in place.

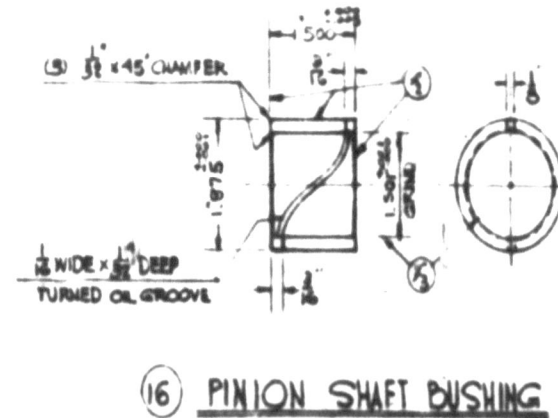

(16) PINION SHAFT BUSHING

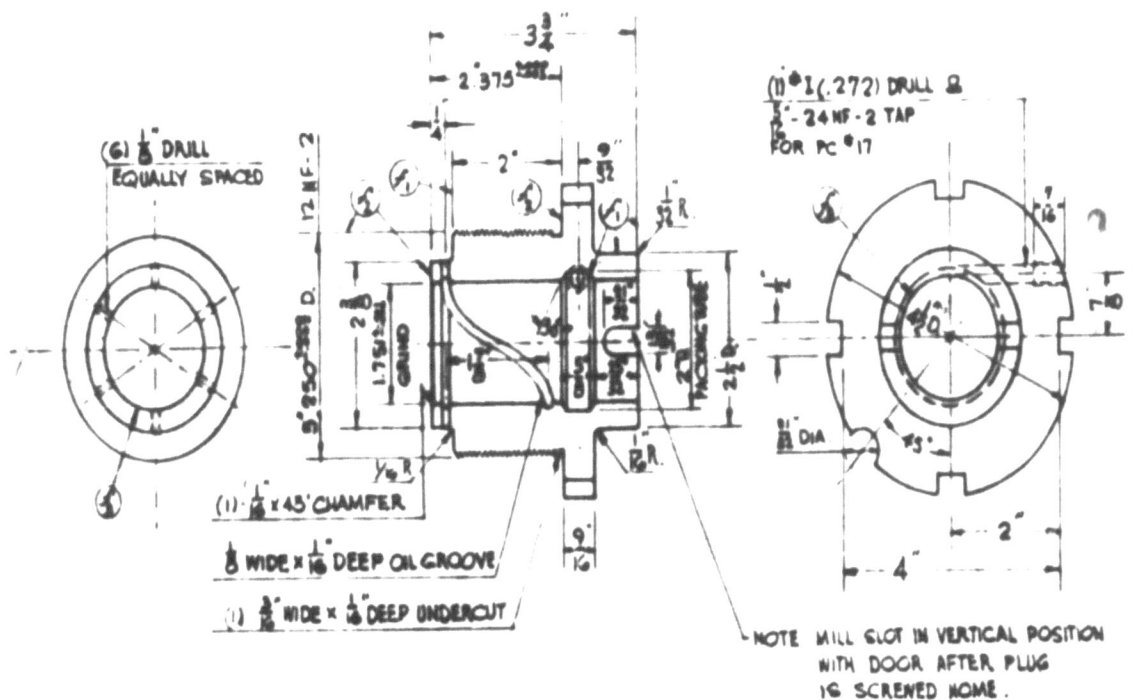

Photograph 7.4: Pinion Shaft Plug. This a pinion shaft plug with pinion shaft protruding on the Flag Level door of *USS Iowa*. The final orientation of the plug depended upon how the hole was tapped. In this example the location of packing plunger ended up at the bottom. The slots for the bolt toggle pin were machined by the yard after the vertical position was determined. The slots had to be vertical in order to match the hole for the pin in the pinion shard. A grease fitting is above the plug. (Jim Kurrasch)

Plan 7.3: (1305BN) The Pinion Shaft Plug was screwed into the door to hold the end of the Pinion Shaft in place. When in its final position, a hole was drilled and tapped into the door face at the curved indentation. A Packing Plunger was screwed into this hole to keep the plug from rotating loose. A slot was machined at the top and bottom of the end of the plug. A Bolt Toggle Pin could be inserted through the slots and a hole in the Pinion Shaft to lock the bolt in position.

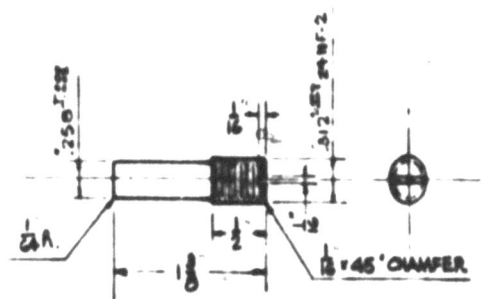

Plan 7.5: (1305BN) These views are of the crank used to extend and retract the door bolts and to manually open the door. One crank was provided for each level of the conning tower.

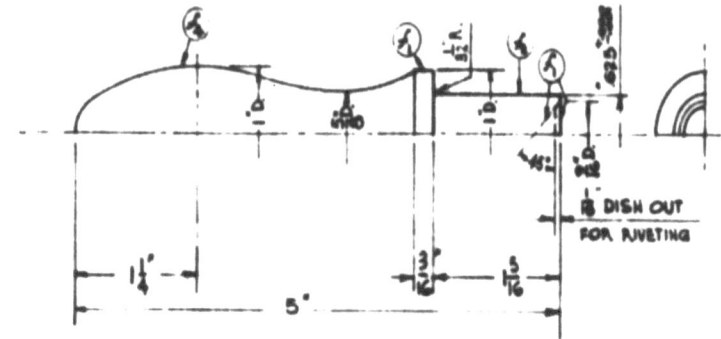

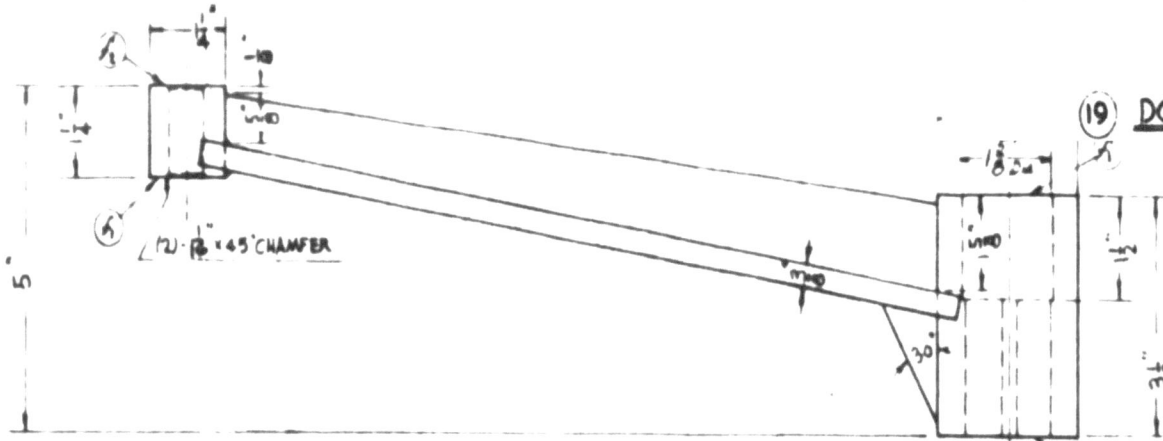

DISH OUT FOR RIVETING

(19) DOOR BOLT CRANK HANDLE

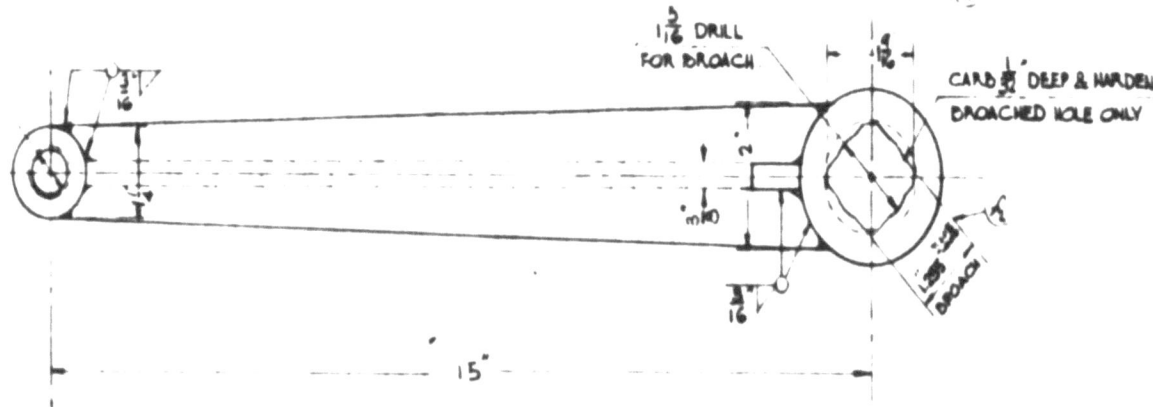

(18) DOOR BOLT CRANK

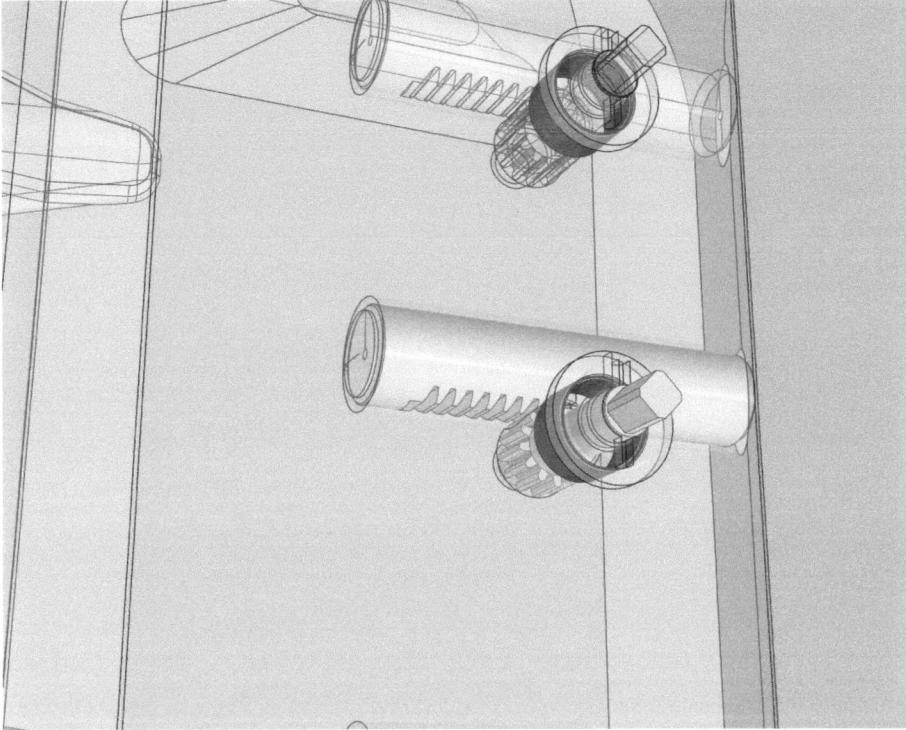

Figure 7.85: Ghosted Rendering Showing How the Door Mechanism Extended and Retracted the Bolts. The crank was placed on the square end of the pinion shaft and was used to rotate the pinion shaft. The gears on the pinion shaft moved the bolt.

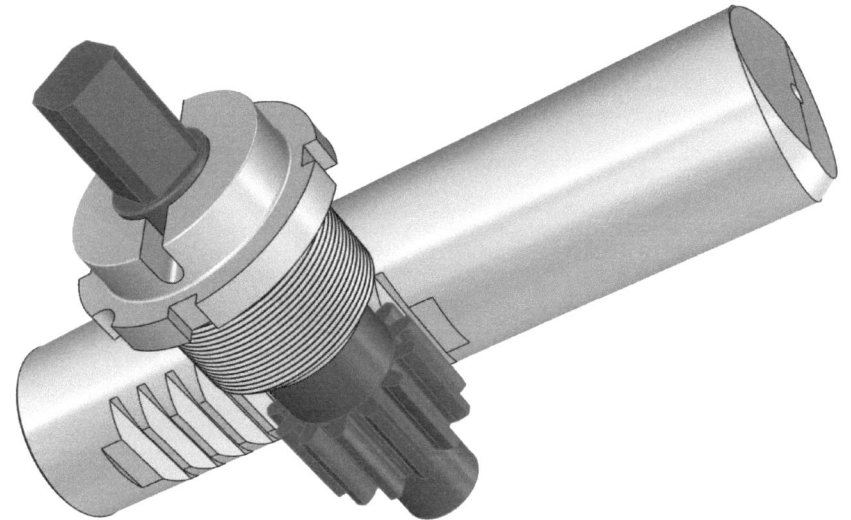

Figure 7.86: Door Opening Mechanism. This view is from below, looking up. The three major components are the bolt (running horizontally), pinion shaft (darkened), and Pinion Shaft Plug (around the pinion shaft). The end of the bolt is angled to clear the door opening. The base of the pinion shaft plug has four rectangular recesses for screwing it in to the door. There is a circular recess for the packing plunger at the lower right. The vertical recess is for inserting a bolt toggle pin through a hole in the pinion shaft. The pinion shaft and bolt could not move with the pin in place.

Photograph 7.7: Door Restraint on USS New Jersey. The holes running through the bolts allow air to escape when opening are visible.

Plan 7.6: (1305BN) Brackets were welded next to each door at the center bolt to hold the door open. The bracket shown at the top was used on USS Iowa at the Flag Level. The bracket at the bottom was used for all the other doors.

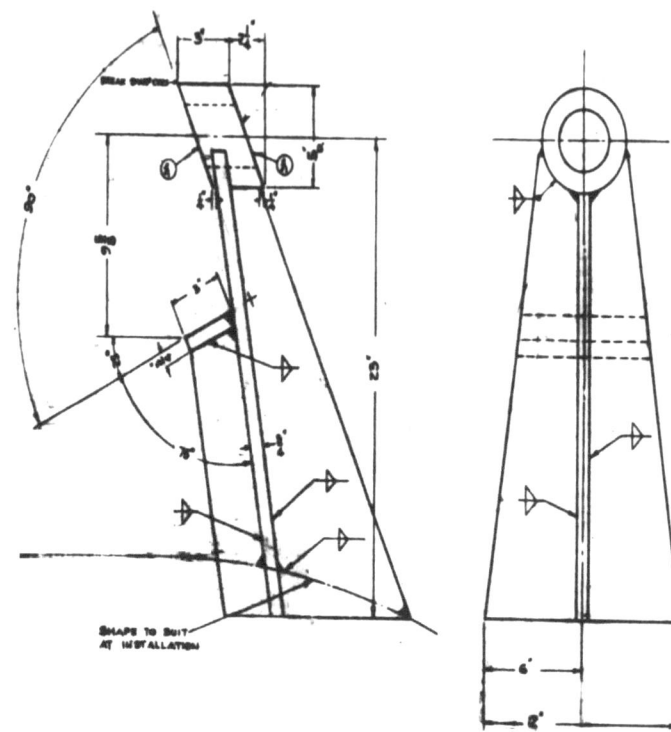

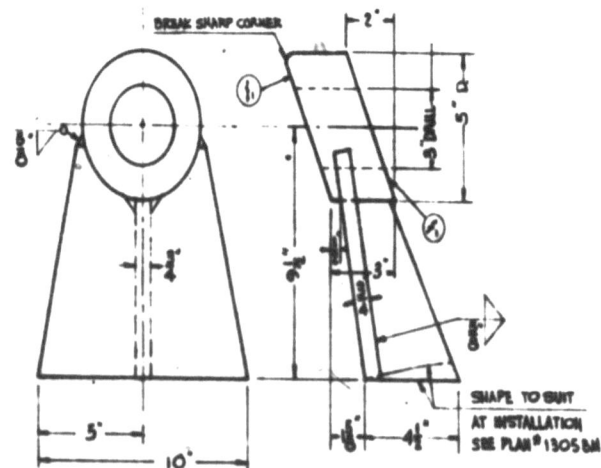

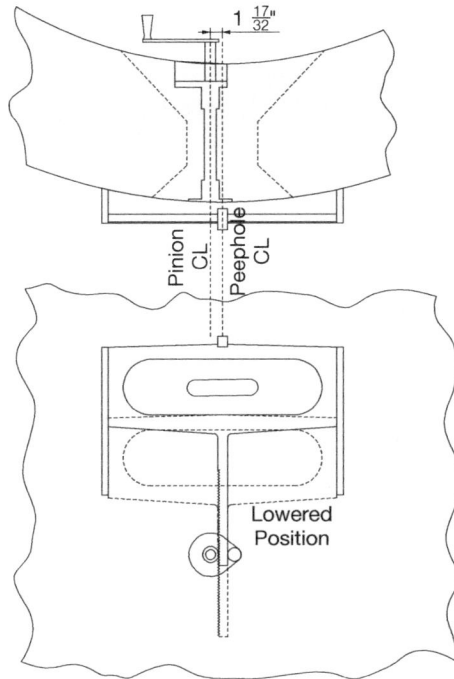

Figure 7.87: Airtight Peephole Cover (1:24). The airtight covers for the peepholes are raised and lowered using a crank. The crank turns a geared pinion shaft meshed with a rack gear attached to the cover. The pinion shaft is attached to a ratchet mechanism that prevents gravity from opening the cover. The pinion shaft is offset from the centerline of the peephole in order to align with the rack gear.

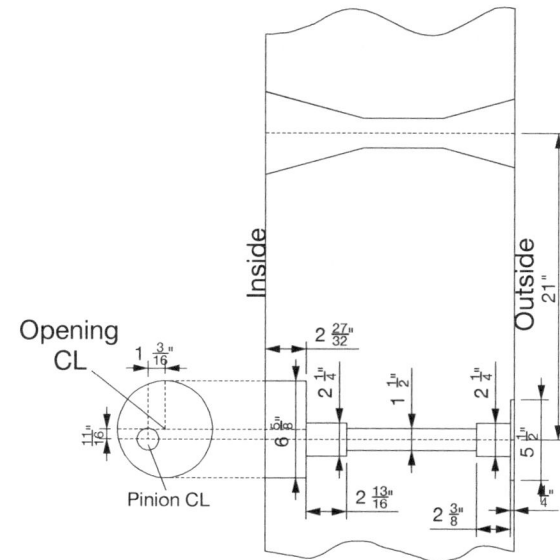

Figure 7.88: Opening for Peephole Cover Mechanism (1:12). The opening for the pinion shaft was drilled in four countersunk segments. The three outer segments are centered on the pinion axis. The center of the innermost segment is offset from that axis in order to accommodate the ratchet mechanism.

Photograph 7.8: An airtight cover over a peephole at the Fire Control Level of USS New Jersey. The rack and pinion gears that raise and lower the cover are visible at the center. The structures that extend below the cover on either side are the tracks that guide the cover when it moves. A joint between side plate passes through the peephole and the weld is very prominent.

Chapter 8: Turret Armor

The turret armor encloses the gunhouse. The Shelf Plate acts as the gunhouse foundation and is the upper structural level of a turret. The shelf plate also protects the underside of the gunhouse. However, the Navy did not consider the Shelf Plate to be part of the ships' armor. The shelf plate rotates above the barbettes but there is no structural connection between them. The Shelf Plate is 80# STS. The gunhouse has a bustle at the rear that offsets the torque caused by the weight of the guns at the front. A 2-inch STS doubling plate is connected to underside of the shelf plate under the bustle using rivets at the edge and quilting rivets at the interior. The area between the doubling plate and main shelf plate was painted on both sides before assembly with red lead to prevent rust. The shelf plate is flat for most of its length but slopes upward at rear.

The final size of the shelf plate was templated from the armor delivered to the yard. The upper shelf plate is divided into front and rear sections. The STS plates making up each section are welded together. The front and rear sections are riveted together along a scarf joint. The main hatch and emergency hatch are located on the slope. The upward slope of the shelf plate orients the hatch in such a way that crew could bend over to enter, rather than having to crawl on the deck to access the hatch..

Armor attachments are located on the top face of the shelf plate. A raised key bar runs along the front edge of the shelf plate to align the bottom of the front armor plate. A scalloped boundary angle runs along the sides and rear of the shelf plate. The boundary angles were machined at the yard out of angle bar so that one edge of the angle was scalloped. The straight edge side of the angle was riveted to the Shelf Plate. The scalloped side of the angle was oriented vertically and was welded to the shelf plate. Backing bulkheads are welded to the side boundary angle and the rear armor plate is welded to the rear boundary angle.

8.27. Backing Bulkheads

There are backing bulkheads at the front and sides. The side backing bulkheads were assembled on the shelf plate before the turret was lifted in place. The front backing bulkhead was installed after the guns were in place. The front backing bulkhead is part of a weldment. The front backing plates are 2.5" Class B armor. In addition to the backing plate, this weldment has a kick plate at the base, a riveting bar at the top, and reinforcements along the inside bottom. The front backing plate was riveted to the shelf plate using four rows of rivets. The side backing plates are 30# STS. The bottom $5\frac{3}{8}$" of the backing bulkhead is vertical. Above the vertical section, the backing bulkhead follows the angle of the Side Plates, offset by one inch. The side backing bulkheads are attached to the Shelf Plate through boundary angles that run along each side, inside the backing bulkheads.

8.28. Armor Plates

Each gunhouse has eleven armor plates: Front Plate (T-1), Front Side Plates, Left and Right (T-2 & T-3), Rear Side Plates, Left and Right (T-4 & T-5), Rear Plate (T6), and Top Plates (T-7, T-8, T-9, T-10, and T-11). The mill machined the plate joints and drilled and tapped all holes at the top and sides. However, the yards drilled and tapped all the holes at the bottom of the plates to connect to the shelf plate. The turret bolts for the completed *Iowa*-class battleships were made by the New York Navy Yard. Bolts for USS *Illinois* and USS *Kentucky* were made by the Navy Yard Norfolk. The New York Navy Yard supplied the mills with gauges for tapping bolt holes and loaned a supply of bolts so that the mills could assemble the turrets for inspection. The bolt threads were coated with a mixture of 5% graphite and 95% tallow for lubrication as they were inserted and removed. The mills were required to fill the bolt holes with temporary plugs to protect the threads during shipping.

8.28.1. Front Plate

The Front Plates (T-1) for the three turrets were the heaviest plates (236 tons) used in the *Iowa*-class battleships. The Front Plates are 17-inches of Class B armor. The Front Plates were machined to sit at a slope of 36° 10' 18" (106" rise over $77\frac{1}{2}$" run). The Front Plates are unusual in that they are the only Class B armor that are mounted on backing bulkheads. This was a design decision made in the *North Carolina*-class that was carried over to the *Iowa*-class.

The front plate has three gunport openings. The openings are cylindrical at the top and bottom with their axes running through the trunnion axis and allow the gun to elevate 45 degrees and depress 2 degrees. However, the machinery for turret no. 2 only allowed its guns to depress to zero degrees.

The edges of the front plate have joints machined into them for mechanical connections. Along the top edge there is a hooked key joint. The tongue of the joint was machined for bolts to hold the Roof Plate (T-7) in place. At each side the rear face has a groove that matches a tongue in the side plates. Along the bottom there is a keyway that matches a key in the Shelf Plate. The mill drilled and tapped connecting bolt holes on the inside face.

8.28.2. Side Plates

The gunhouse has two facets at the side so the each side's armor consists of a Front Side Plate and Rear Side Plate (T-2 & T-4, T-3 & T-5). The side plates are 9.5" of Class A armor that are angled along the centerline and turret axis. The side plates have openings for the rangefinder, pointer, and trainer periscopes. Cast hoods were bolted over these openings after the gunhouse armor was installed. The rangefinders in Turret 1 shipped water, had poor visibility, and added weight. As part of preparation for the Korean War, the rangefinder and hoods were removed from Turret 1 on all four ships and the opening was plated over.

The side plates have a different type of joint machined at each edge. Along the joint between the two side plates, the mill machined an hour-glass keyway. The mill machined a tongue into the forward edge of the front side plates that match the groove at the side of the front plate. The aft side plates have groove along the inside face that mate with a tongue on the rear plate. There is a rabbet along the lower edge. Along the top there is a tongue that matches grooves in the Top Plates. The mill drilled and tapped holes in that tongue to secure the Top Plates with bolts.

The mill also drilled and tapped bolt holes on the inside face for mounting to the backing plate.

8.28.3. Rear Plate

The Rear Plate (T-6) is twelve inches of Class A armor and is the only plate on the gunhouse oriented vertically. It is also the only turret plate that is not designed to be flat. The center of the plate is circular and the outer ends are straight. Presumably the rear plate was made thicker than the sides in order to balance the turret. The upper edges and sides have tongues that fit the roof plate (T-11) and Rear Side plates. The mill drilled and tapped bolt holes along the upper tongue to secure the Roof Plate (T-11). The bottom edges have a rabbet that forms a recess for the shelf plate. The rear plate is an example of Class A armor that was used structurally.

8.28.4. Top Plates

Five plates make up the top of the gunhouse armor (T-7, T-8, T-9, T-10, & T-11). The Top Plates meet at scarf joints. The scarf joints are oriented so that the center plate (T-9) can be removed to allow the yoke and breach assemblies to be lifted out of the turret for gun replacement. Along the outer edges, except over the rangefinder opening, the Top Plates have a groove that fits the tongue at the top of the front, side, and rear plates. The mill machined holes through the Top Plates along the groove so that they could be bolted to the side and rear plates below. The rearmost Top Plate (T-11) has two openings for periscopes for spotting the fall of shots. One turret difference is that the rearmost Top Plate (T-11) for Turret Nos. 2 and 3 had an additional hole machined to run a power cable from inside the turret to a quad 40mm mount. USS *Iowa* lacked the mount and hole on Turret No. 2. These holes were filled in when the 40mm mounts were removed.

8.29. Backing Bulkheads

The Front and Side Plates of the gunhouse are mounted on backing bulkheads. They have openings that correspond to the openings in the armor plate and have through holes that match the bolt holes on the inside face of the armor plates. These bolt openings were templated from the armor plates as delivered from the mill. The backing plates were offset one inch from the designed inside face of the armor plates. The gap was intended to compensate for any irregularities on the inside face.

Backing bulkheads were erected at the sides and front of the shelf plate. The forward backing bulkhead is 2.5" of Class B plate. The forward backing weldment incorporates a kick plate that was riveted to the shelf plate. The upper edge of the weldment has a plate that serves to a rivet point for the forward roof plate. The side backing bulkheads are 30# STS. The side plates were welded to the kick plate on the shelf plate along the scalloped vertical edge. To account for any irregularities in the armor plates, the backing bulkheads were offset one inch from the designed location of the armor. The plans specified that rivets used to connect to the shelf plate should have a bead weld around the edge.

8.30. Plate Installation

The side plates were installed first. After they were in position, the yard hammered a keyway into the joint, permanently locking the two plates into a single unit. The side plates were bolted to the backing plates and tap riveted to the shelf plate. Presumably, roof plates were put in place temporarily to align the sides. The front plate was slipped over the gun barrels and over the key along the front of the shelf plate. The front plate was bolted to the front backing plate and tap riveted to the shelf plate. Collars were installed at the openings in the front and side plates to retain cement. Covers were welded to the inside of the backing bulkhead over the bolts to restrain any bolt that broke off due to an impact. The rear plate was thick enough to be welded on the inside face.

The rear plates were attached to the shelf plate differently from the front and side plates.

There is no backing plate at the rear. Instead, the rear plate was welded at the bottom directly to the scallops on the kick plate at the rear of the shelf plate. The rear plate was welded to the side backing bulkheads and roof plate along scalloped angles. Like the front and side plates, tap rivets were driven up through the shelf plate into the bottom of the armor.

A mixture of one part cement and two parts haydite filled the gap between the armor and plate so that the force of an impact would be spread across the entire face of the backing bulkheads rather than just to the bolts. The collars prevented cement from leaking out of the openings. In places where he gap between the armor

and backing plate would not allow haydite to flow, the plans specify that a mixture of cement, pine tar, and shellac was to be gunned into the gap.

The top plates could then be put in place. The top plates were bolted together and to front, rear, and side plates. The rear top plate (T-11) was welded to the rear plate along a scalloped angle. The four outer top plates (T-7, T-8, T-10, and T-11) were welded to the side backing bulkhead. The forward top plate was tap riveted to the front backing bulkhead. The moveable center plate (T-9) was bolted to the side backing bulkhead. Finally a key was hammered into the joint between the front plate and the forward top plate and was welded in place.

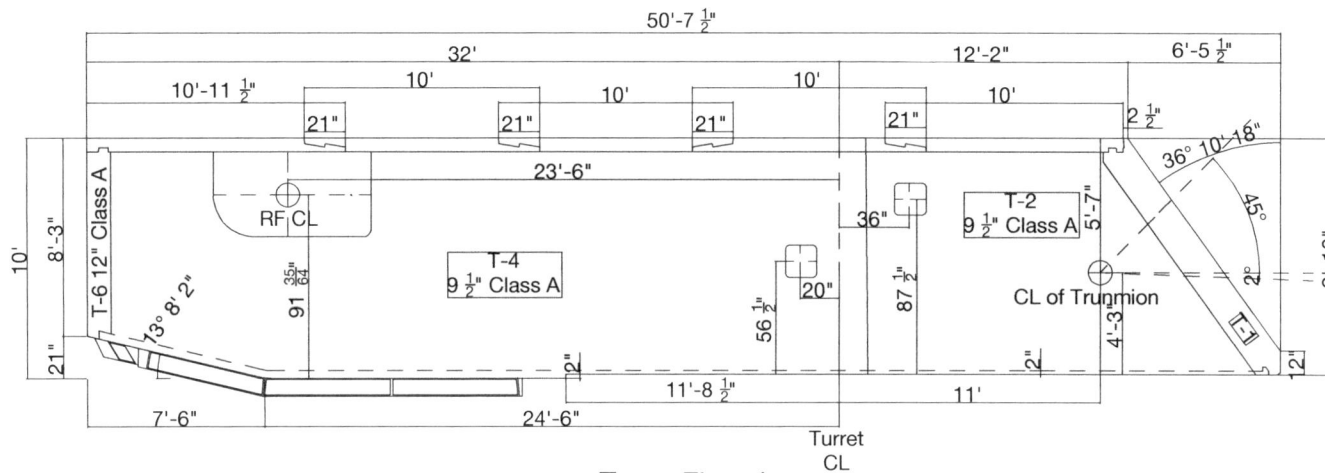

Figure 8.1: Turret Side Elevation (1:96)

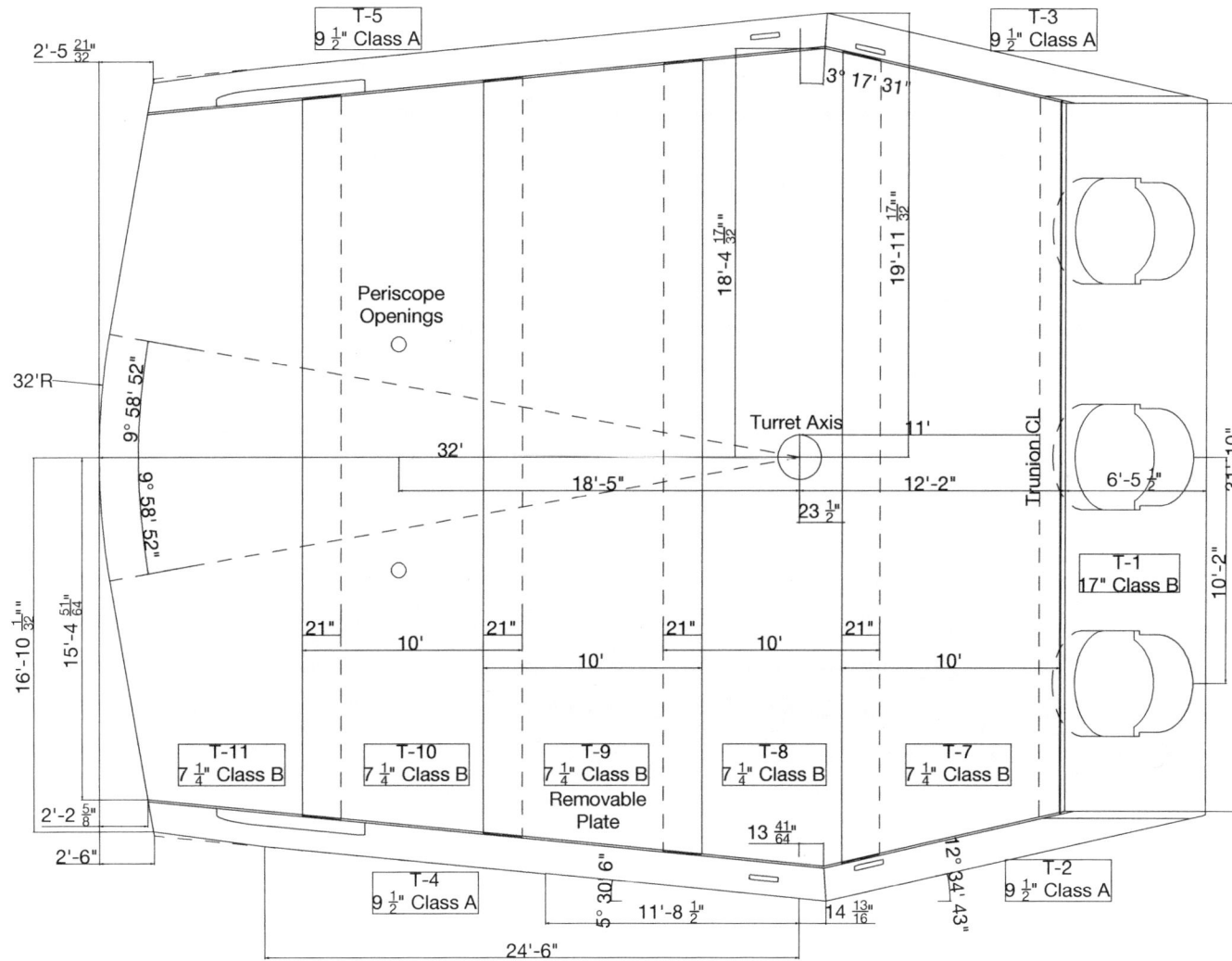

Figure 8.2: Turret Plan (1:96)

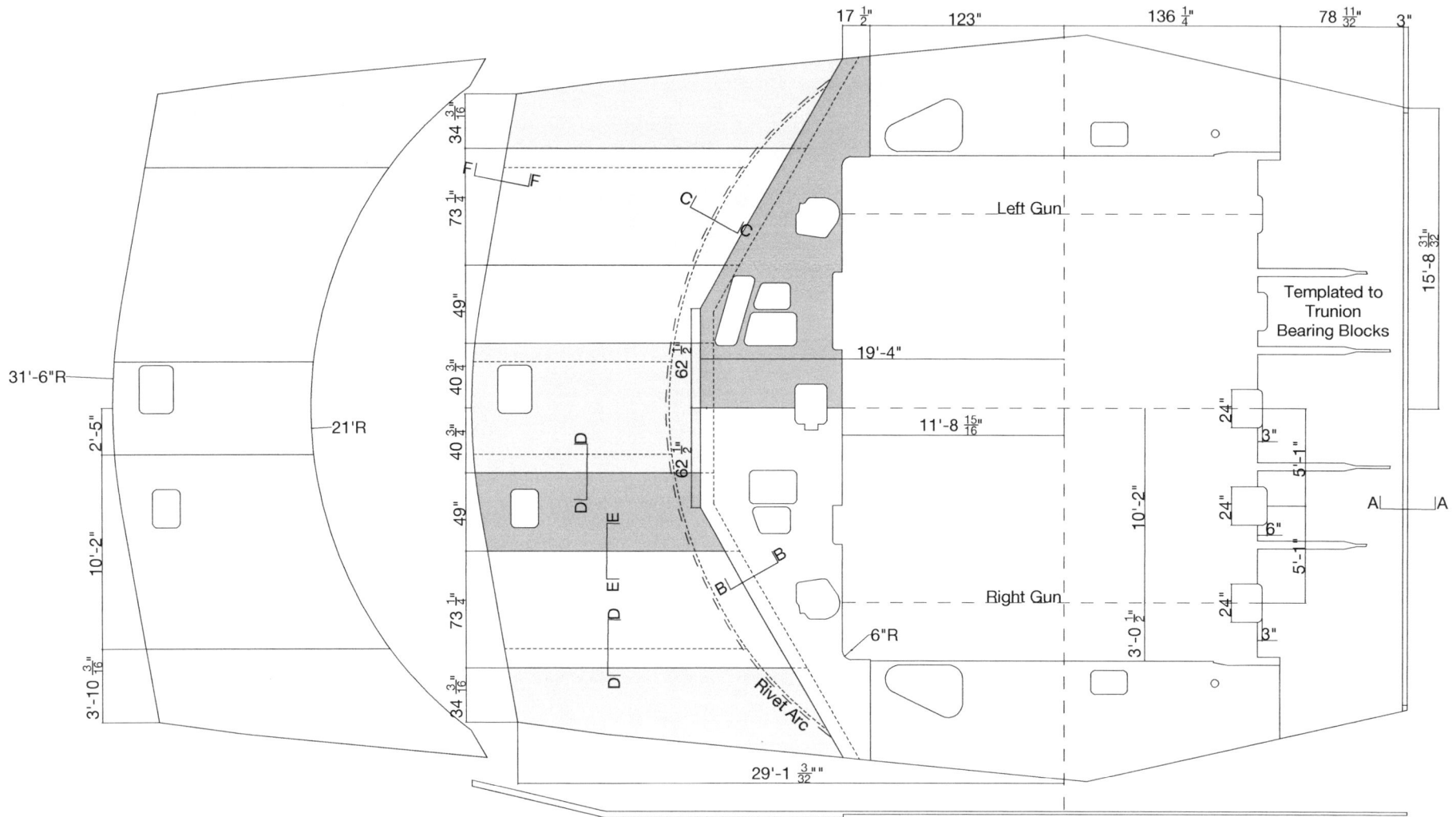

Figure 8.3: Shelf Plate and Doubler (1:96).
The doubler is at the left and the upper shelf plate at the right. The profile is shown below. The individual plates are indicated with shading. The shelf plate is 80# STS. Under the turret bustle the doubling plate adds another 80# STS to that thickness.

Plan 8.1: (7201ZT) Section A–Key at Forward Edge of Shelf Plate. An original plan showing the key bar in the shelf plate that aligns the front armor plate.

Plan 8.3: (7201ZT) Section C–Connection of Doubling Plate. The doubling plate was attached using rivets.

Plan 8.2: (7201ZT) Section B—Scarf Joint in Shelf Plate. The section of the shelf plate under the turret bustle was connected using a scarf joint.

Plan 8.4: (7201ZT) Section D–Connection of Doubler to Shelf Plate. This plan shows how the shelf plate and doubler were attached where their plate seams were close together.

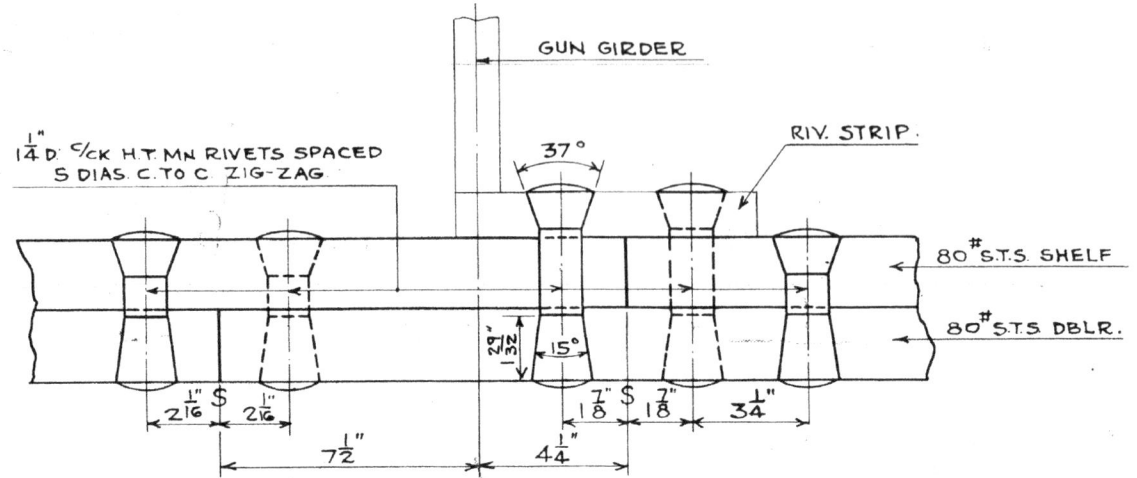

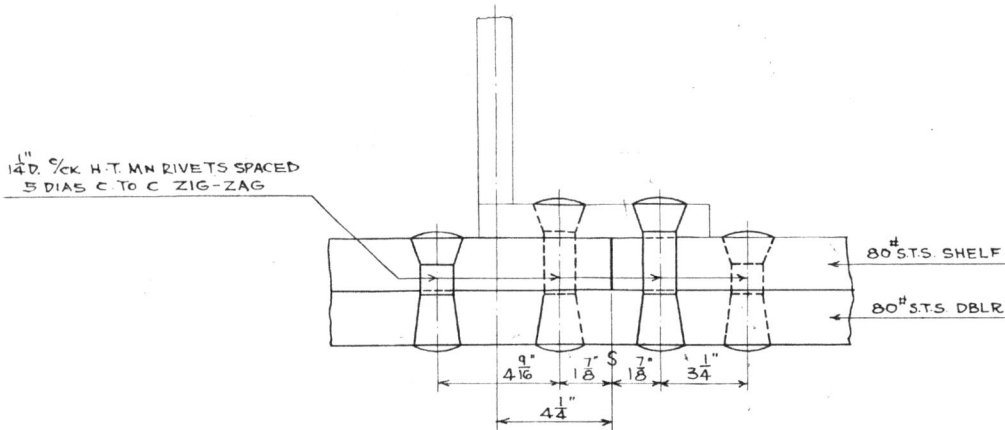

Plan 8.5: (7201ZT) **Section E–Connection of Doubler to Shelf Plate.** This plan shows how the doubler was connected near a seam in the shelf plate.

Photograph 8.1: Start of the Doubling Plate. This is Turret No. 1 on USS New Jersey. Construction resulted in misalignment between the doubling plate and the side plate. The two plates are aligned on the right side.

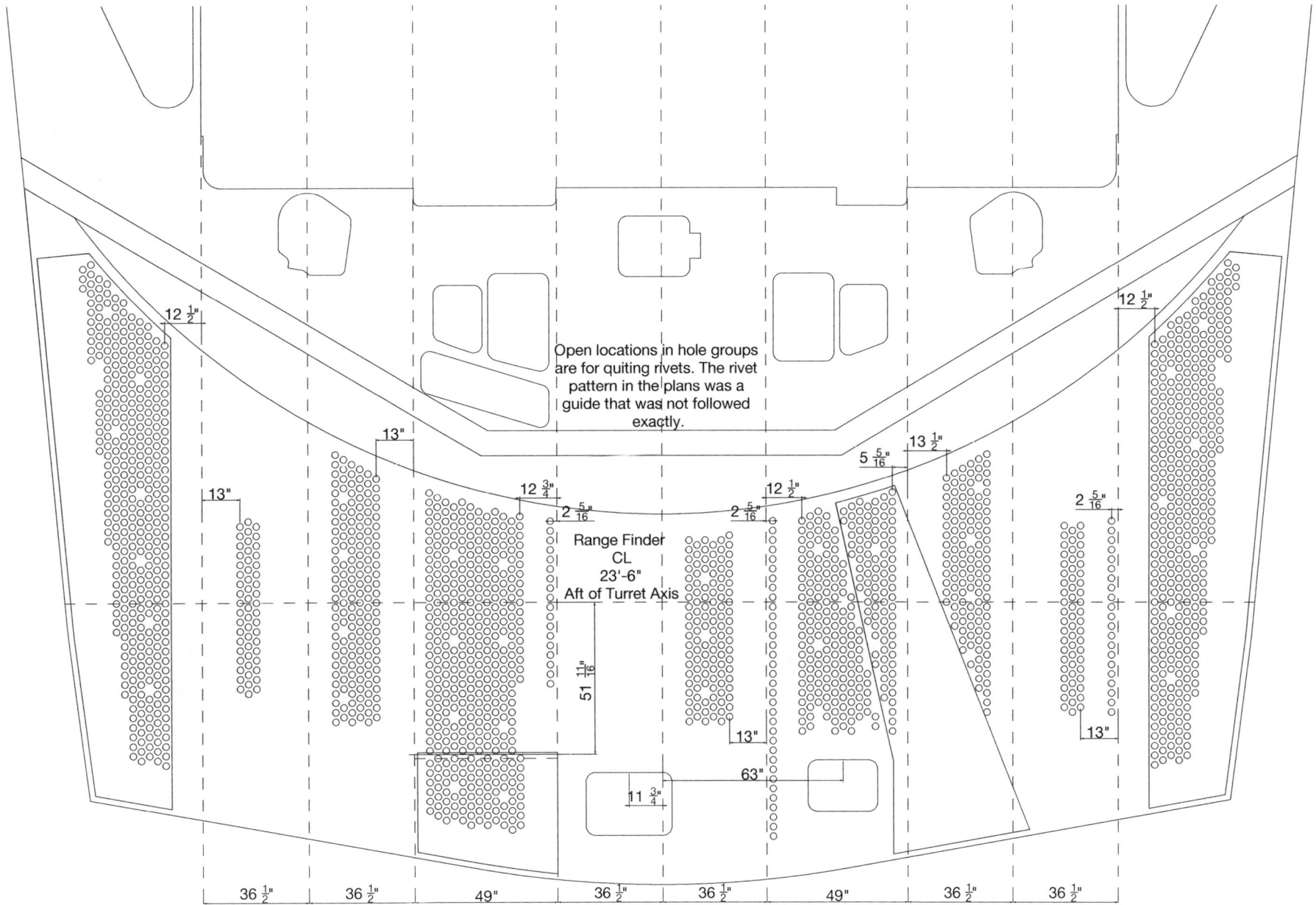

Figure 8.4: Vent Holes in Shelf Plate (1:48). Ventilation air was drawn into the turret through holes in the shelf plate. This top plan view shows how the vent holes were arranged on the original plans.

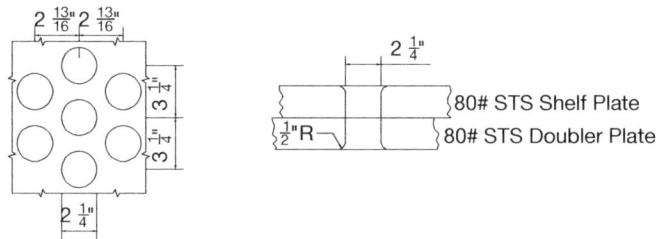

Figure 8.5: Vent Holes in Shelf Plate (1:12). A plan and cross section showing the arrangement of vent holes in the shelf plate.

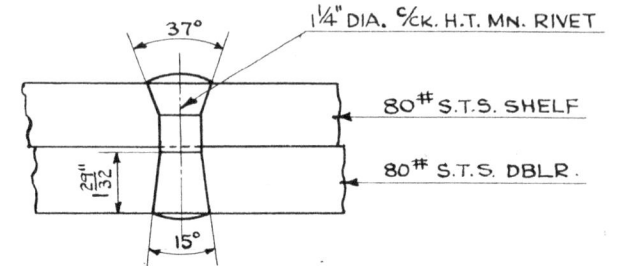

Plan 8.2: (7201ZT) Quilting rivets secured the doubling plates to the shelf plate away from the edge.

Photograph 8.3: The bottom of the shelf plate of turret no. 1 on USS New Jersey. The main hatch is at the upper center and the escape hatch is to the right. Two of the four external vent trunks are visible. The hole pattern is similar to that shown in Figure 8.4 but not identical. (Nat'l Archives Phila.)

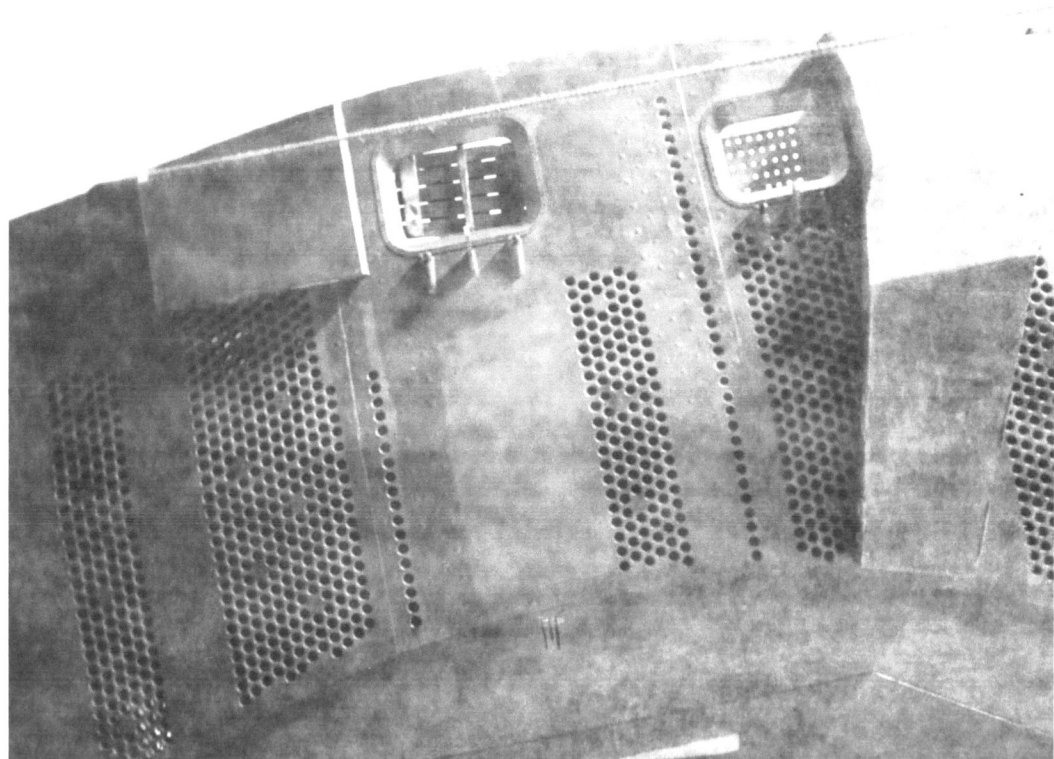

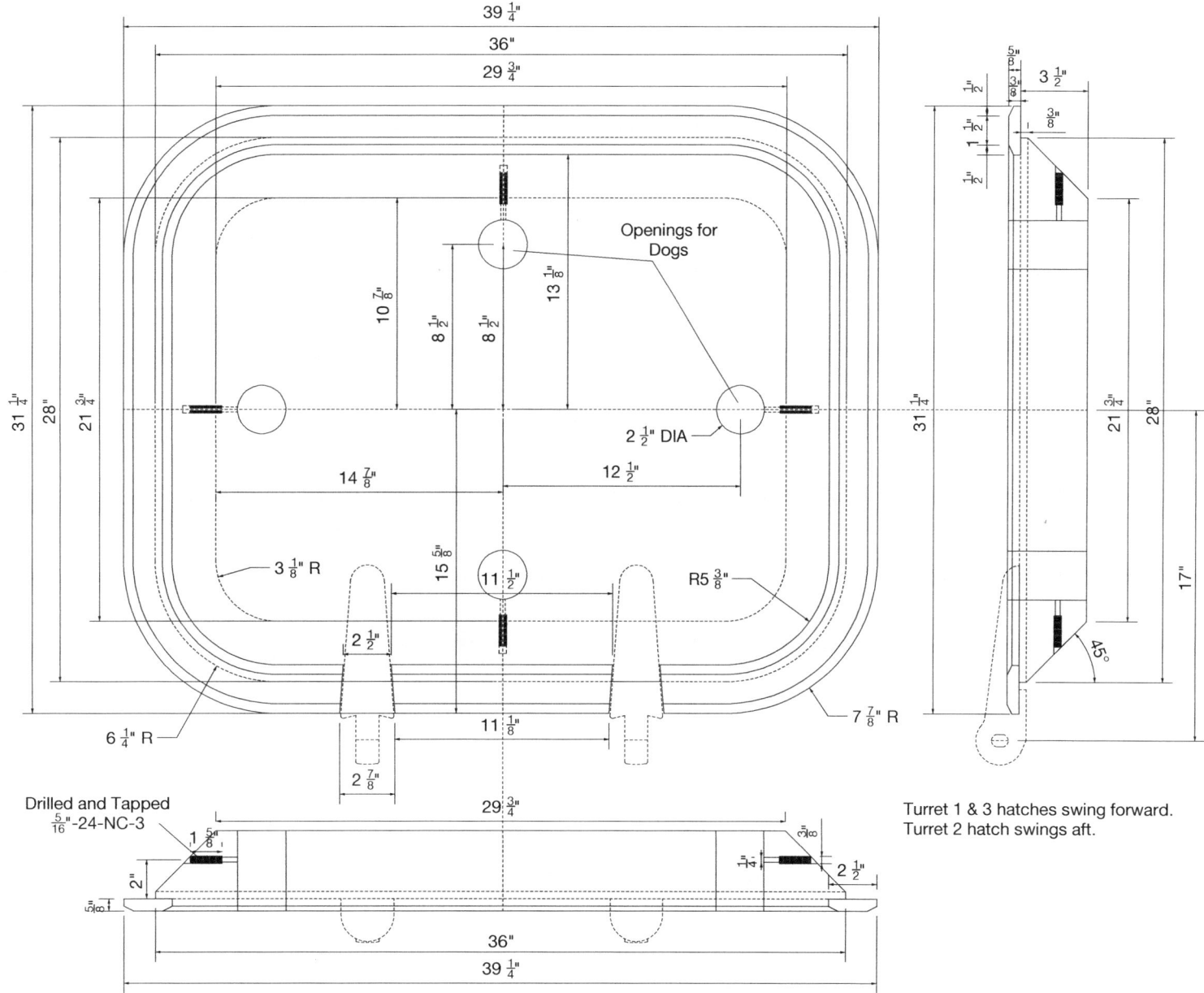

Figure 8.6: 22" x 30" Hatch (1:8). This is the main turret access located under the turret bustle.

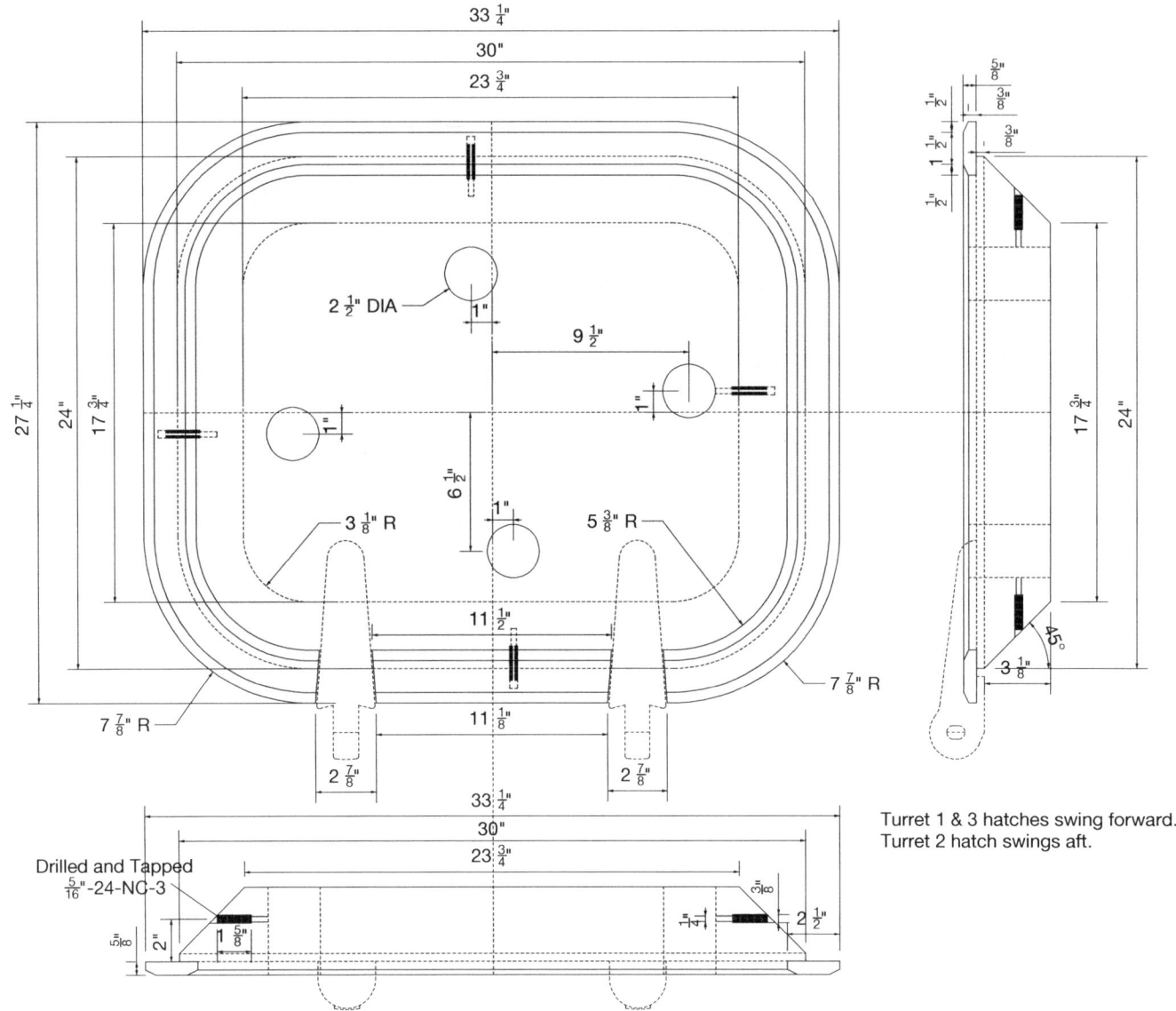

Figure 8.7: 18"x24" Turret Hatch (1:8). This hatch was available in case of emergencies if the main hatch was made unusuable. These hatches were rarely used.

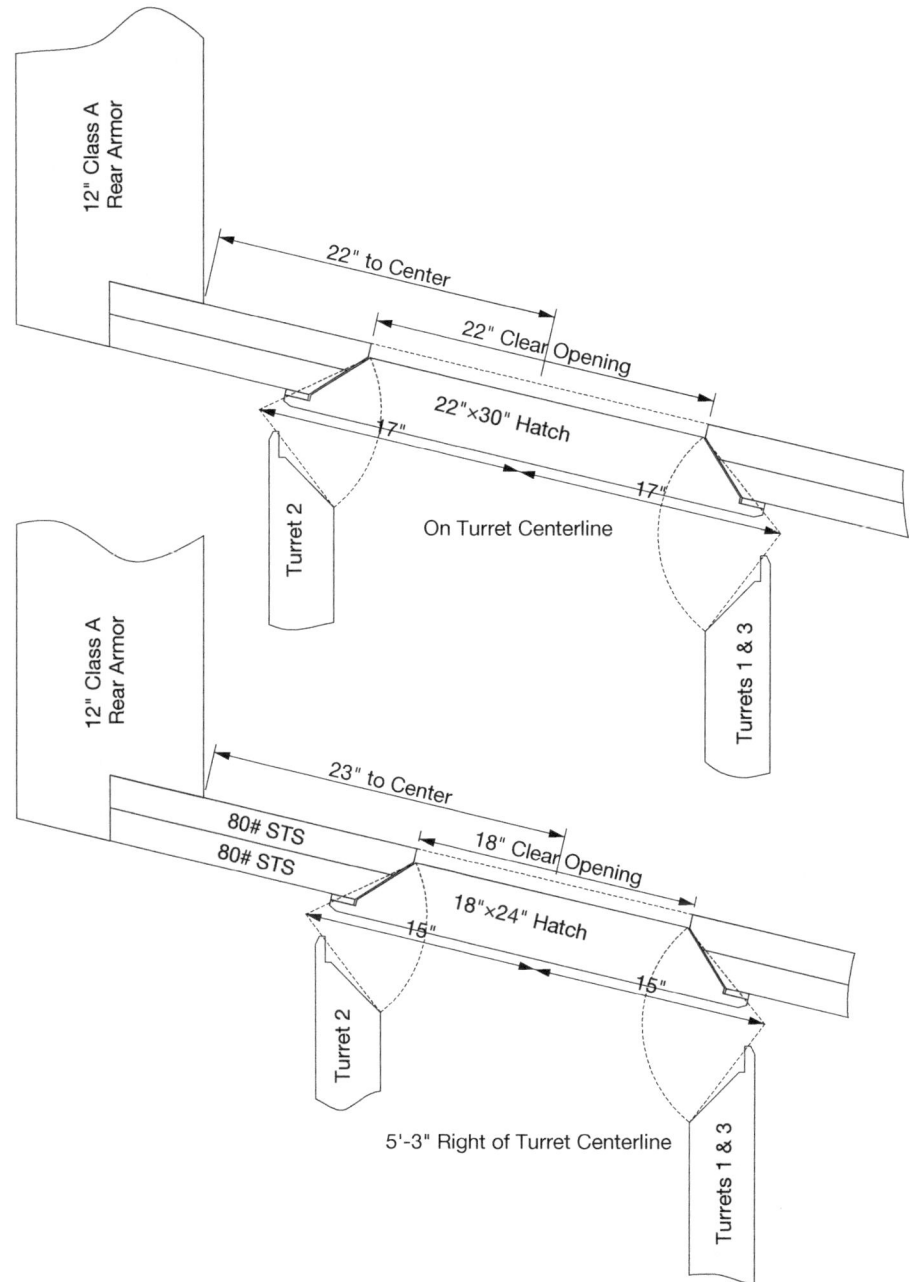

Figure 8.8: Locations of Turret Hatch Openings (1:12). The hatches for for Turret 2 open to the rear while those for the other turrets swing forward.

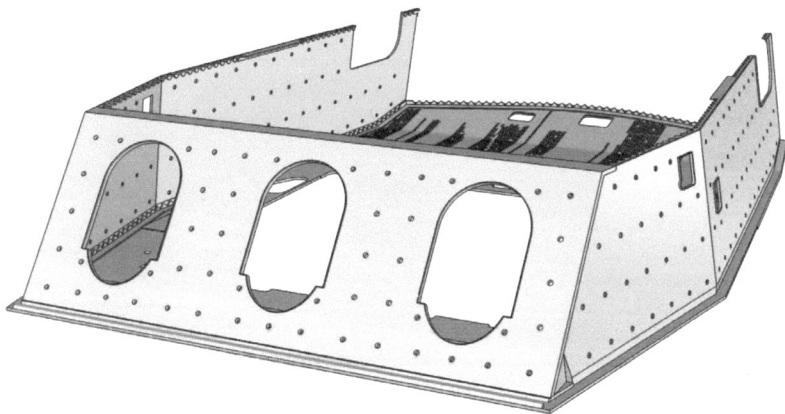

Figure 8.9: **Forward Rendering of the Backing Plates**.

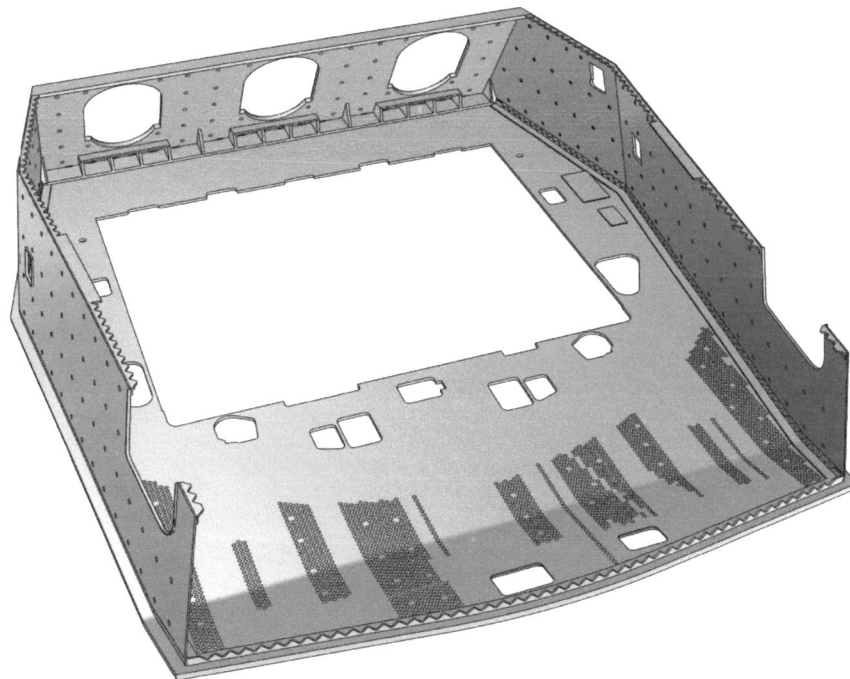

Figure 8.10: **Renderings Showing Structure Supporting the Inside Face of the Forward Backing Plate**. The scallops along the upper sides break where the removable roof plate is bolted.

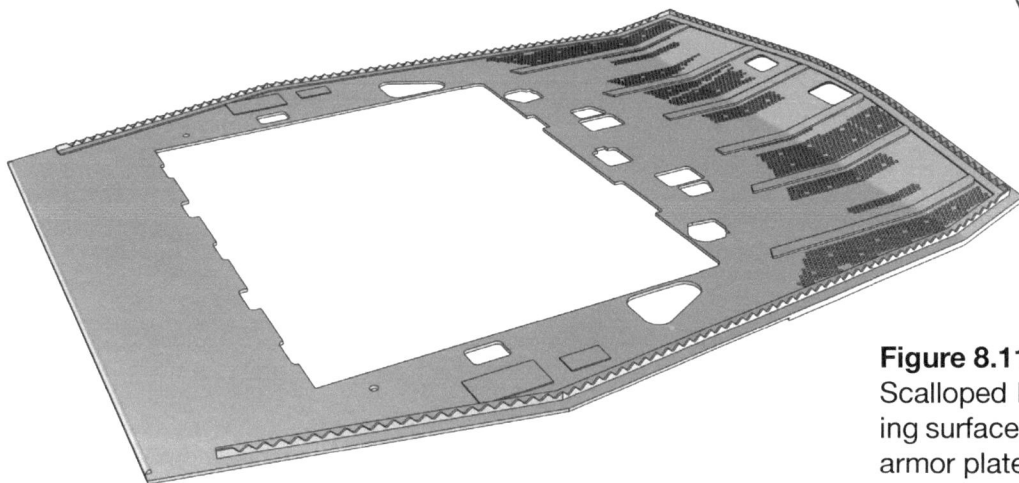

Figure 8.11: **Rendering of the Shelf Plate**. Scalloped boundary angles provide the welding surface for the side backing plates and rear armor plate.

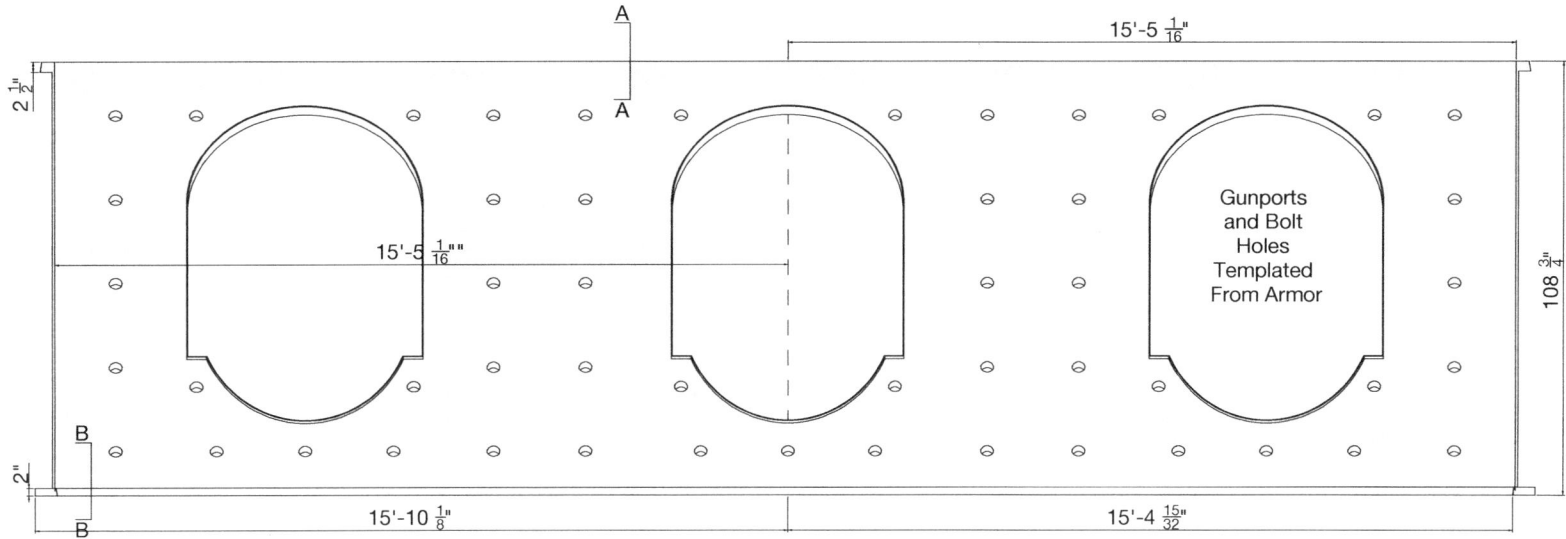

Figure 8.12: Front Backing Plate Front Elevation (1:48)

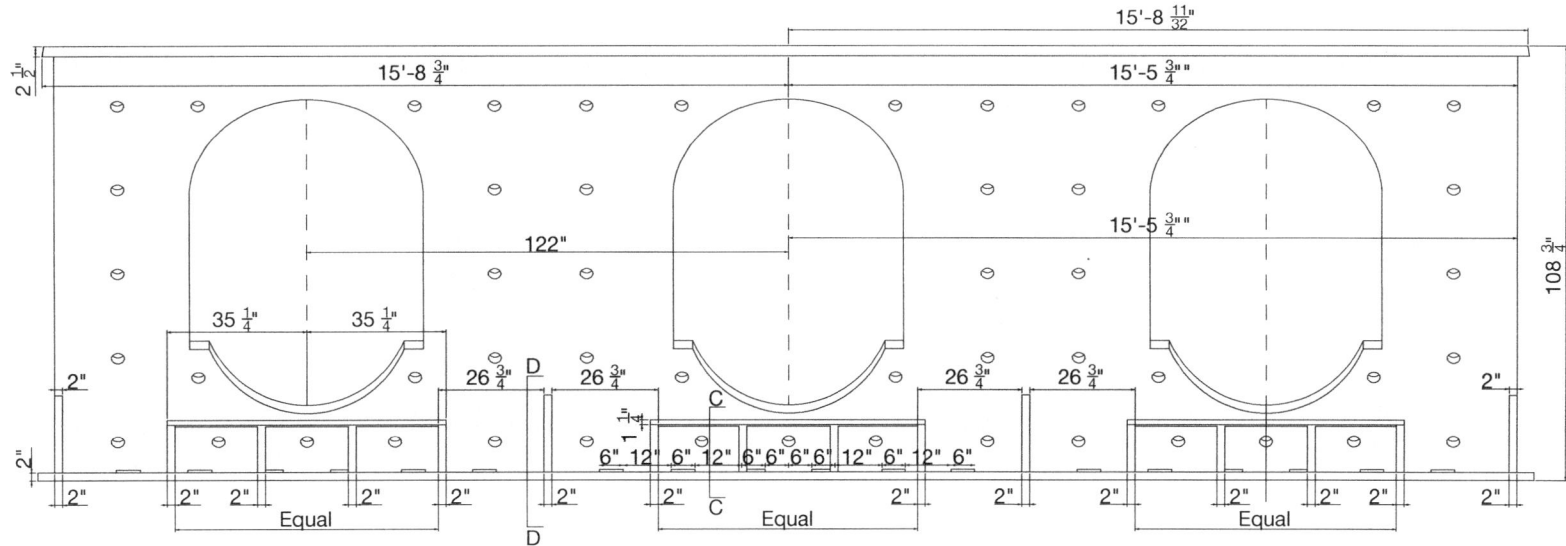

Figure 8.13: Front Backing Plate Inside Elevation (1:48)

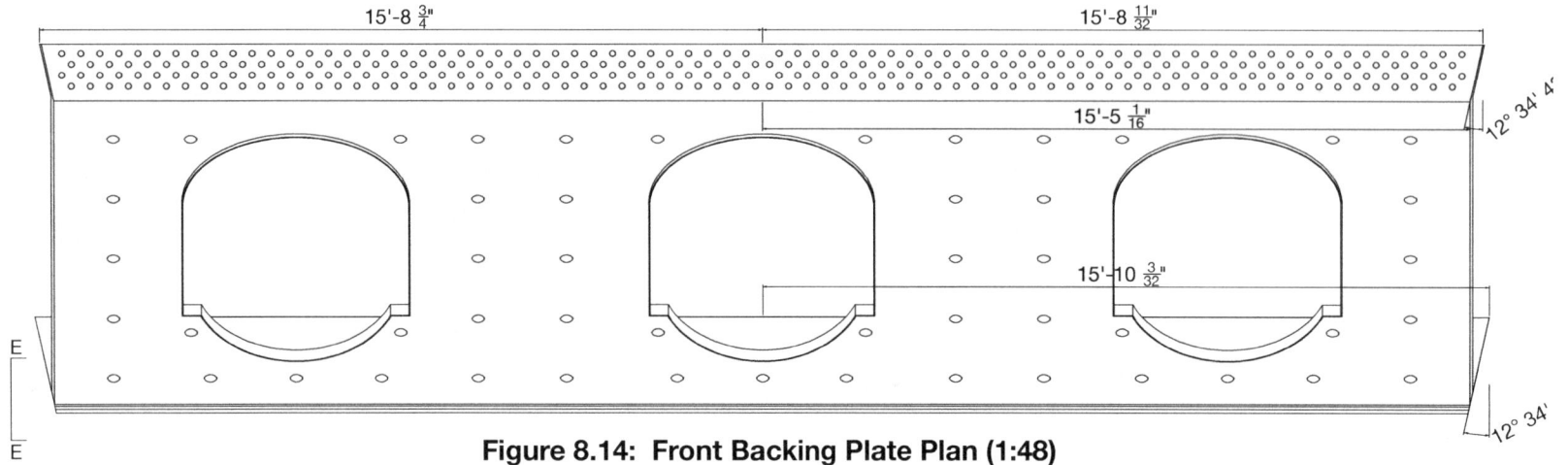

Figure 8.14: Front Backing Plate Plan (1:48)

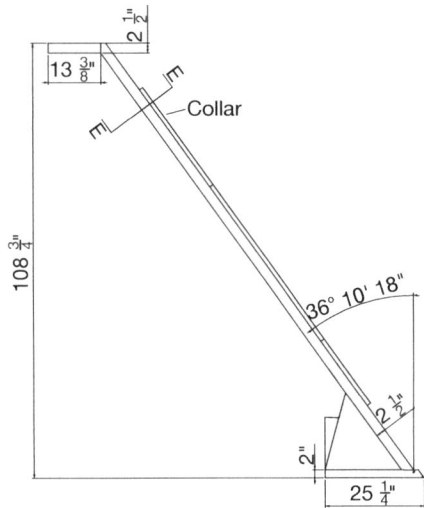

Figure 8.15: Front Backing Plate Side Elevation (1:48)

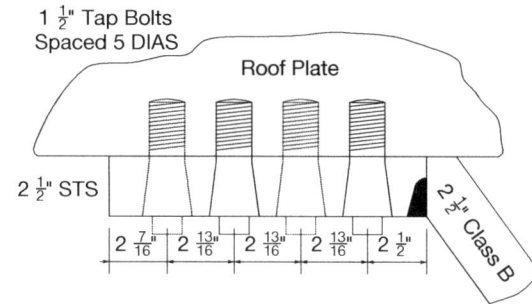

Figure 8.16: Section A–Connection to Roof Plate (1:8). See Figure 8.12.

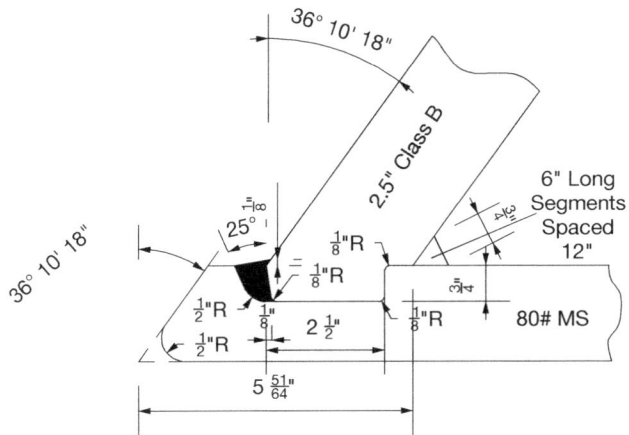

Figure 8.18: Section B–Connection to Base Plate (1:4).

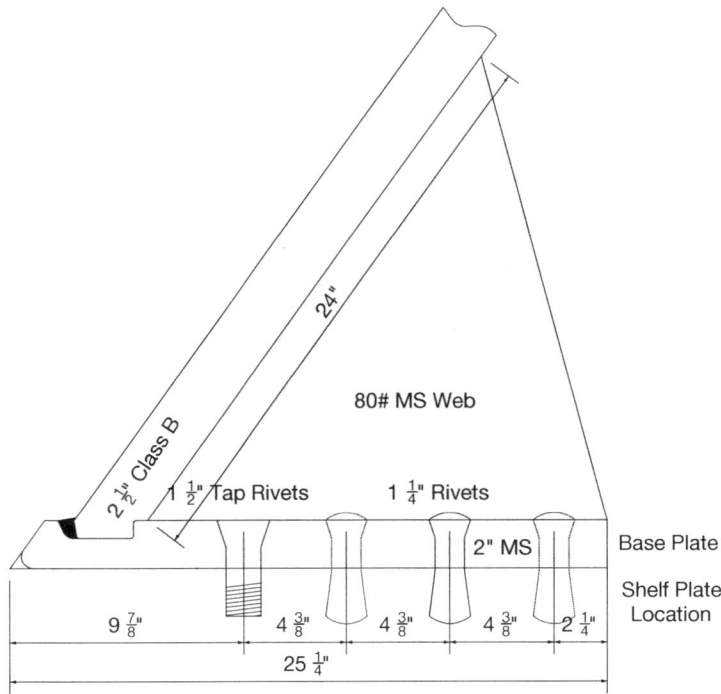

Figure 8.19: Section C–"A" Web (1:8). These webs support horizontal bracing under the gunport openings.

Figure 8.20: Section D–"B" Web (1:8).

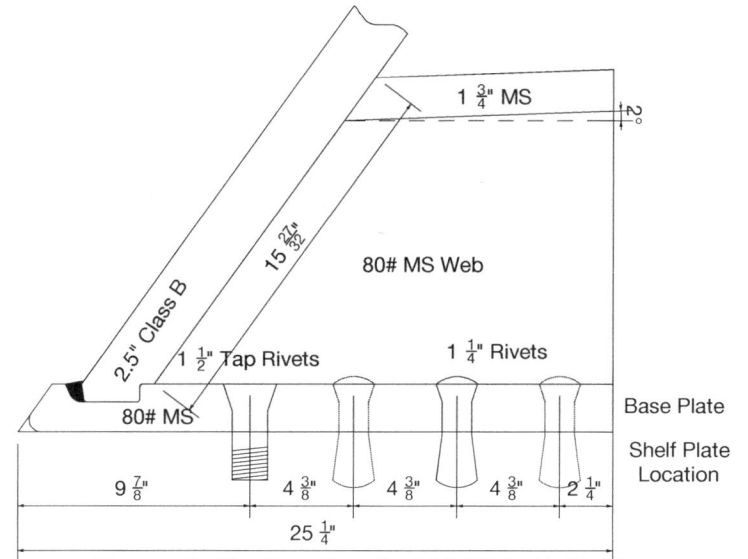

Figure 8.17: The upper section of Turret no. 1 being lowered into place on USS New Jersey. The rotating turret's roller track is visible just above the top edge of the barbette. The bolt holes for securing the armor plates are visible in the backing plate. (Nat'l Archives, Phila.)

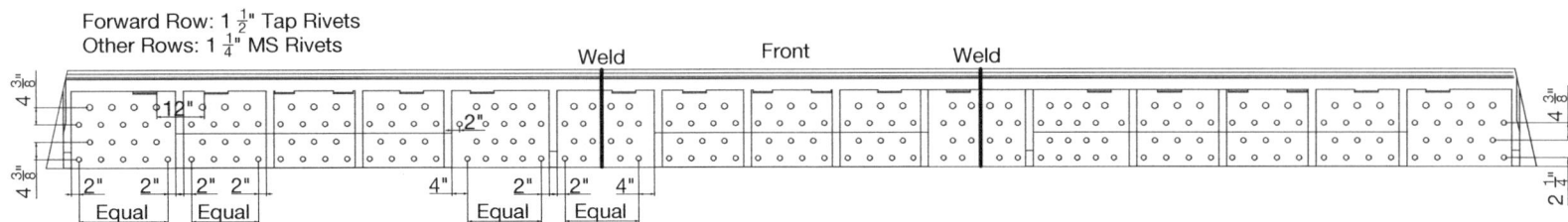

Forward Row: $1\frac{1}{2}$" Tap Rivets
Other Rows: $1\frac{1}{4}$" MS Rivets

Figure 8.21: Section E–Front Base Plate Rivet Locations (1:48).

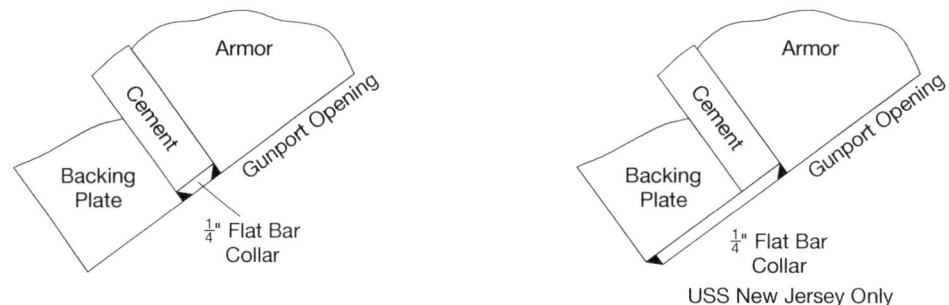

Figure 8.22: Section F–Gun Port collar (1:4). Collars around the backing plate openings prevented cement from flowing out.

Photograph 8.4: The upper section of turret no. 1 waiting to be installed on USS New Jersey. The turrets were too heavy to be lifted so they were installed in upper and lower sections with out the armor. The side backing plates and shelf plate are clearly visible. The transverse bulkhead that separates the officers' booth is also prominent. The working level of the officers' booth is elevated above vent trunks. (Nat'l Archives, Phila.)

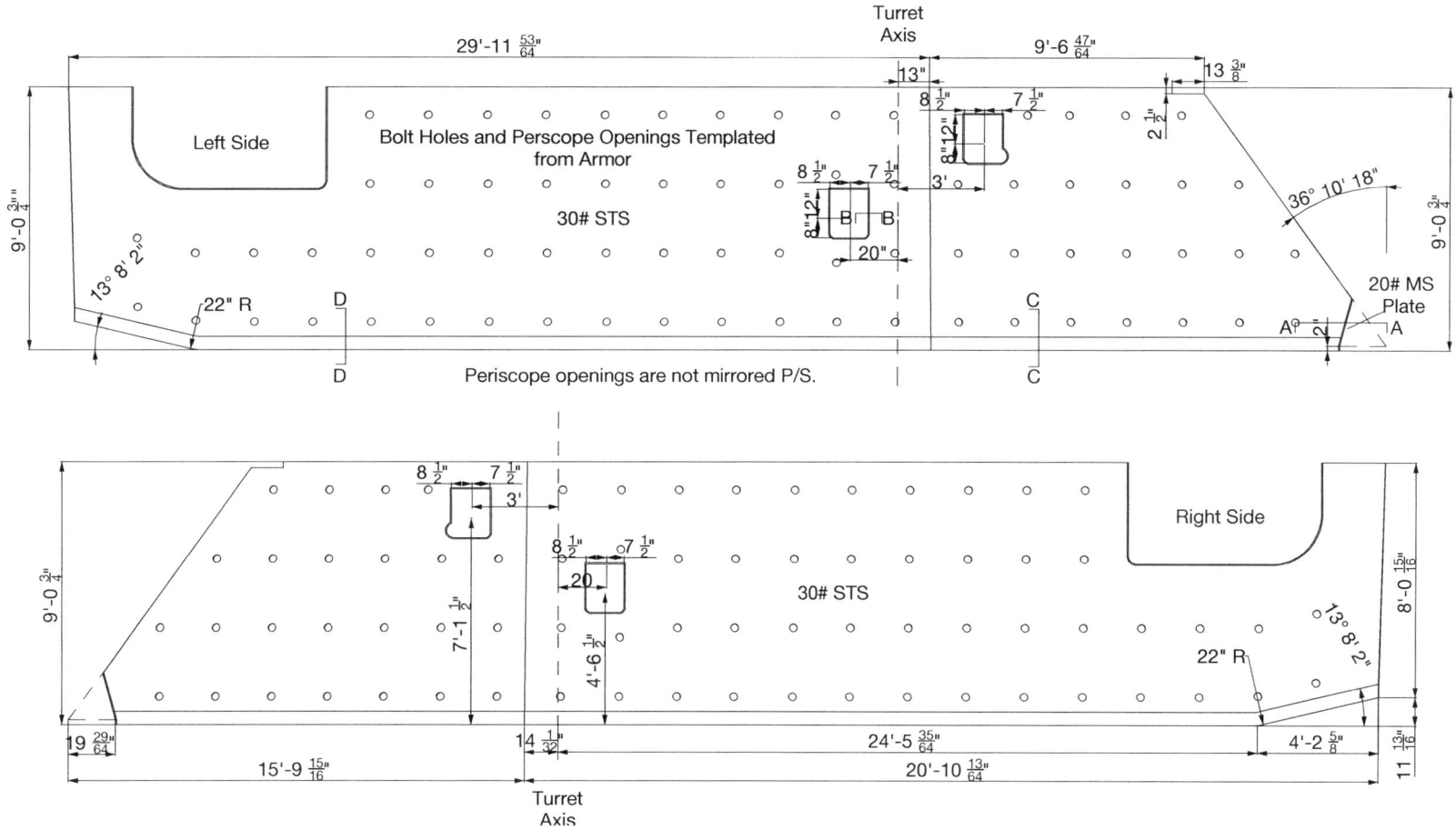

Figure 8.23: **Side Backing Bulkhead Inside Elevations (1:72).**

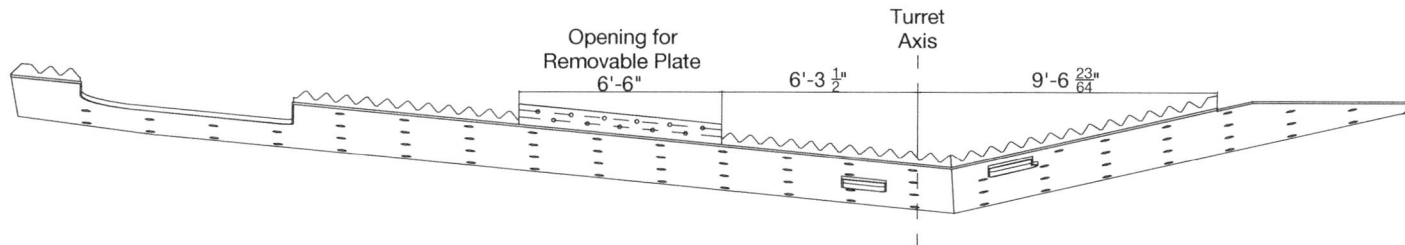

Figure 8.24: **Right Backing Bulkhead Plan (1:72).** The left side is a mirror except for the periscope openings.

80# MS Web on
Front Plate

20# MS
Plate

30# STS
Backing Plate

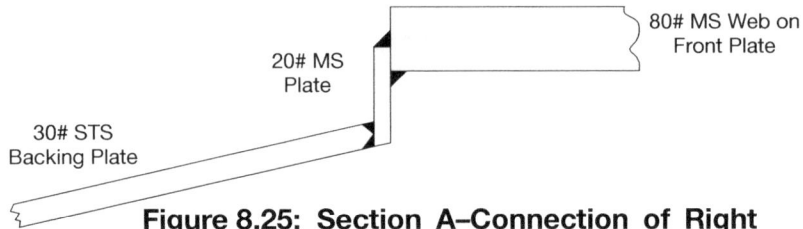

Figure 8.25: Section A–Connection of Right Backing Bulkhead to Front Backing Bulkhead (1:6). See lower corner of Figure 8.9.

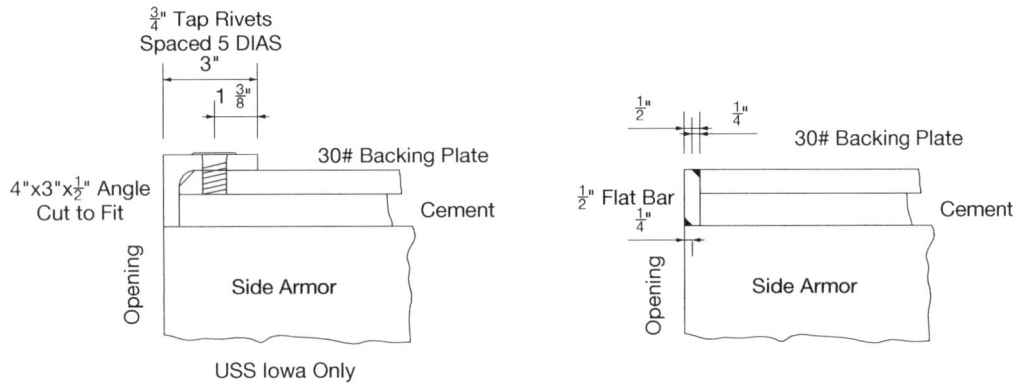

$\frac{3}{4}$" Tap Rivets
Spaced 5 DIAS
3"

$1\frac{3}{8}$"

30# Backing Plate

4"x3"x$\frac{1}{2}$" Angle
Cut to Fit

Cement

Opening

Side Armor

USS Iowa Only

$\frac{1}{2}$" $\frac{1}{4}$"

30# Backing Plate

$\frac{1}{2}$" Flat Bar
$\frac{1}{4}$"

Cement

Opening

Side Armor

Figure 8.26: Section B-Collars Around Side Backing Bulkhead Openings (1:6).

Photograph 8.5: Collar at Trainer Telescope Opening on USS Iowa. A thin metal strip has been welded over the tap rivets that does not appear in the original plans. (Jim Kurrasch)

Photograph 8.6: Collar on a Pointer Telescope Opening on USS New Jersey. The later ships used welded flat bars for the collar. (Ryan Szimanski)

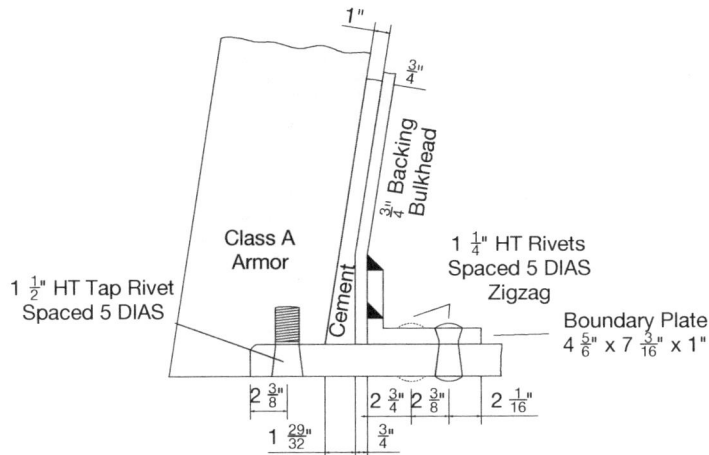

Figure 8.31: Section C–Connection of Side Backing Bulkhead to Shelf Plate (1:12).

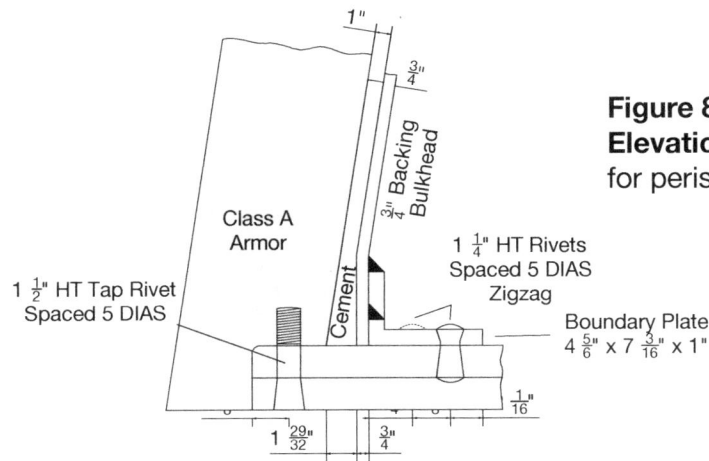

Figure 8.32: Section D–Backing Bulkhead Connection to Doubled Shelf Plate (1:12).

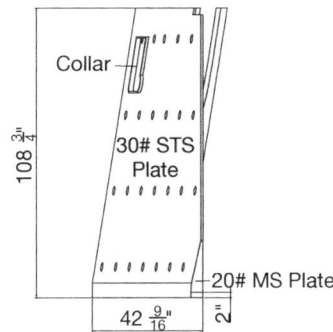

Figure 8.29: Right Backing Bulkhead Front Elevation (1:72). Left side is a mirror except for periscope openings.

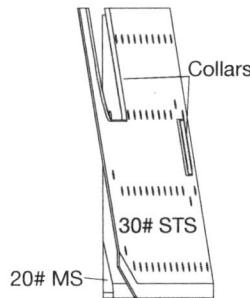

Figure 8.30: Right Backing Plate Rear Elevation (1:72). Left side is a mirror except for periscope openings.

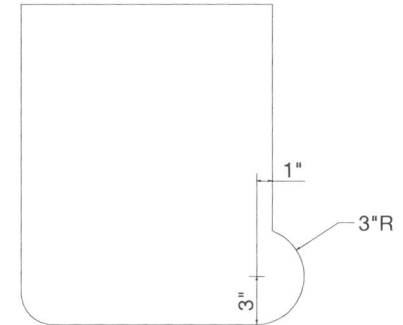

Figure 8.27: Trainer's Handwheel Extension into Backing Plate (1:12). The trainers' handwheels did not fit in the rectangular opening in the bcking plate for the periscope. This extension was cut into the backing plate to allow the handwheel to be operated.

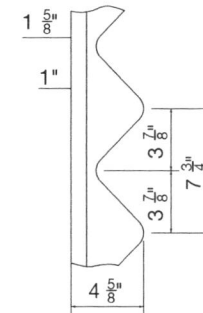

Figure 8.28: Scallop Pattern (1:12). This scallop pattern used for all the backing plate connections.

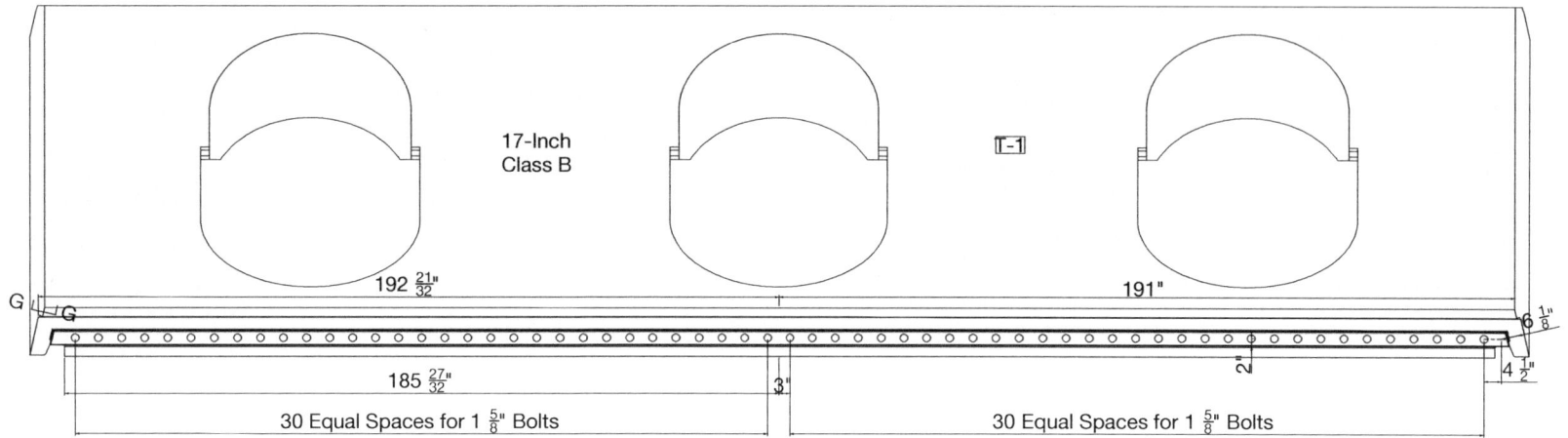

17-Inch
Class B

T-1

192 $\frac{21}{32}$"

191"

G
G

6 $\frac{1}{8}$"

185 $\frac{27}{32}$"

3"

2"

4 $\frac{1}{2}$"

30 Equal Spaces for 1 $\frac{5}{8}$" Bolts

30 Equal Spaces for 1 $\frac{5}{8}$" Bolts

Figure 8.33: Front Plate Top Plan (1:48)

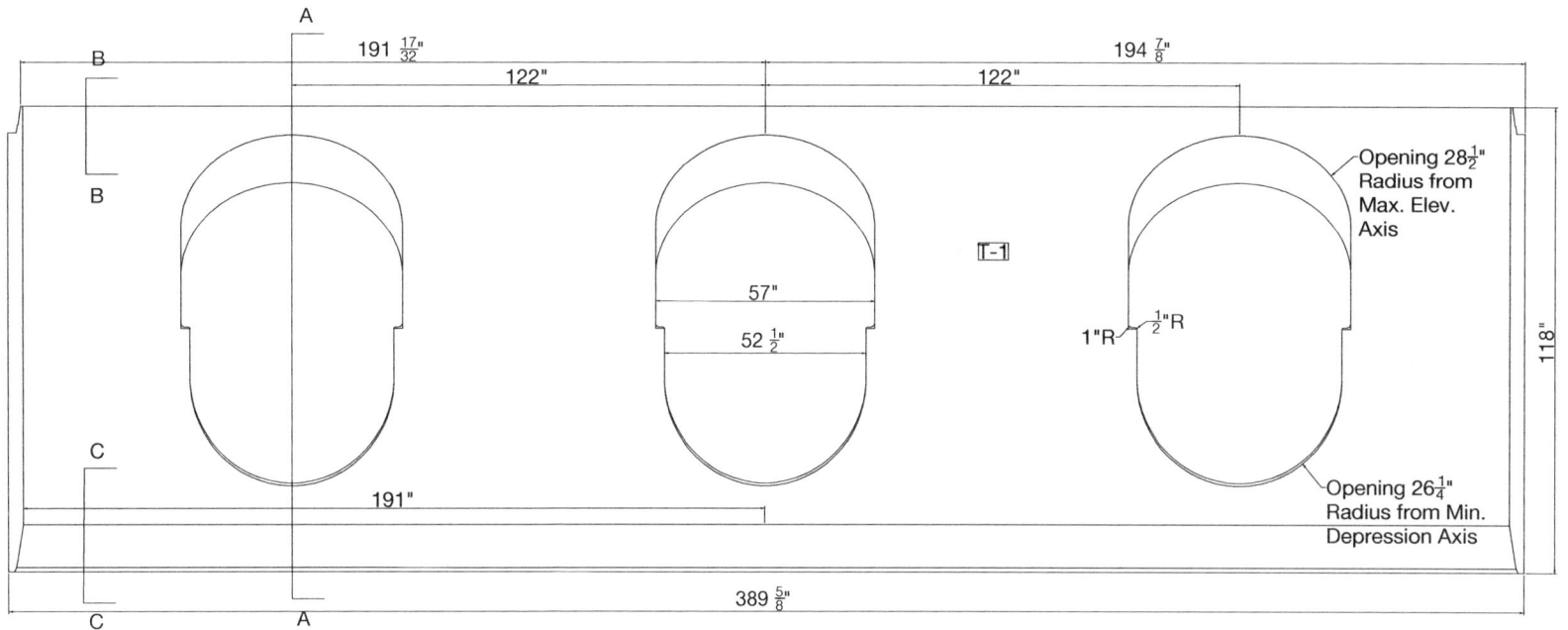

A

191 $\frac{17}{32}$"

194 $\frac{7}{8}$"

B

122"

122"

B

Opening 28 $\frac{1}{2}$"
Radius from
Max. Elev.
Axis

T-1

57"

52 $\frac{1}{2}$"

1"R

$\frac{1}{2}$"R

118"

C

Opening 26 $\frac{1}{4}$"
Radius from Min.
Depression Axis

191"

C

389 $\frac{5}{8}$"

A

Figure 8.34: Front Plate Outside Elevation (1:48)

194 $\frac{13}{16}$"

192 $\frac{21}{32}$"

191 $\frac{17}{32}$"

122"   122"

1"R $\frac{1}{2}$"R

T-1

118"

389 $\frac{5}{8}$"

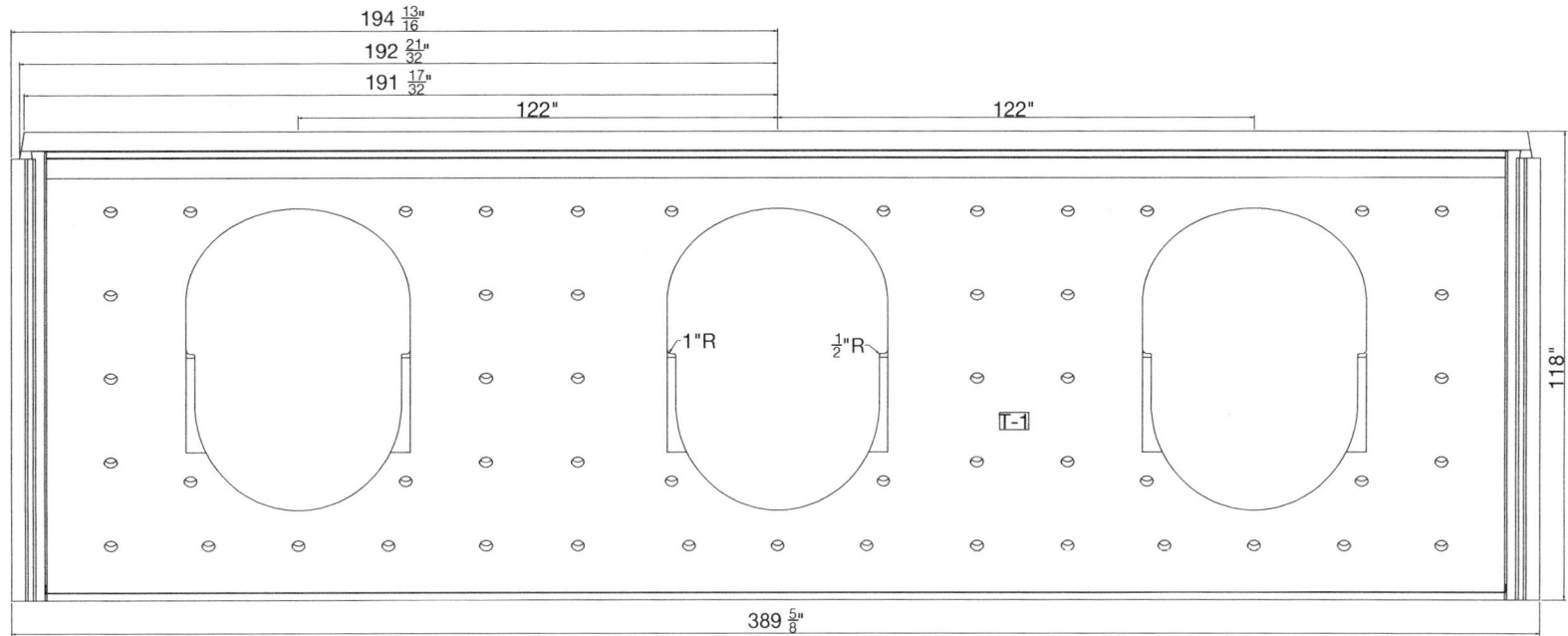

Figure 8.35: Front Plate Inside Elevation (1:48)

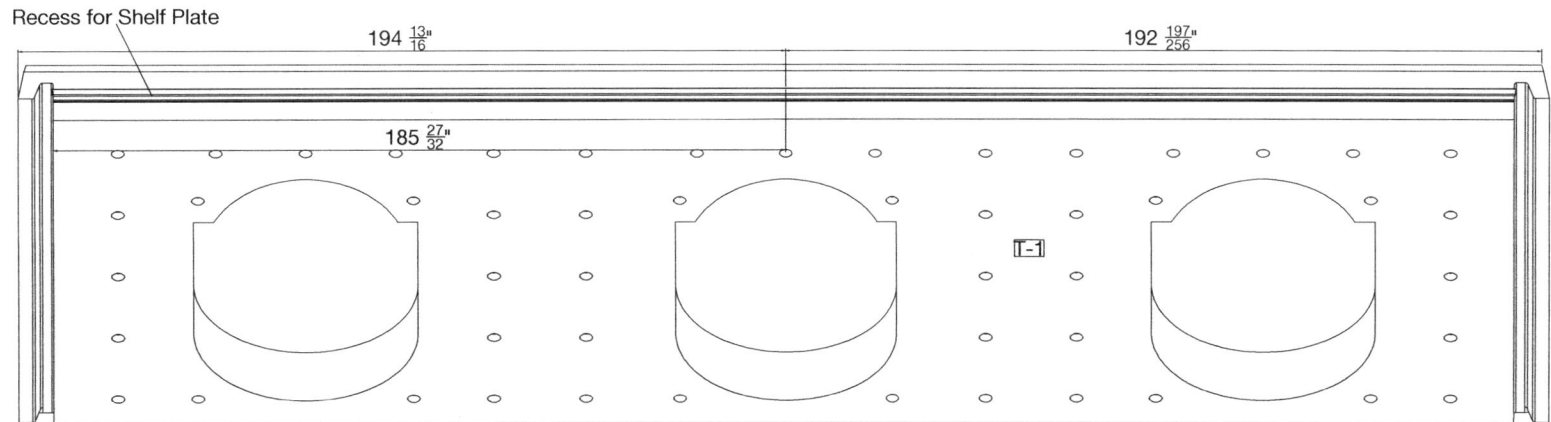

Recess for Shelf Plate

194 $\frac{13}{16}$"   192 $\frac{197}{256}$"

185 $\frac{27}{32}$"

T-1

Figure 8.36: Front Plate Bottom Plan (1:48)

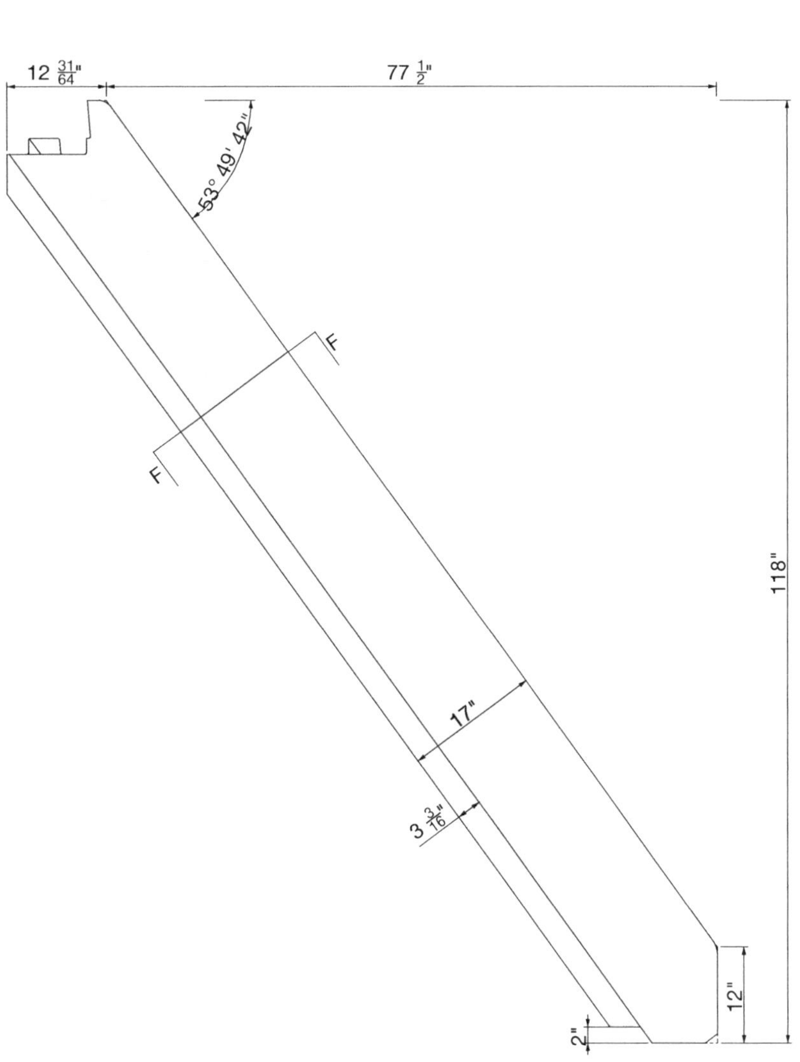

Figure 8.37: Front Plate Side Elevation (1:24)

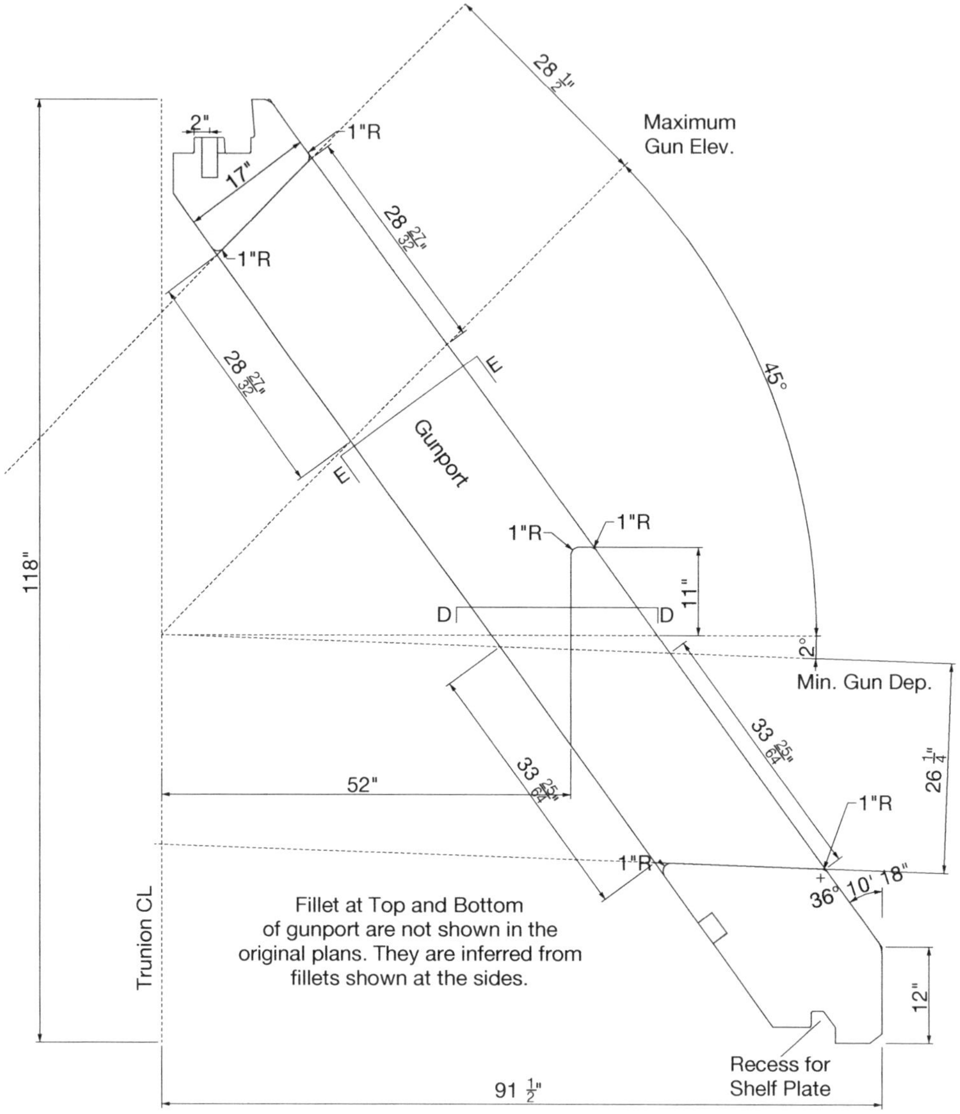

Figure 8.38: Section A–Through Gunport Center (1:24)

Fillet at Top and Bottom
of gunport are not shown in the
original plans. They are inferred from
fillets shown at the sides.

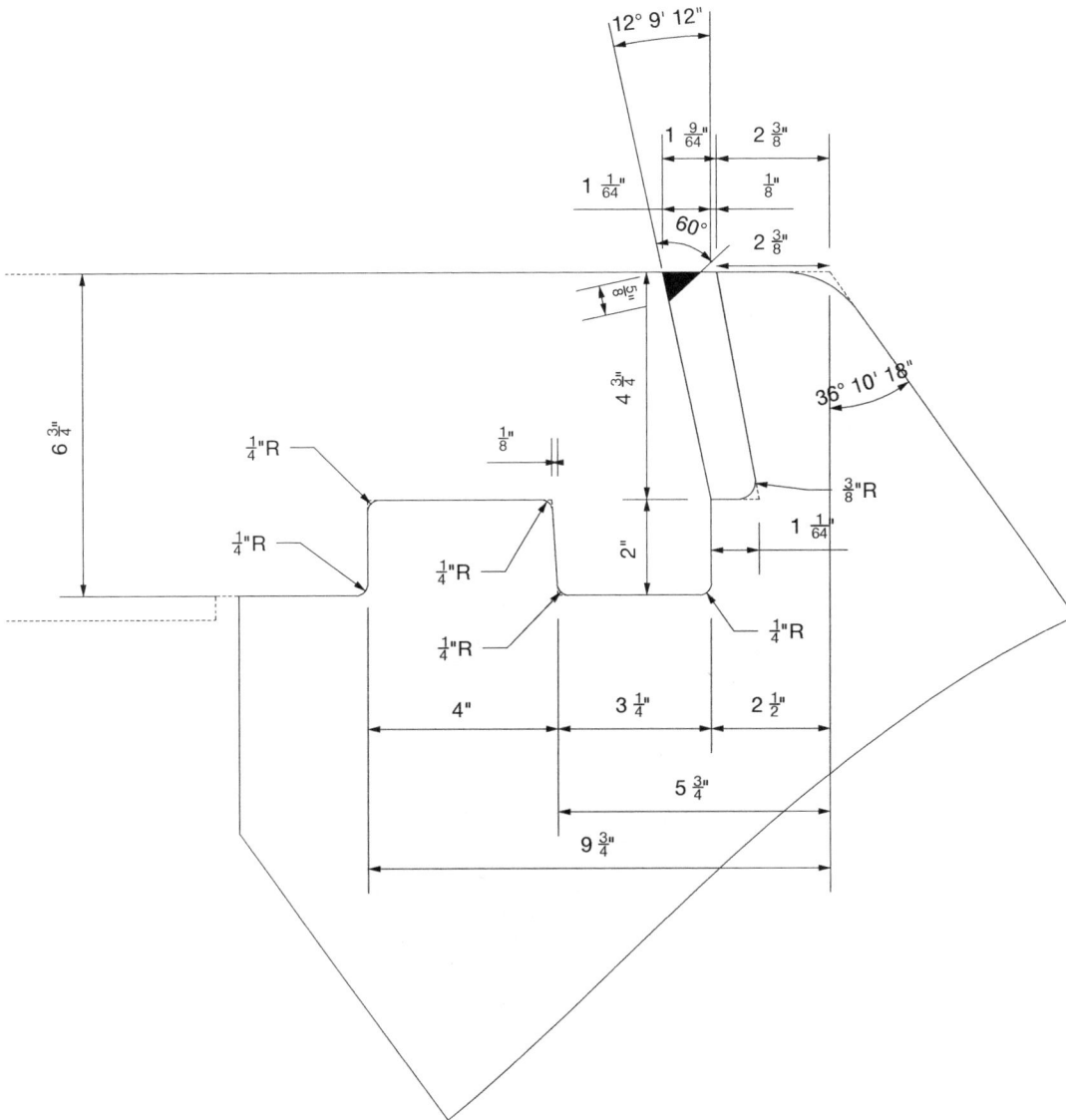

Figure 8.39: Section B–Through Upper Edge (1:4)

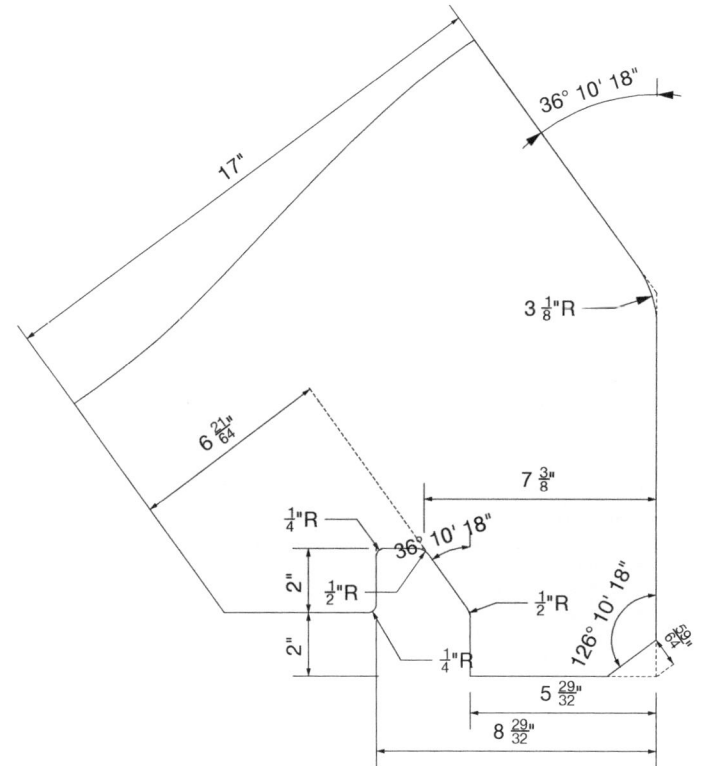

Figure 8.40: Section C–Through Lower Edge (1:6)

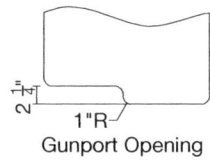

Figure 8.41: Section D–Through Gunport Side (1:24)

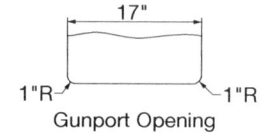

Figure 8.42: Section E–Through Gunport Wall Normal to Faces (1:24)

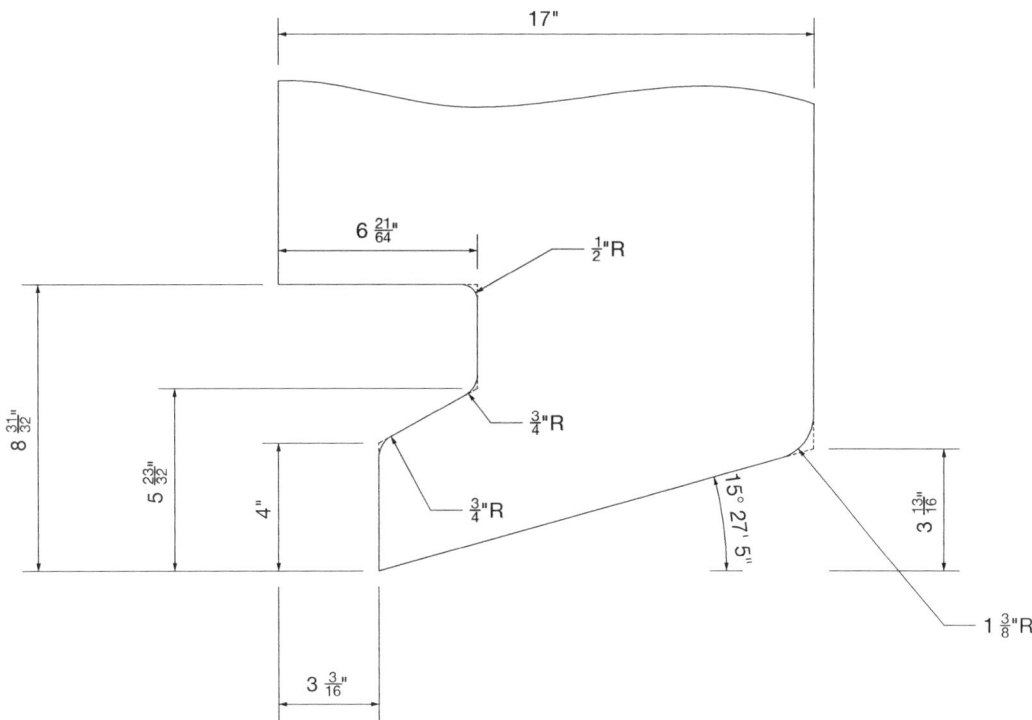

Figure 8.43: Section G–Through Upper Side Edge

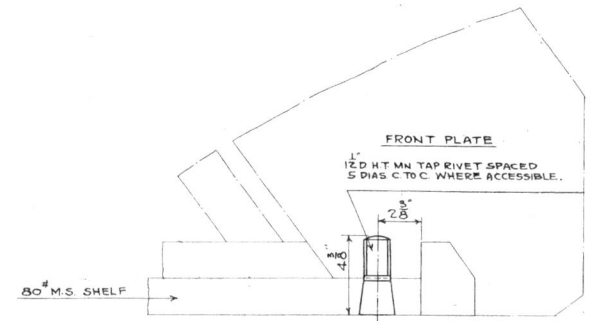

Figure 8.44: Section F–Through Side Edge (1:6). This section is normal to the inside face. It shows the groove that receives the front side plate.

Plan 8.1: (7201T). How the front plate was attached to the shelf plate. See Plan 8.1 and Figure 8.40.

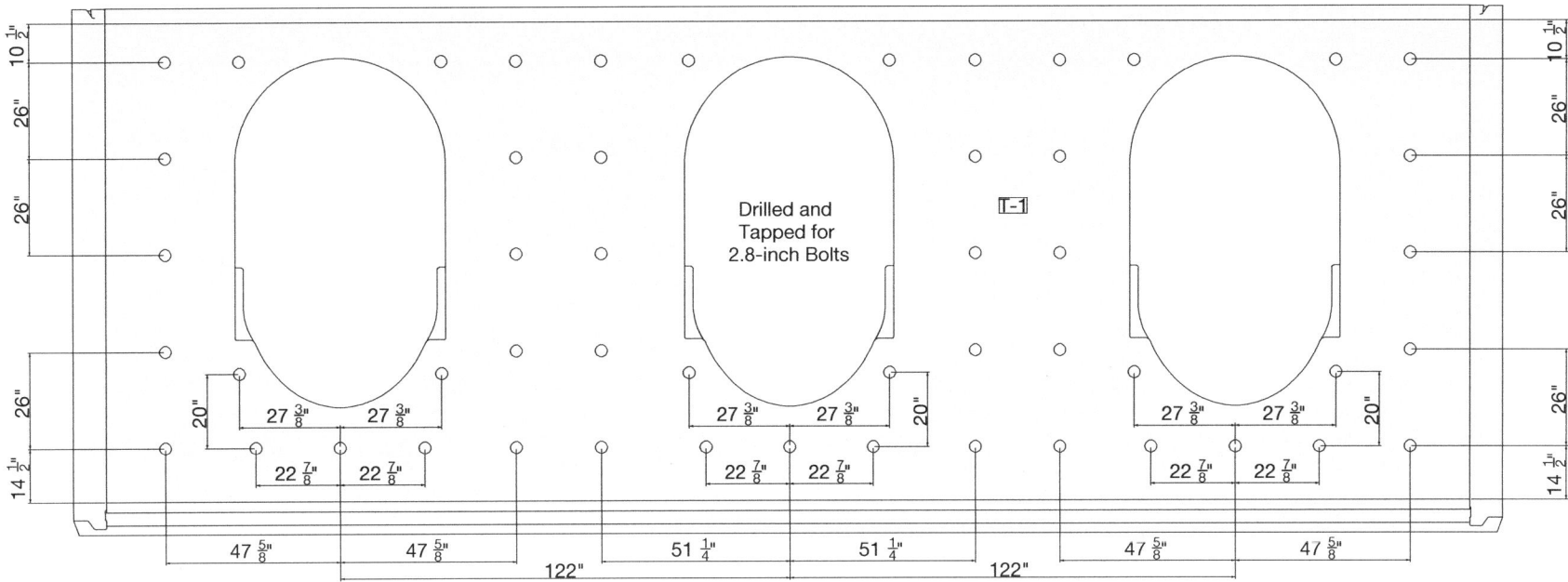

Figure 8.45: Front Plate Inside Expanded View (1:48).
This view shows the bolt locations.

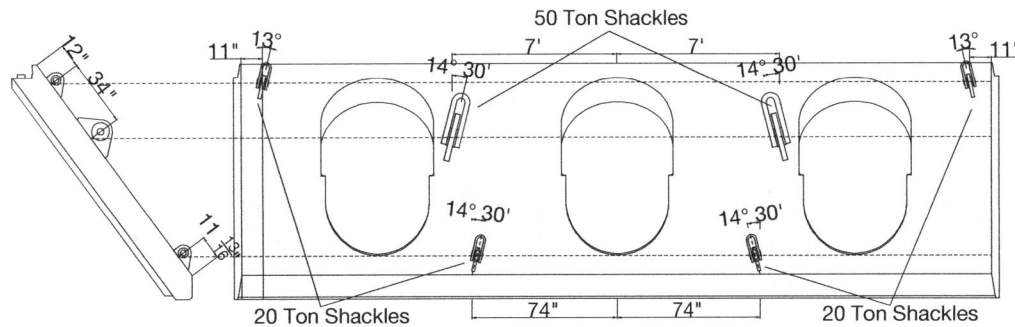

Figure 8.46: Lifting Pad Locations (1:96).
Lifting pads were welded to the front plate.
The pads were grinded off after installation.

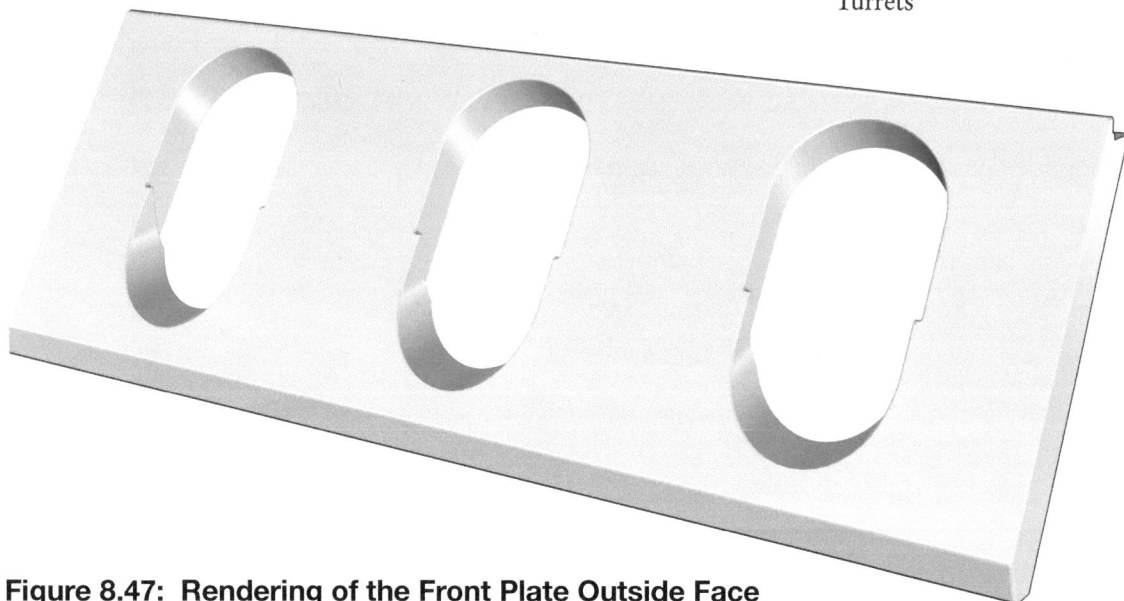

Figure 8.47: Rendering of the Front Plate Outside Face

Photograph 8.2: Front Backing Plate Bolts. This is USS Iowa. The circular caps were welded over the bolts to the backing plate. A trunnion bearing block is to the right. (Jim Kurrasch)

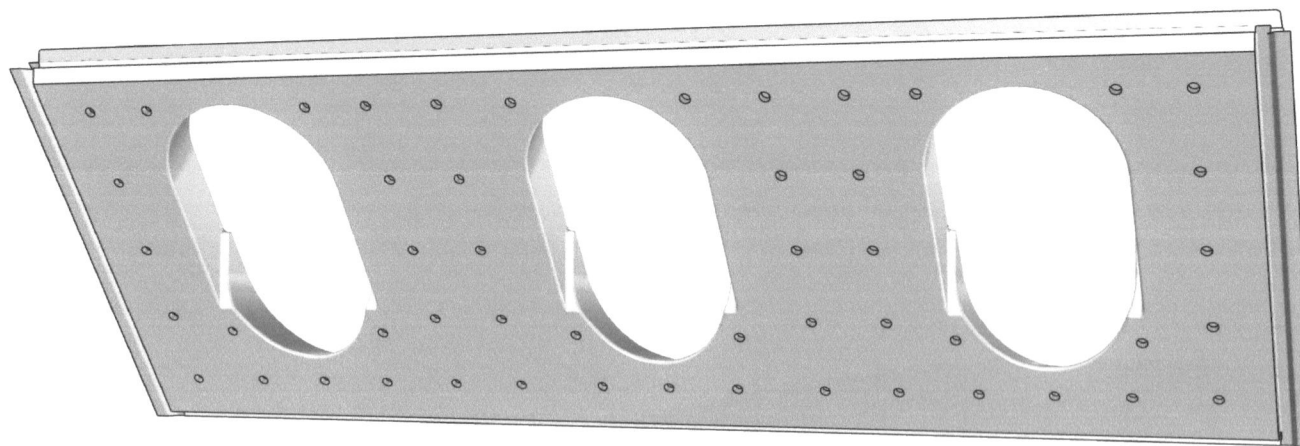

Figure 8.48: Rending of the Front Plate Inside Face

Figure 8.50: Rendering of a gunport

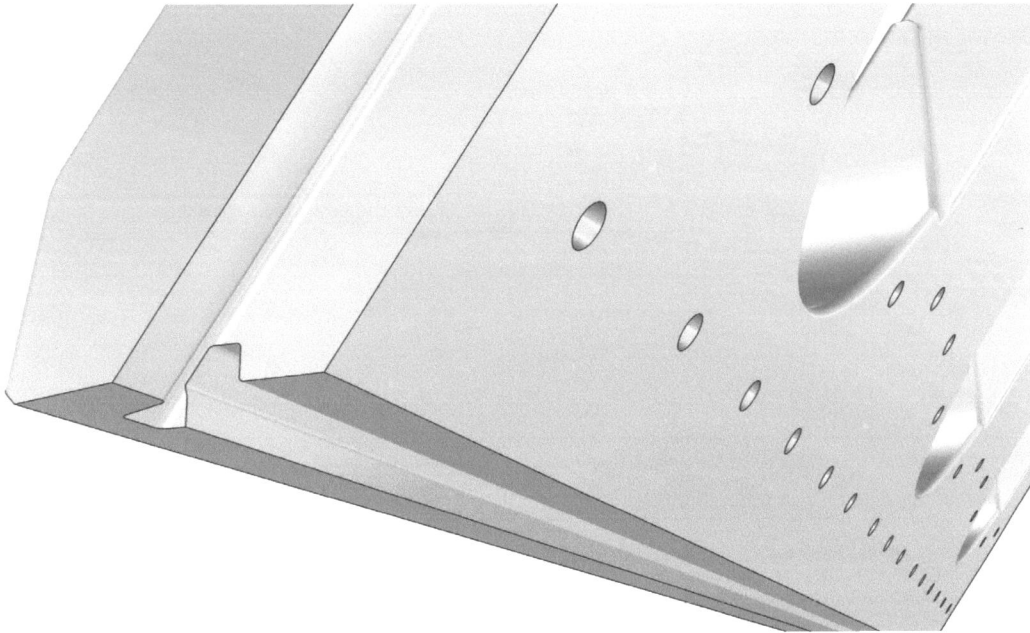

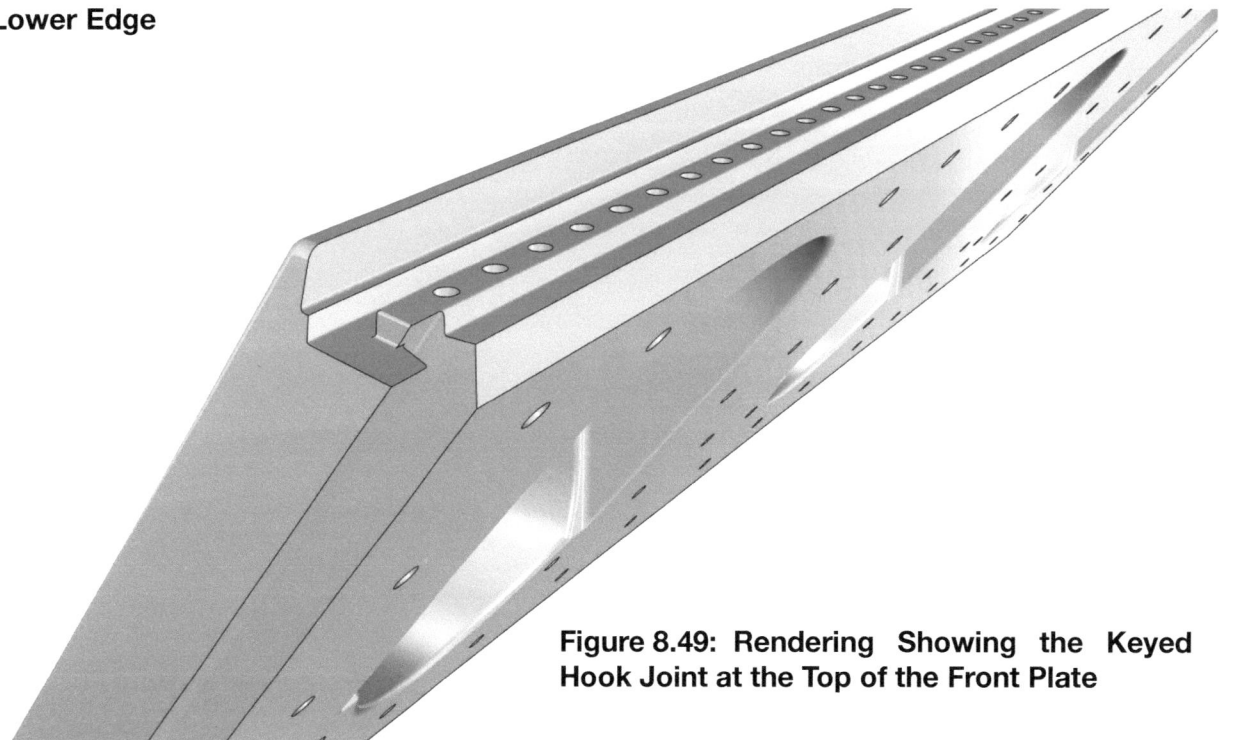

Figure 8.51: Rendering of Lower Edge

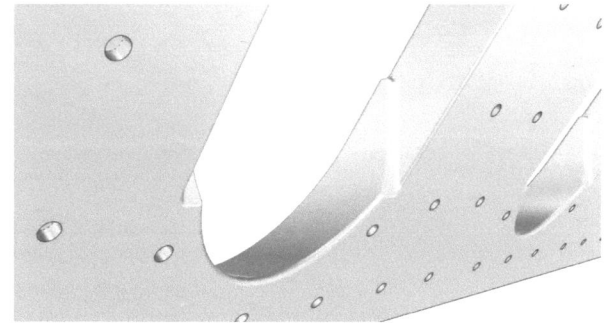

Figure 8.49: Rendering Showing the Keyed Hook Joint at the Top of the Front Plate

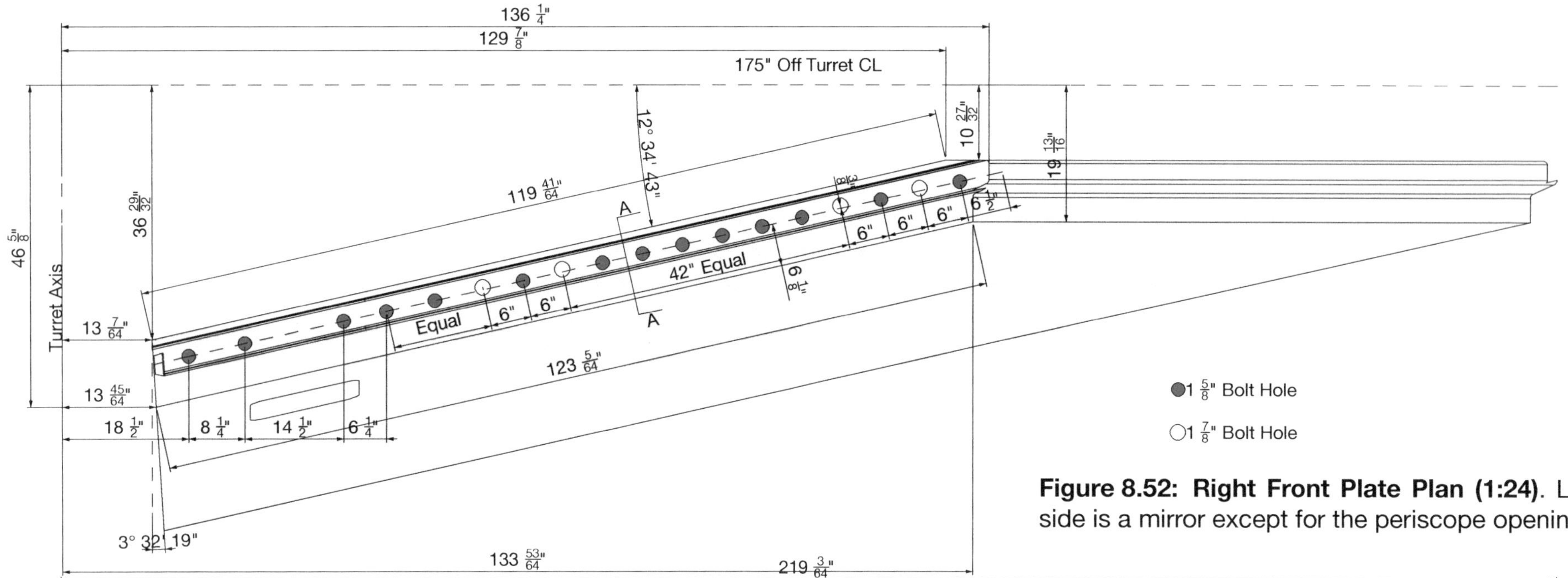

Figure 8.52: Right Front Plate Plan (1:24). Left side is a mirror except for the periscope opening.

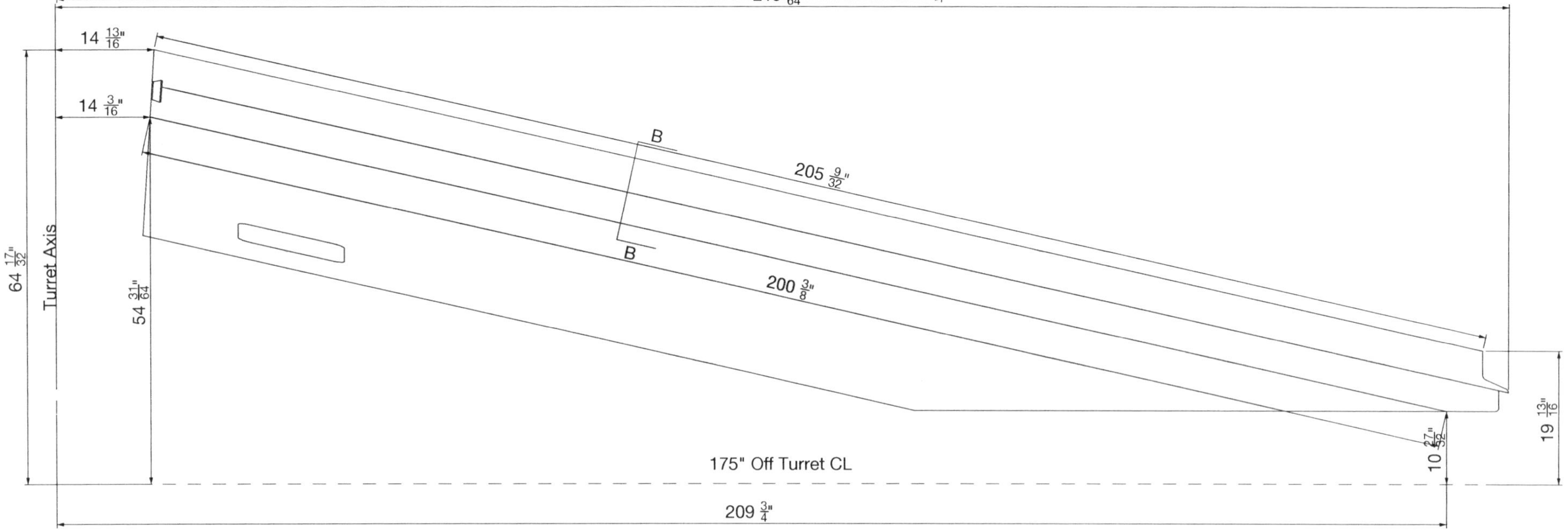

Figure 8.53: Right Front Plate Bottom Plan (1:24).
Left side is a mirror except for the periscope opening

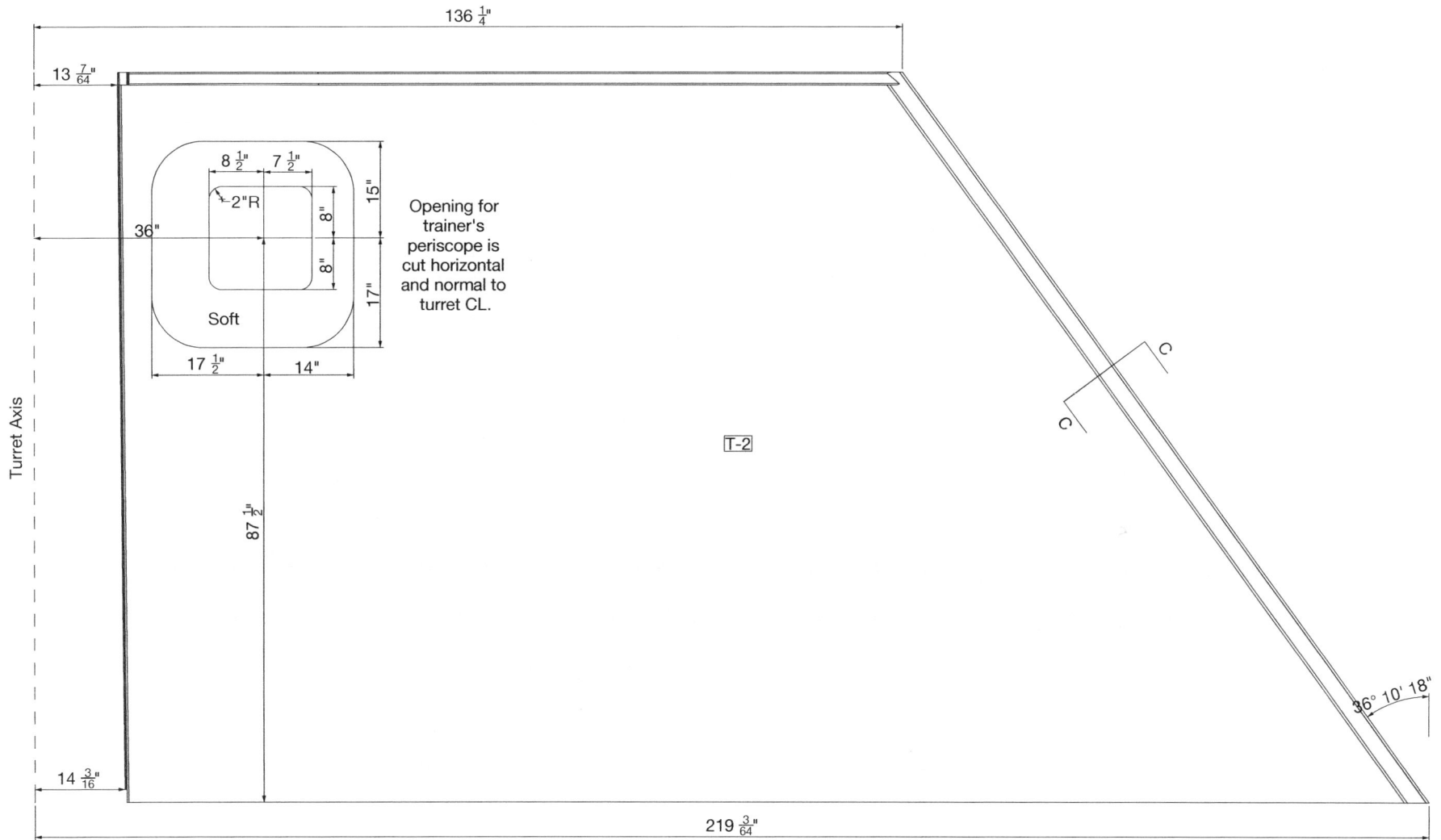

Figure 8.54: Forward Right Plate Outside Elevation (1:24)

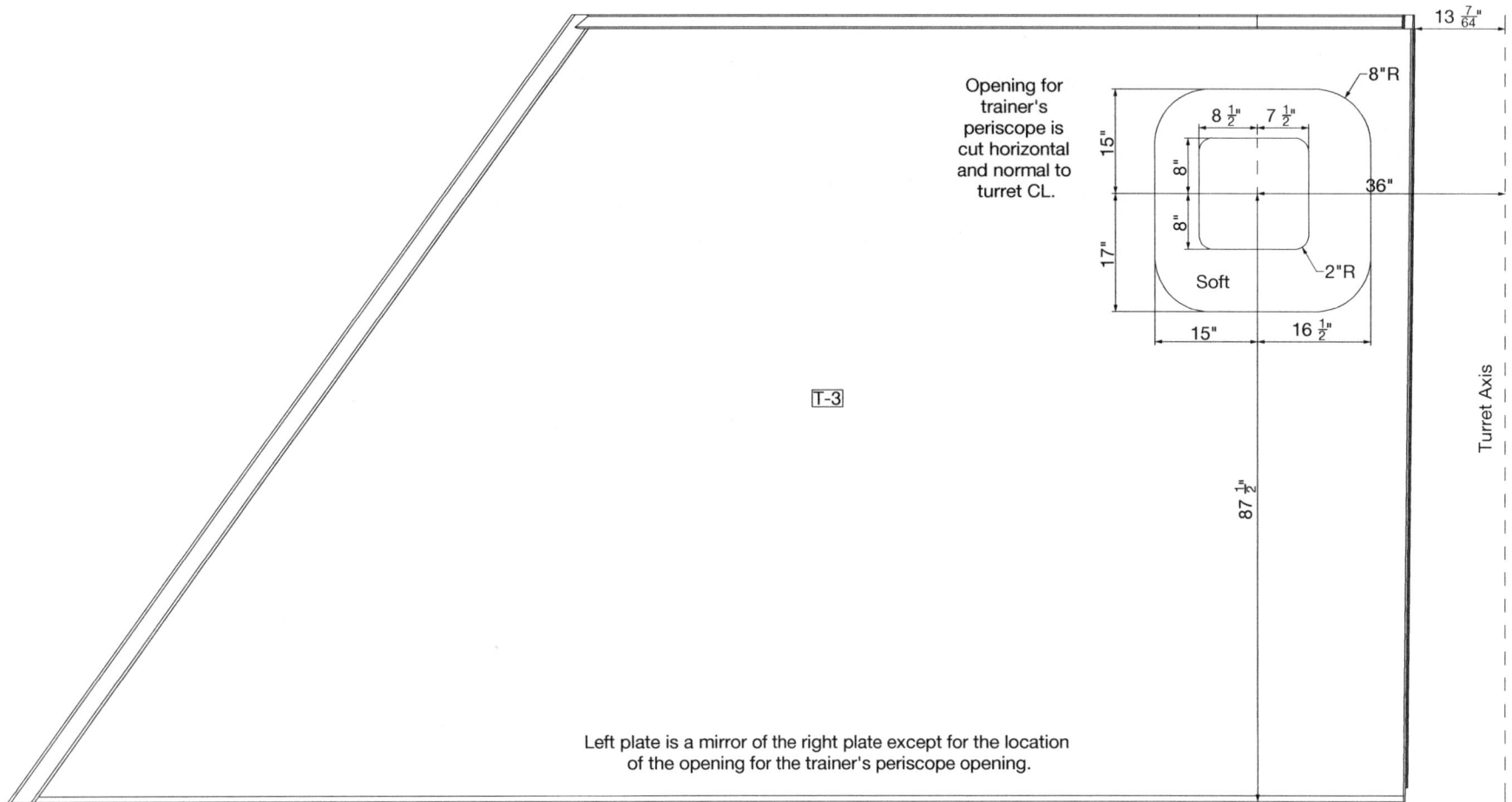

$13\frac{7}{64}"$

Opening for
trainer's
periscope is
cut horizontal
and normal to
turret CL.

8"R

$8\frac{1}{2}"$ $7\frac{1}{2}"$

15"

8"

36"

8"

17"

Soft

2"R

15" $16\frac{1}{2}"$

T-3

Turret Axis

$87\frac{1}{2}"$

Left plate is a mirror of the right plate except for the location
of the opening for the trainer's periscope opening.

**Figure 8.55: Forward Left Plate Outside El-
evation (1:24)**

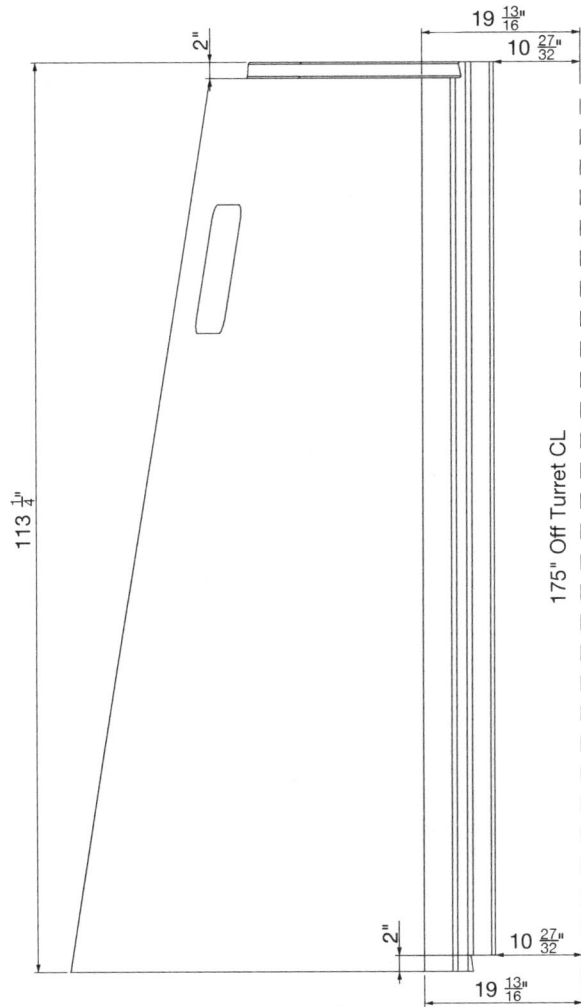

Figure 8.56: **Right Front Plate Rear Elevation (1:24).** Left side is a mirror except for periscope opening

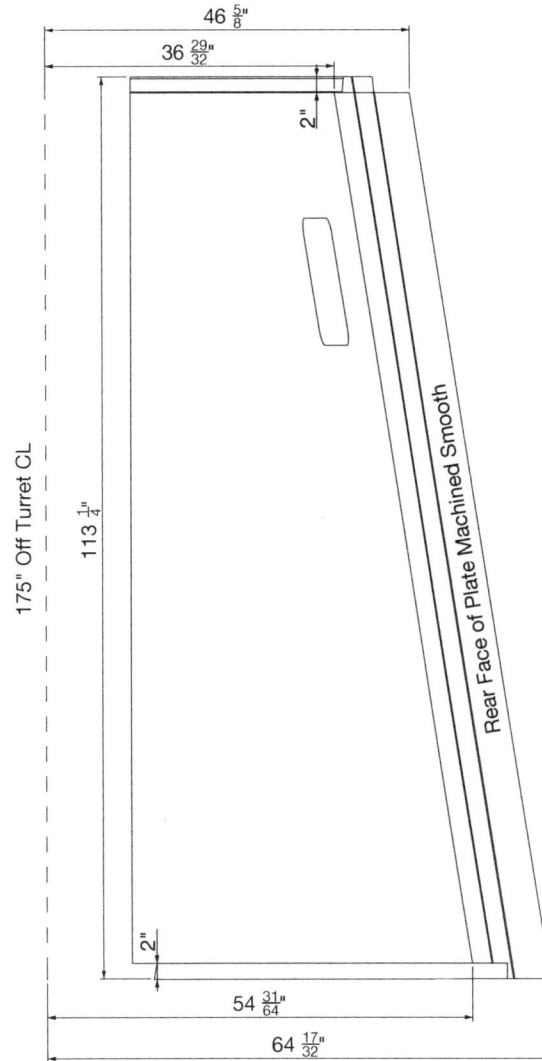

Figure 8.57: **Right Front Plate Front Elevation (1:24).** Left side is a mirror except for periscope opening

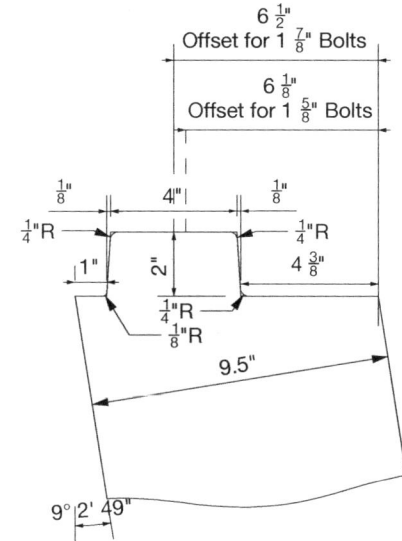

Figure 8.58: **Section A–Upper Edge (1:6).** See Figure 8.52.

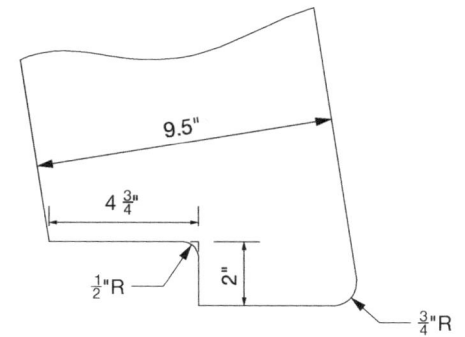

Figure 8.59: **Section B–Lower Edge (1:6).** See Figure 8.53.

Figure 8.62: **Section Normal to Keyway (1:6)**.

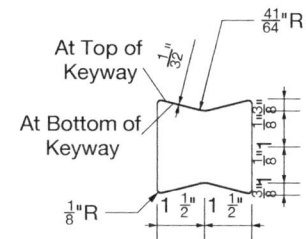

Figure 8.63: **Cross Section of Key (1:6)**

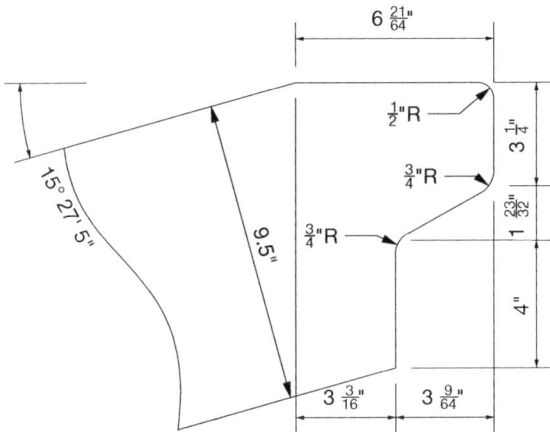

Figure 8.60: Section C–Forward Edge (1:6).
See Figure 8.54

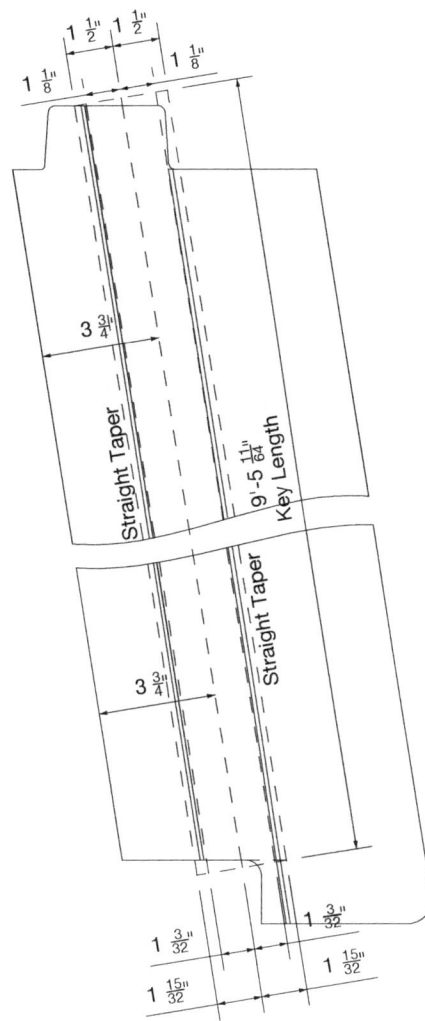

Figure 8.61: Elevation of Keyway Between Side Plates (1:6). The front and rear side plates are joined by an hourglass key hammered into a tapered keyway.

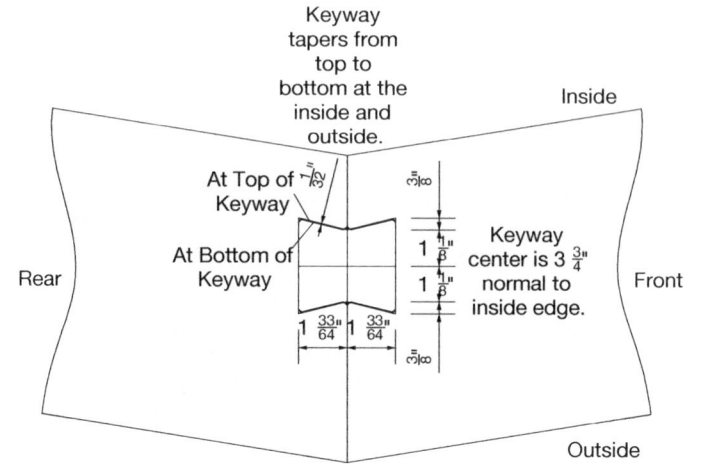

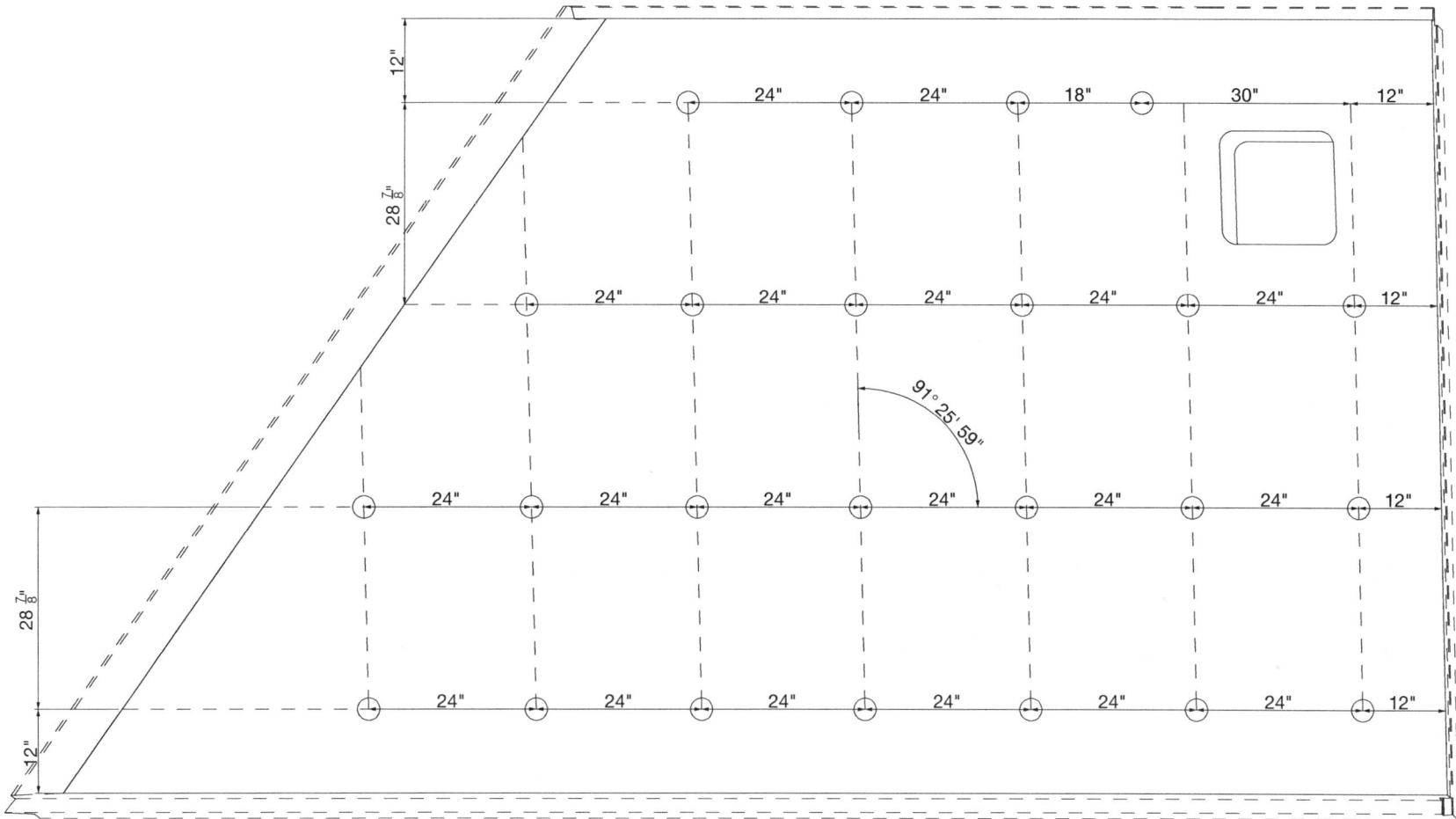

Figure 8.64: Expansion of Forward Right Plate Inner Face (1:24).
This diagram shows the location of bolt holes for mounting the armor plate to the backing plate.

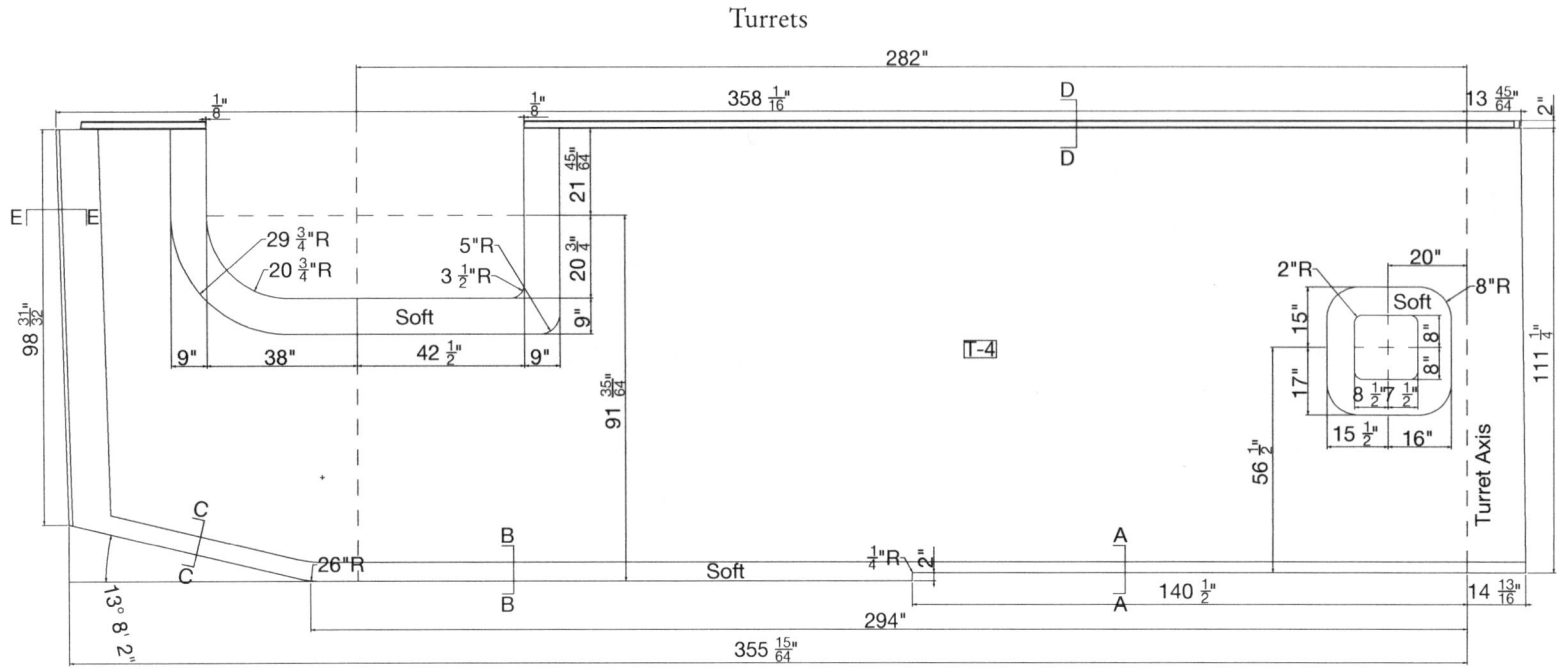

Figure 8.65: Right Rear Plate Elevation (1:48)

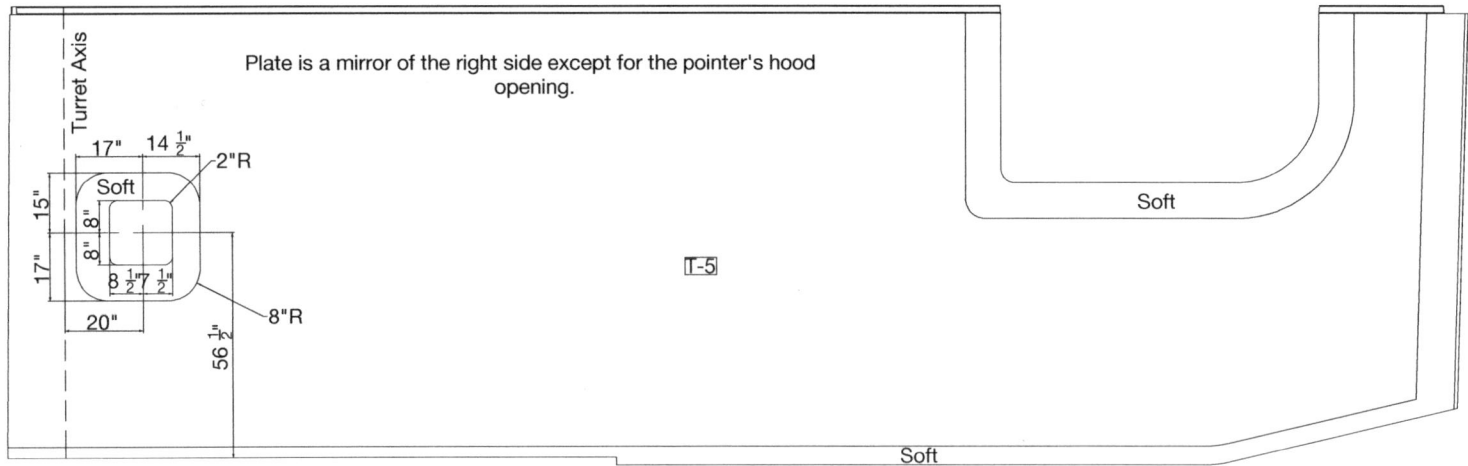

Plate is a mirror of the right side except for the pointer's hood opening.

Figure 8.66: Left Rear Plate Elevation (1:48)

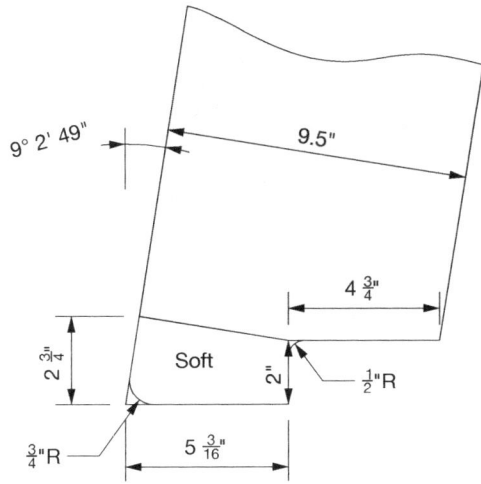

Figure 8.67: Section A–Lower Edge (1:6). This section is normal to the faces.

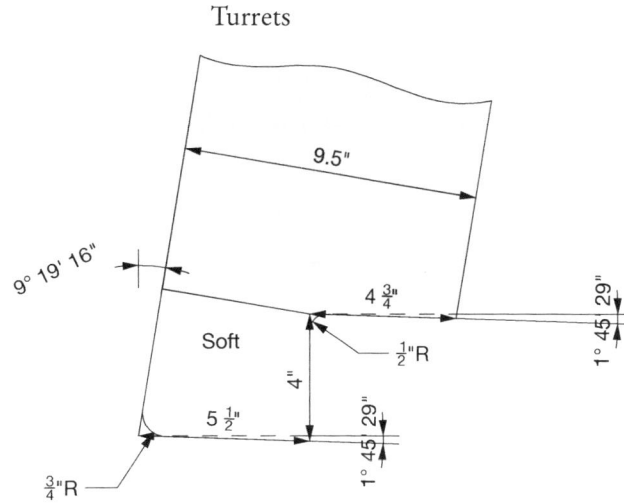

Figure 8.68: Section C–Lower Edge at Slope (1:6). Section normal to faces.

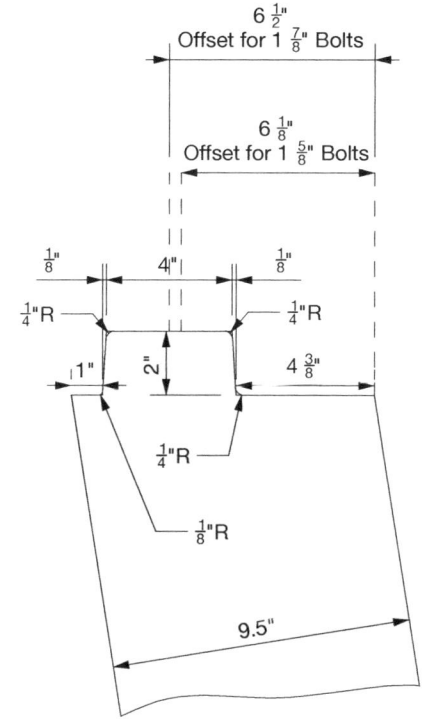

Figure 8.70: Section D–Upper Edge (1:6). Normal to faces.

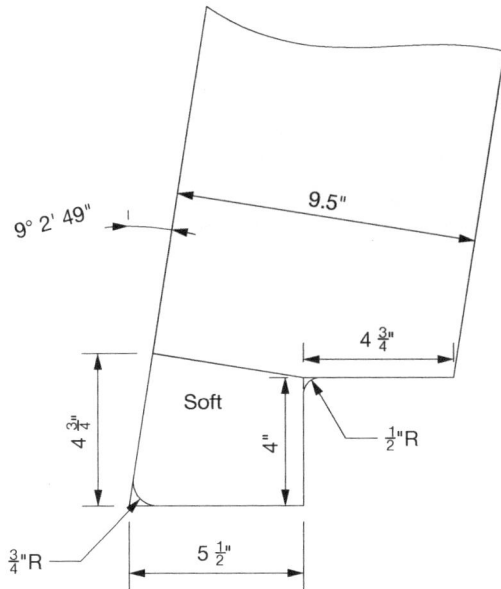

Figure 8.69: Section B–Lower Edge at Doubled Shelf Plate (1:6). Section Normal to Faces.

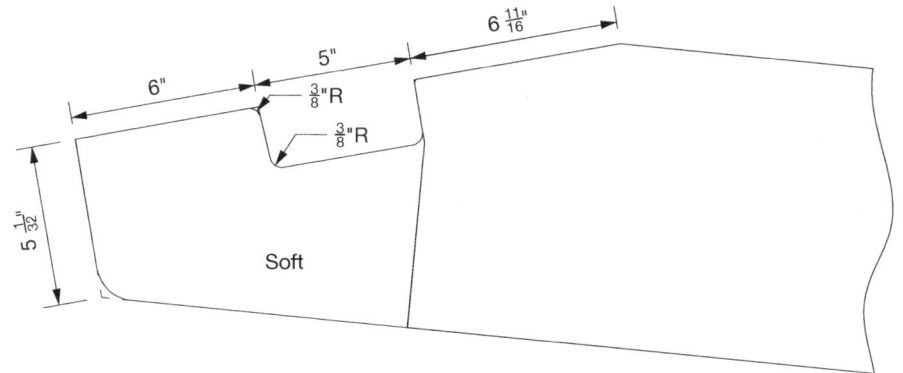

Figure 8.71: Section E–Rear Edge (1:6). Normal to Faces.

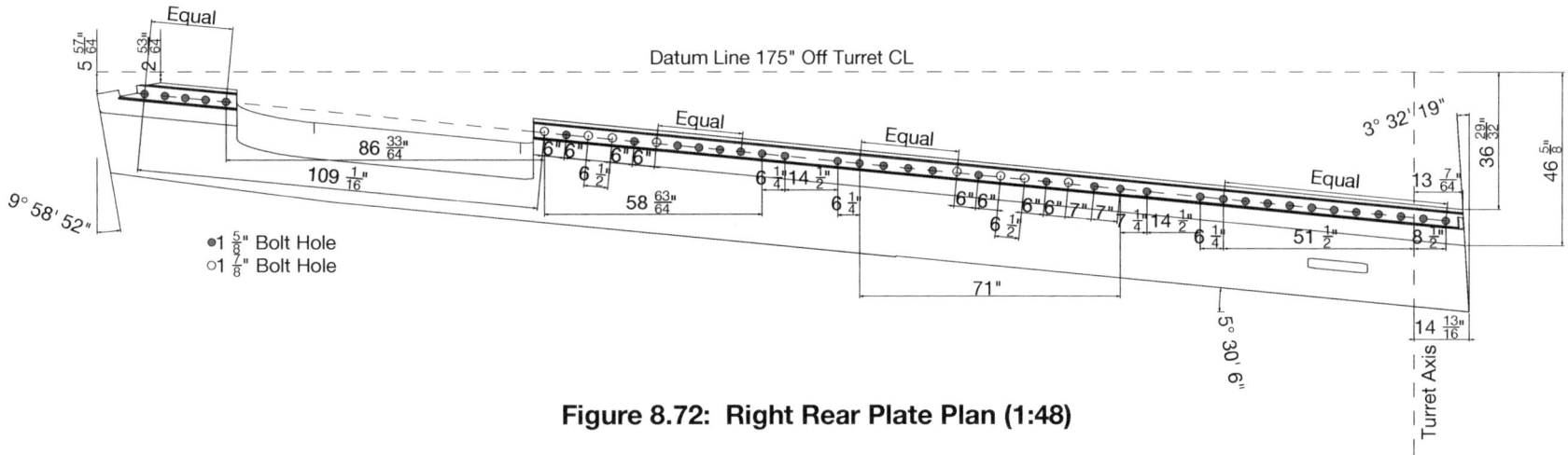

Figure 8.72: **Right Rear Plate Plan (1:48)**

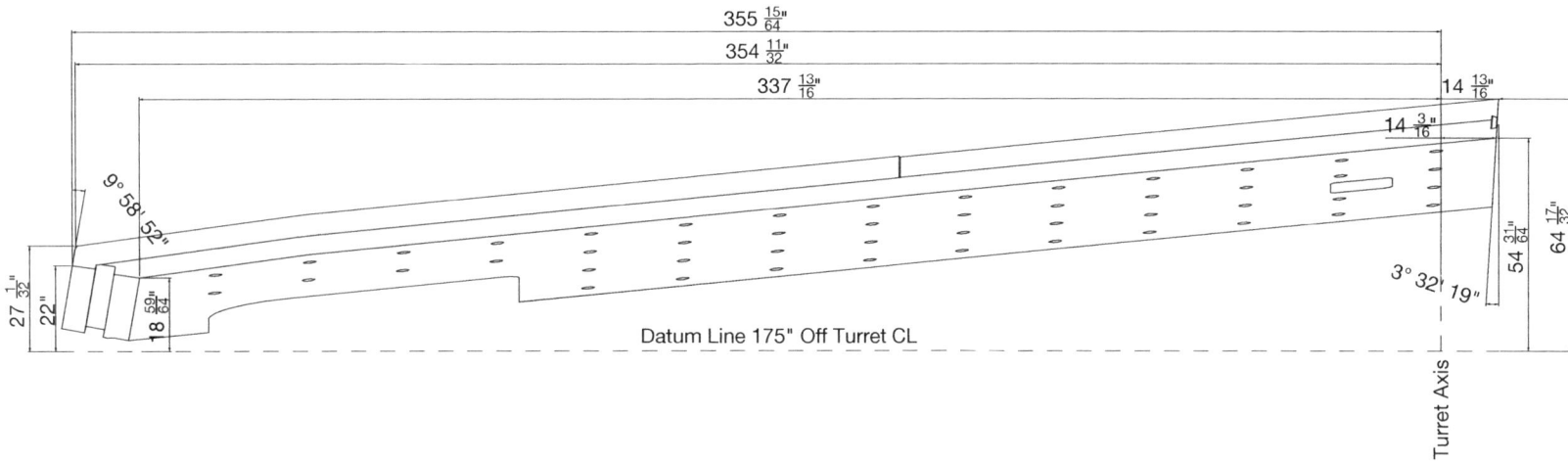

●1 $\frac{5}{32}$" Bolt Hole
○1 $\frac{7}{8}$" Bolt Hole

Datum Line 175" Off Turret CL

Equal

Figure 8.73: **Right Rear Plate Bottom Plan (1:48)**

Datum Line 175" Off Turret CL

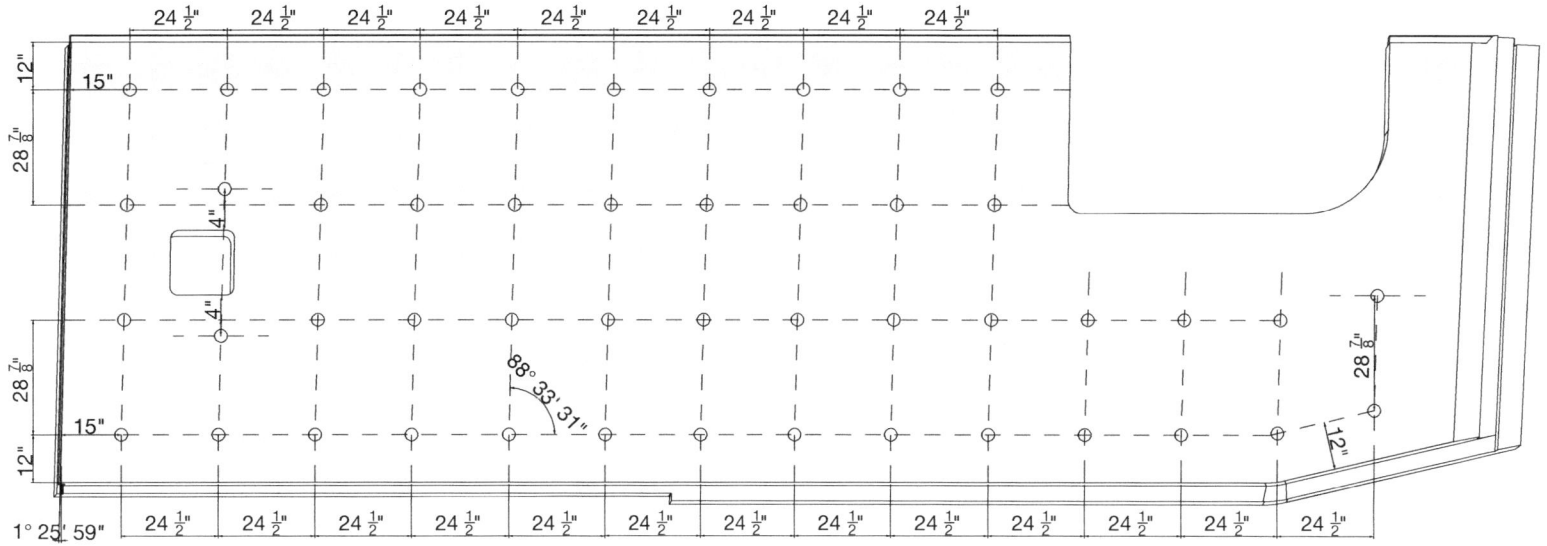

Figure 8.74: Expansion of Inside Right Rear Plate (1:48).
Showing the location of bolt holes. Left side is a mirror except for periscope opening,

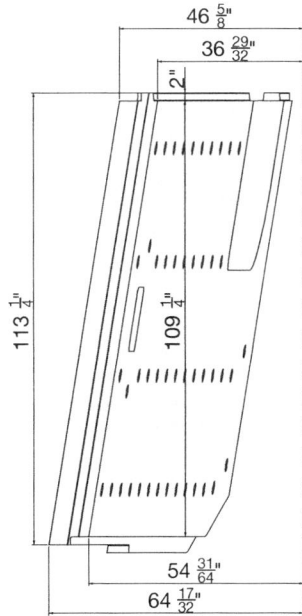

Figure 8.75: Right Rear Plate Front Elevation (1:48)

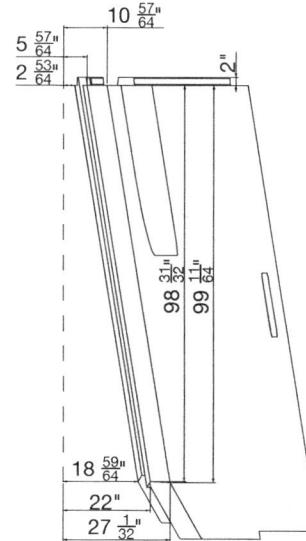

Figure 8.76: Right Rear Plate Rear Elevation (1:48)

178 $\frac{13}{16}$"

8 $\frac{1}{2}$"

Finished Smooth

C | C

T-6

12"

102 $\frac{11}{32}$"

97 $\frac{39}{64}$"

22 $\frac{1}{8}$"

55 $\frac{1}{4}$"

66 $\frac{3}{4}$"

27 $\frac{5}{8}$"

8 $\frac{3}{4}$"

8 $\frac{3}{4}$"

8 $\frac{3}{4}$"

8 $\frac{3}{4}$"

8 $\frac{3}{4}$"

8 $\frac{3}{4}$"

8 $\frac{1}{4}$"

44"

Finished Smooth

194 $\frac{11}{16}$"

196 $\frac{3}{16}$"

Datum Line 4'-3" BL Trunion CL

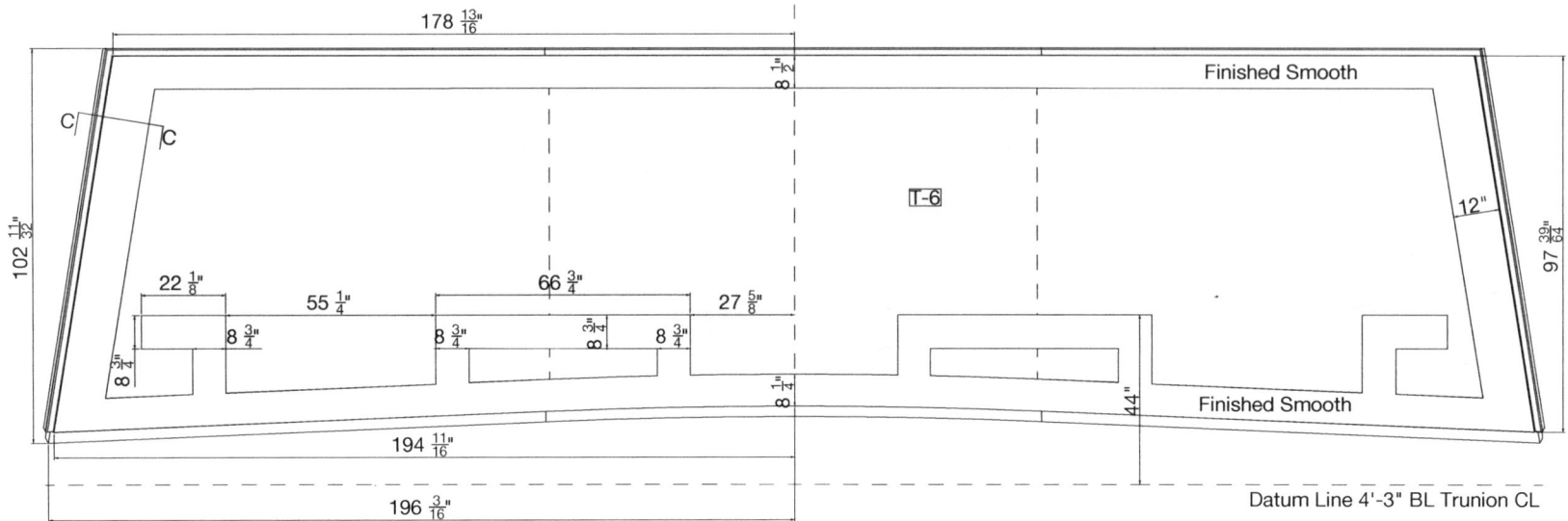

Figure 8.77: Rear Plate Inside Elevation (1:48). Smoother areas around the edge are for welding angled scallops to connect the plate to the turret. The other smooth areas are for welding the gun girders.

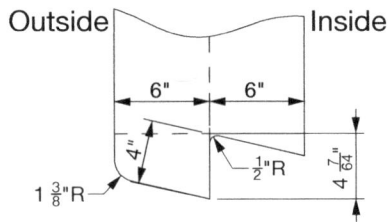

Outside   Inside

6"   6"

4"

$\frac{1}{2}$"R

4 $\frac{7}{64}$"

1 $\frac{3}{8}$"R

Figure 8.78: Section A–Lower Edge (1:12)

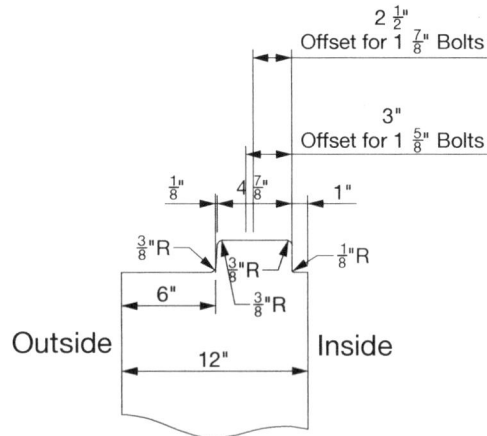

2 $\frac{1}{2}$"
Offset for 1 $\frac{7}{8}$" Bolts

3"
Offset for 1 $\frac{5}{8}$" Bolts

$\frac{1}{8}$"   4 $\frac{7}{8}$"   1"

$\frac{3}{8}$"R $\frac{1}{8}$"R

$\frac{3}{8}$"R

$\frac{3}{8}$"R

6"

Outside   Inside

12"

Figure 8.80: Section B–Upper Edge (1:12)

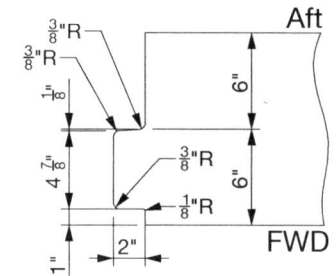

Aft

$\frac{3}{8}$"R
$\frac{3}{8}$"R

$\frac{1}{8}$"

6"

4 $\frac{7}{8}$"

$\frac{3}{8}$"R

6"

$\frac{1}{8}$"R

1"   2"

FWD

Figure 8.79: Section C–Side Edge (1:12). Section Normal to side,

Photograph 8.3: Turret Rear Plate. This is turret 1 on USS New Jersey. Milling variations caused edges of the side and rear plates to be out of alignment. The cementing process has left noticeable irregularities in the surface of the rear plate.

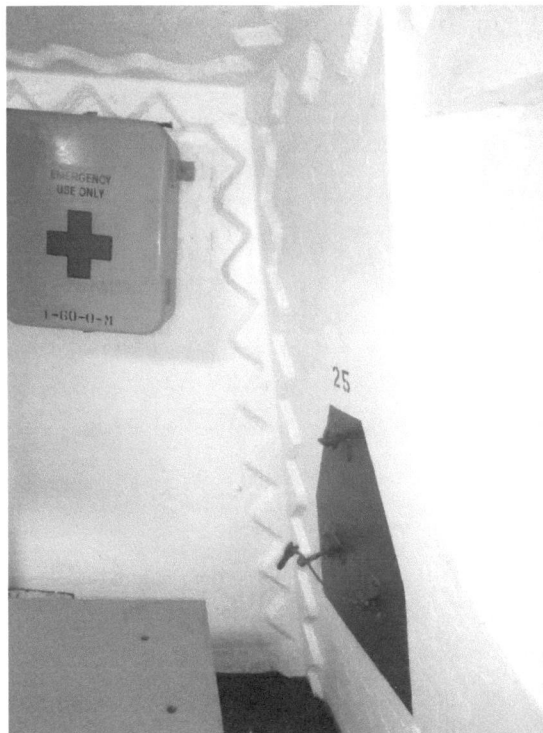

Photograph 8.4: Rear Plate Connections. This is the left rear corner of Turret 1 on USS New Jersey. The rear plate at the left was welded to the side backing plate at the right and to the top plate along scalloped angles.

66 $\frac{9}{16}$"

32'R

Equal

Equal

116 $\frac{3}{32}$"

B

1 $\frac{45}{64}$"

B

● 1 $\frac{5}{8}$" Bolt Hole
○ 1 $\frac{7}{8}$" Bolt Hole

3"

66"

6" 6" 6" 1$\frac{1}{2}$" 6" 6"

19° 57' 44"

180 $\frac{7}{16}$"

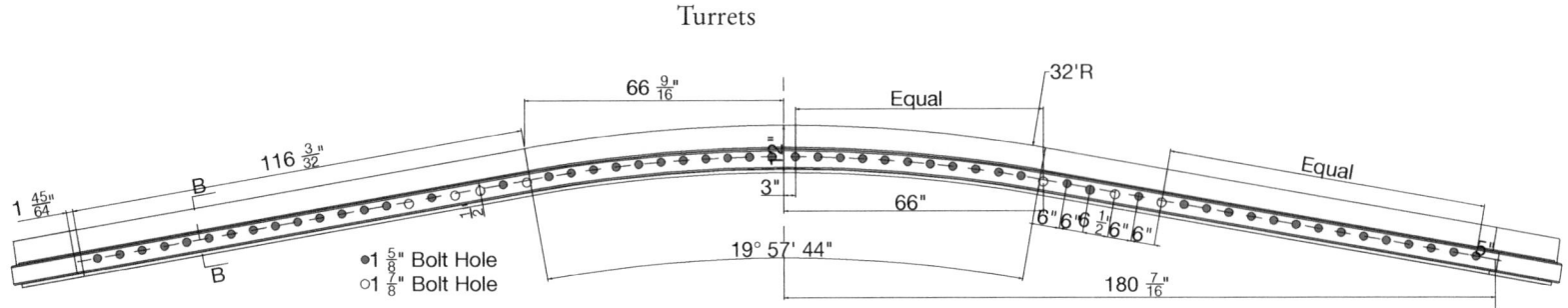

Figure 8.81: Top Plan (1:48)

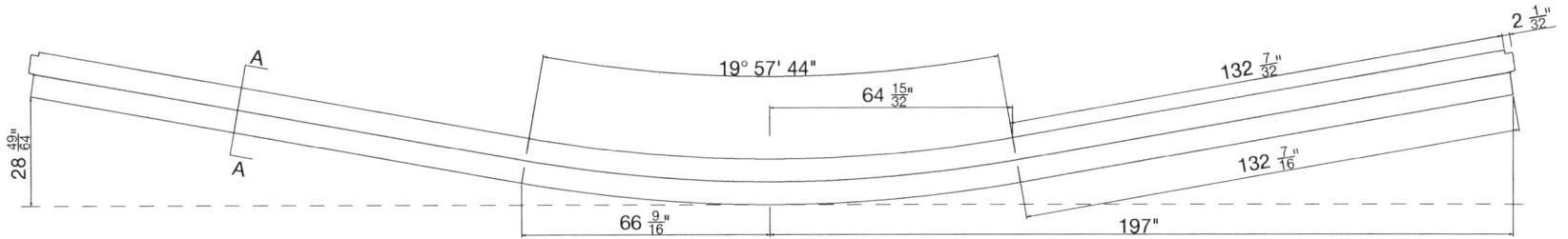

A

19° 57' 44"

132 $\frac{7}{32}$"

2 $\frac{1}{32}$"

64 $\frac{15}{32}$"

132 $\frac{7}{16}$"

28 $\frac{49}{64}$"

A

66 $\frac{9}{16}$"

197"

Figure 8.82: Bottom Plan (1:48)

Figure 8.83: Right Elevation (1:48). Left side is a mirror.

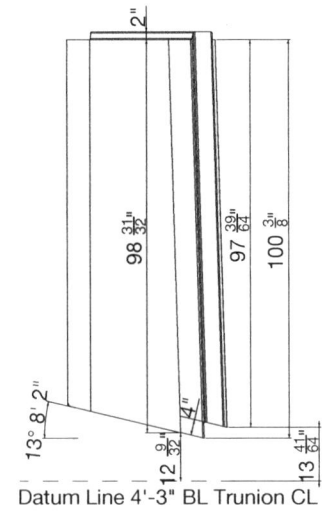

2"

98 $\frac{31}{32}$"

97 $\frac{39}{64}$"

100 $\frac{3}{8}$"

13° 8' 2"

4"

12 $\frac{9}{32}$"

13 $\frac{41}{64}$"

Datum Line 4'-3" BL Trunion CL

Plan 8.5: (7201ZT) Welding of Rear Plate to Roof Plate.

Plan 8.7: (7201ZT) Welding and Bolting of Rear Plate to Shelf Plate

Plan 8.6: (7201ZT) Welding of Rear Plate to Side Plates

Plan 8.8: (7201ZT) Detail of Tap Bolts Connecting Rear Plate to Shelf Plate

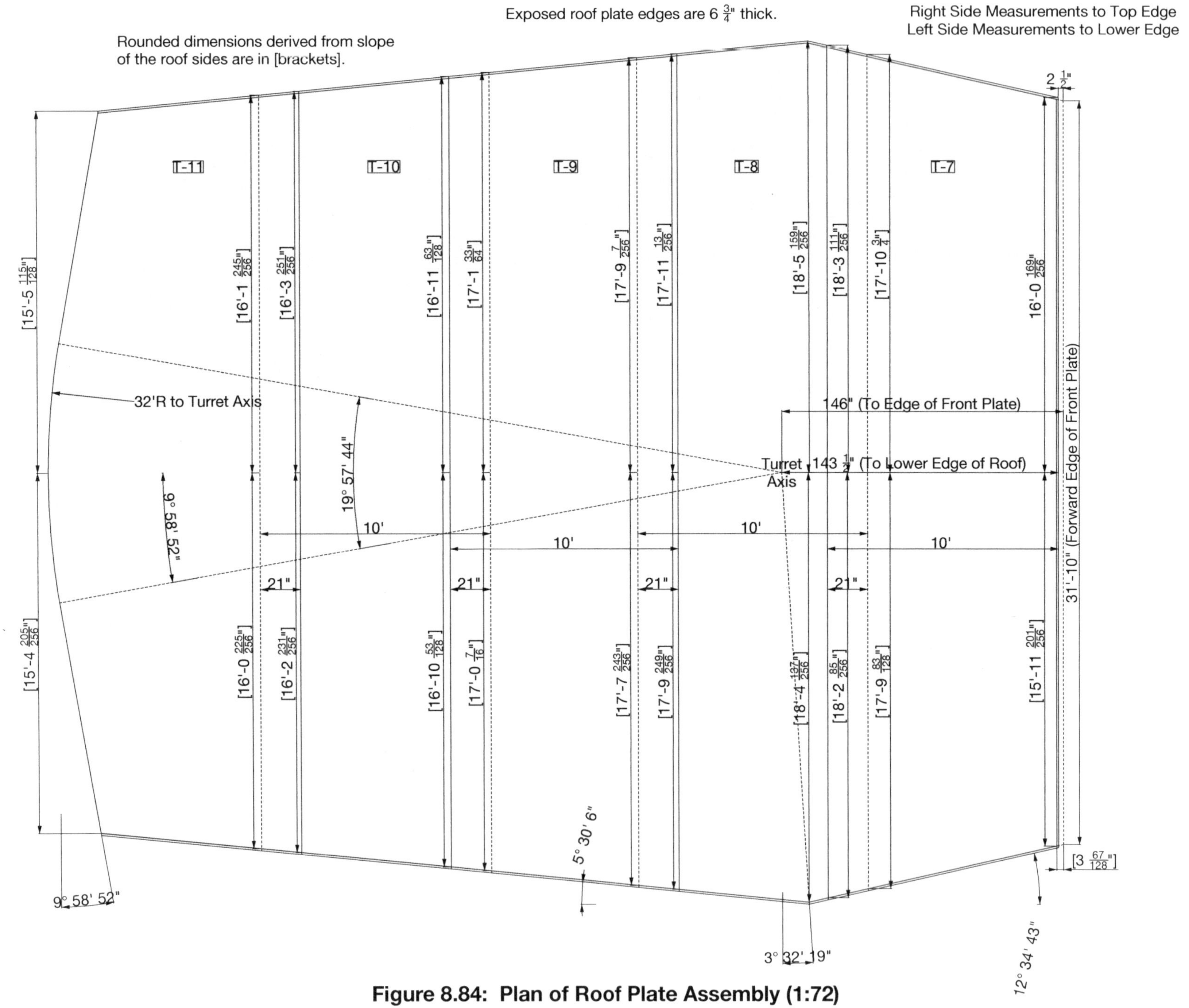

Turrets

Figure 8.84: Plan of Roof Plate Assembly (1:72)

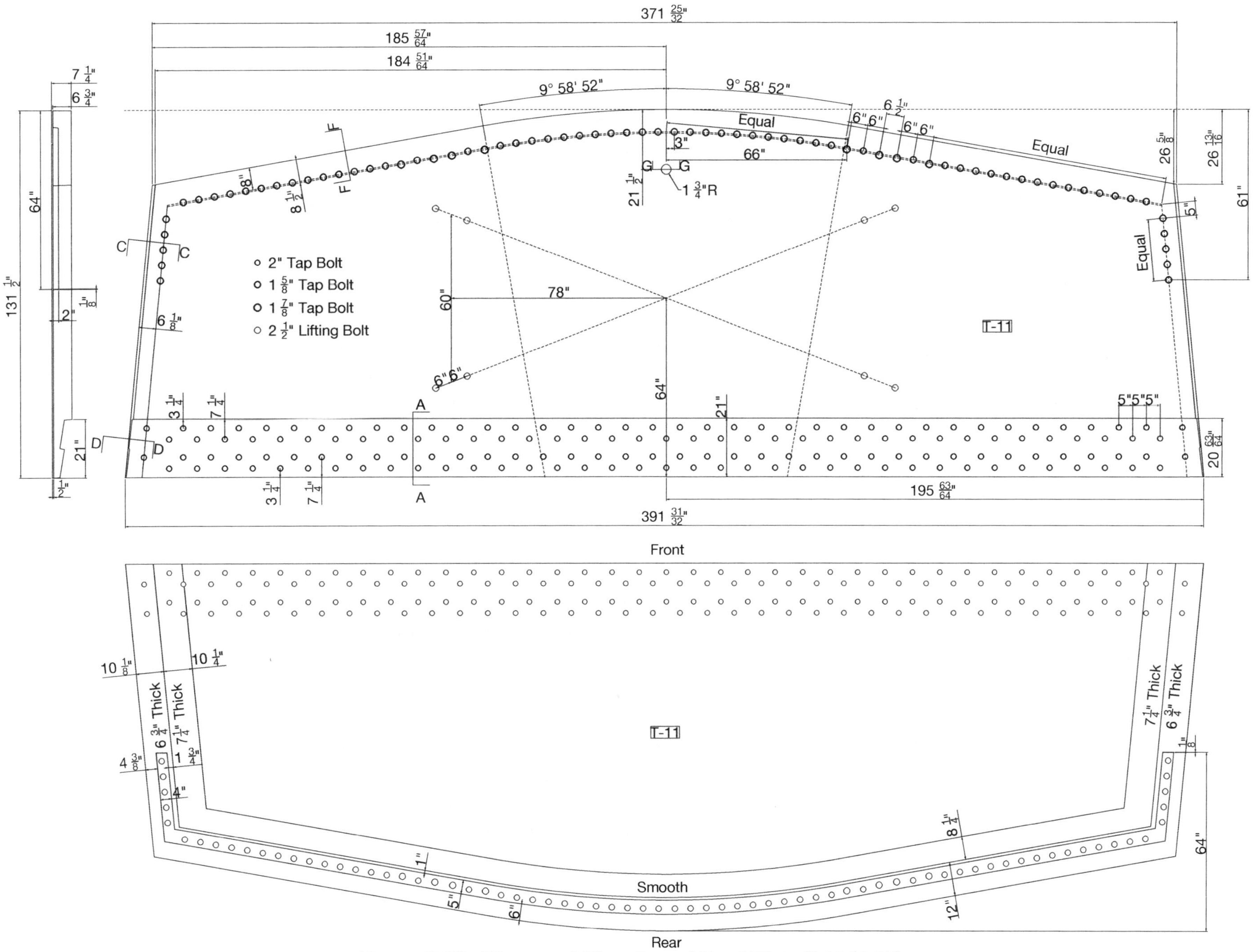

Figure 8.85: Plans and Elevation of Roof Plate T-11 (1:48)

Legend:
- ○ 2" Tap Bolt
- ○ 1 5/8" Tap Bolt
- ○ 1 7/8" Tap Bolt
- ○ 2 1/2" Lifting Bolt

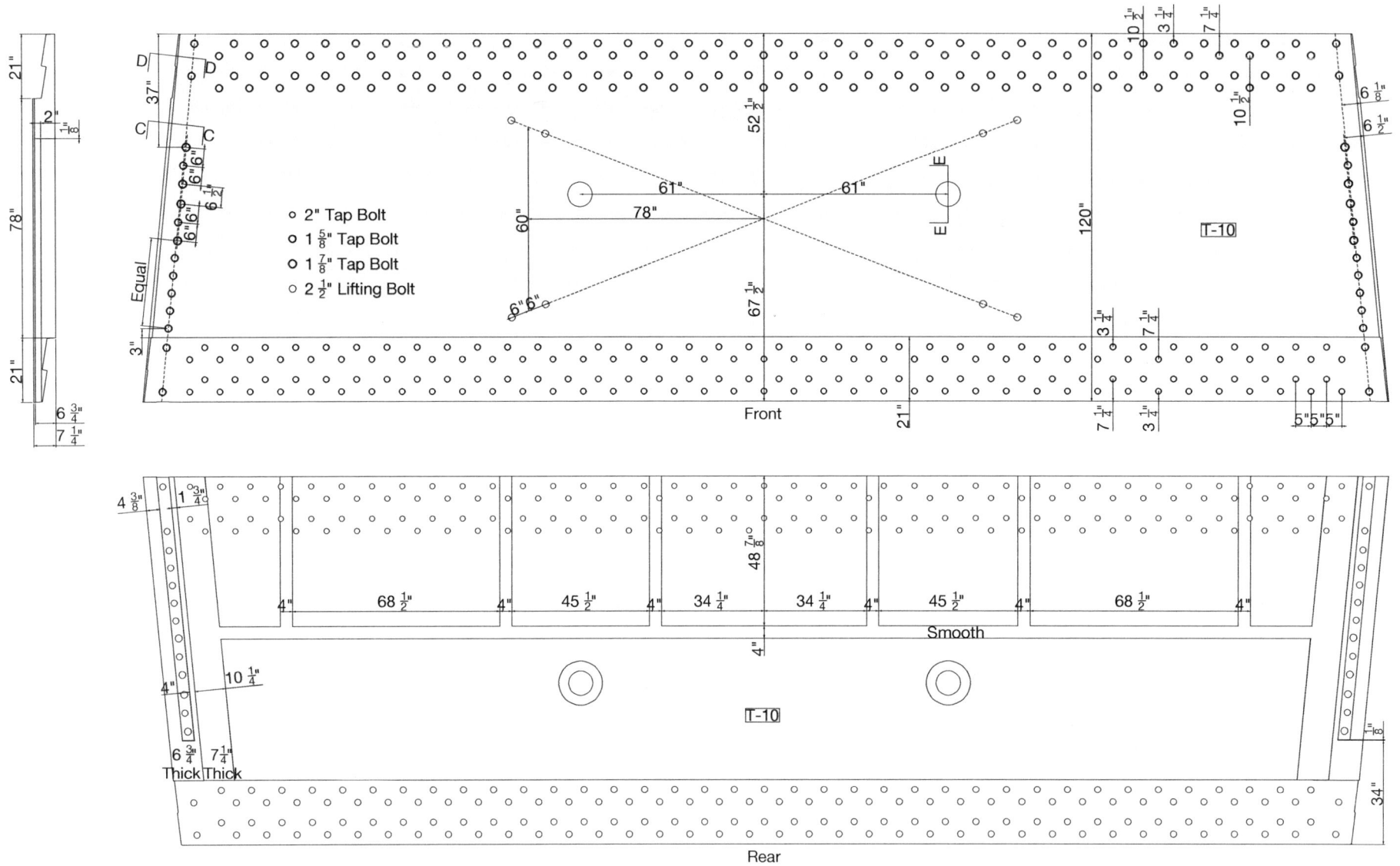

Figure 8.86: Plans and Elevation of Roof Plate T-10 (1:48)

o 2" Tap Bolt
o 1 5/8" Tap Bolt
o 1 7/8" Tap Bolt
o 2 1/2" Lifting Bolt

Front

Rear

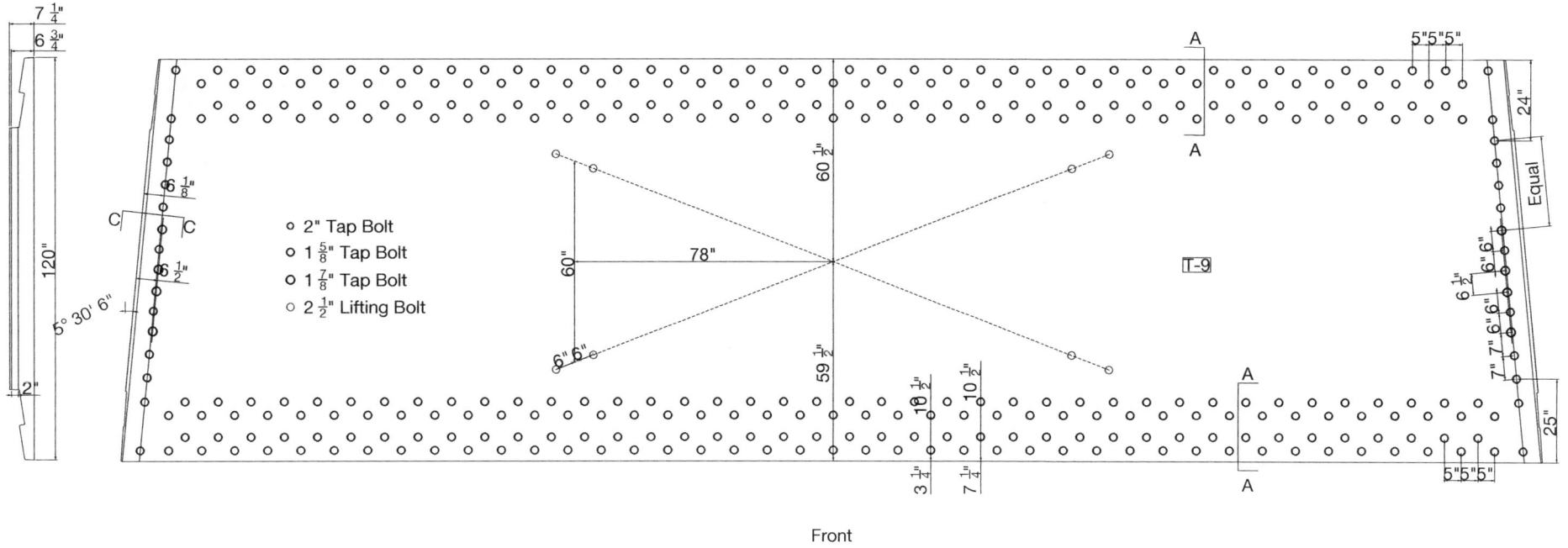

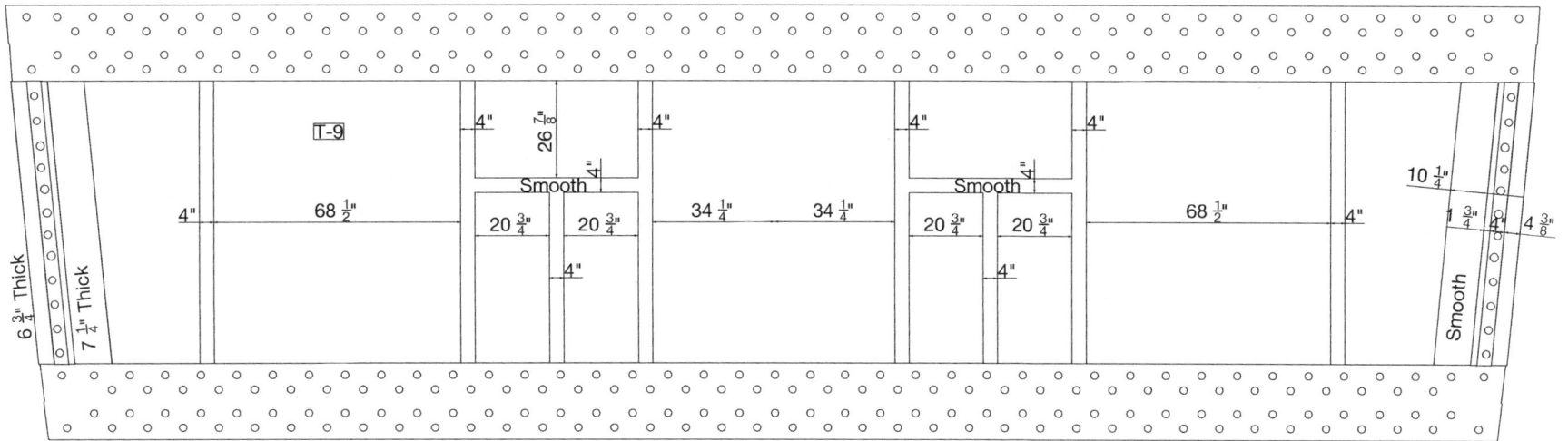

Figure 8.87: Plans and Elevation of Roof Plate T-9 (1:48).
This plate is removable.

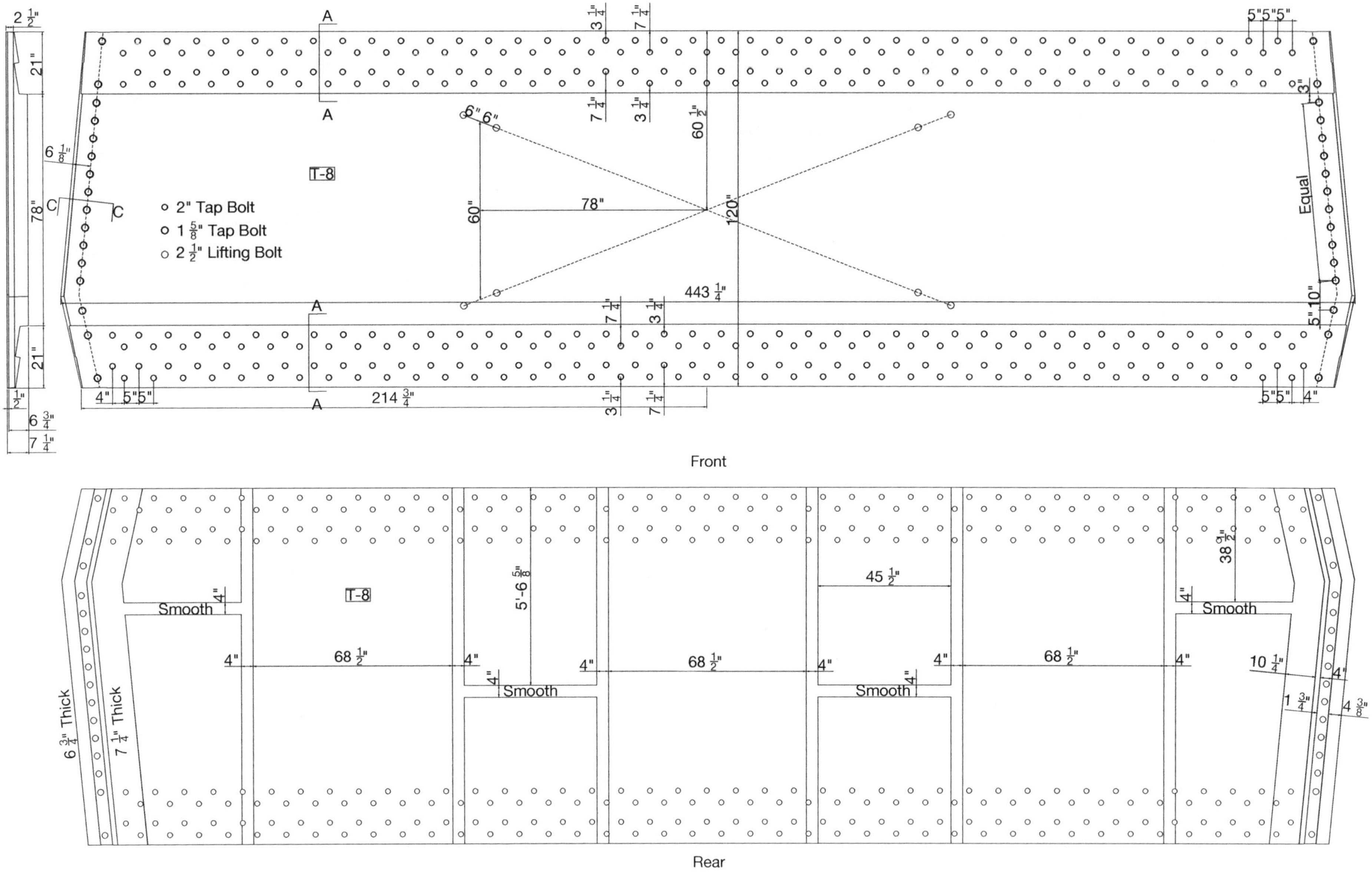

Figure 8.88: Plans and Elevation of Roof Plate T-8 (1:48)

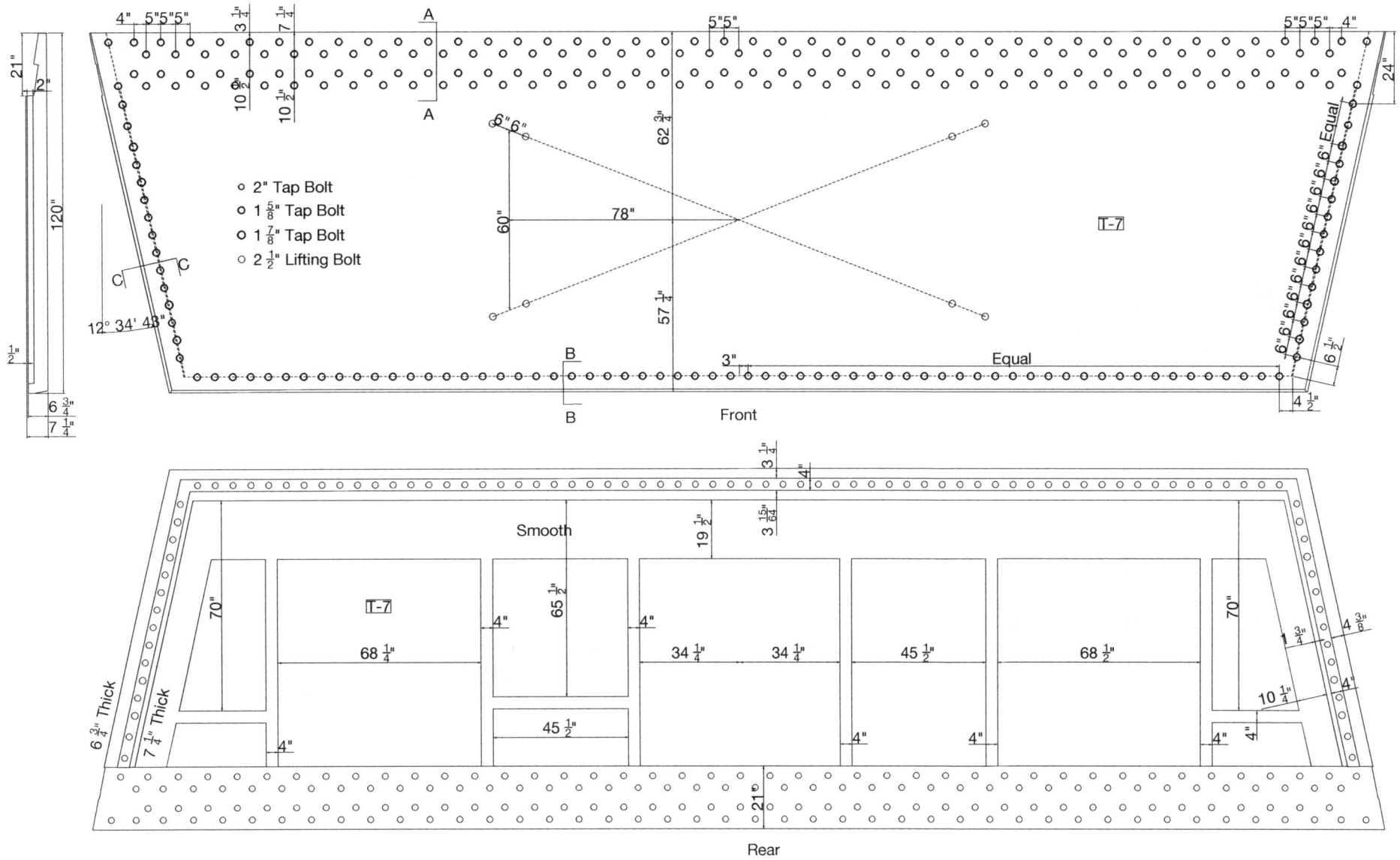

Figure 8.89: Plans and Elevation of Roof Plate T-7 (7201ZT)

Photograph 8.9: Turret Roof Plates. This view of the roof of Turret 3 on USS New Jersey show shows the bolts through the scarf joint connecting the roof plates.

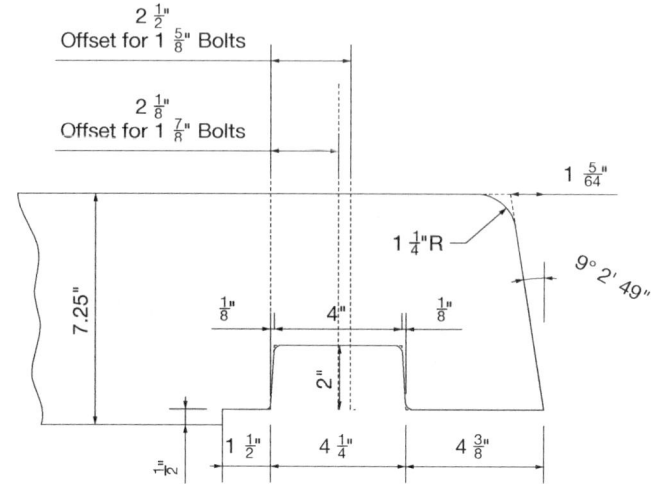

Figure 8.90: Section C–Side Edge (1:6)

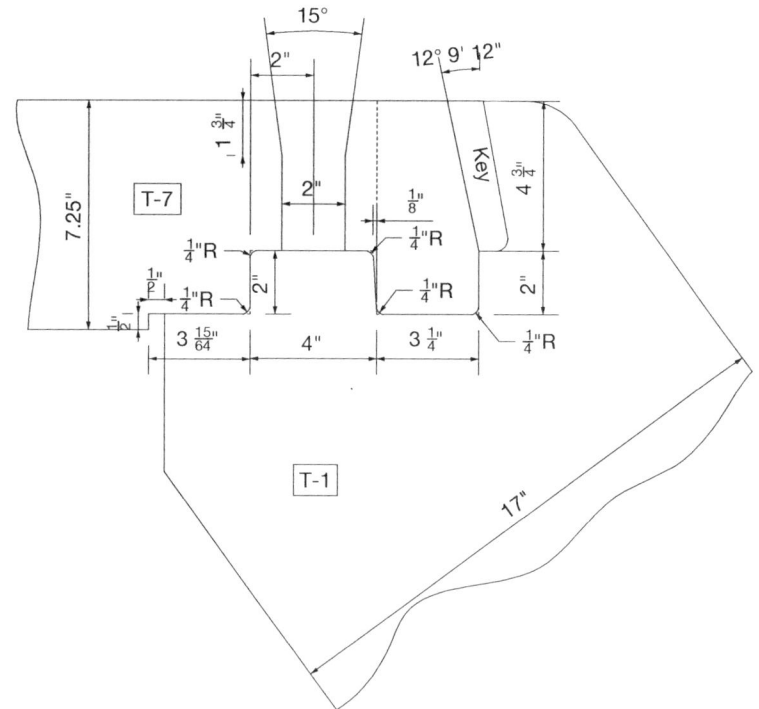

Figure 8.91: Section B–Forward Edge (1:6)

Photograph 8.10: Root of Turret No. 1 on USS Missouri.
The scarf joint between roof plates is covered by a 5# mantlet plate. USS Iowa is similar. (Franklin Clay)

Photograph 8.11: Roof of Turret No. 1 on USS New Jersey.
USS Wisconsin and USS New Jersey lack the mantlet plate over
the root scarf joints.

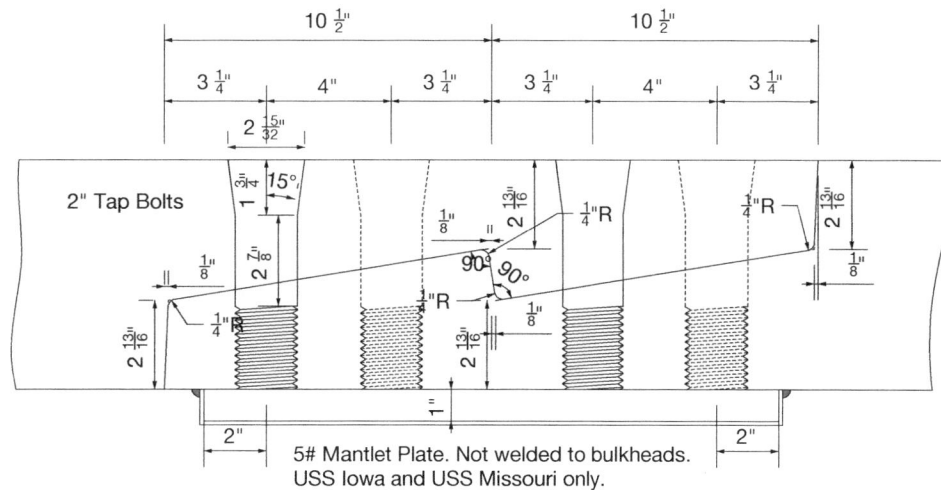

Figure 8.92: Section A–Scarf Joint Between Roof Plates (1:6)

3"
Offset for $1\frac{5}{8}$" Bolts

$2\frac{1}{2}$"
Offset for $1\frac{7}{8}$" Bolts

1"R

T-11

7.25"

$4\frac{7}{8}$"

$\frac{1}{8}$"

2"

$2\frac{1}{2}$"

$\frac{1}{2}$"

$1\frac{1}{2}$"

5"

6"

Figure 8.95: Section F–Rear Edge (1:6)

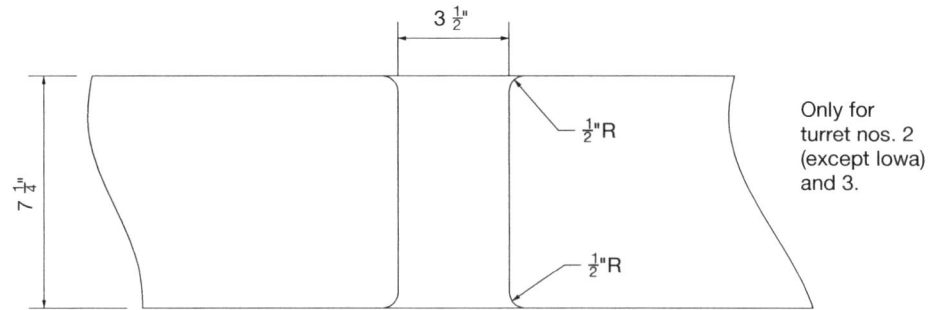

2" Tap
Bolt Hole
At Scarf Joint

$1\frac{5}{64}$"

$9°\,2'\,49''$

7.25"

$\frac{1}{2}$"

4"

$6\frac{1}{8}$"

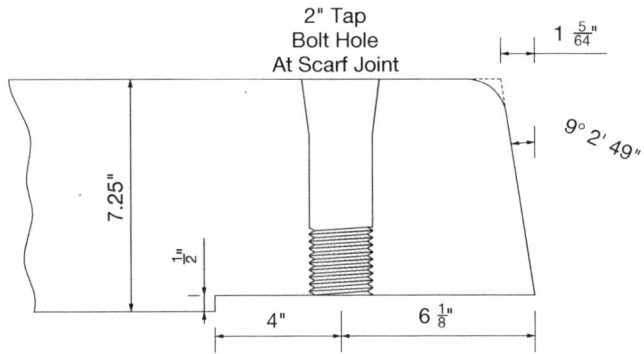

Figure 8.93: Section D–Side Edge Over Rangefinder Opening (1:6)

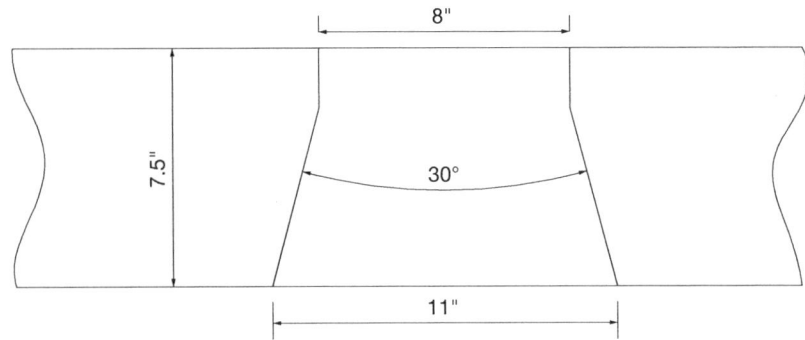

$3\frac{1}{2}$"

$\frac{1}{2}$"R

$7\frac{1}{4}$"

$\frac{1}{2}$"R

Only for
turret nos. 2
(except Iowa)
and 3.

Figure 8.96: Section G–Hole for 40mm Power Cable (1:6)

8"

7.5"

30°

11"

Figure 8.94: Section E–Periscope Openings (1:6)

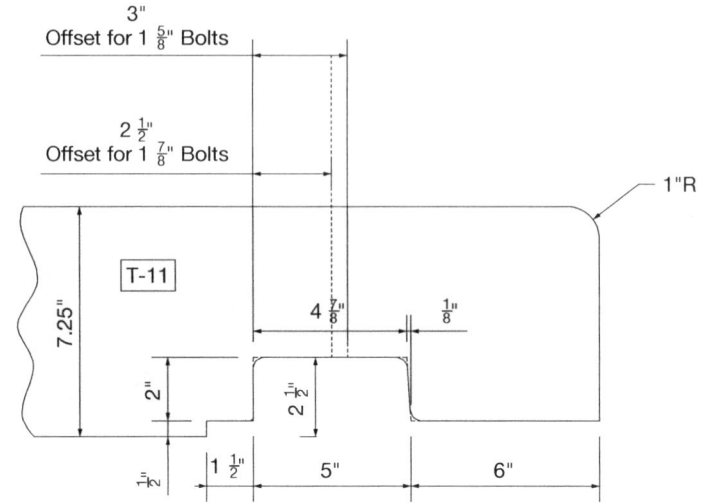

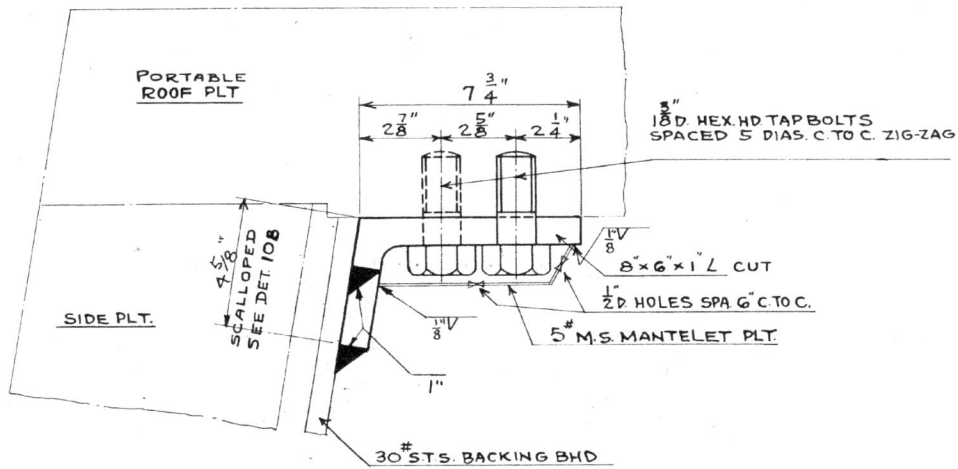

Plan 8.12: (7201ZT). Connection of Removable Roof Plate to Backing Bulkhead.

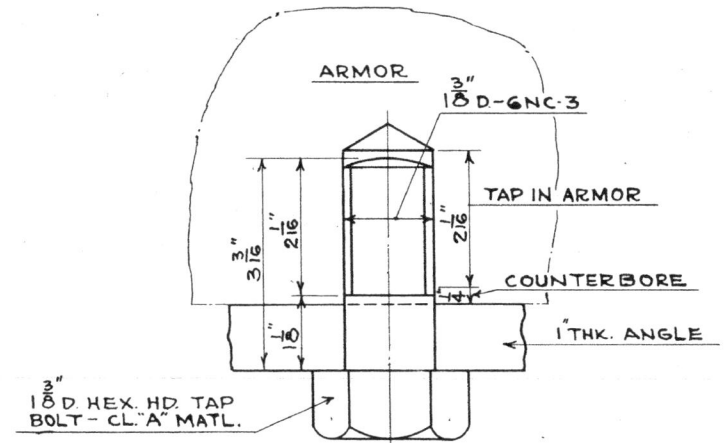

Plan 8.14: (7201ZT). Detail of Bolts Attaching Removable Roof Plate.

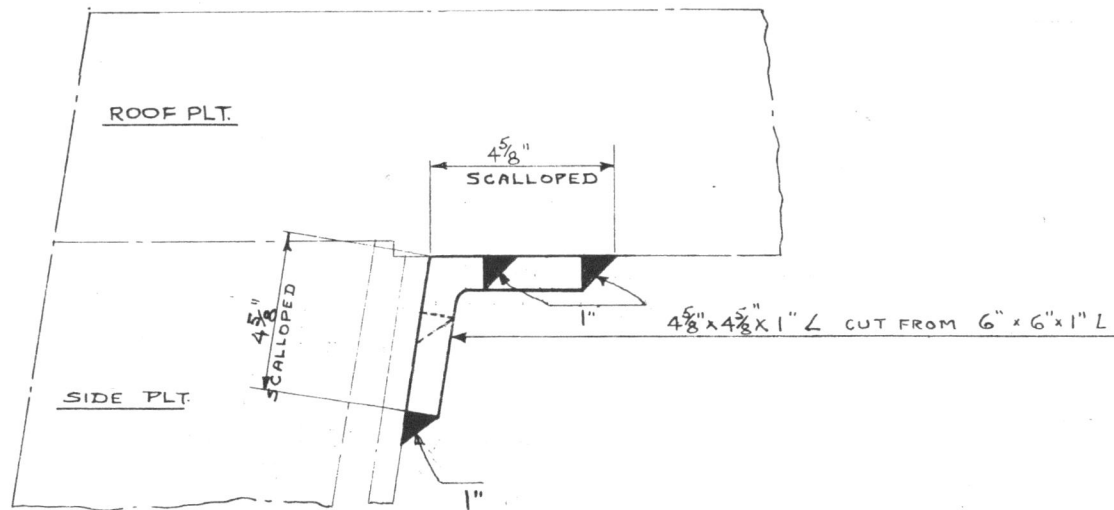

Plan 8.13: (7201ZT). Welding of Fixed Roof Plates to Backing Bulkhead.

Reference point Is 38.55" ABV Trunnion CL, 23'-6" Aft of Turret CL, and is
aligned with upper edge of the turret armor ABT 16'-0 $\frac{1}{16}$" off the CL

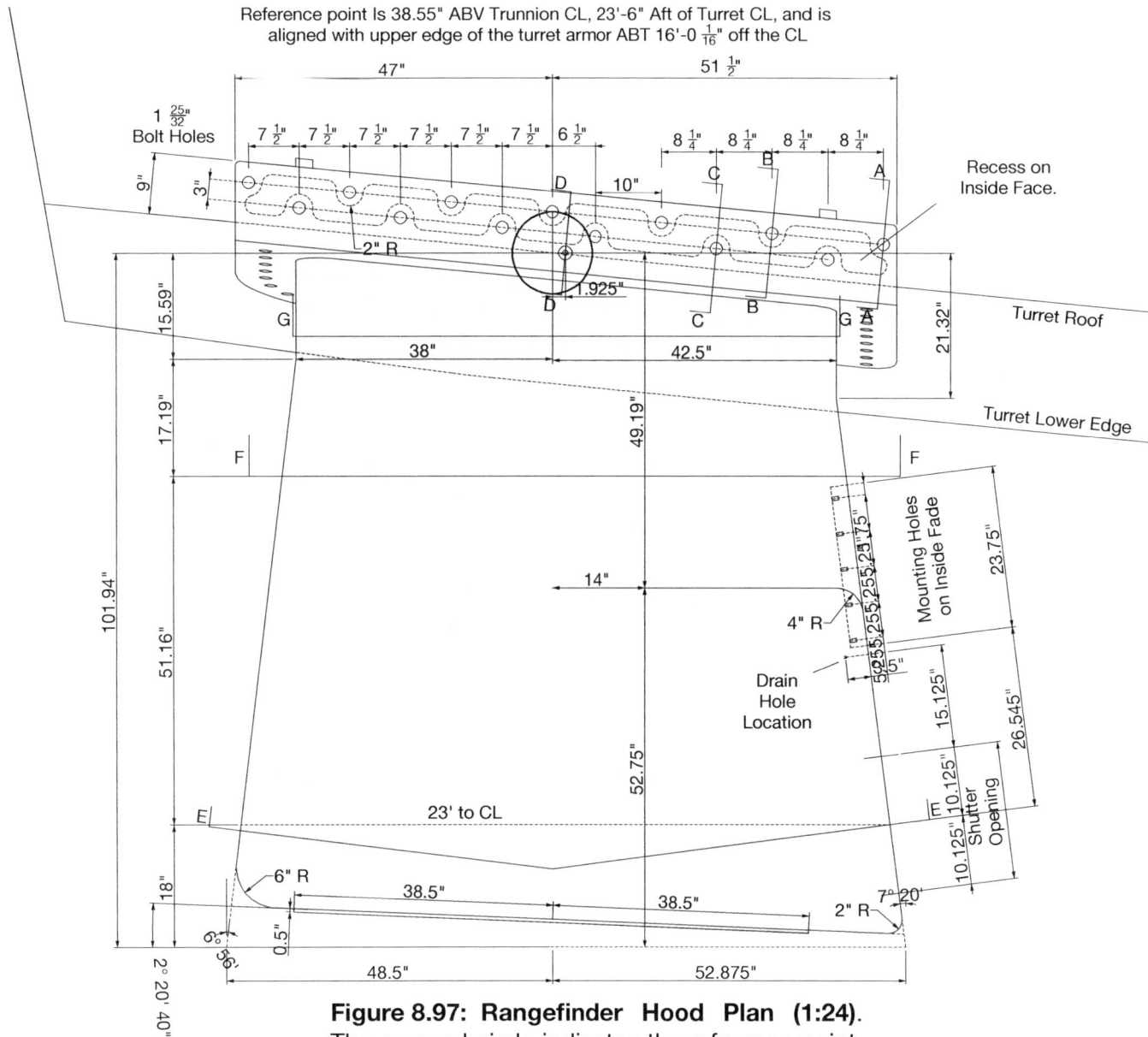

Figure 8.97: Rangefinder Hood Plan (1:24).
The crossed circle indicates the reference point.
This is the intersection of the rangefinder centerline and the upper edge of the turret. Note that
the front and rear faces are perpendicular to the
centerline where they meet the baseplate.

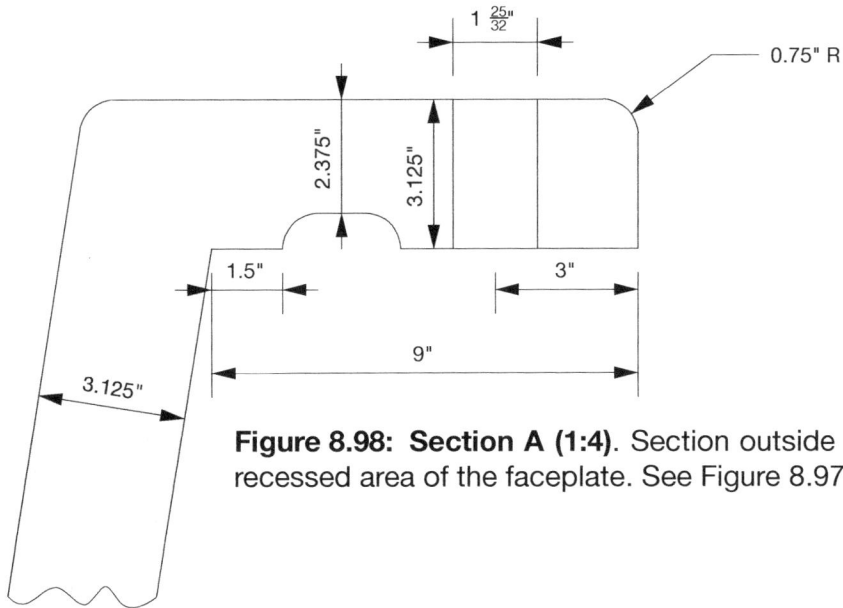

Figure 8.98: Section A (1:4). Section outside the recessed area of the faceplate. See Figure 8.97.

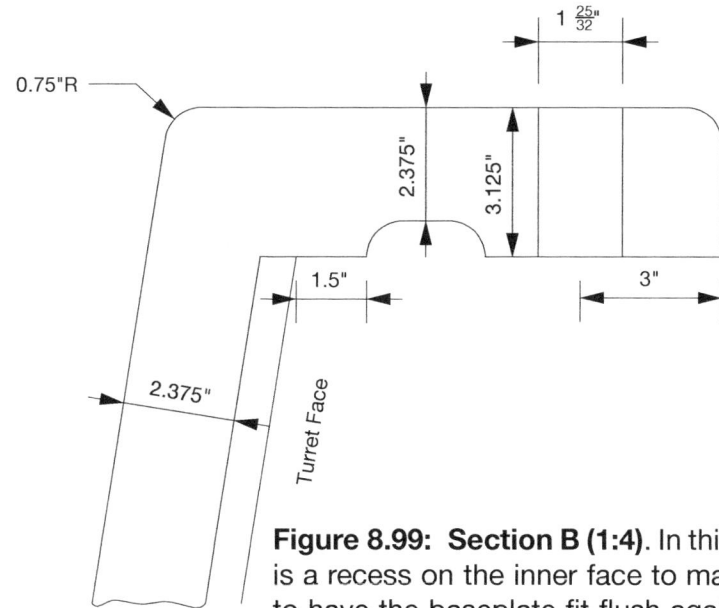

Figure 8.99: Section B (1:4). In this area there is a recess on the inner face to make it easier to have the baseplate fit flush against the turret. See Figure 8.97.

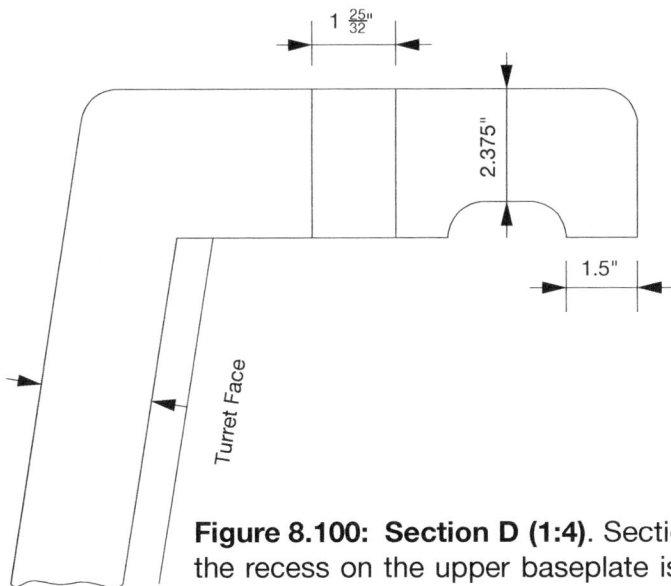

Figure 8.100: Section D (1:4). Section where the recess on the upper baseplate is inboard of a bolt hole. See Figure 8.97.

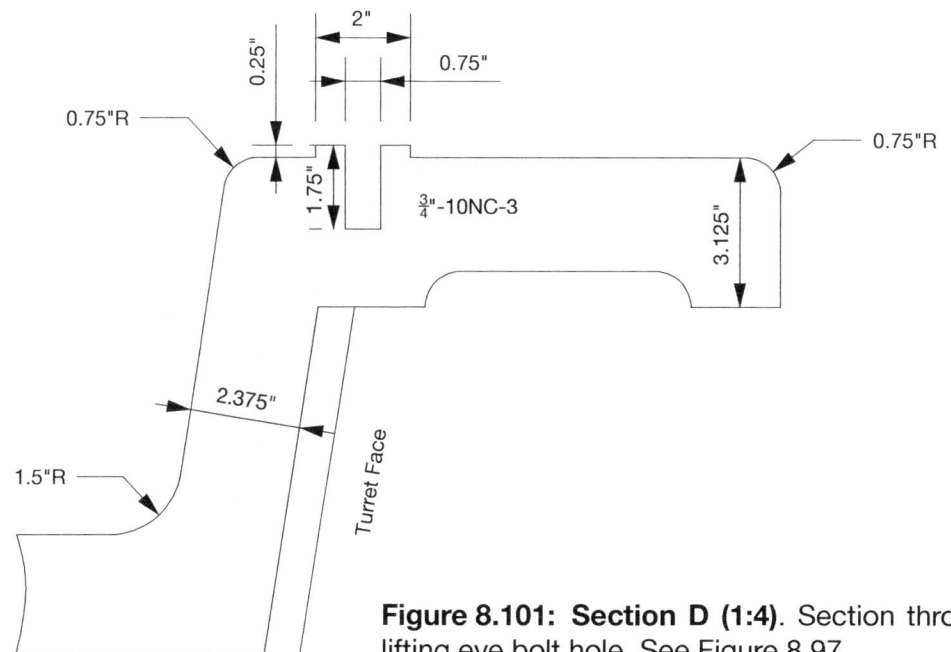

Figure 8.101: Section D (1:4). Section through lifting eye bolt hole. See Figure 8.97.

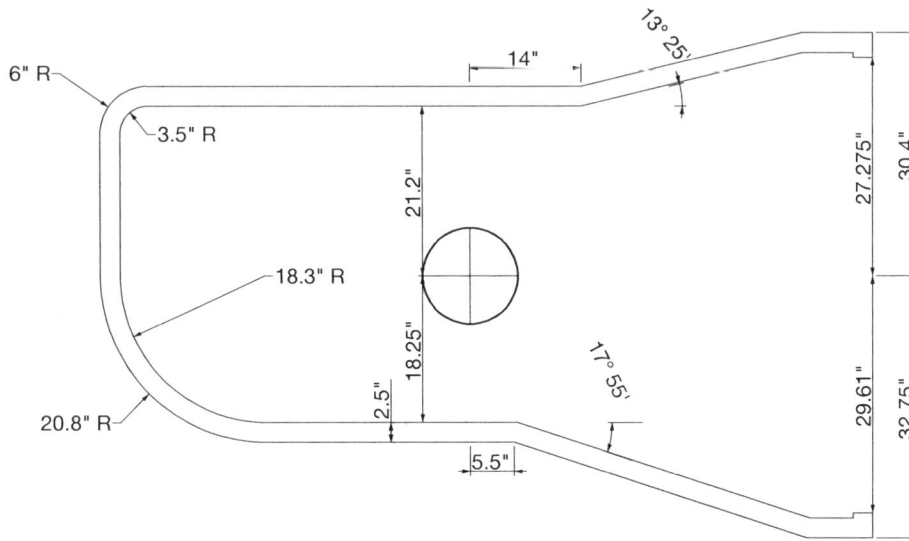

Figure 8.102: Section E–Through Rangefinder Opening (1:24). See Figure 8.97.

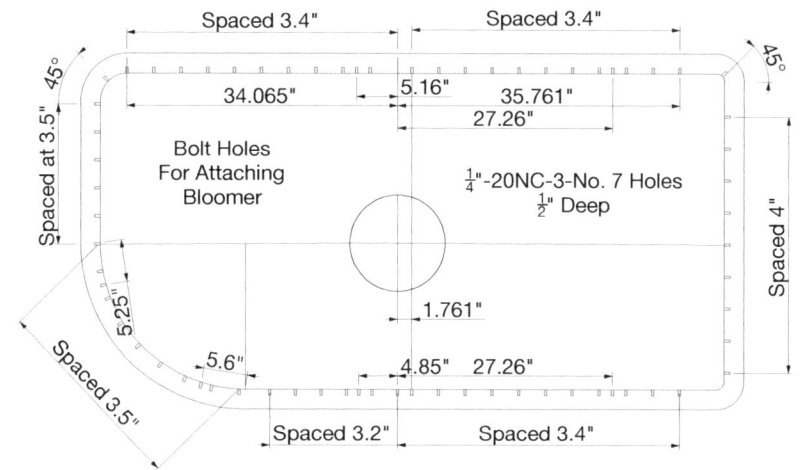

Figure 8.103: Section F (1:24). This section is through the bolt holes for attaching the bloomer. See Figure 8.97.

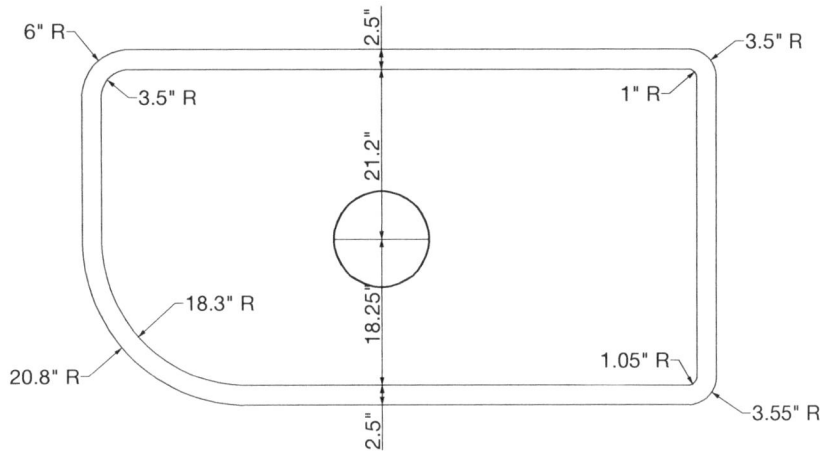

Figure 8.104: Section G (1:24). The section is at a point where the front and back faces are perpendicular to the turret centerline. See Figure 8.97.

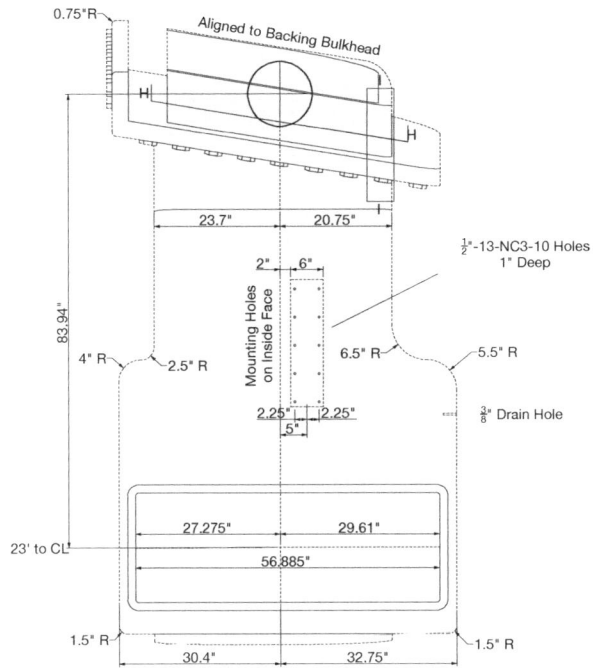

0.75"R
Aligned to Backing Bulkhead
23.7" 20.75"
$\frac{1}{2}$"-13-NC3-10 Holes
1" Deep
2" 6"
Mounting Holes on Inside Face
83.94"
4" R 2.5" R 6.5" R 5.5" R
$\frac{3}{8}$" Drain Hole
2.25" 2.25"
.5"
23' to CL
27.275" 29.61"
56.885"
1.5" R 1.5" R
30.4" 32.75"

Figure 8.105: Rangefinder Hood Front Elevation (1:24)

The inside face of the extension from the mounting plate aligns with the backing plate.

2.5"
21.2"
18.3"R
18.25"
1.05"R
20.8"R
3.55"R

Figure 8.106: Section H–Inboard of Mounting Plate (1:24). See Figure 8.105.

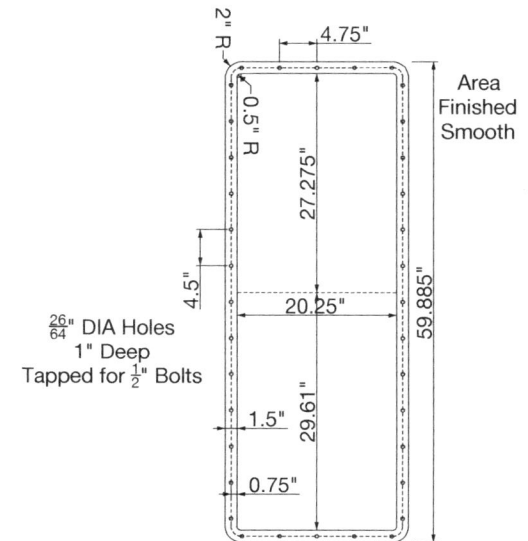

Photograph 8.15: Rangefinder hood on Turret No. 3 of USS New Jersey. Casting produced a rough surface.

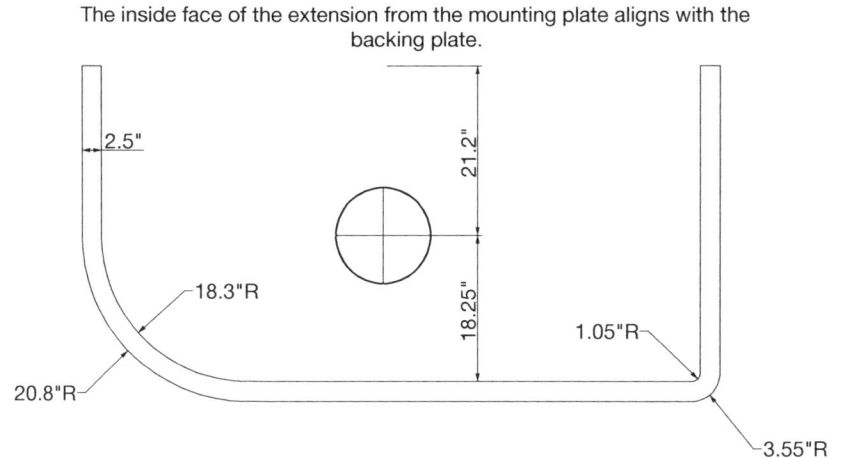

2" R
4.75"
Area Finished Smooth
0.5" R
27.275"
$\frac{26}{64}$" DIA Holes
1" Deep
Tapped for $\frac{1}{2}$" Bolts
4.5"
20.25"
59.885"
29.61"
1.5"
0.75"

Figure 8.107: Door Opening (1:24)

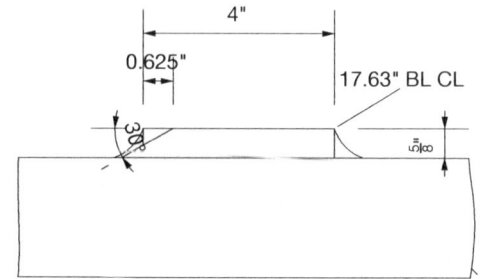

Figure 8.108: Section J (1:4). See Figure 8.109.

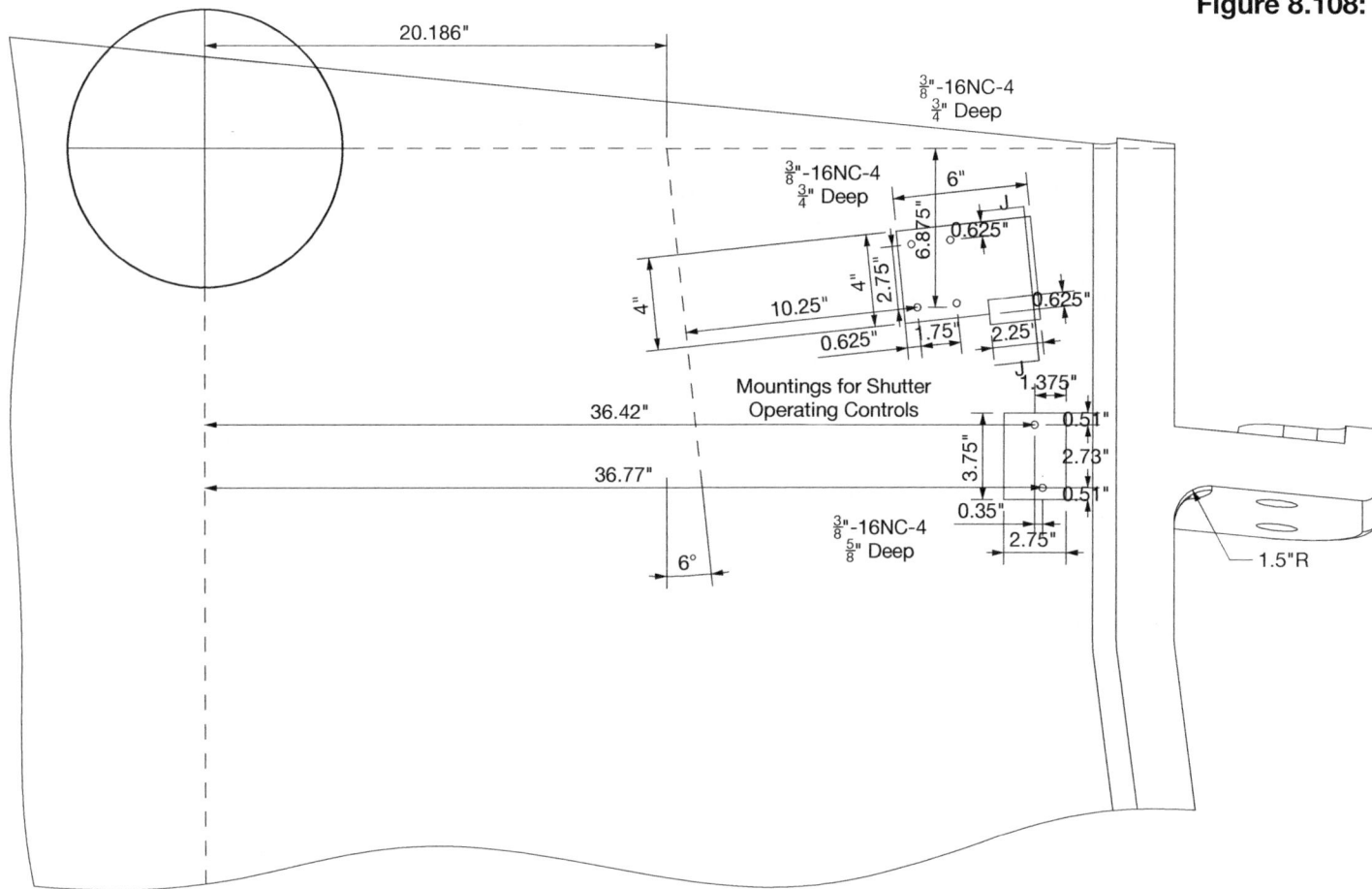

Figure 8.109: Section H (1:8)–Shutter Opening Mechanism Mounting (1:8). See Figure 8.105.

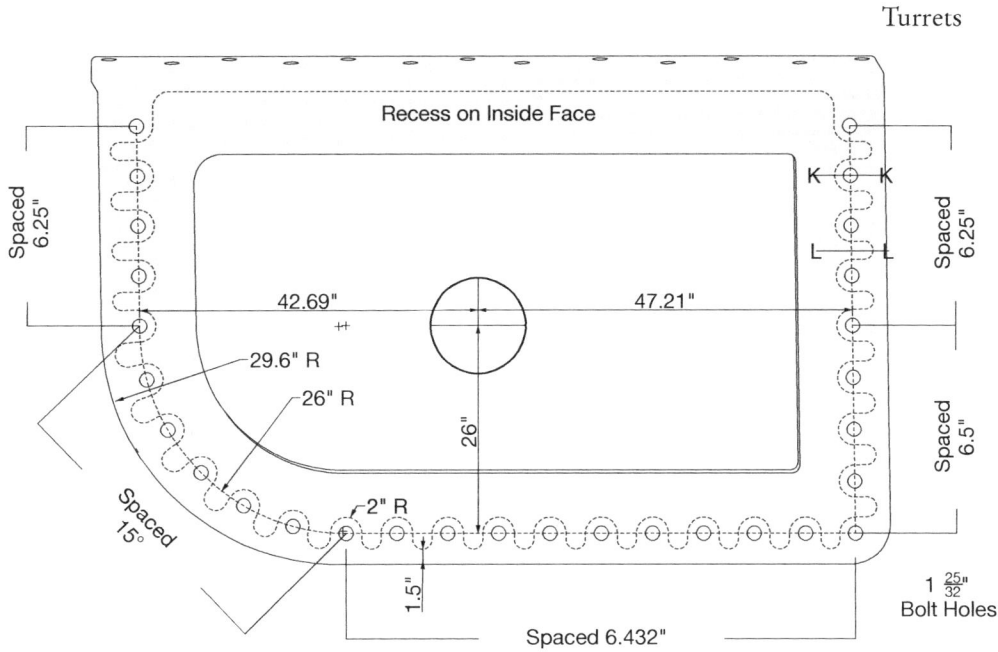

Figure 8.110: Baseplate Expansion (1:24).
The dashed line denotes a 0.75" recess on the inside face.

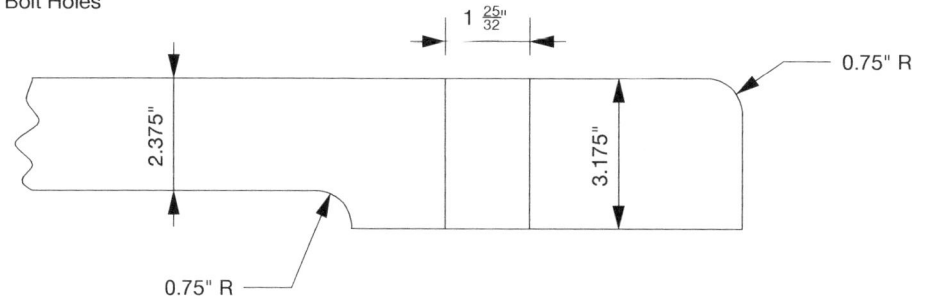

Figure 8.111: Section K–Through Vertical Baseplate (1:4).
See Figure 8.110.

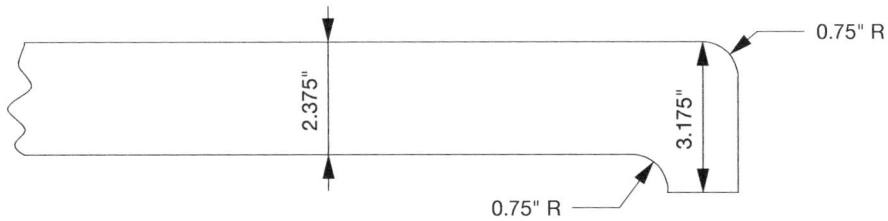

Figure 8.112: Section L–Through Vertical Baseplate (1:4).
See Figure 8.110.

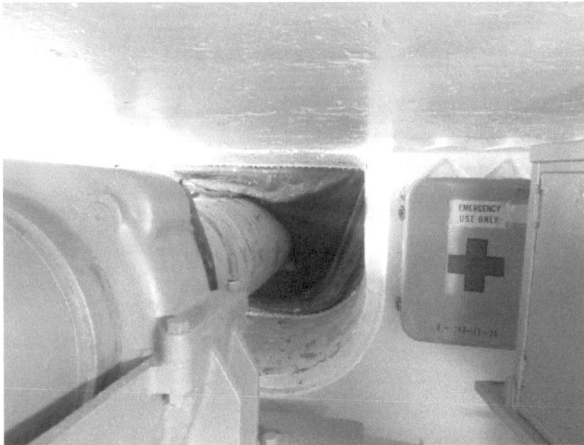

Photograph 8.16: Rangefinder on Turret No. 2 of USS New Jersey Entering the Right Hood, The rubber bloomer creates a gas seal.

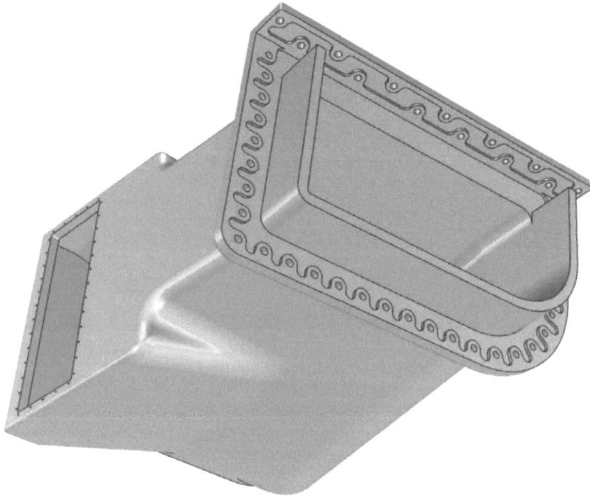

Figure 8.113: Expansion of End of Rangefinder Hood (1:24). The opening has a 2.5" thick cover.

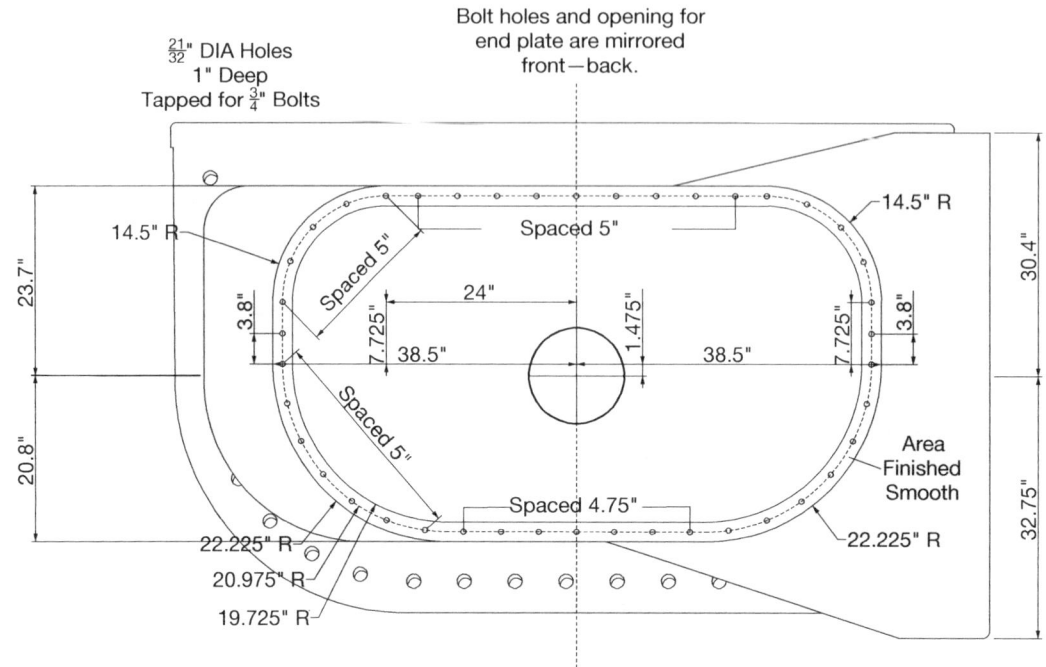

Figure 8.114: Rangefinder Hood Rendering. This view shows the recesses on the base plate.

Photograph 8.17: Turret 1 on USS New Jersey. The coincidence rangefinders in turret 1 were removed from the Iowa-class prior to the Korean War and the opening was plated over. The plate is 5-inches thick. There is a 5×9" brace at the top of the opening. The scalloped angles are 1-inch thick. The round caps are welded to the side backing plate over the bolts holding the armor.

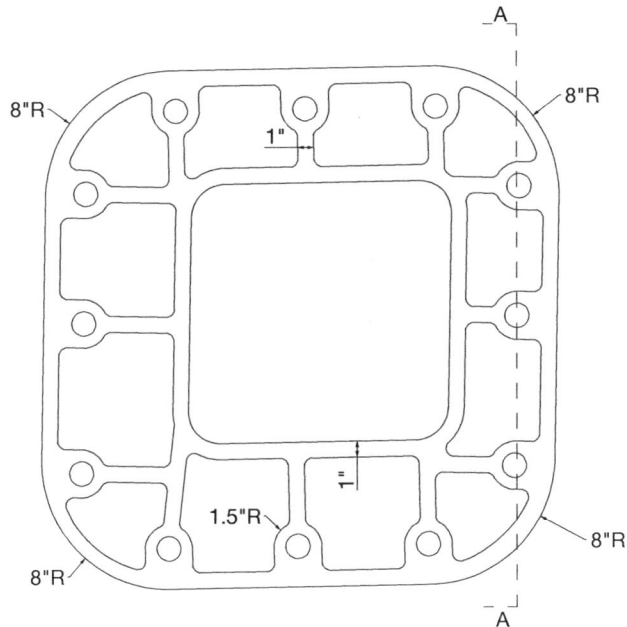

Left Trainer Hood is shown. The bases of the other hoods are similar.

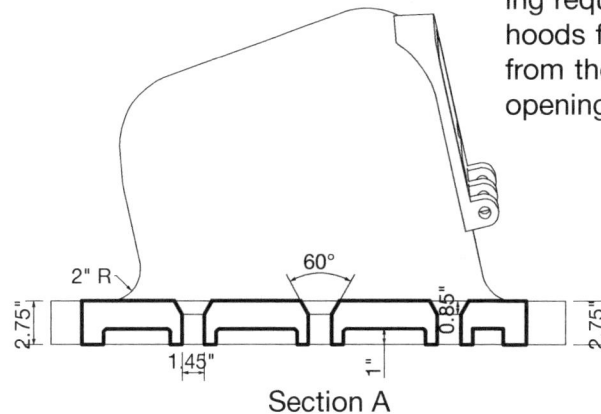

Figure 8.115: Hood Base (1:12). The inside faces of the pointer and trainer hoods are not flat in order to reduce the amount of grinding required to make them sit flush. All of the hoods follow this pattern of 1" ribs extending from the bolt holes and wrapping around the opening for the telescope.

Section A

Base was delivered 3.25" thick and machined at the navy yard to fit armor. Hood is attached using $1\frac{1}{2}$" standard countersunk head screw bolts.

Photograph 8.18: Trainer and Pointer Hoods. There is a trainer and pointer hood on each side of all the turrets. The trainer hood is forward and above the pointer hood. All four hoods are shaped differently. This is the left side of turret 1 on USS New Jersey. The yard had to install a shutter that closed the periscope opening. On the left hoods the opening is centered on the periscope objective. On the right hoods the opening center is 0.75" outboard of the periscope objective.

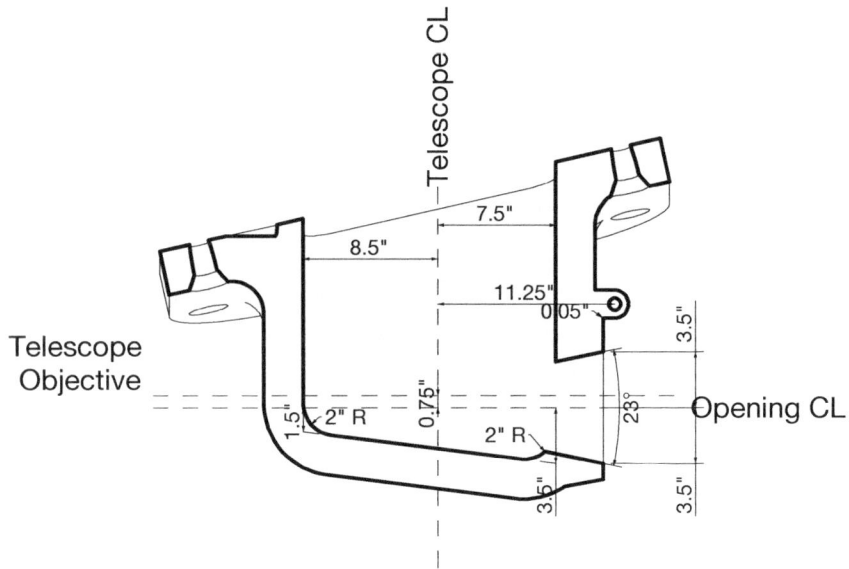

Figure 8.116: Right Trainer Hood Top Section Through Centerline (1:12).

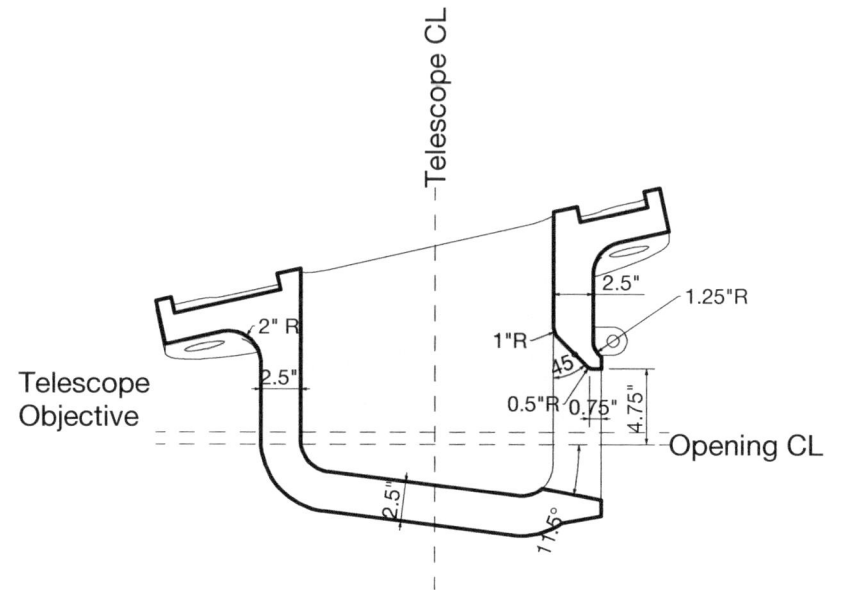

Figure 8.117: Right Trainer Hood Section Through Opening Mechanism Widening (1:12)

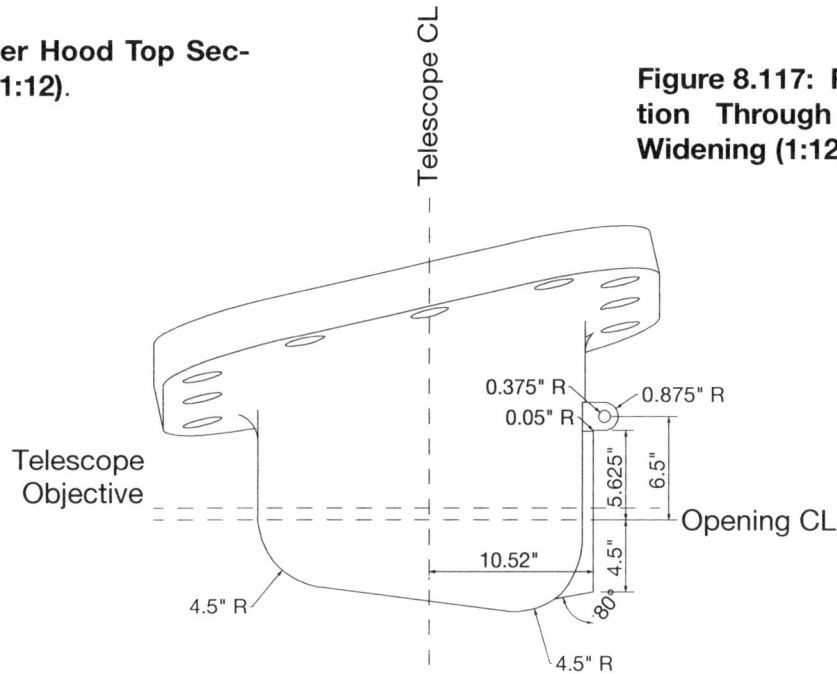

Figure 8.118: Right Trainer Hood Plan (1:12).
Each of the four periscope hoods is different.

Figure 8.119: Right Trainer Hood Front Elevation (1:12)

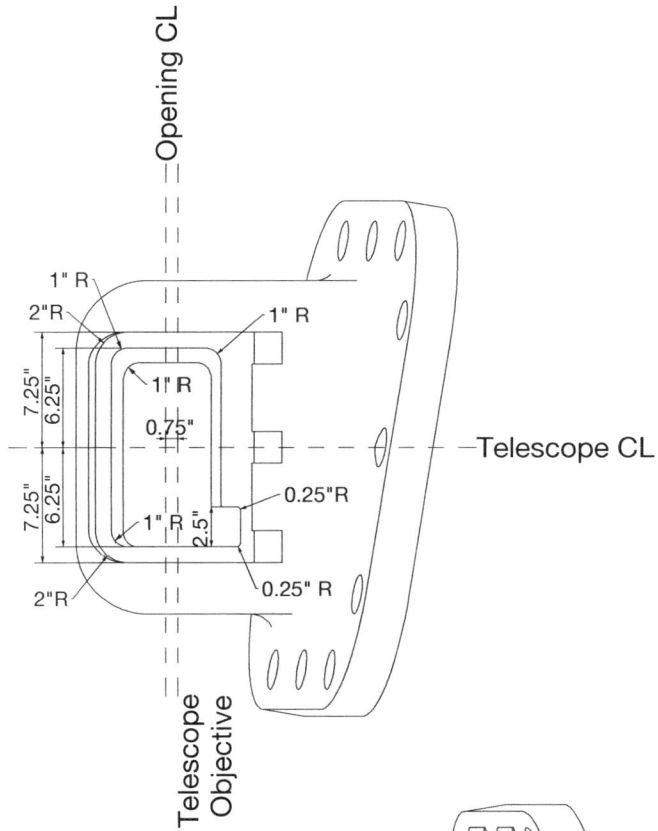

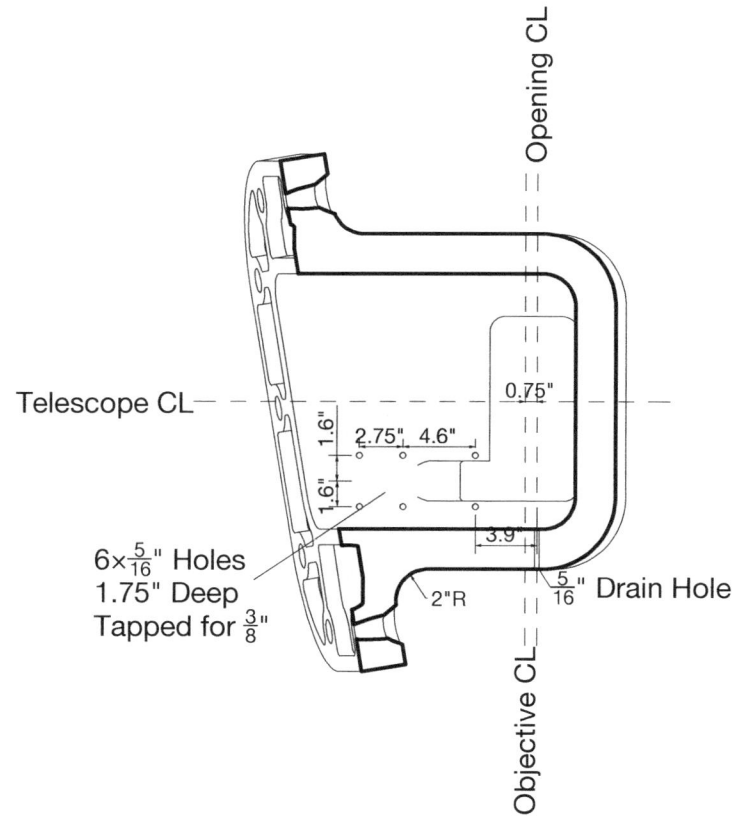

Opening CL

1" R
2"R
1" R
1" R
7.25"
6.25"
0.75"
Telescope CL
7.25"
6.25"
0.25"R
1" R
2.5"
0.25" R
2"R

Telescope Objective

Figure 8.120: Right Trainer Hood Rear Section (1:12)

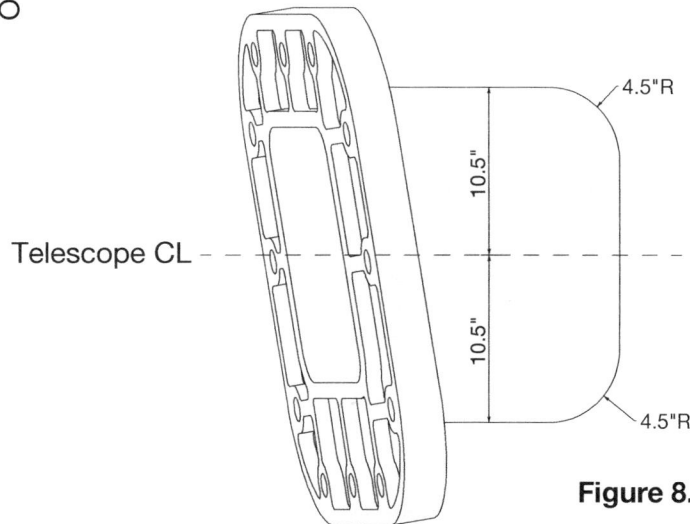

Opening CL

Telescope CL
0.75"
1.6"
2.75" 4.6"
1.6"
3.9"
$6 \times \frac{5}{16}$" Holes
1.75" Deep
Tapped for $\frac{3}{8}$"
2"R
$\frac{5}{16}$" Drain Hole

Objective CL

Figure 8.121: Right Trainer Hood Rear Elevation (1:12)

4.5"R
10.5"
Telescope CL
10.5"
4.5"R

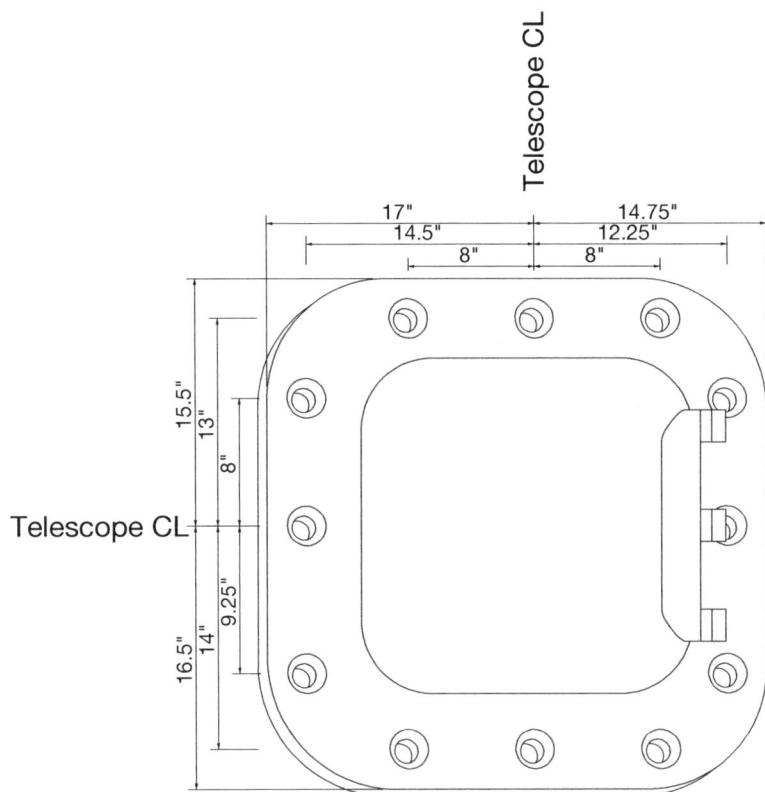

Figure 8.122: Right Trainer Hood Right Elevation (1:12)

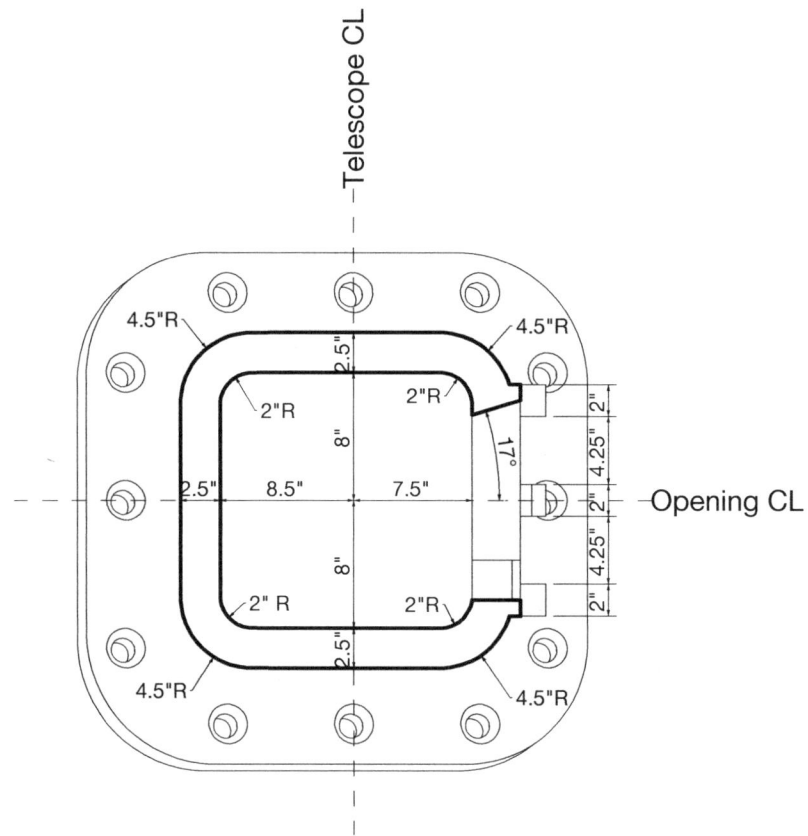

Figure 8.123: Right Trainer Hood Right Section (1:12)

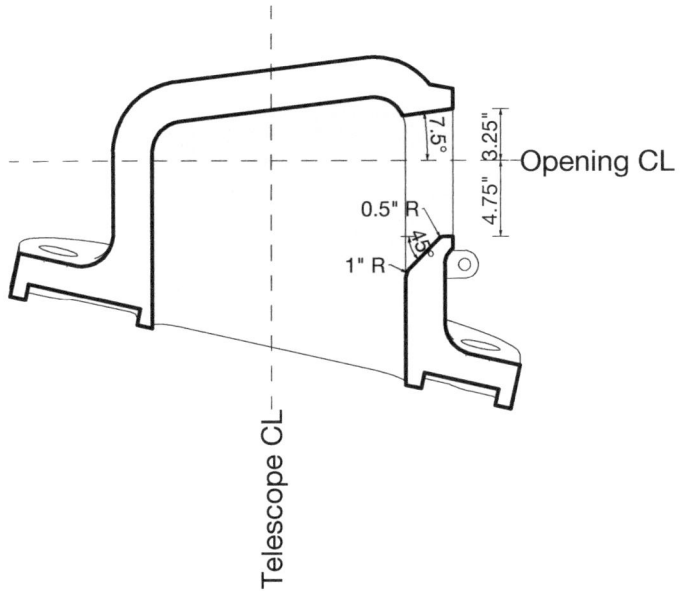

Figure 8.124: Left Trainer Hood Section Through Opening Mechanism Widening (1:12)

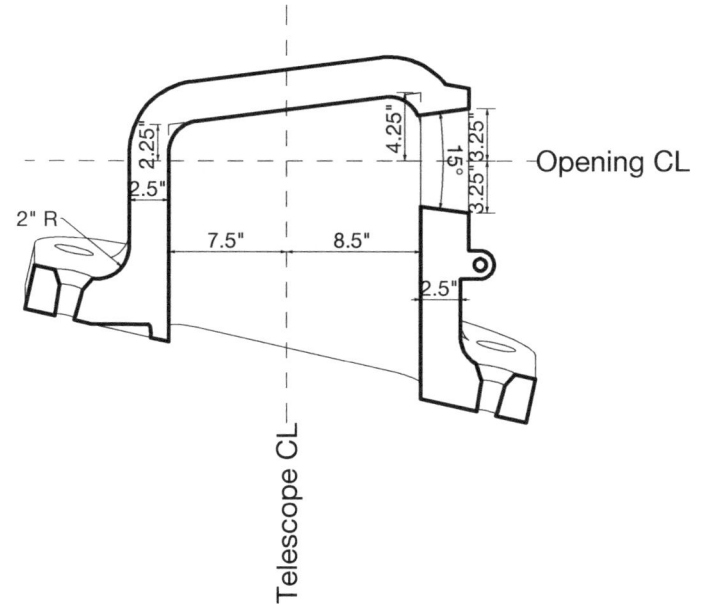

Figure 8.125: Left Trainer Hood Top Section Through Centerline (1:12).

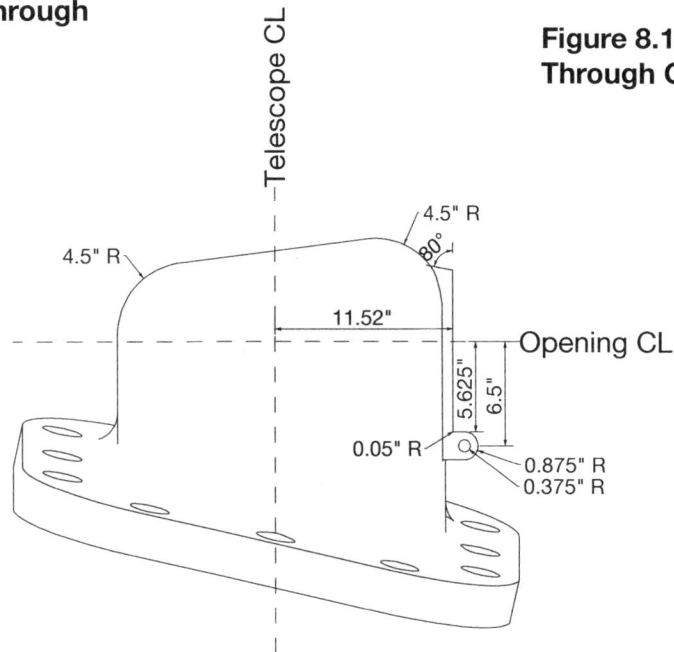

Figure 8.126: Left Trainer Hood Plan (1:12)

Figure 8.127: Left Trainer Hood Front Elevation (1:12)

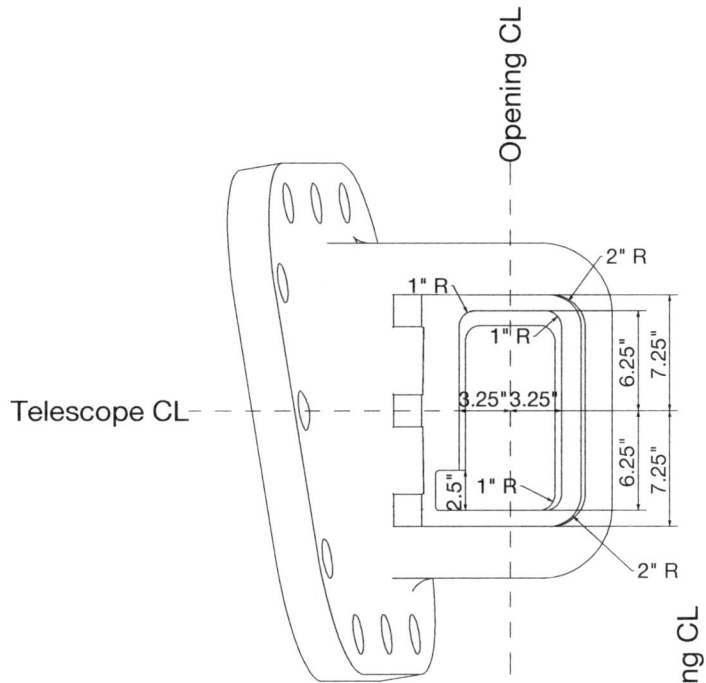

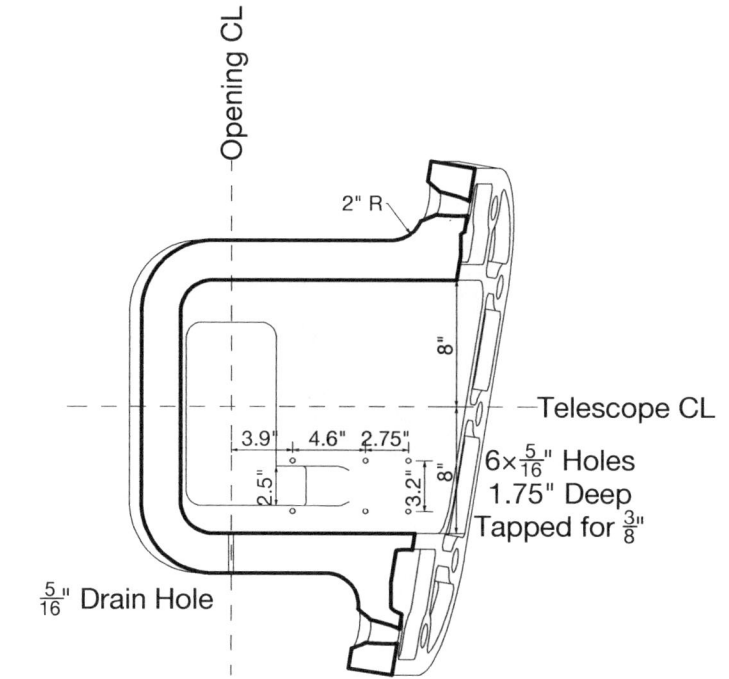

Figure 8.128: Left Trainer Hood Rear Section (1:12)

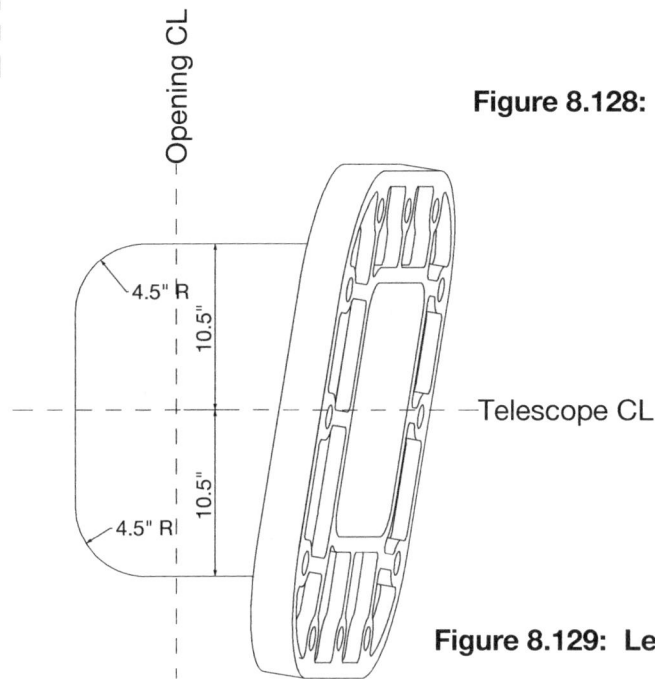

Figure 8.129: Left Trainer Hood Rear Elevation (1:12)

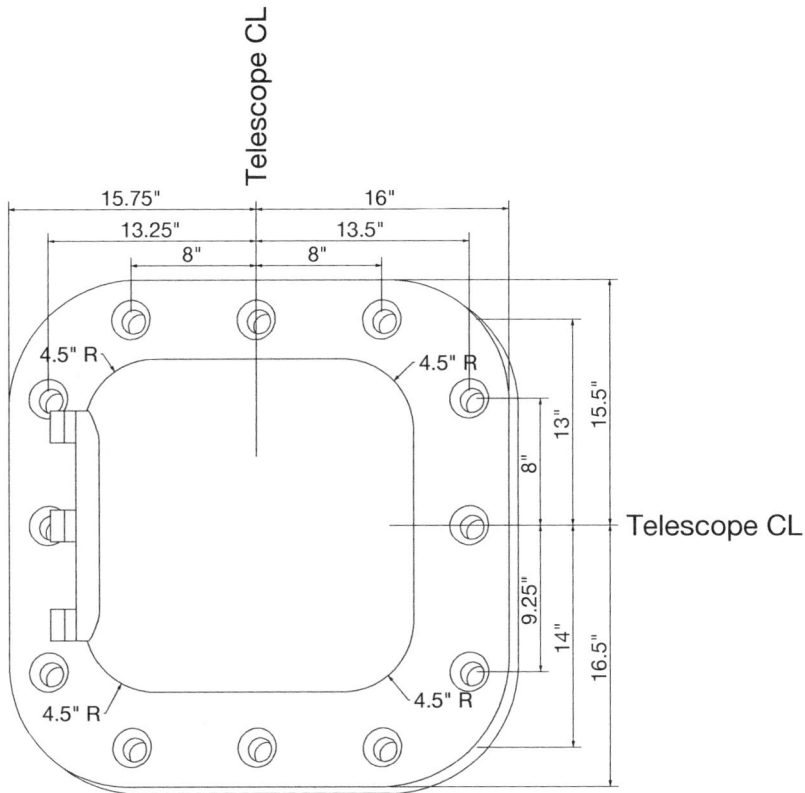

Figure 8.130: Left Trainer Hood Right Elevation (1:12)

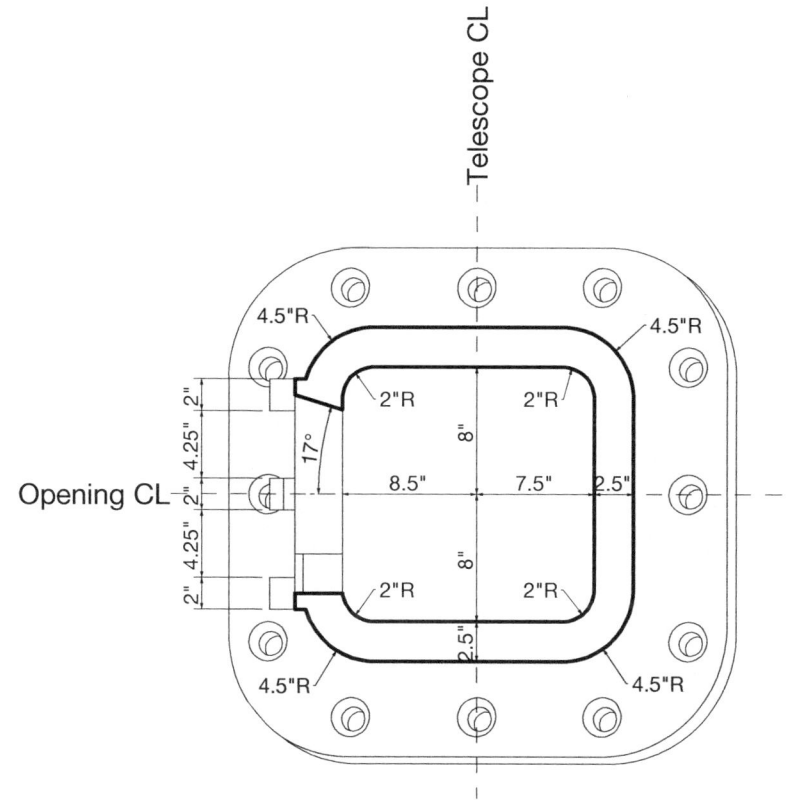

Figure 8.131: Left Trainer Hood Right Section (1:12)

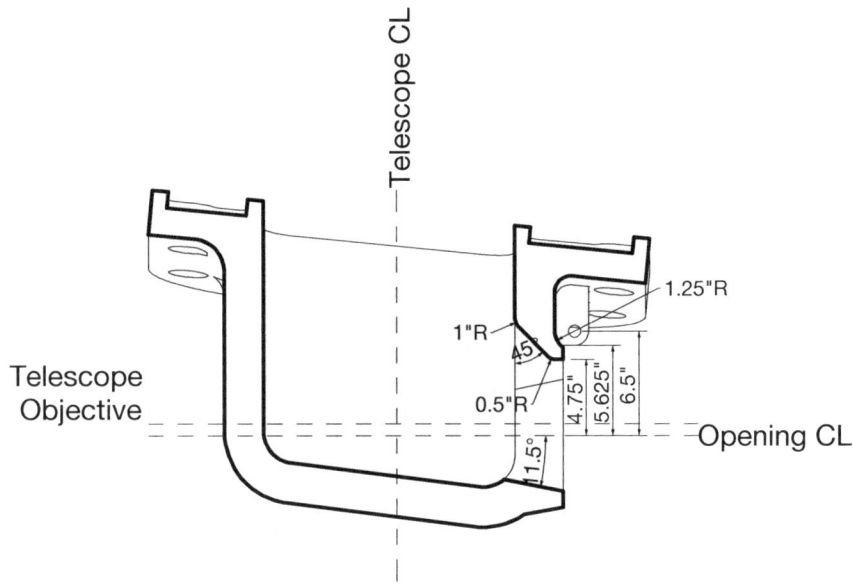

Figure 8.132: Right Pointer Hood Section Through Opening Mechanism Widening (1:12)

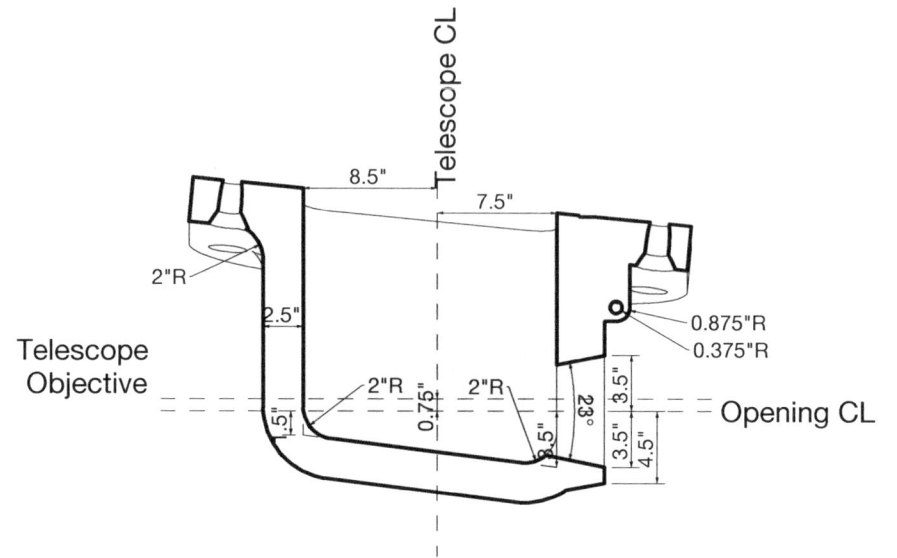

Figure 8.133: Right Pointer Hood Top Section Through Centerline (1:12).

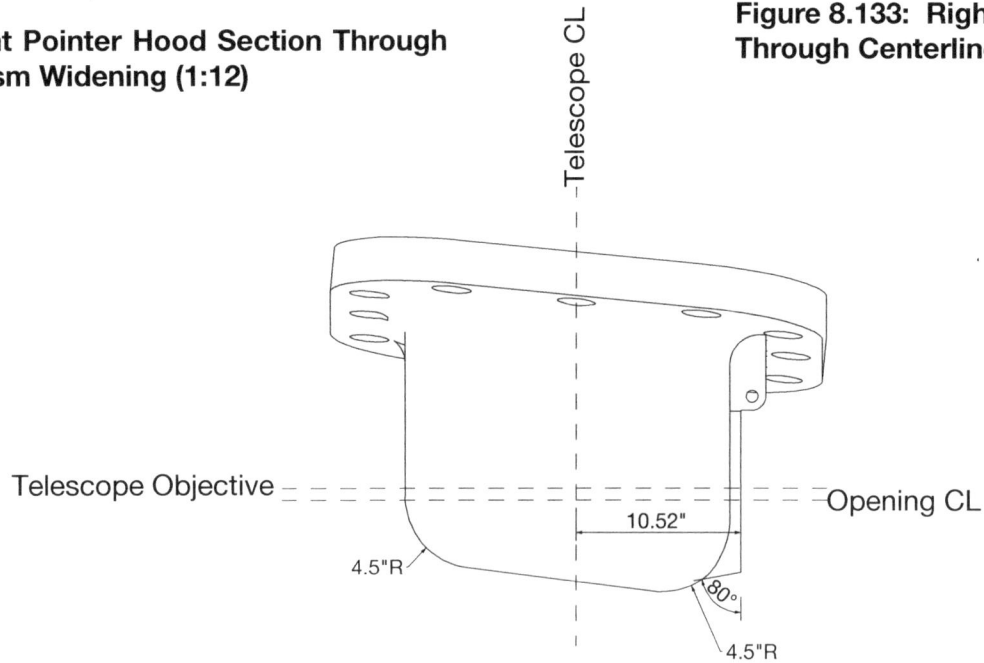

Figure 8.134: Right Pointer Hood Plan (1:12)

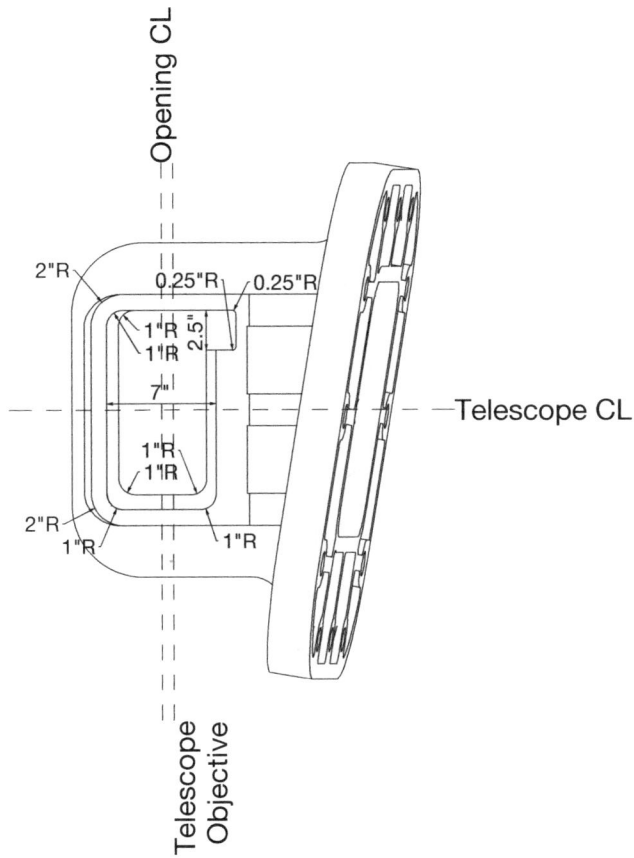

Figure 8.135: Right Pointer Hood Front Elevation (1:12)

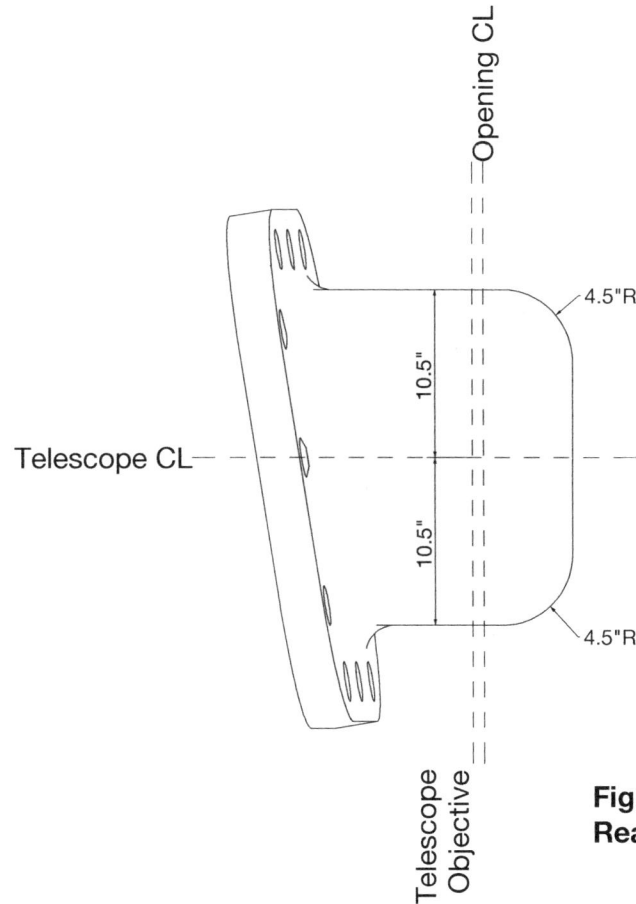

Figure 8.137: Right Pointer Hood Rear Elevation (1:12)

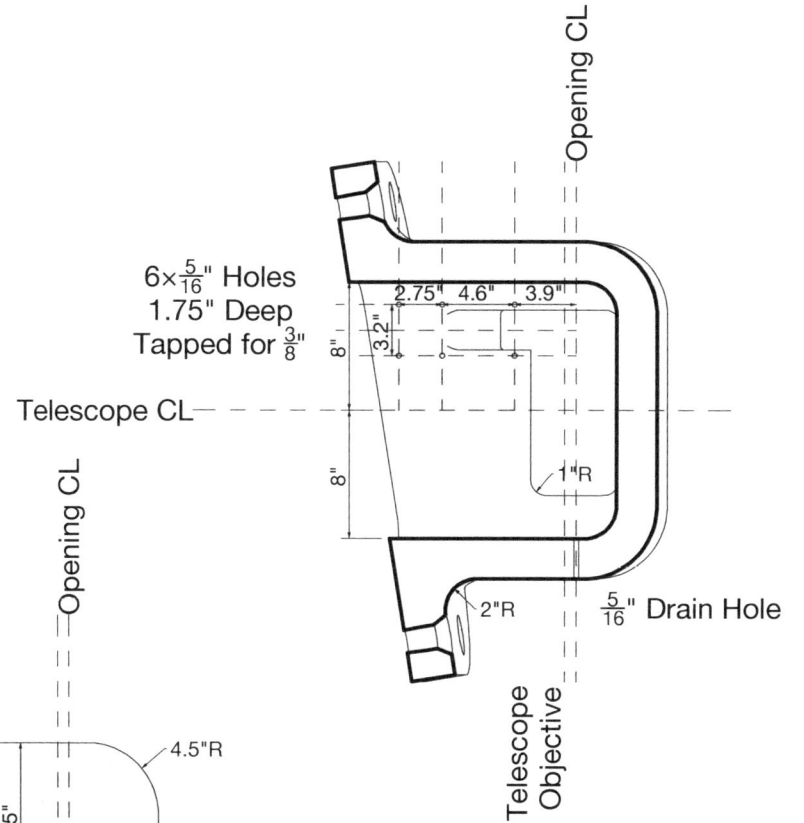

Figure 8.136: Right Pointer Hood Rear Section (1:12)

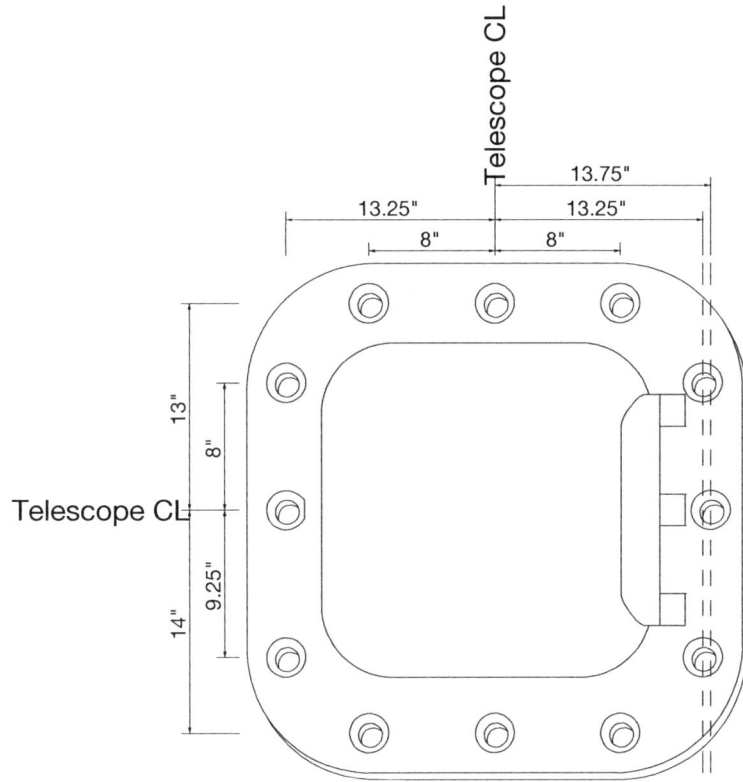

Figure 8.138: Right Pointer Hood Right Elevation (1:12)

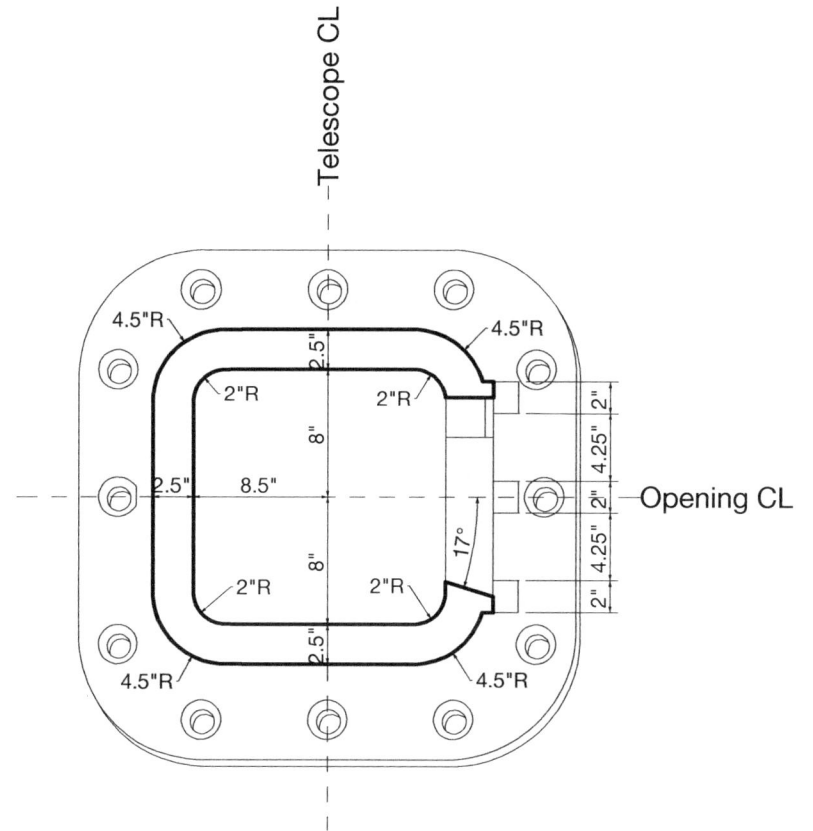

Figure 8.139: Right Pointer Hood Right Section (1:12)

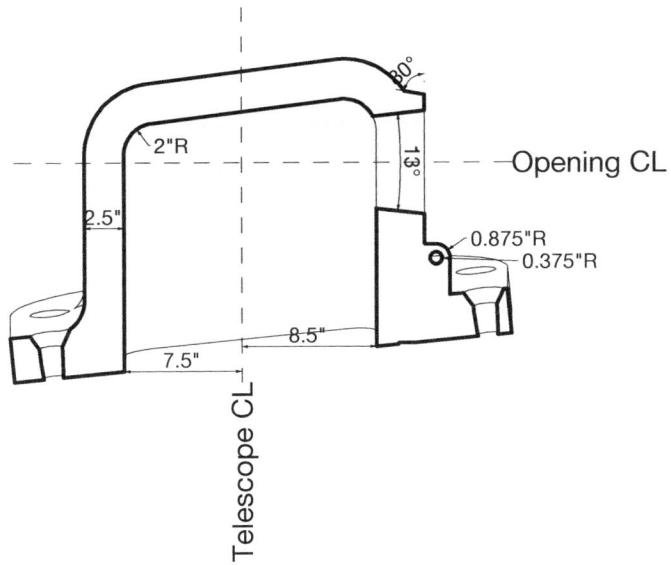

Figure 8.140: Left Pointer Hood Section Through Centerline (1:12)

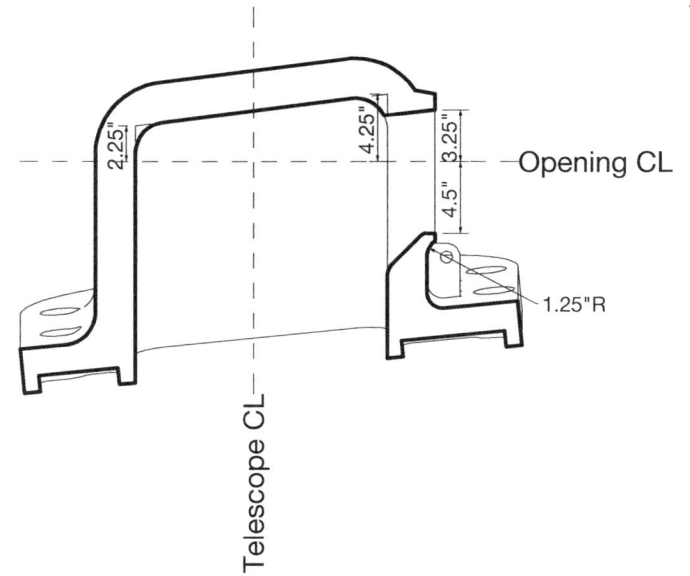

Figure 8.141: Left Pointer Hood Top Section Through Mechanism Widening (1:12)

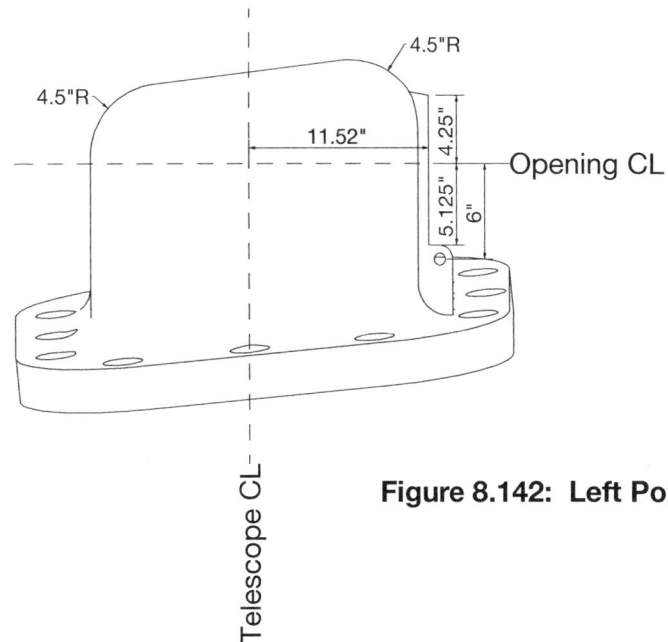

Figure 8.142: Left Pointer Hood Plan (1:12)

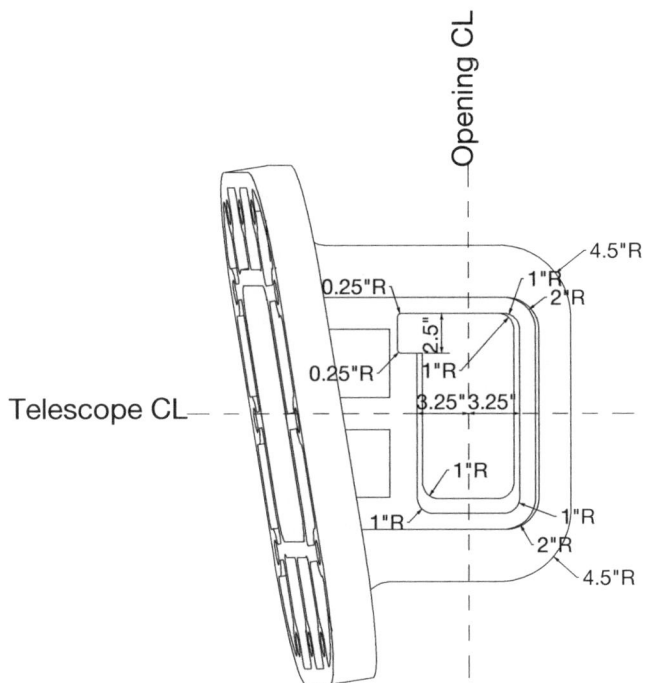

Figure 8.143: Left Pointer Hood Front Elevation (1:12)

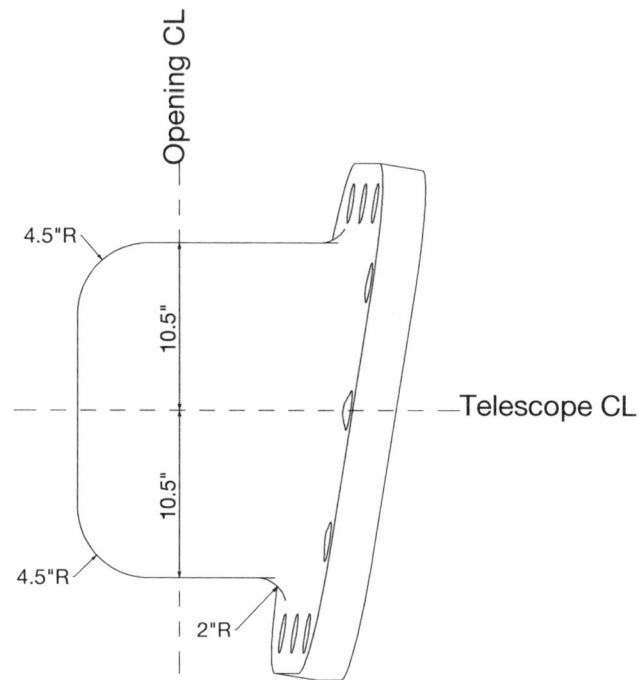

Figure 8.144: Left Pointer Hood Rear Elevation (1:12)

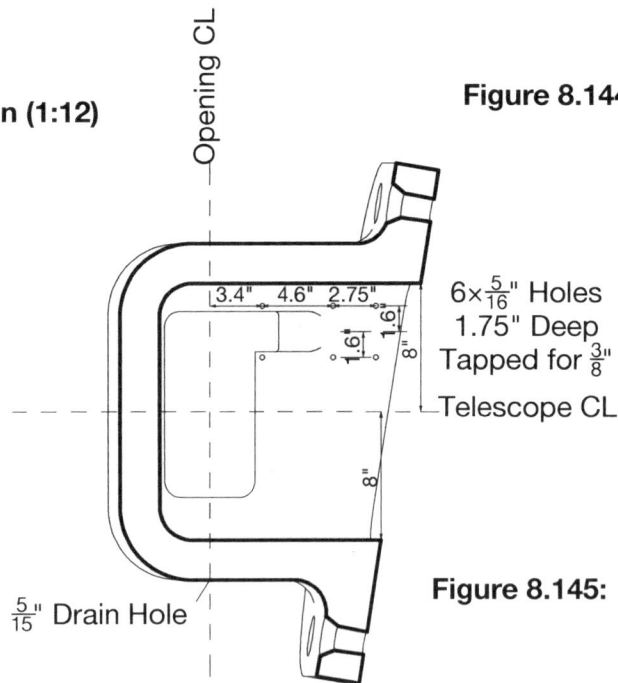

Figure 8.145: Left Pointer Hood Rear Section (1:12)

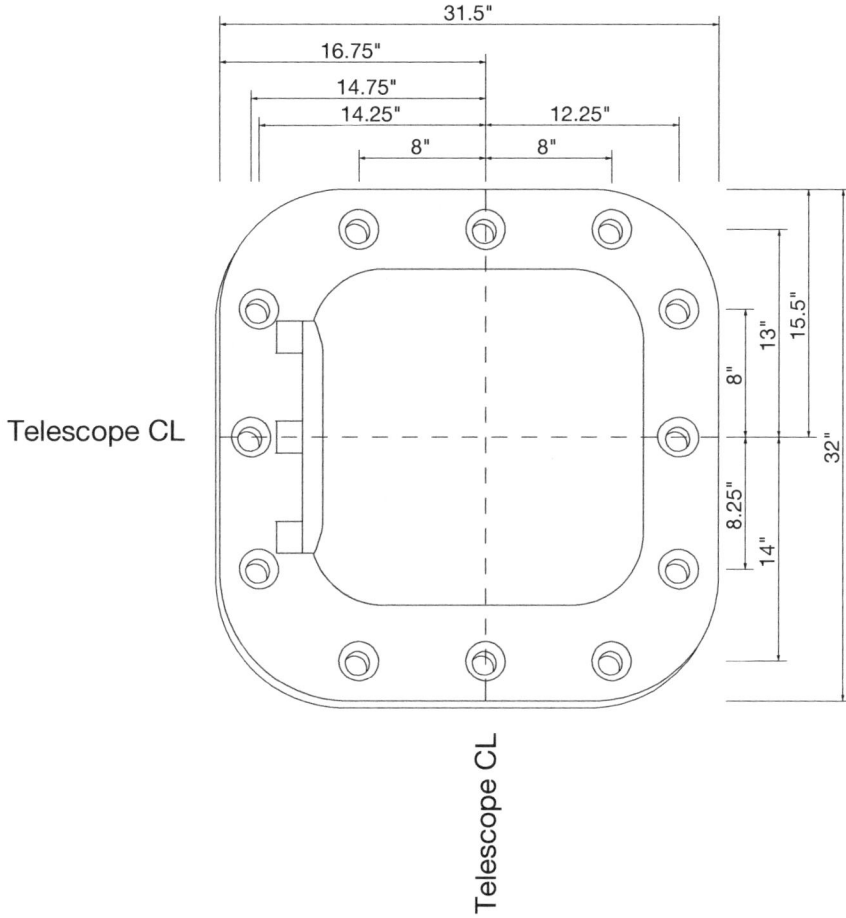

Figure 8.146: Left Pointer Hood Left Elevation (1:12)

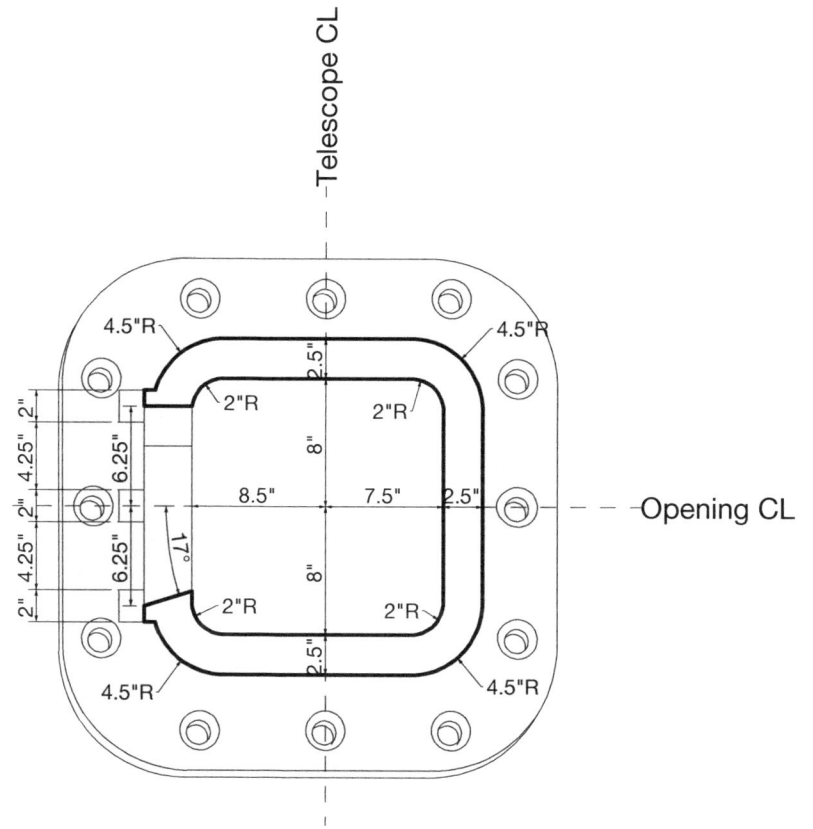

Figure 8.147: Left Pointer Hood Left Section (1:12)

Chapter 9: Armored Grates

The *Iowa*-class designers tried to minimize the number of openings in the citadel armor. However, they could not be eliminated entirely. Openings for crew, wiring, ammunition, air, water, and exhaust were essential for the operation of the ship. In some cases, such as water pipes and crew ventilation, the openings could be kept small enough that they did not reduce armor protection. Even such small openings were kept to a minimum. Ammunition and wiring openings (to be addressed in a later volume) were protected with armored trunks. The largest openings were for air and exhaust for the engine rooms, boiler rooms, and emergency diesel generator rooms. These openings are protected by armored grates. These grates have small holes but are thicker than the surrounding armor to compensate for the weakness the holes create. These grates are primarily located at main deck and second deck. Grates are positioned such that it would be nearly impossible for an incoming projectile to strike more than one grate.

There are four types of openings protected by grates: uptake, intake, supply and exhaust. The uptakes routed hot gases from the boilers through the smoke pipes. Intakes provided air to the boilers for combustion. Intake air enters through vents at the side of the superstructure and flows unrestrained through the uptake spaces to second deck. This allowed the intake air flow to assist in cooling the uptake trunks. Below splinter deck, intake air enters a plenum

chamber for each boiler. Other areas requiring a large air flow, are connected to supply trunks. Supply air flows provided cooling for machinery space and air for diesel motors. Air enters supply trunks at various deck vents and vents at the side of the superstructure. The exhaust flows took cooling air out of the machinery spaces. The exhaust from the diesel generator rooms flowed to deck vents, exhaust flow from the engine and boiler rooms entered the intake air for the boilers, trash burner exhaust passed up the aft smoke pipe.

Most of the armored grates at each deck have some common features. All main deck grates are 4.3 inches thick, second deck grates are 15.2 inches, 2.8 inches at third deck. For the most part all the grates at the same deck have the same profile and method of connection to the deck. The exceptions are the uptake grates and the trash burner exhaust which have unique connection methods and profiles. In addition, the trash burner exhaust has an armored grate oriented vertically in FR143.

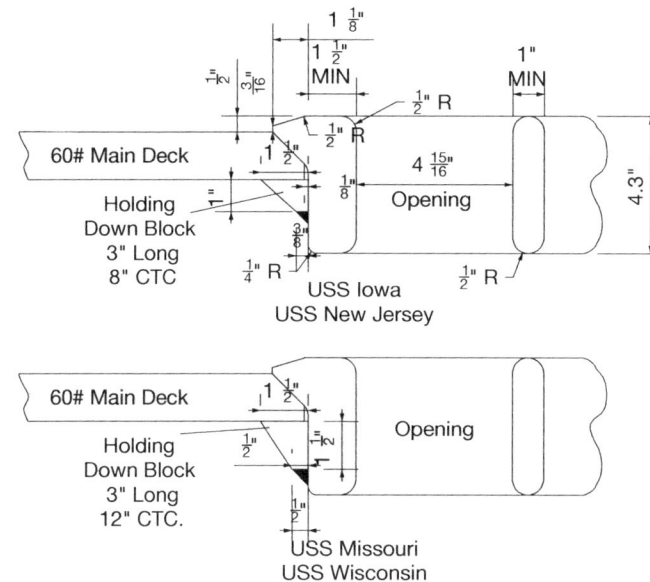

Figure 9.1: Connection of Grates to Main Deck (1:6). This diagram shows the connecion and cross section of main deck grates. This applies to all main deck grates except for the uptake and trash burner grates. The grate overlaps the deck opening by one inch.

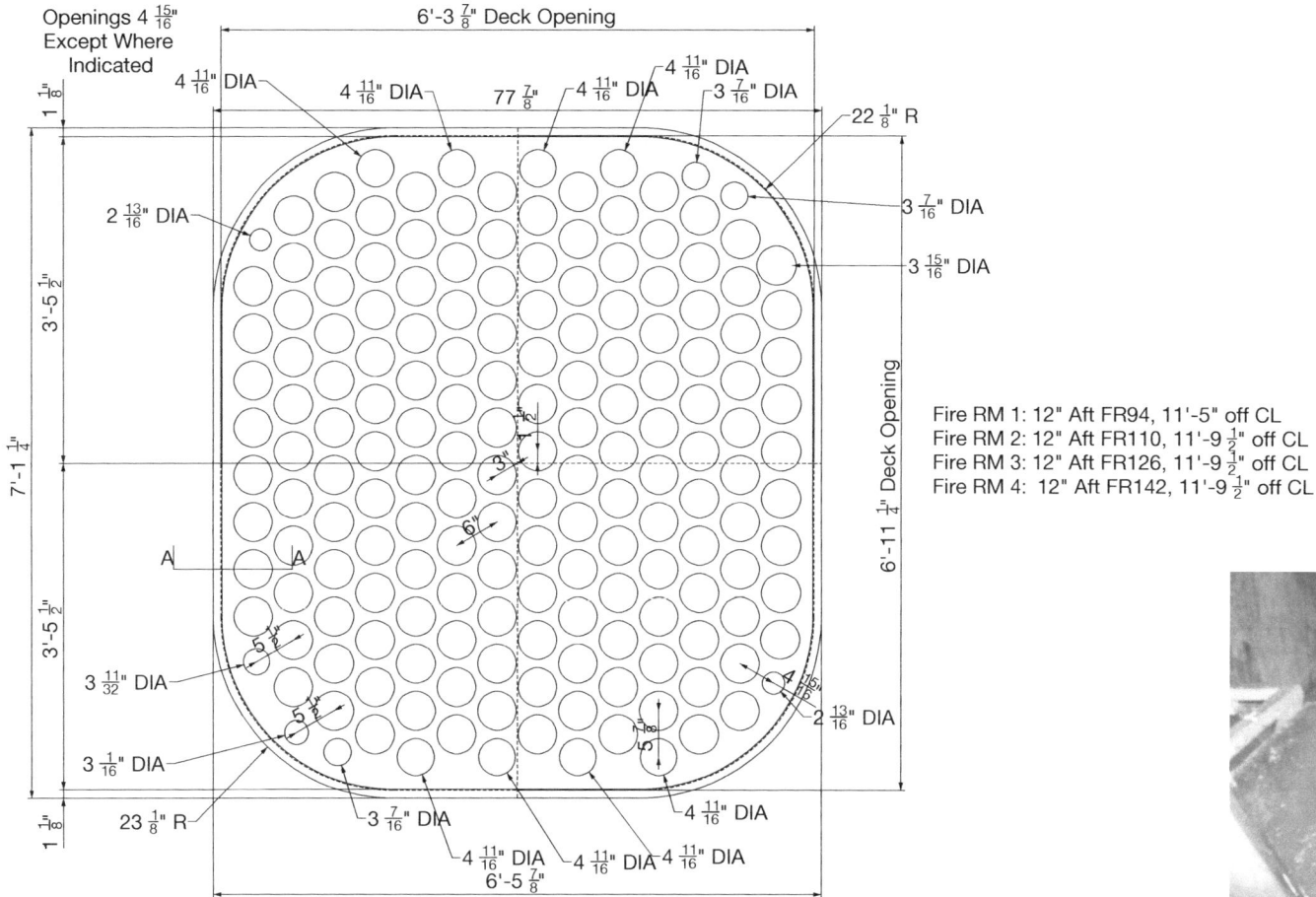

Figure 9.2: Main Deck Intake Grates (1:24). There are eight intake grates (two for each fire room) located in the uptake spaces. Intake air flows freely through the uptake spaces until second deck so the air flow through these grates is not directed towards a specific boiler. These are the only boiler room grates that have the same edge profile and deck connection as the other armored grates.

Fire RM 1: 12" Aft FR94, 11'-5" off CL
Fire RM 2: 12" Aft FR110, 11'-9 $\frac{1}{2}$" off CL
Fire RM 3: 12" Aft FR126, 11'-9 $\frac{1}{2}$" off CL
Fire RM 4: 12" Aft FR142, 11'-9 $\frac{1}{2}$" off CL

Photograph 9.1: Main deck intake grate on USS New Jersey. This is the port grate for fire room 2.

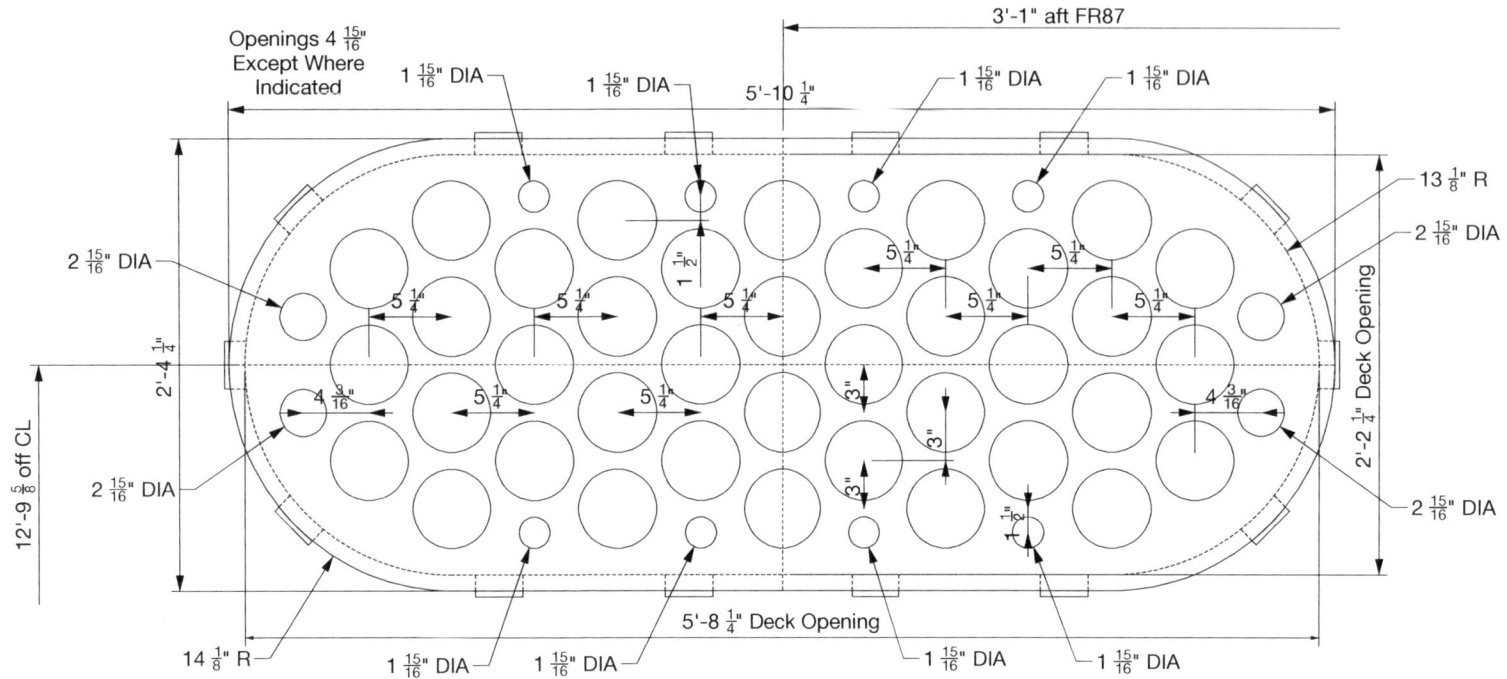

Figure 9.3: Port Main Deck Supply Grate for Fire Room 1 (1:12). Each fire room has a port and starboard supply trunk. This grate is located in the port aft corner of the conning tower support. The air intake is at the side of the conning tower support at the 02 level, below the flag bridge. At times, this intake was protected from water by a shroud. The 02 level of the conning tower support functions as a plenum chamber. On USS Iowa, this area only contains armored door opening machinery. On the other ships it is empty.

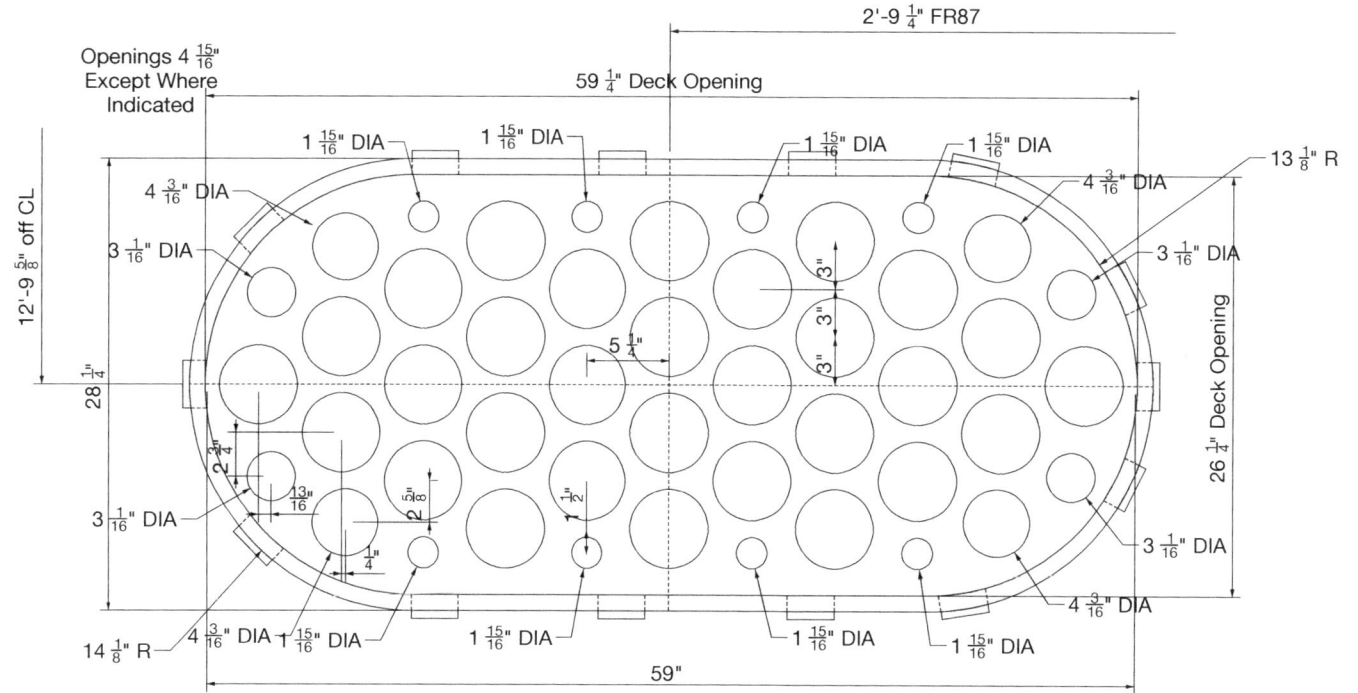

Figure 9.4: Starboard Main Deck Supply Grate for Fire Room 1 (1:12). This grate is located at the starboard aft corner of the conning tower support. This air supply draws air from the 02 level of the conning tower support through an opening at the starboard side. This opening was also shrouded at times during the ships' history.

Photograph 9.2: A View Down the Intake Trunk on USS New Jersey. This is from the 02 level of the conning tower support looking down to Main Deck.

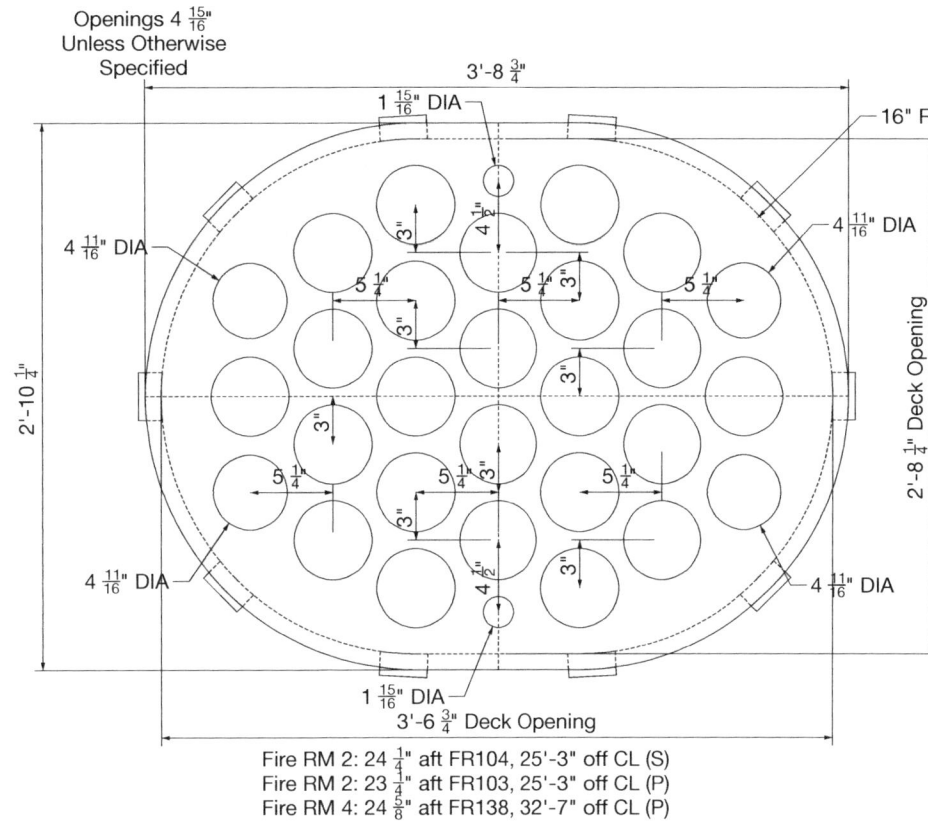

Figure 9.5: Main Deck Supply Grate for First Room 2 and Fire Room 4 (Port) (1:12). Three grates follow this plan. The trunks for fire room two open at the sides of the 01 deckhouse behind five-inch gun mounts 3 and 4. The trunk for fire room 4 opens at the starboard side of the 01 level aft of the former five-inch gun mount 9 and behind the current fueling at sea control station platform.

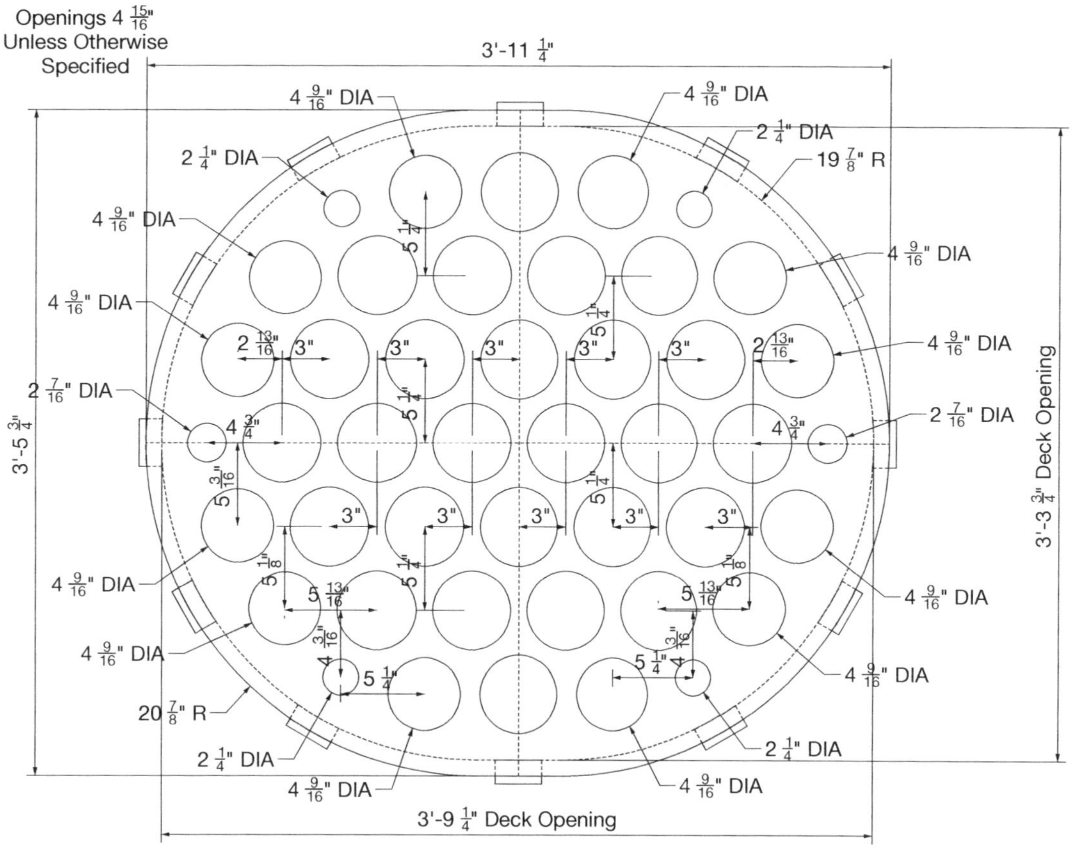

Fire RM 3 (P): 26" aft FR119, 9'5- 1/2" off CL
Fire MR 4 (S): 24" aft FR138, 32'-0 1/2" off CL (rotated)

Figure 9.6: Main Deck Supply Grate for Fire Room 3 Starboard and Fire Room 4 Starboard (1:12). The starboard trunk for Fire Room 3 opens in the 03 level near the centerline. This is now covered by the Tomahawk missile platform, which created a plenum chamber below. The port trunk for fire room 4 opens at the 01 level aft of the former five-inch gun mount 10.

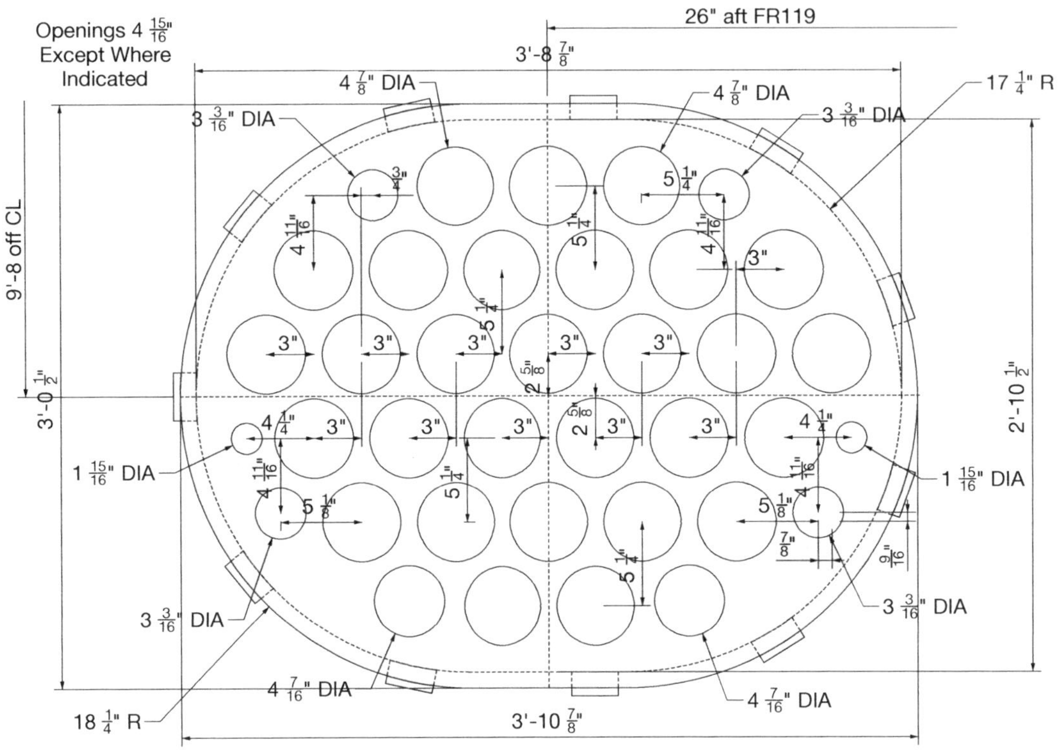

Figure 9.7: Main Deck Supply Grate for Fire Room 3 Starboard (1:12). This trunk opens at the 03 level. Currently it is covered by the Tomahawk missile platform that created a plenum chamber below.

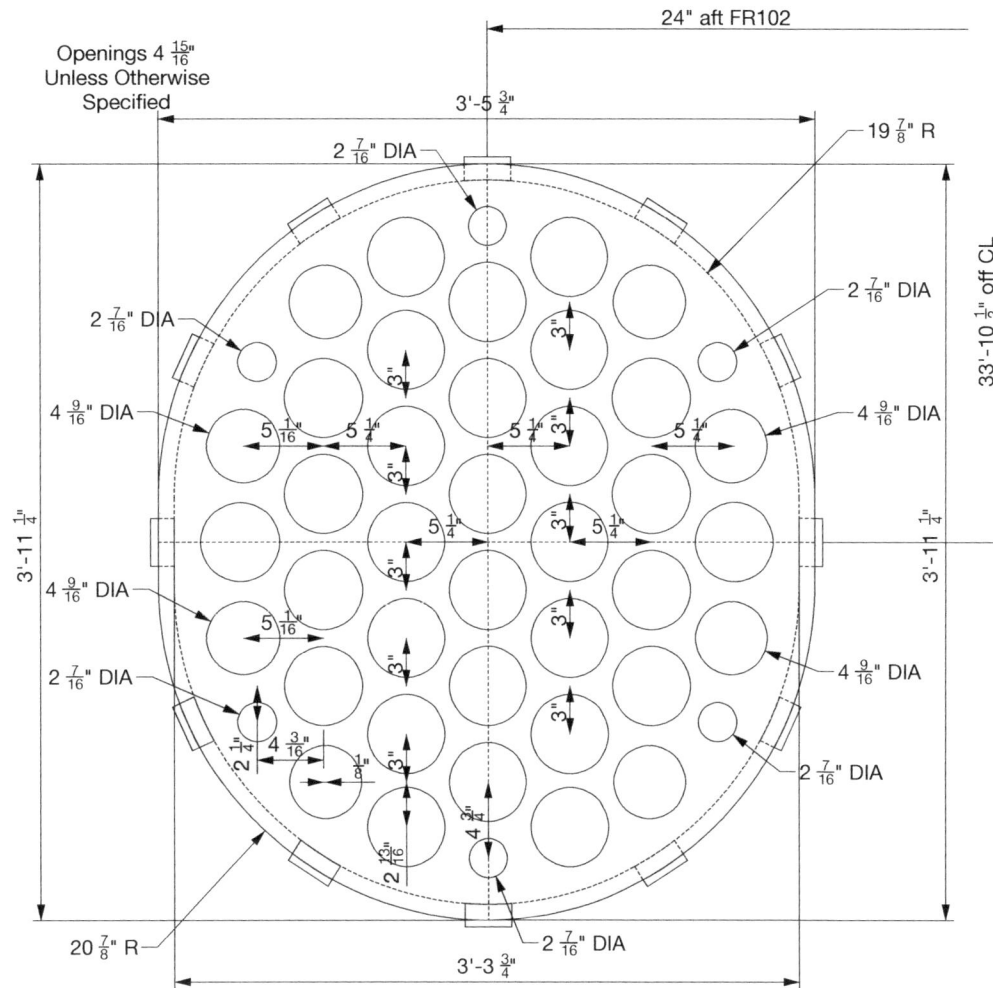

Figure 9.8: Main Deck Engine Room 1 Supply Grate (1:12). The engine rooms have one supply supply trunk. The intake for engine room 1 is at the port side of the deckhouse at 01, aft of five-inch gun mount 1.

Openings $4\frac{15}{16}$"
Unless Otherwise
Specified

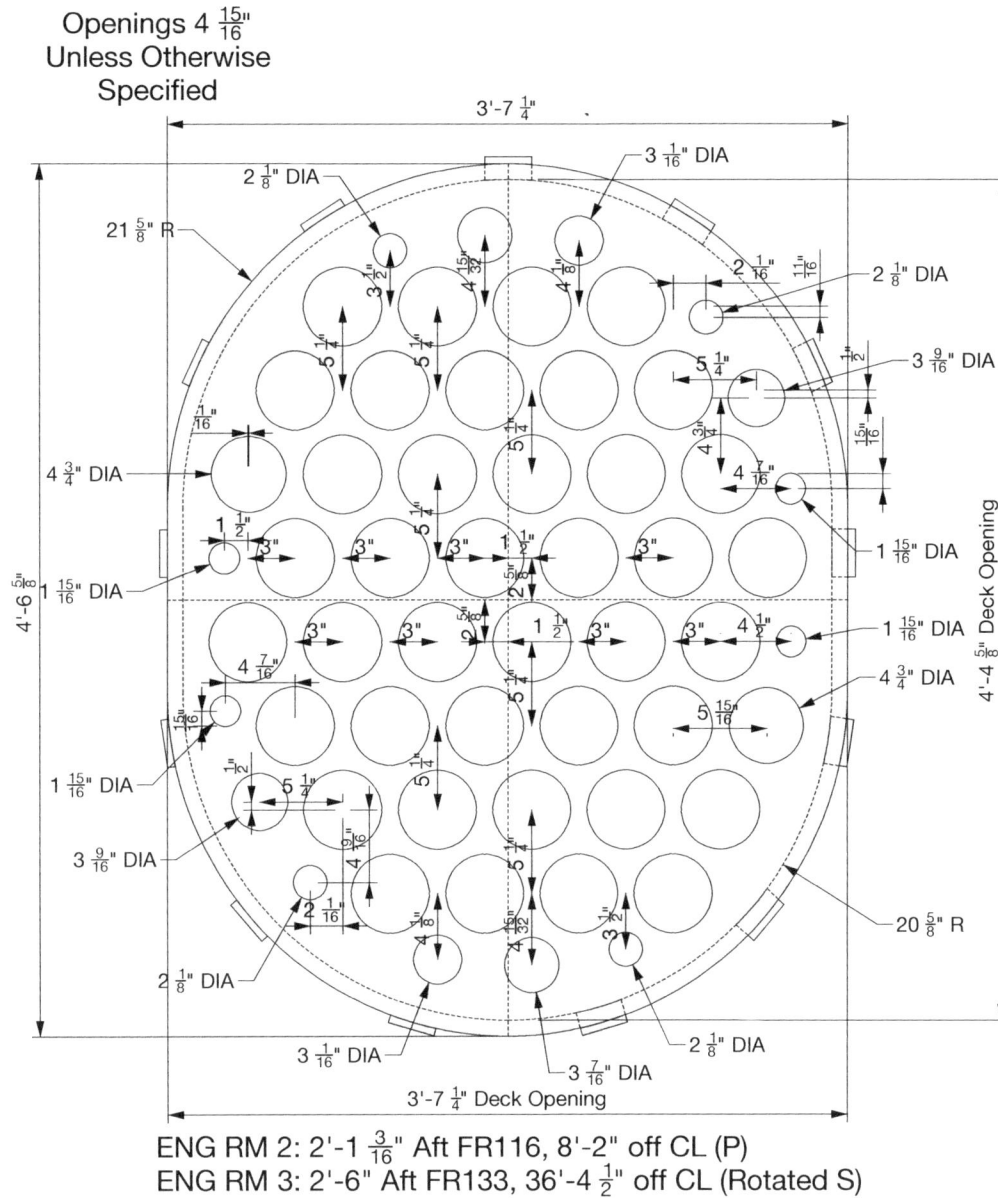

Figure 9.9: Main Deck Supply Grates for Engine Rooms 2 & 3 (1:12). The intake for for engine supply trunk room 3 is in the port side of the deckhouse at the 02 level, aft of the break between the fore and aft sections of the superstructure. The intake for the engine room 3 supply trunk is the large structure at the starboard side of the deckhouse against the handling room for the former five-inch gun mount 9.

ENG RM 2: $2'\text{-}1\frac{3}{16}$" Aft FR116, $8'\text{-}2$" off CL (P)
ENG RM 3: $2'\text{-}6$" Aft FR133, $36'\text{-}4\frac{1}{2}$" off CL (Rotated S)

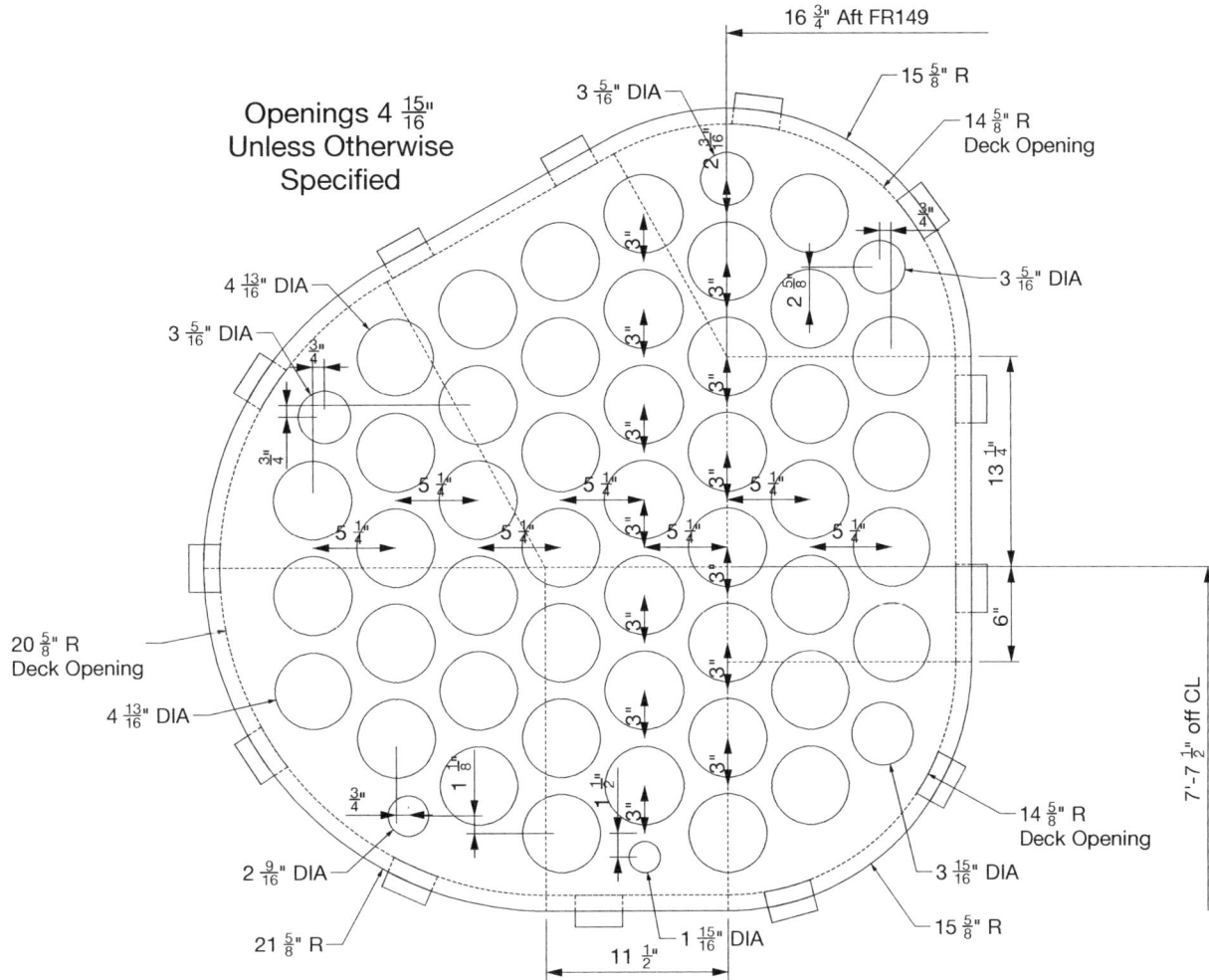

Figure 9.10: Main Deck Supply Grate for Engine Room 1 (1:12). The intake for this supply trunk is at the 01 level in the port side of the deckhouse, near the aft end of the superstructure.

Photograph 9.3: Enginer Room 1 Supply Vent. The odd shape of the supply grate for engine room 1 is a result of it having to fit within the the angle of the deckhouse.

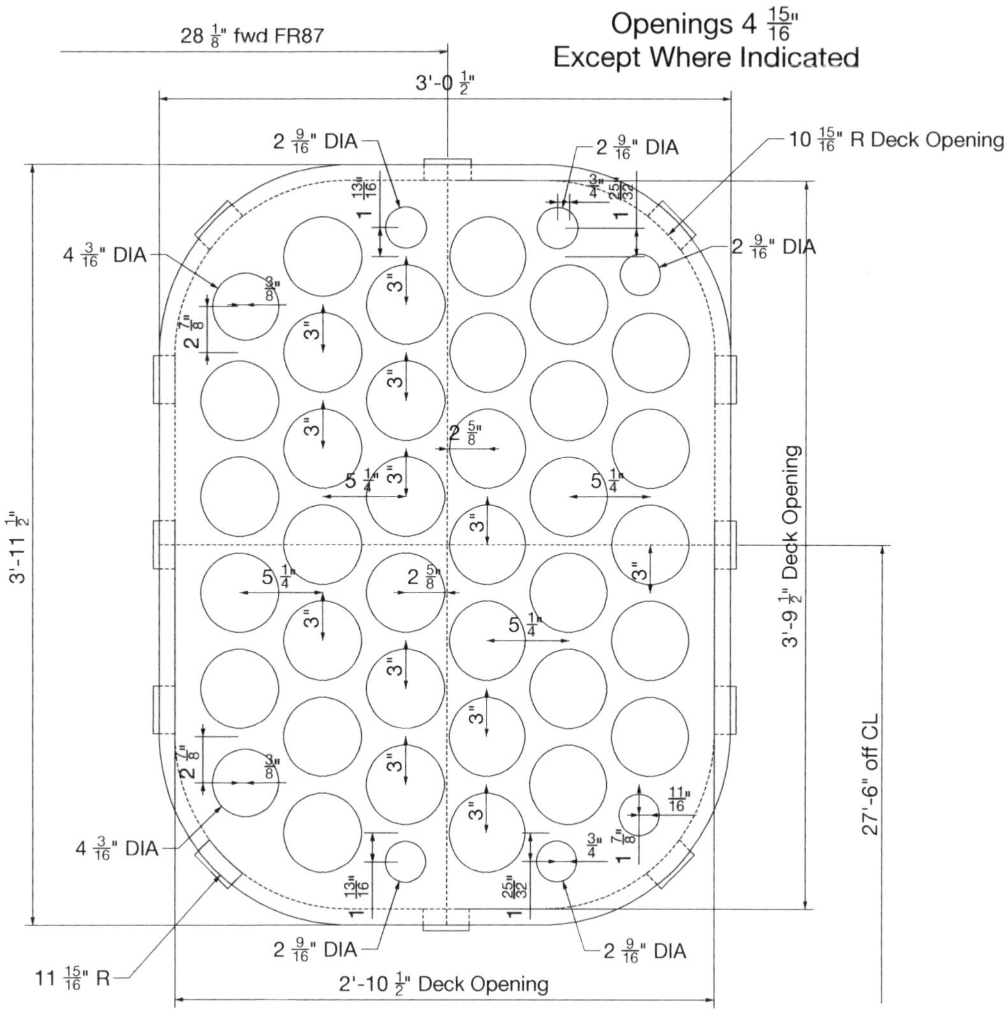

Figure 9.11: Forward Emergency Diesel Generator Room Main Deck Supply Grate (1:12). The intake for this trunk is a mushroom vent on the port side of the 01 level opposite the conning tower support.

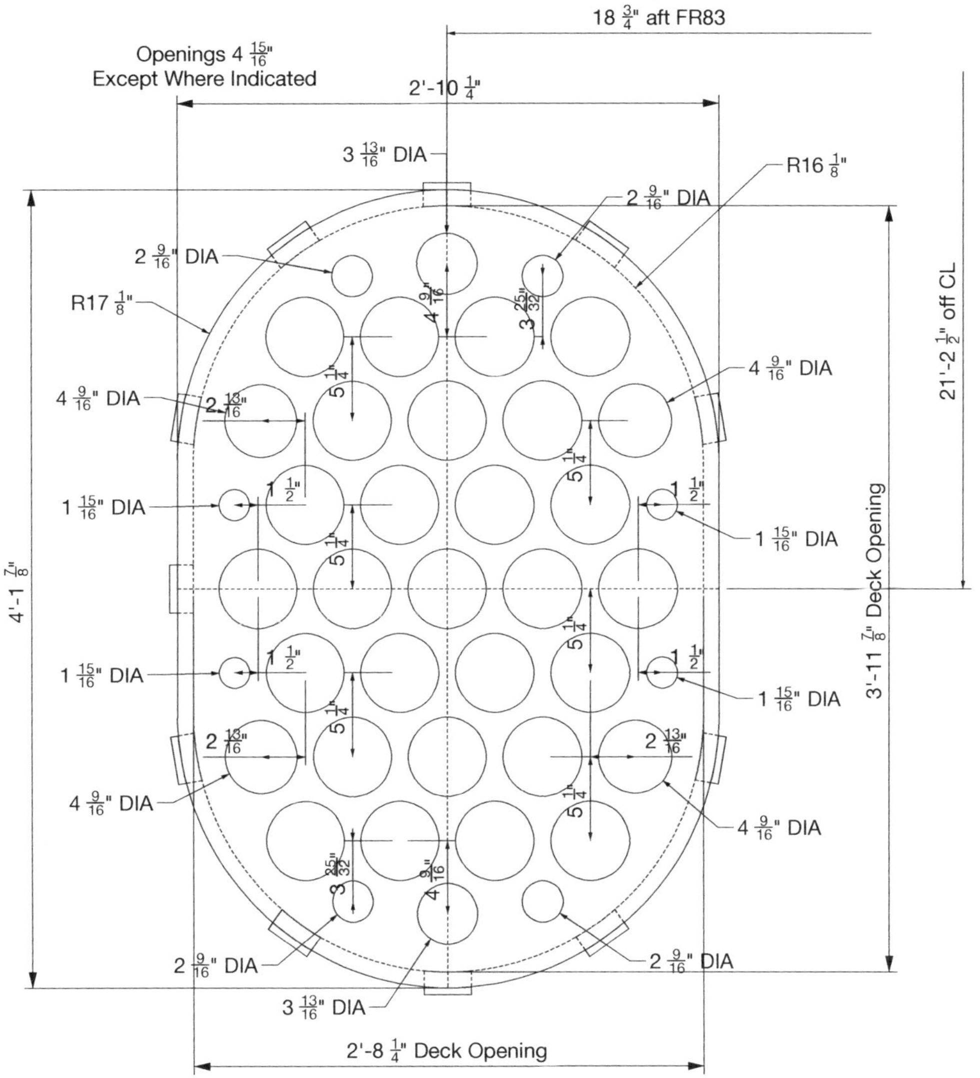

Figure 9.12: Forward Emergency Diesel Generator Room Main Deck Exhaust Grate (1:12). This trunk exits at the mushroom vent at the starboard side of the 01 level.

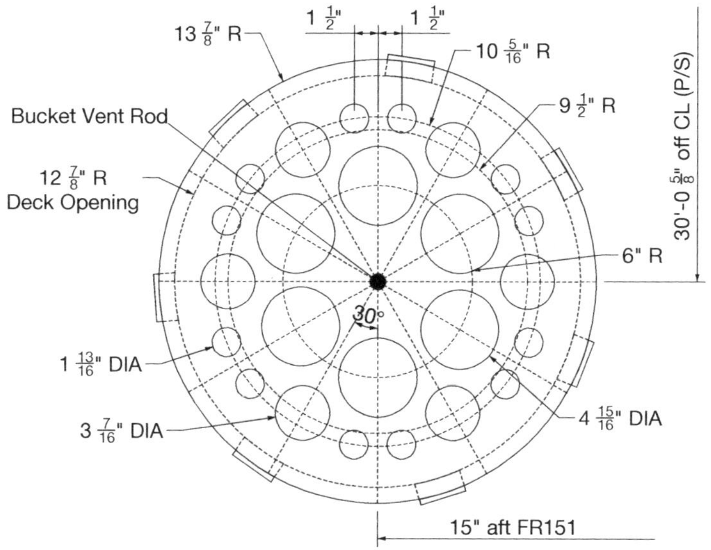

Figure 9.13: Aft Emergency Diesel Generator Main Deck Supply and Exhaust Grates (1:12). The intake for the supply trunk is a bucket vent at the starboard side of the main deck at the gap between turret 3 and the aft end of the superstructure. The exhaust is another bucket vent at the port side.

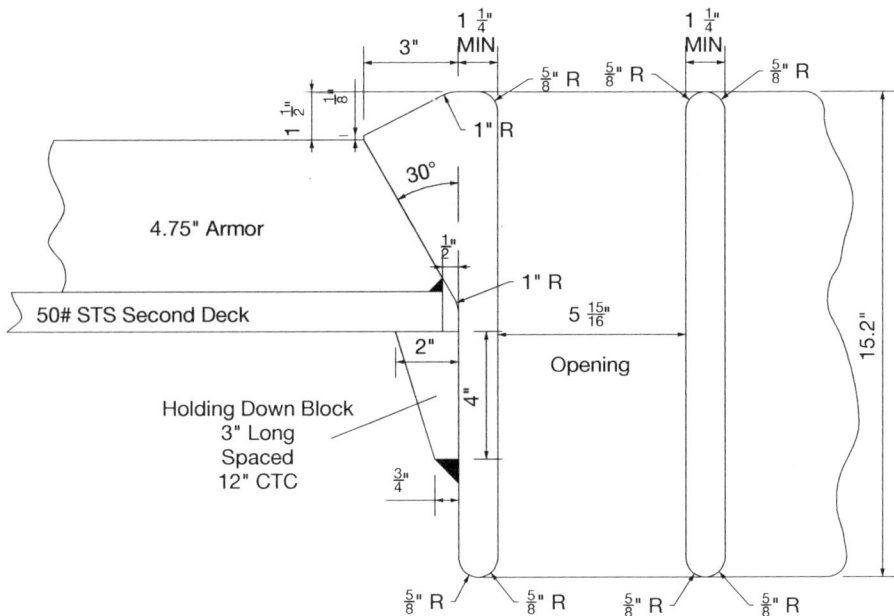

Figure 9.14: Connection of Grates to Second Deck (1:6). The edge cross section and deck connections apply to all grates except for uptake grates. The grate overlaps the deck opening by $2\frac{1}{2}$".

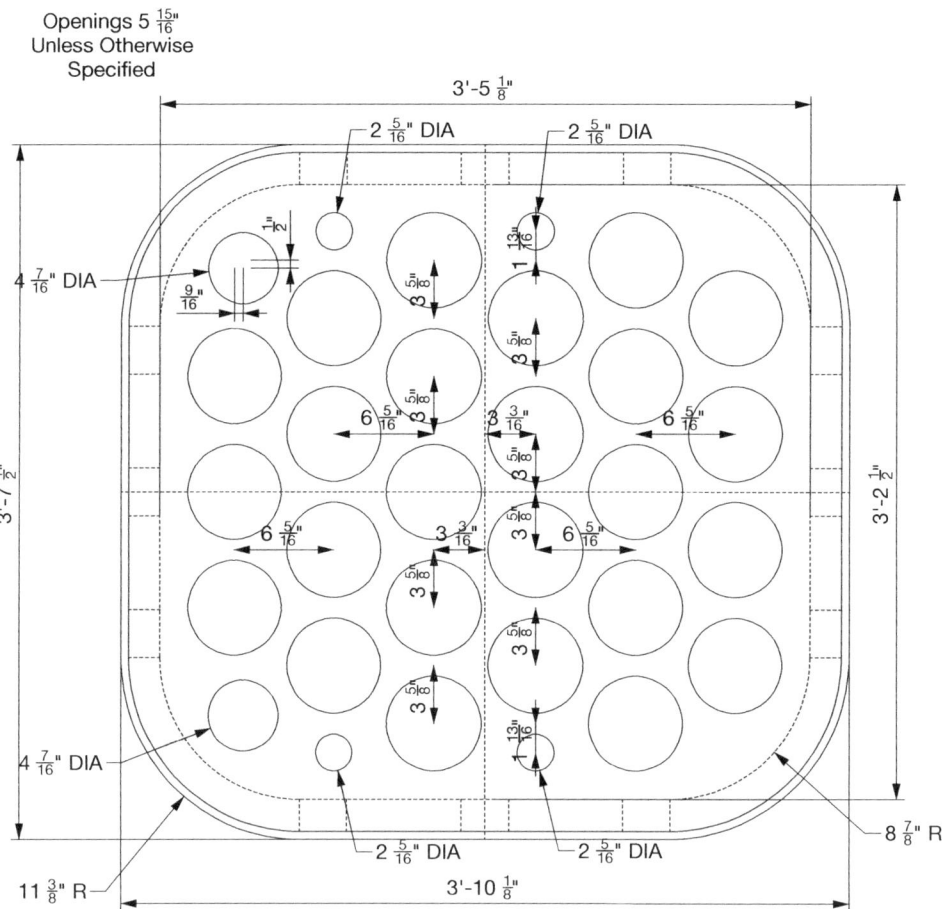

Figure 9.15: Engine Room Second Deck Supply Grates (1:12). The second deck supply grates for the engine rooms are all the same.

ENG RM 1: 24" Aft FR101, 14'-6$\frac{1}{2}$" off CL
ENG RM 2: 24" Aft FR117, 14'-10$\frac{1}{2}$" off CL
ENG RM 3: 24" Aft FR133, 14'-10" off Cl
ENG RM 4: 24" Aft FR149, 14'-10" off CL

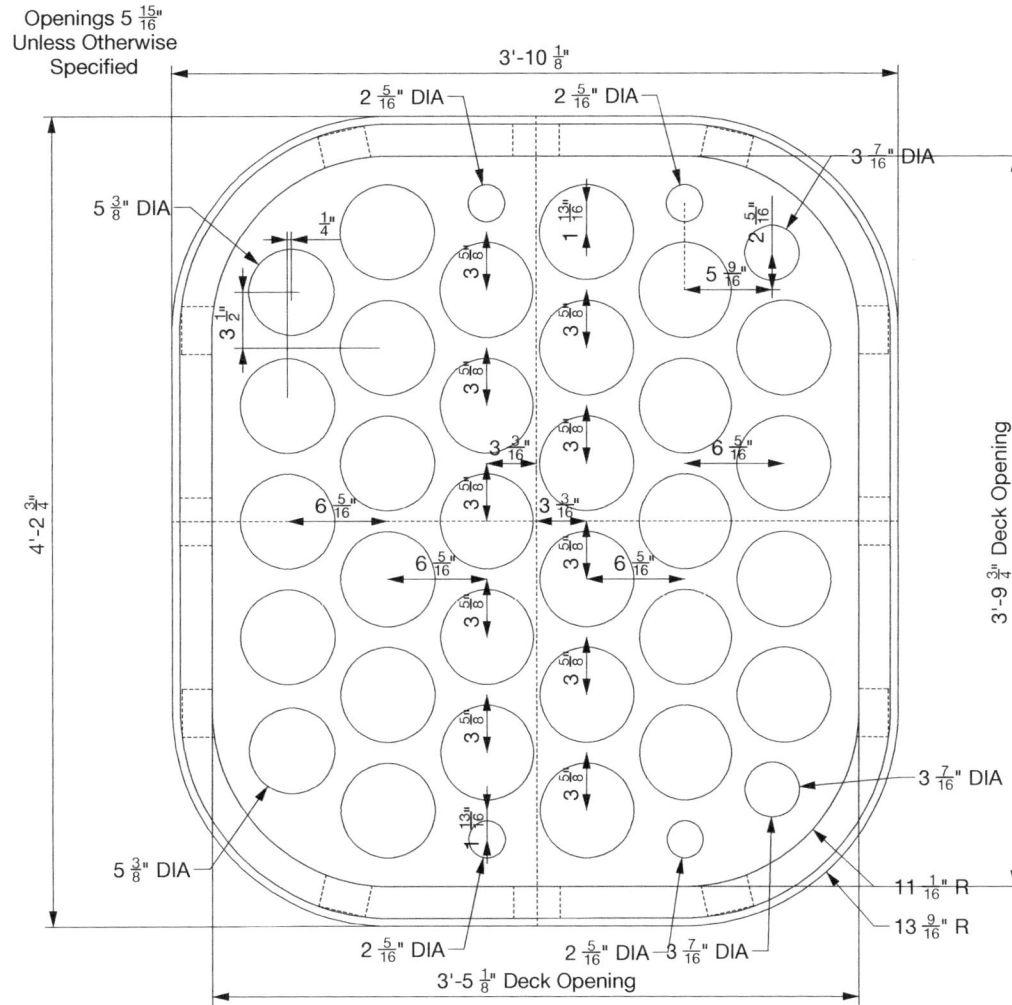

Figure 9.16: Engine Room Second Deck Exhaust Grates (1:12). The second deck exhaust grates for the engine rooms are all the same. The engine room exhaust emerges from these grates in the uptake spaces where it merges with the intake airflow and enters the boilers.

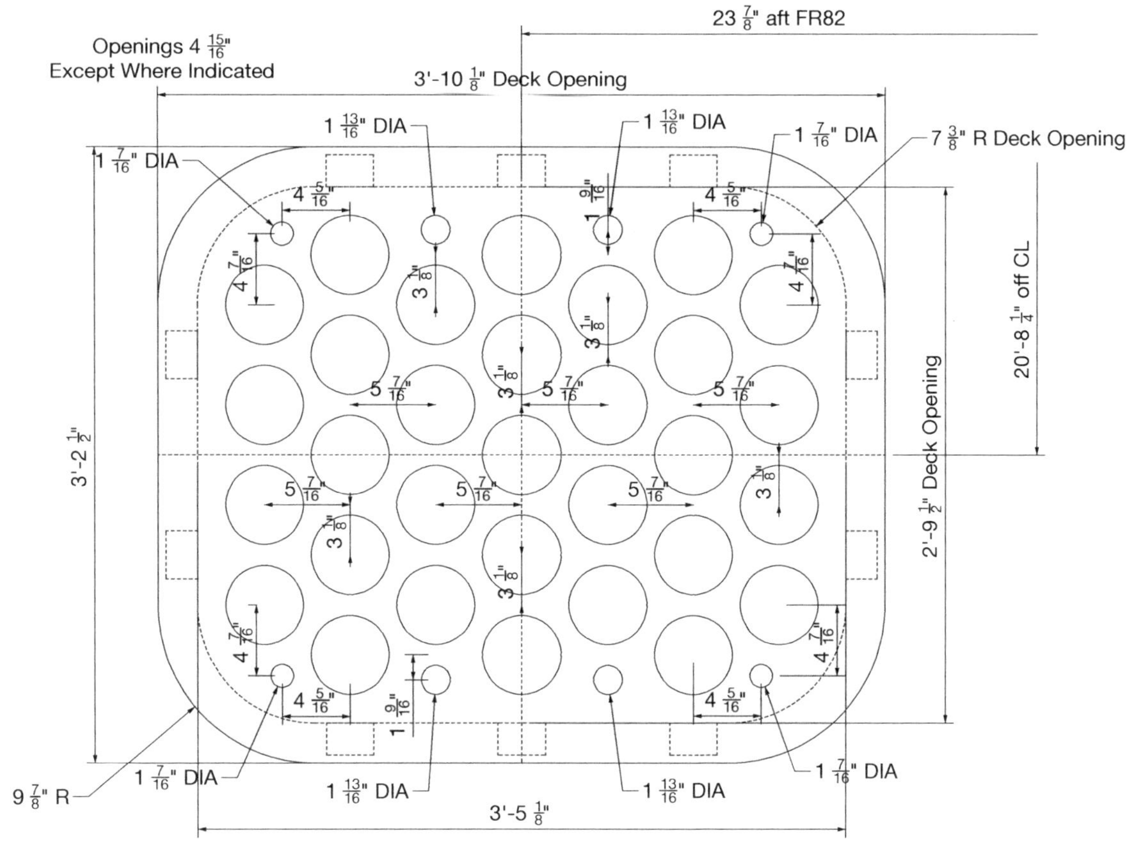

Figure 9.17: Second Deck Forward Emergency Diesel Generator Room Exhaust Grate (1:12). This grate is on on the port side.

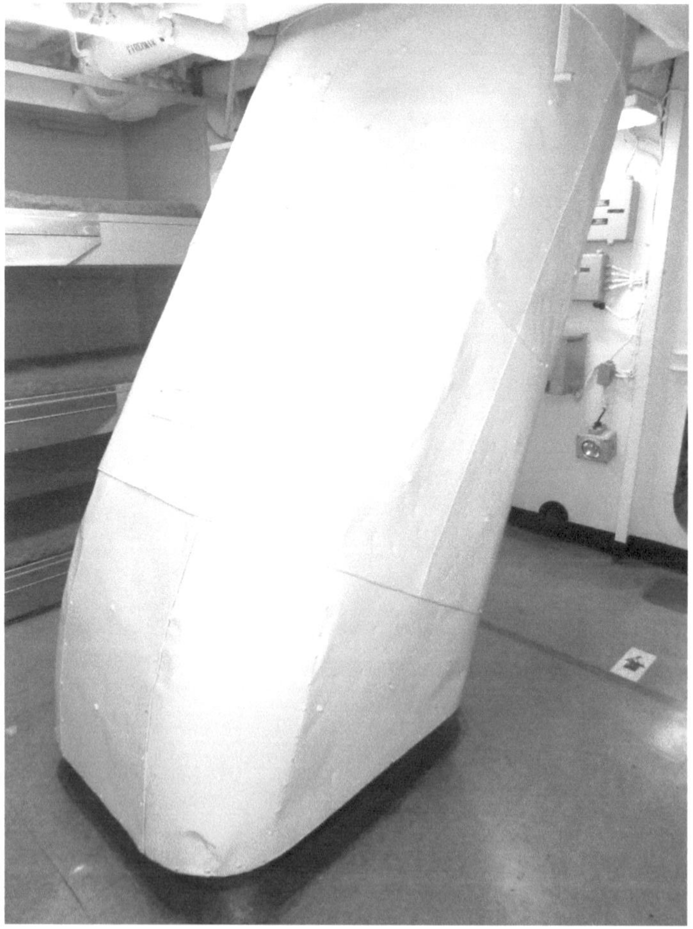

Photograph 9.4: Forward Emergency Diesel Generator Room Exhaust. This is Second Deck on USS New Jersey. The armored grate is visible at the deck.

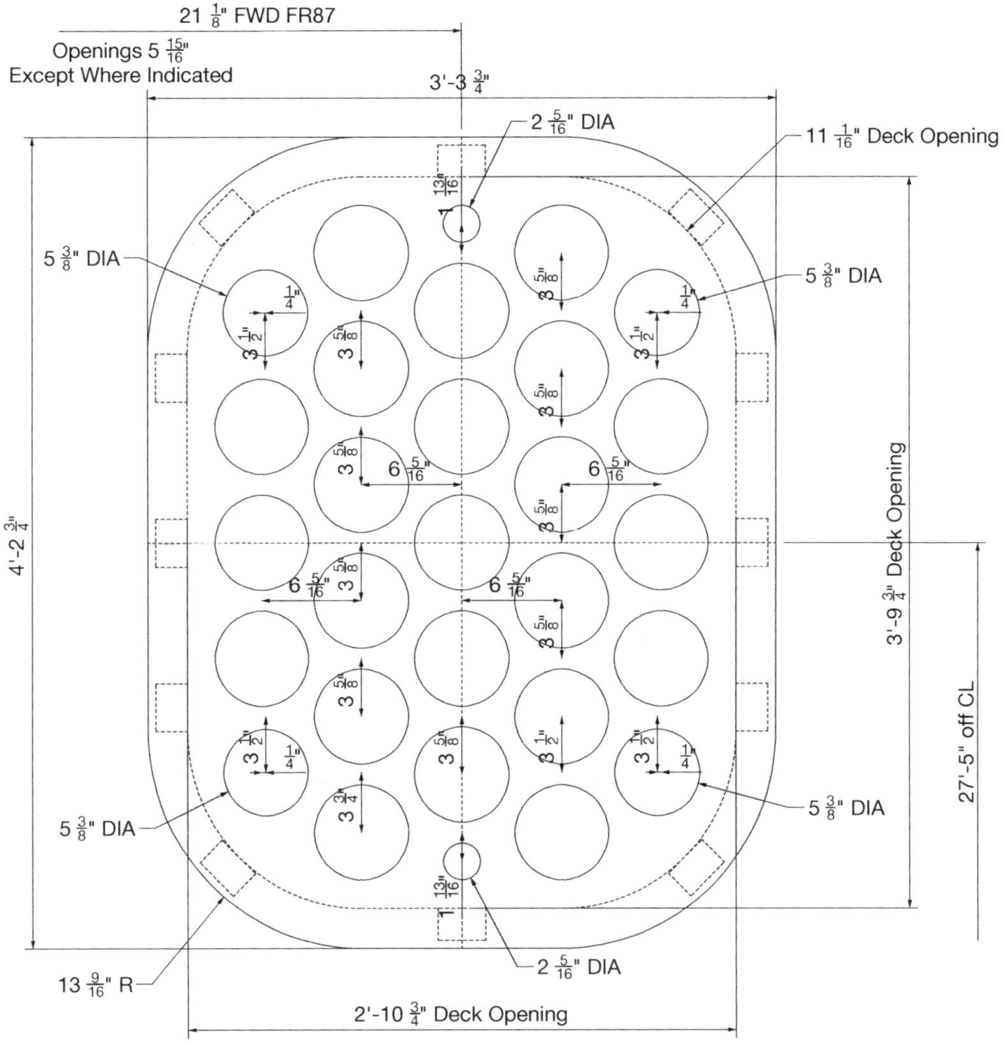

Figure 9.18: Second Deck Forward Emergency Diesel Generator Room Supply Grate (1:12). This grate is on the port side.

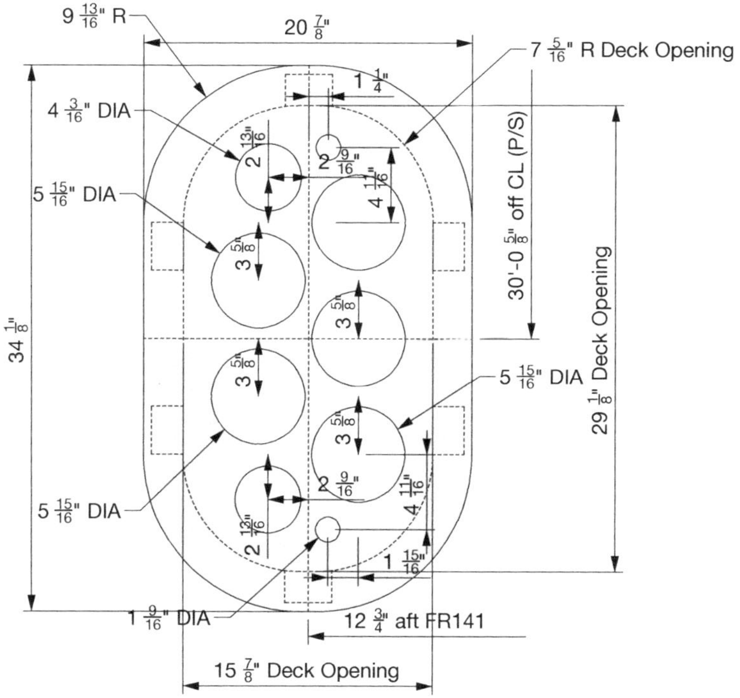

Figure 9.19: Second Deck Aft Emergency Diesel Generator Room Supply and Exhaust Grates (1:12). The supply grate is on the starboard side and the exhaust grate is on the port side.

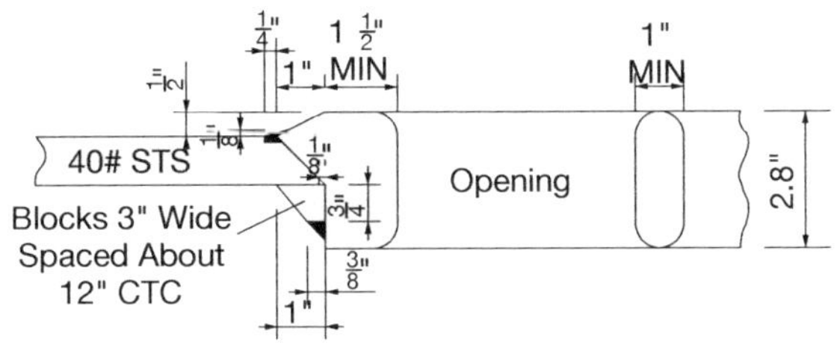

Figure 9.21: Connection of Grates to Third Deck (1:6). This applies to all third deck grates. The grate overlaps the deck opening by $\frac{7}{8}$".

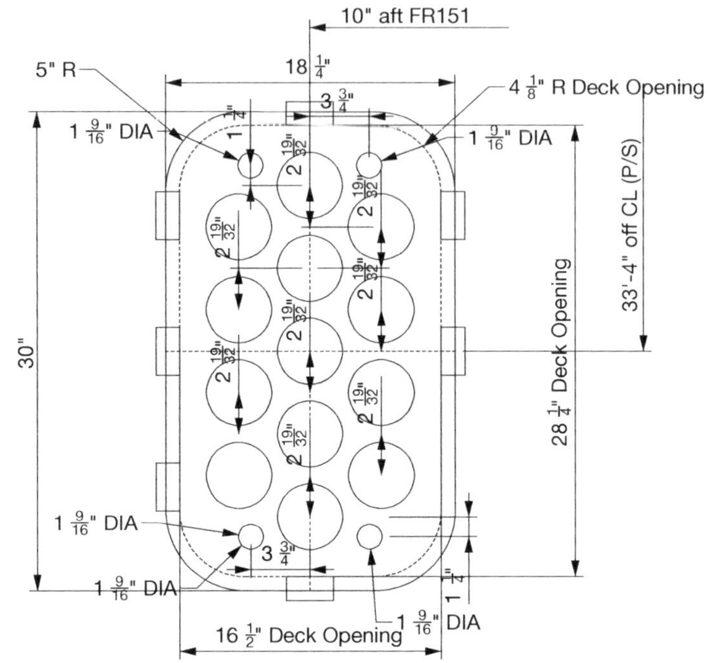

Figure 9.20: Third Deck Aft Emerency Diesel Generator Room Supply and Exhaust Grates (1:12). The supply grate at the starboard side and exhaust grate is at the port side are the same.

Photograph 9.5: Aft Emergency Generator Room Exhaust. This is second deck on USS New Jersey. The armored grate is visible at the deck. This trunk is for exhaust air. The diesel engine exhaust exits at the top of the smoke pipe.

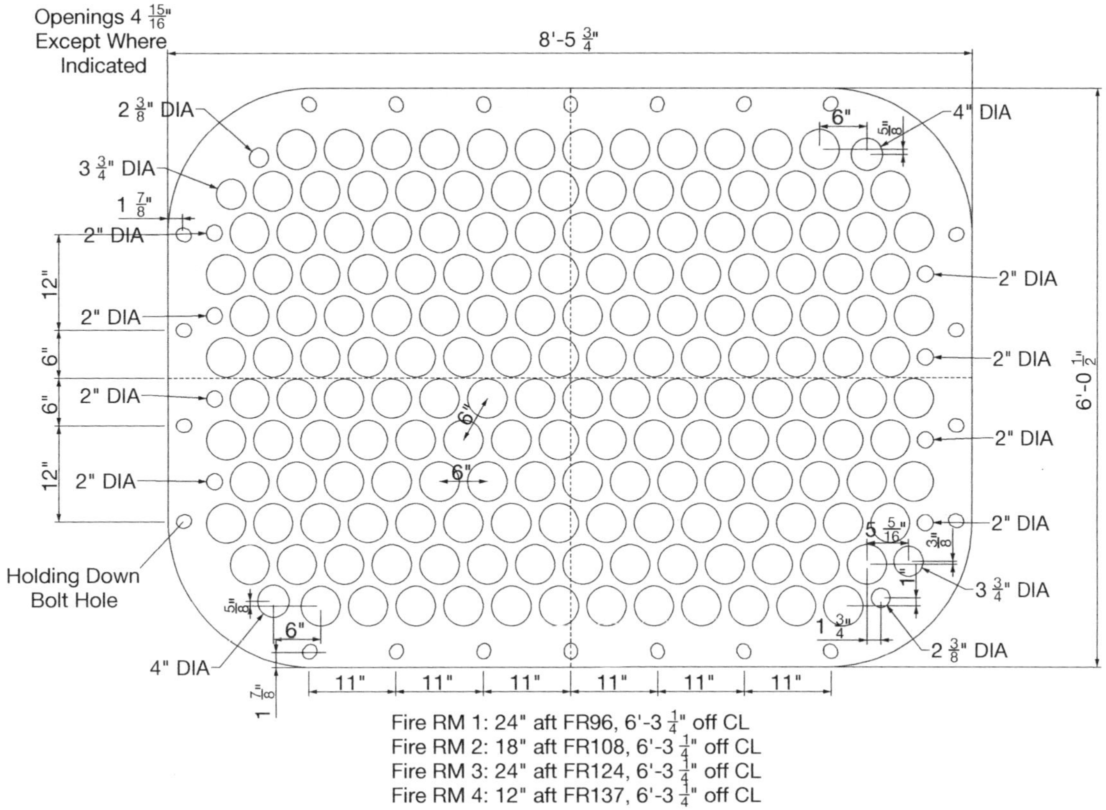

Openings 4 $\frac{15}{16}$" Except Where Indicated

8'-5 $\frac{3}{4}$"

2 $\frac{3}{8}$" DIA

3 $\frac{3}{4}$" DIA

1 $\frac{7}{8}$"

2" DIA

6"

5 $\frac{5}{8}$"

4" DIA

12"

2" DIA

6"

6"

2" DIA

6"

2" DIA

12"

2" DIA

6"

6"

2" DIA

Holding Down Bolt Hole

5 $\frac{5}{8}$"

2" DIA

6"

4" DIA

1 $\frac{7}{8}$"

11" 11" 11" 11" 11" 11"

6'-0 $\frac{1}{2}$"

2" DIA

2" DIA

5 $\frac{5}{16}$" 3 $\frac{5}{8}$"

1"

3 $\frac{3}{4}$" DIA

1 $\frac{3}{4}$"

2 $\frac{3}{8}$" DIA

Fire RM 1: 24" aft FR96, 6'-3 $\frac{1}{4}$" off CL
Fire RM 2: 18" aft FR108, 6'-3 $\frac{1}{4}$" off CL
Fire RM 3: 24" aft FR124, 6'-3 $\frac{1}{4}$" off CL
Fire RM 4: 12" aft FR137, 6'-3 $\frac{1}{4}$" off CL

Figure 9.22: Plan of Main Deck Uptake Grates (1:24). There is an uptake for each of the eight boilers. The grates are mirrored along the centerline. These grates are uniformly 4.3" thick. The bolt holes are the edge are oval shaped and oriented towards the center.

Opening Axis Oriented to Center of Grate

$\frac{13}{16}$" R

$\frac{5}{16}$"

$\frac{13}{16}$" R

1" DIA Bolt
Reference Point
$\frac{1}{4}$" Asbestos

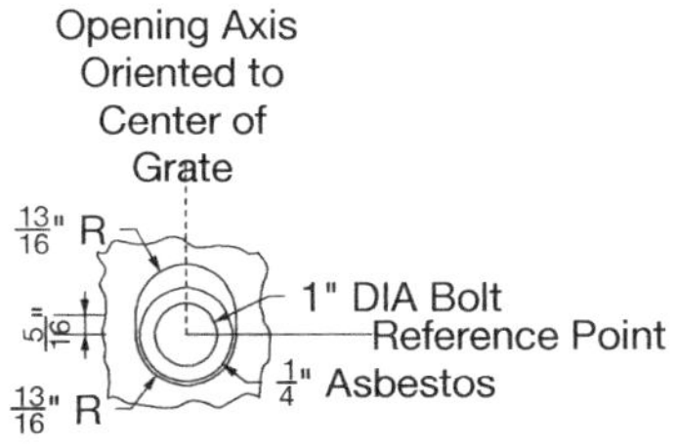

Figure 9.23: Detail of Bolt Holes in Uptake Grate (1:6). The bolt holes are ovals oriented towards the center of the grate. This allowed the grate to expand without damaging the bolts.

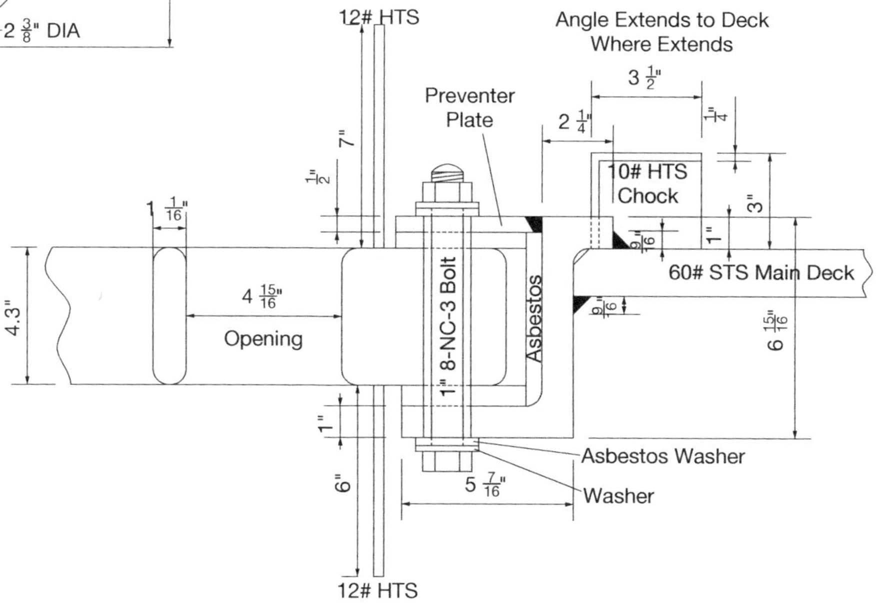

12# HTS

Angle Extends to Deck Where Extends

3 $\frac{1}{2}$"

Preventer Plate

2 $\frac{1}{4}$"

$\frac{1}{4}$"

1 $\frac{1}{16}$"

$\frac{1}{2}$"

7"

10# HTS Chock

3"

4.3"

4 $\frac{15}{16}$"

Opening

1"-8-NC-3 Bolt

Asbestos

$\frac{9}{16}$"

1"

$\frac{9}{16}$"

60# STS Main Deck

6 $\frac{15}{16}$"

1"

6"

5 $\frac{7}{16}$"

Asbestos Washer

Washer

12# HTS

Figure 9.24: Connection of Uptakes Grates to Main Deck (1:6). The angle at the upper right supports the uptake leading to the smoke pipe.

$21 \frac{11}{16}$" $5 \frac{13}{16}$" $16 \frac{1}{2}$"   22" — Outboard Side Boiler 2

$24 \frac{1}{16}$" $5 \frac{7}{16}$" $16 \frac{1}{2}$"   22" — Outboard Side Boiler 8

$25 \frac{5}{16}$" $2 \frac{3}{16}$" $16 \frac{1}{2}$"   22" — Inboard Side Boilers 5 & 6

22" $5 \frac{1}{2}$" $16 \frac{1}{2}$"   22" — Outboard Side Boilers 3 & 4

— Everywhere Else

Chocks

108"

$107 \frac{1}{8}$"

The original plans are inconsistent in the size of the grate and fail to account for plate thickness. The measurements here are corrected.

12"

12"

12"

12"

12"

12"

12"

$64 \frac{1}{8}$"

$76 \frac{3}{4}$" Deck Opening

$78 \frac{3}{4}$"

$77 \frac{7}{8}$"

$-14 \frac{1}{16}$" R

Preventer Plates

11"   11"   11"   11"   11"   11"

$93 \frac{3}{8}$"

106" Deck Opening

Outline of Grate

Outline of Bracket Upper Edge

Outline of Deck Opening

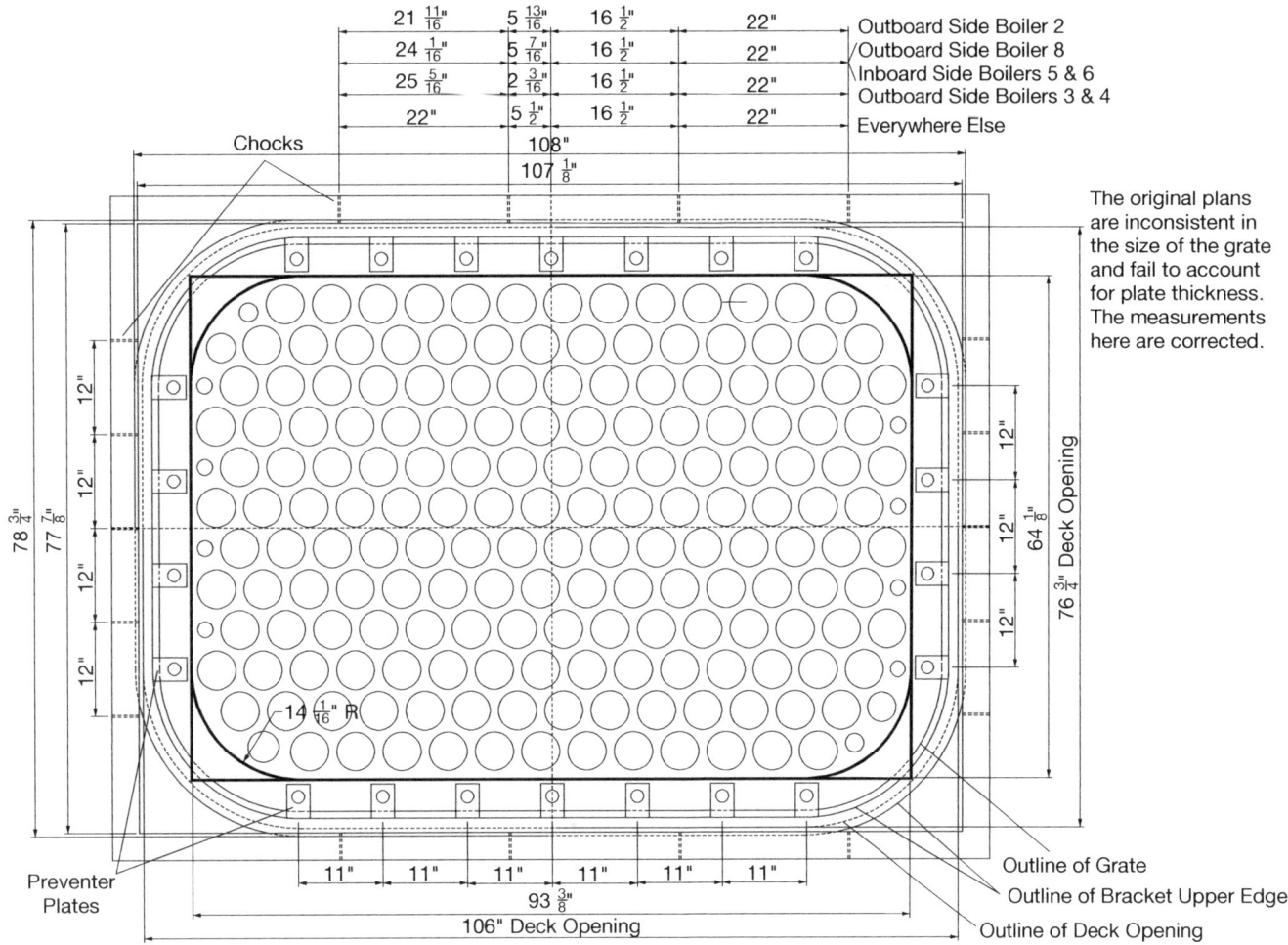

Figure 9.25: Plan of Uptake Grate Connect-sion to Main Deck (1:24).

Figure 9.26: Plan View of Preventer Plate (1:6).

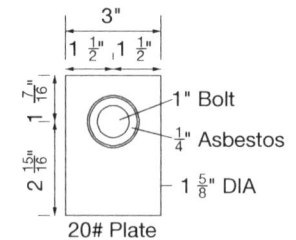

3"

$1 \frac{1}{2}$" $1 \frac{1}{2}$"

$1 \frac{7}{16}$"

$2 \frac{15}{16}$"

1" Bolt

$\frac{1}{4}$" Asbestos

$1 \frac{5}{8}$" DIA

20# Plate

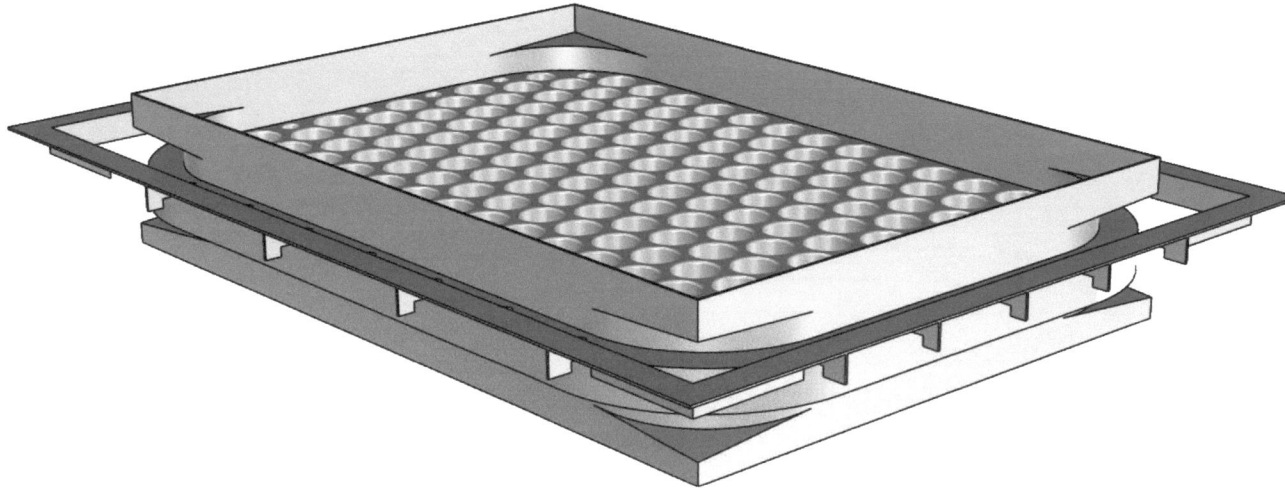

Figure 9.28: Rendering of Uptake Grate Corner. This view shows how the corners are filled in. The rectangular coaming shown is welded to the surrounding deck.

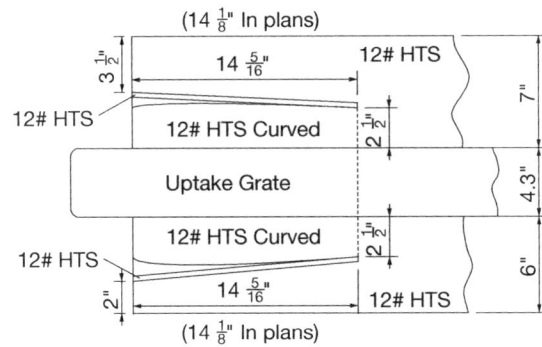

Figure 9.27: Main Deck Uptake Grate Corner (1:12). The overhangs at the corner are closed off with sloping fill.

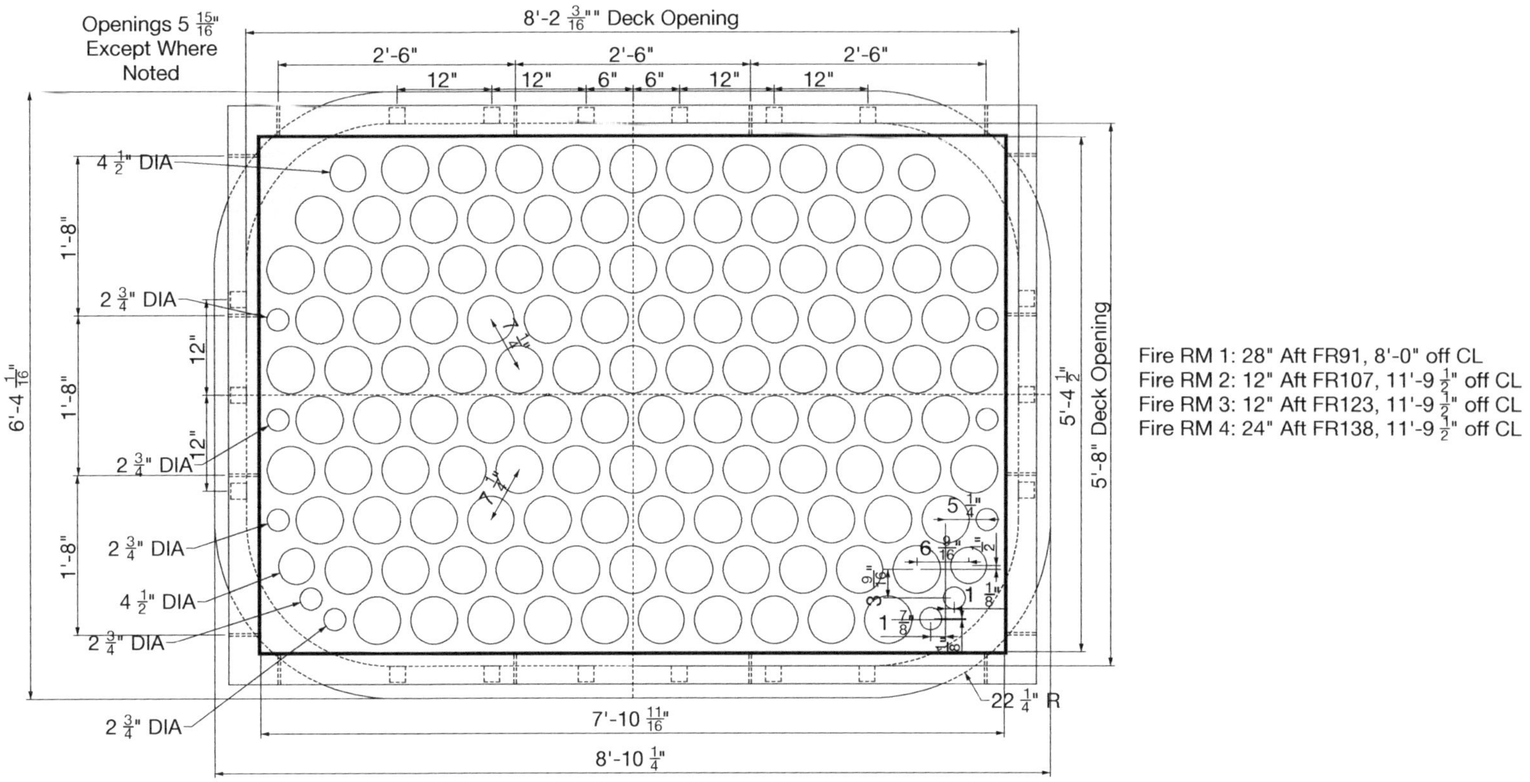

Figure 9.29: Plan of Second Deck Uptake Grate (1:24).

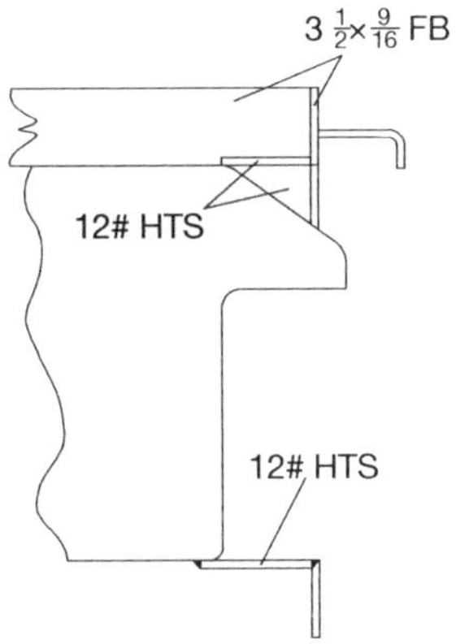

Fire RM 1: 28" Aft FR91, 8'-0" off CL
Fire RM 2: 12" Aft FR107, 11'-9½" off CL
Fire RM 3: 12" Aft FR123, 11'-9½" off CL
Fire RM 4: 24" Aft FR138, 11'-9½" off CL

Figure 9.30: Closing of the Corner Gap of Second Deck Uptake Grate (1:12).

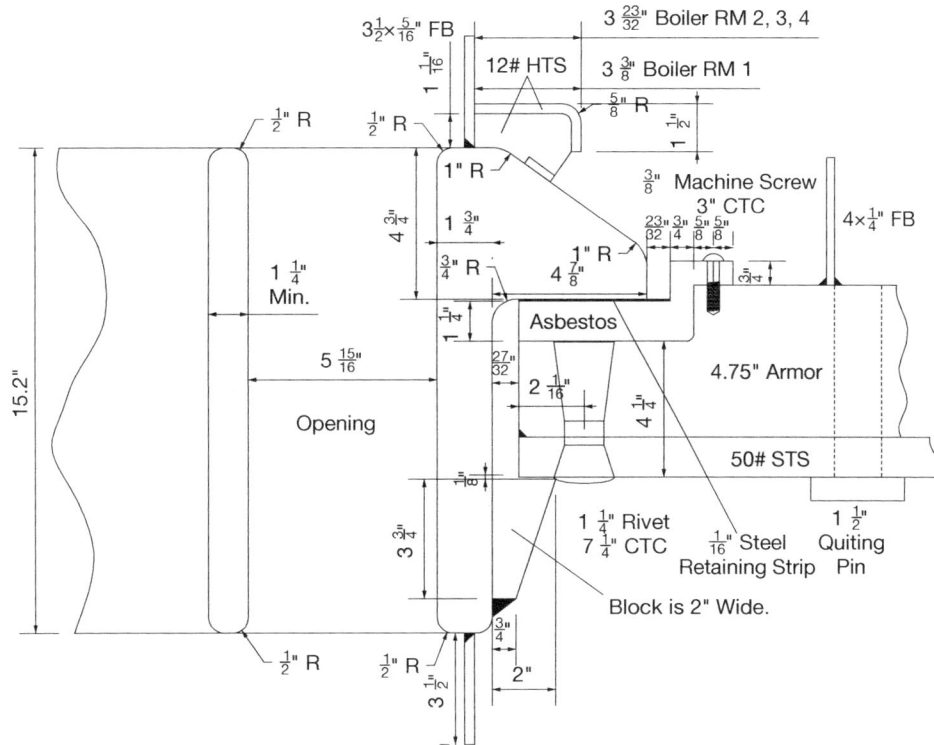

Figure 9.32: Connection of Uptake Grate to Second Deck (1:6).

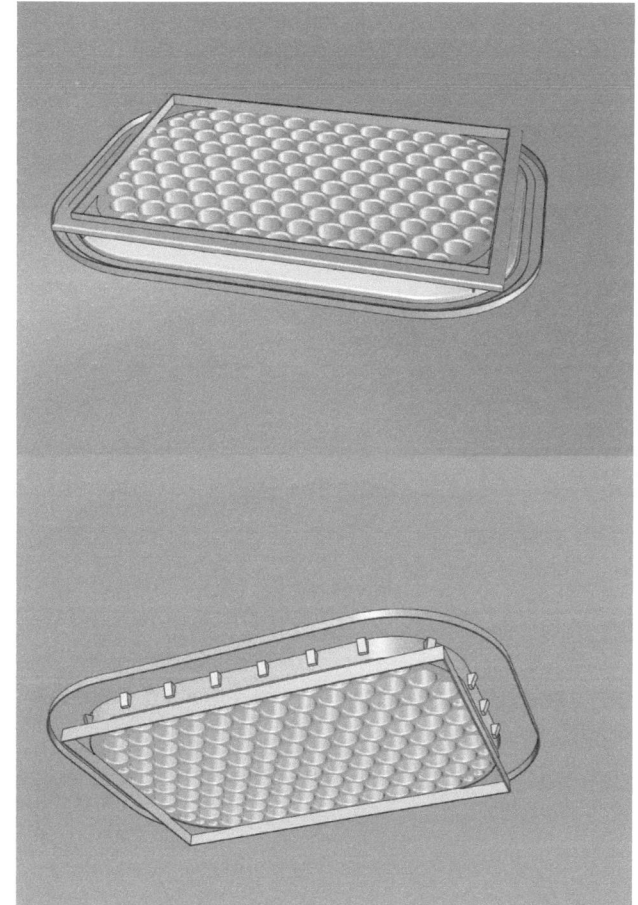

Figure 9.31: Renderings of the Second Deck Uptake Grate. This views show the grate sitting in the deck and how the blocks hold the grate in position from the bottom.

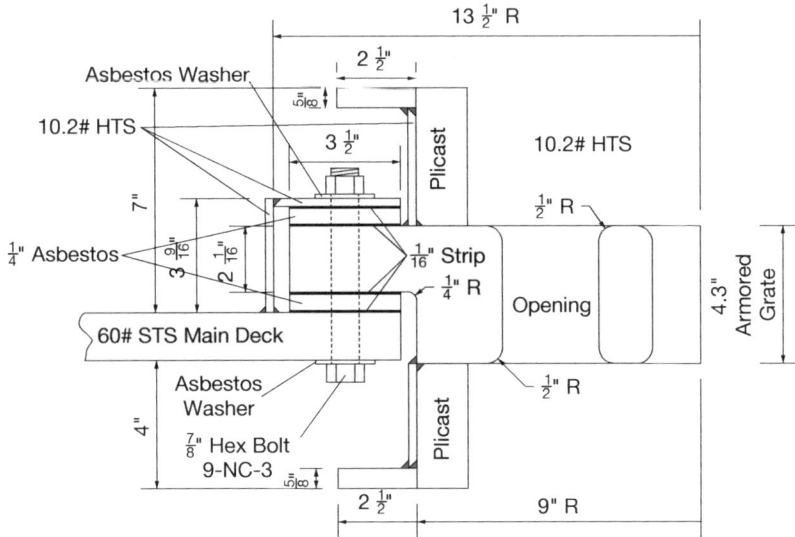

Figure 9.33: Connection of Trash Burner Grate to Main Deck (1:6). The trash burner exhuast runs from Second Deck to the aft smoke pipe. The armored grates for the trash burner exhaust are positioned horizontally at main deck and vertically in the bulkhead at FR143. The grates have the same plan but differ in thickness as shown. The grates and their connections are circular. This view shows one half of the grate.

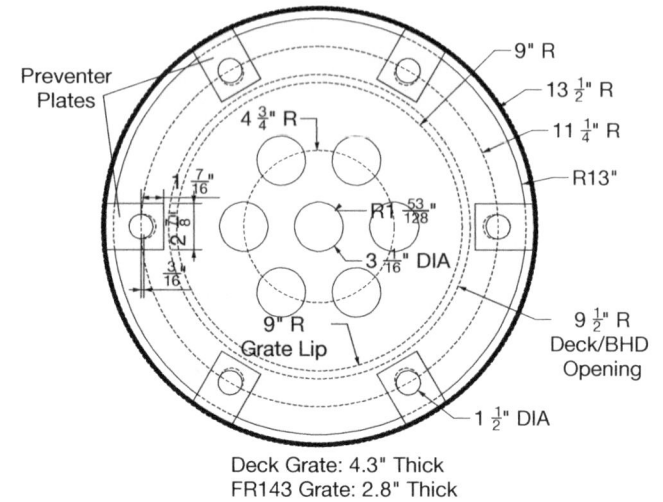

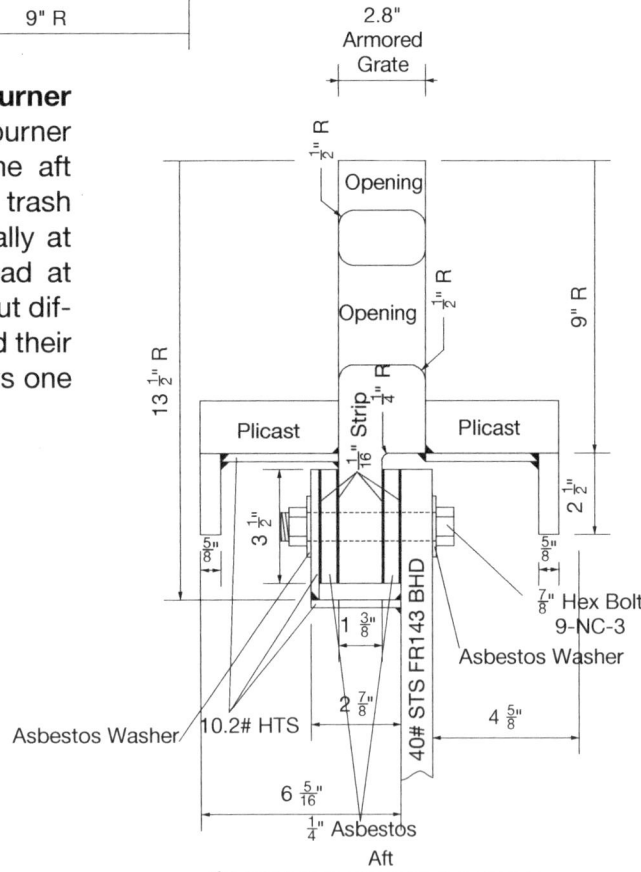

Deck Grate: 4.3" Thick
FR143 Grate: 2.8" Thick

Figure 9.34: Plan of Trash Burner Armored Grates (1:12). The bolt holes in the preventer plates are circular while the holes in the grate are oval. The oval openings allow the grate to expand without damaging the bolts.

Figure 9.35: Connection of Trash Burner Grate to FR143 Bulkhead (1:6). This grate is thinner than the main deck grate and is oriented vertically. This figure shows the lower half of the grate.

Chapter 10: Hatch Covers

There are thirteen hatch openings in the second deck armor and four in the third deck armor. The covers for these openings are Class B armor. The third deck hatches are all in the 5.6-inch thick deck armor between FR166 and FR189. The second deck hatches are 5.625 inches thick and the third deck hatches are 6.125 inches thick. The hatch openings were machined at a 45-degree angle so the openings are wider at the top than at the bottom. The specified hatch sizes measure the clear opening. The sizes on the plans give the opening length parallel to the centerline first then the transverse width. The various hatches open in all four directions and some open along their short length and others open along their long length.

The three forwardmost hatch covers (FR50, FR61 P/S) were made from the plugs cut for 36x42 hatches in third deck. The other armored hatch covers were ordered from the mills at their final thicknesses. One of the second deck hatch openings is a circular ammunition loading hatch at FR68. This hatch was ordered as an octagonal plate. All the other hatch openings are rectangular with rounded corners for personel and stores (including ammunition) access. Their covers were ordered as rectangular plates.

The yard had to machine the hatch covers to their final shape and add fixtures. The corners had to be rounded and the sides were machined to 45 degrees. The yard welded a gasket strip around the upper edge so that a watertight gas-

ket could be added. The yard also drilled holes for dogs in the face and holes for lubricating the dogs. The hatches weighed between three quarters of a ton and nearly two tons. They would be impossible for one person to open so the yard added hinge arms that were counter-balanced by a pair of springs on the opposite side from the hinges. The yard also welded staples to the upper face to create attachment points for opening with

block and tackle if the springs failed.

The FR68 ammunition hatch was a special case. It has no hinge mechanism. To open this hatch, three removable lifting eyes had to be screwed into the upper face and the hatch had to be lifted out and away. Three centering pads aligned the hatch when the hatch was closed so that the dogs would be opposite their wedges for sealing.

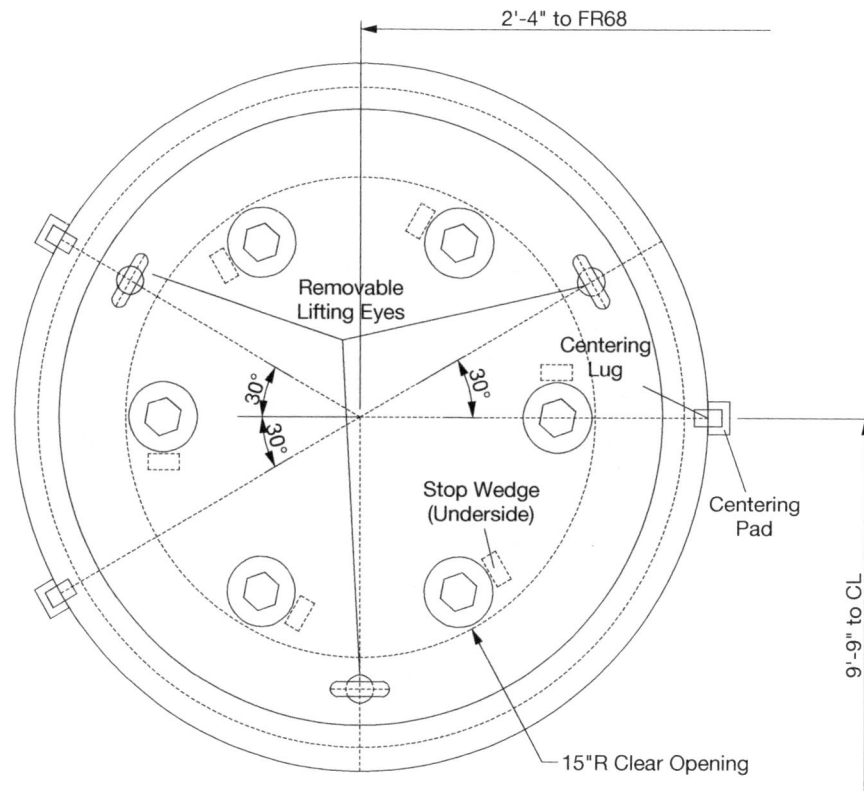

Figure 10.1: Ammunition Hatch Cover FR68 (1:12). There is a series of vertically aligned hatches from main deck to first platform. This is the hatch within the armored second deck. This hatch has to be lifted clear to be opened.

Location	Size	USS *Iowa*	USS *New Jersey*	USS *Missouri*	USS *Wisconsin*	USS *Illinois*	USS *Kentucky*
FR68	30×36	Bureau of Ordnance	Bureau of Ordnance	The Midvale Company	Carnegie-Illinois Steel	The Midvale Company	Carnegie-Illinois Steel
FR68	30 DIA	Carnegie-Illinois Steel	Carnegie-Illinois Steel	The Midvale Company	Carnegie-Illinois Steel	The Midvale Company	Carnegie-Illinois Steel
FR79	30×36	The Midvale Company	The Midvale Company	The Midvale Company	Carnegie-Illinois Steel	Bethlehem Steel	Carnegie-Illinois Steel
FR85	46×42	The Midvale Company	The Midvale Company	The Midvale Company	Carnegie-Illinois Steel	Bethlehem Steel	Carnegie-Illinois Steel
FR101	30×36	The Midvale Company	The Midvale Company	The Midvale Company	Carnegie-Illinois Steel	Bethlehem Steel	Carnegie-Illinois Steel
FR117	30×36	The Midvale Company	The Midvale Company	The Midvale Company	Carnegie-Illinois Steel	Bethlehem Steel	Carnegie-Illinois Steel
FR133	52×48	Carnegie-Illinois Steel	Carnegie-Illinois Steel	The Midvale Company	Carnegie-Illinois Steel	Bethlehem Steel	Carnegie-Illinois Steel
FR149	36×30	The Midvale Company	The Midvale Company	The Midvale Company	Carnegie-Illinois Steel	Bethlehem Steel	Carnegie-Illinois Steel
FR155	36×30	Carnegie-Illinois Steel	Carnegie-Illinois Steel	The Midvale Company	Carnegie-Illinois Steel	The Midvale Company	Carnegie-Illinois Steel
FR164	36×30	Carnegie-Illinois Steel	Carnegie-Illinois Steel	The Midvale Company	Carnegie-Illinois Steel	The Midvale Company	Carnegie-Illinois Steel
FR168	42×36	Bethlehem Steel	Bethlehem Steel	The Midvale Company	Carnegie-Illinois Steel	The Midvale Company	Carnegie-Illinois Steel
FR177	36×42	Bethlehem Steel	Bethlehem Steel	The Midvale Company	Carnegie-Illinois Steel	The Midvale Company	Carnegie-Illinois Steel
FR185	42×36	Bethlehem Steel	Bethlehem Steel	The Midvale Company	Carnegie-Illinois Steel	The Midvale Company	Carnegie-Illinois Steel
FR188	32×40	Carnegie-Illinois Steel	Carnegie-Illinois Steel	The Midvale Company	Carnegie-Illinois Steel	The Midvale Company	Carnegie-Illinois Steel

Table 10.1: Armored Hatch Cover Suppliers
Others Used Third Deck Armor Hatch Opening Plugs

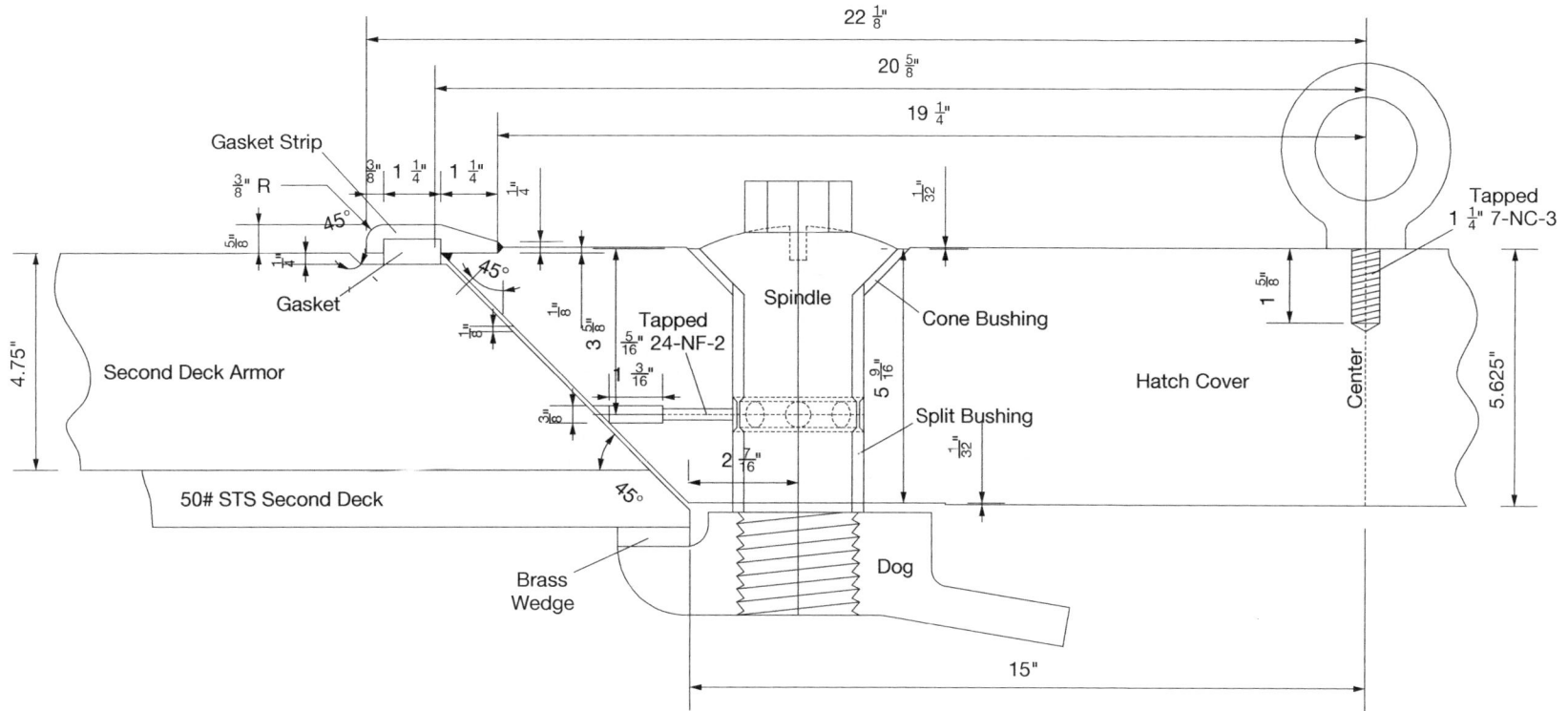

Figure 10.2: Cross section of the Second Deck 30-Inch Ammunition Hatch at FR68 (1:4).

Photograph 10.1: Ammunition Hatch on USS Iowa. On USS Iowa the armored hatch covers have dogs opened by with a blade tool. (Jim Kurrasch)

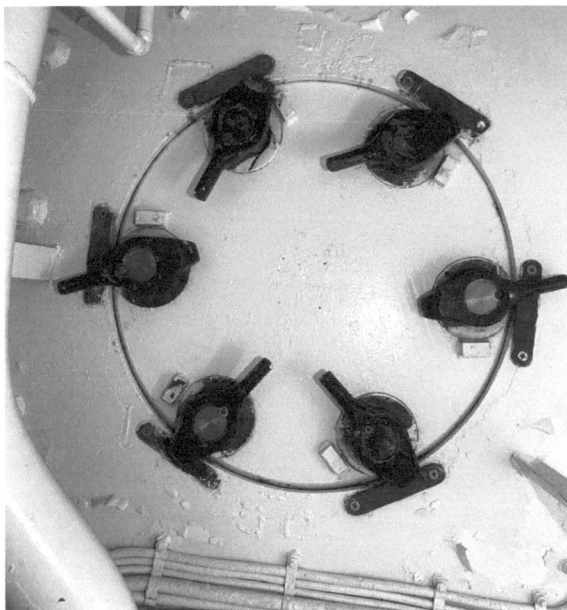

Photograph 10.2: Ammunition Hatch on USS New Jersey. The later ships used a hex wrench to open the dogs. The plans indicate that USS New Jersey was to be the same as USS Iowa. USS New Jersey's construction either deviated from the plans or it was modified later to match USS Missouri and USS Wisconsin.

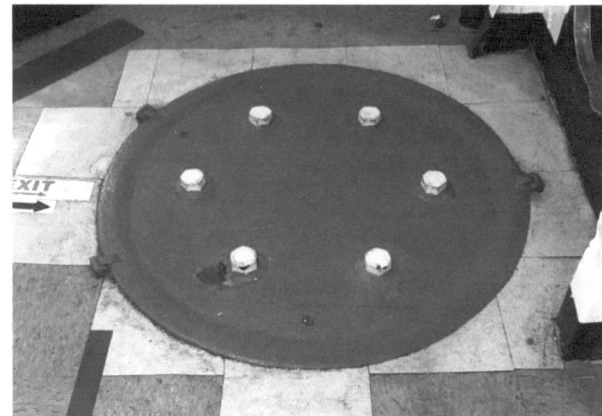

Photograph 10.3: Ammunition Hatch on USS Iowa Viewed from Third Deck. Wedges ring the hatch opening. There is a stop wedge next to each dog to keep the dog from rotating too far to open the hatch. (Jim Kurrasch)

Plan 10.1: (1623CD) The operation of an armored hatch cover. This is the second deck hatch at FR85. The others are similar. There is a spring connected by a cable to each of the two hinges at the opposite side. With effort, one man can open the hatch. A catch rod locks the hatch against either a latch or hole to hold it in the open position.

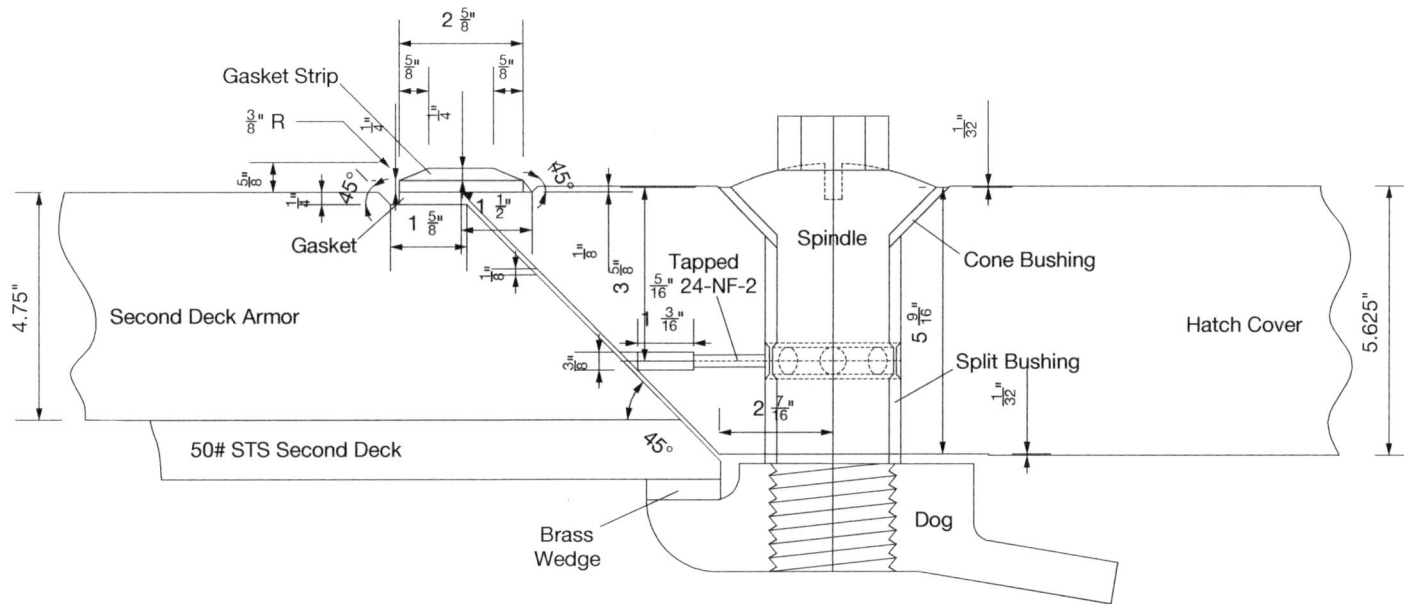

Figure 10.3: Cross Section of Second Deck Armored Hatch Cover (1:4). All of these hatch covers follow this pattern. There is an opening for lubricating each of the dog spindles. The plans indicate that USS Iowa and USS New Jersey use a slotted spindle head and USS Missouri and USS Wisconsin use hex heads. USS New Jersey currently has hex heads.

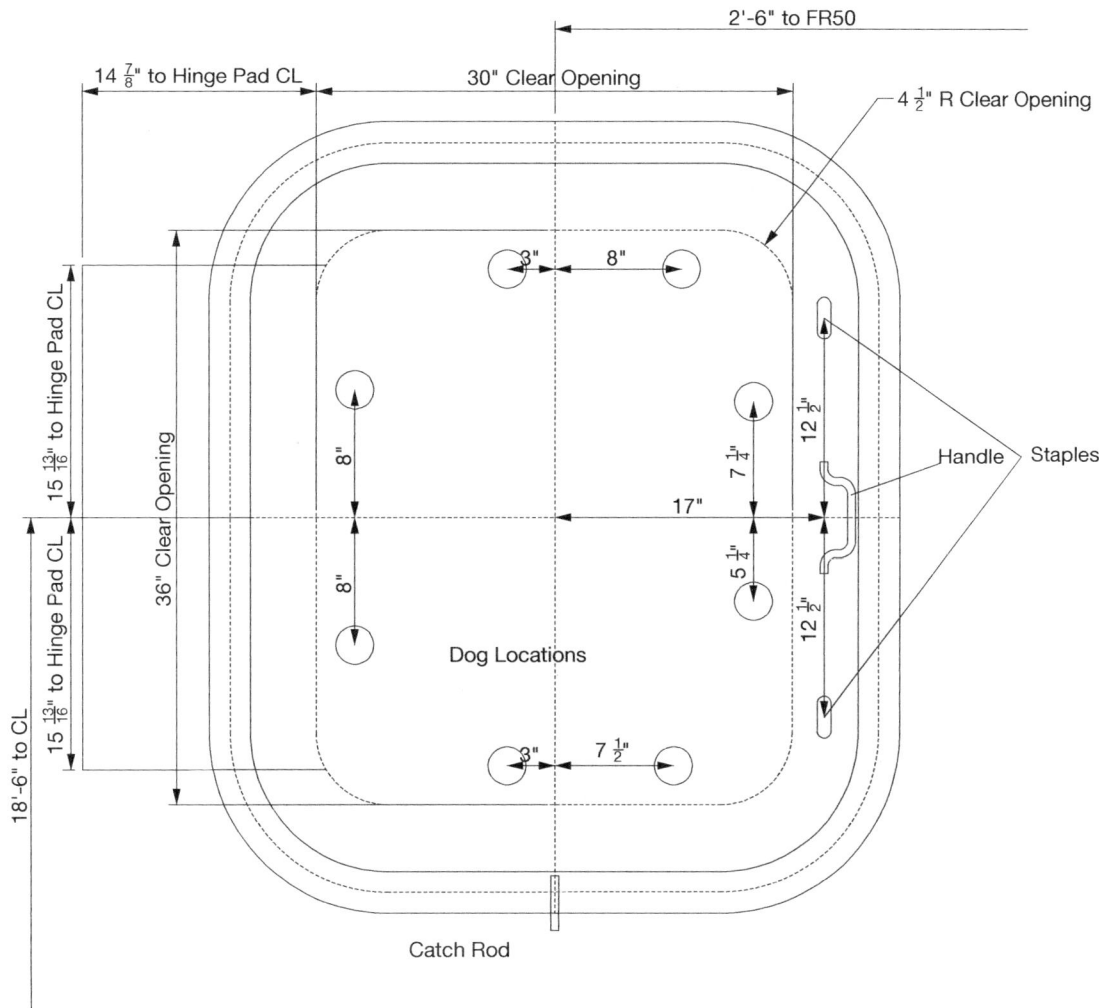

Figure 10.4: Second Deck 30x36 FR50 Hatch Cover (1:12). This is the only armored hatch cover that opens aft. This hatch leads to a watertight trunk that extends to Second Platform. The primer magazine was at Second Platform and 20mm and 40mm magazines were at First platform. This cover was made from a plug cut from one of the 36x42 third deck armor hatch openings.

Photograph 10.2: FR50 Hatch Cover on USS New Jersey. The catch rod locks into the latch at the left to hold the cover in place when the hatch is open. The staples at the right have a link for attaching a tackle to open the hatch if the spring fails. The hatch cover weights over 1500 lbs. However, the spring counter-balance allows one person pulling hard to open it.

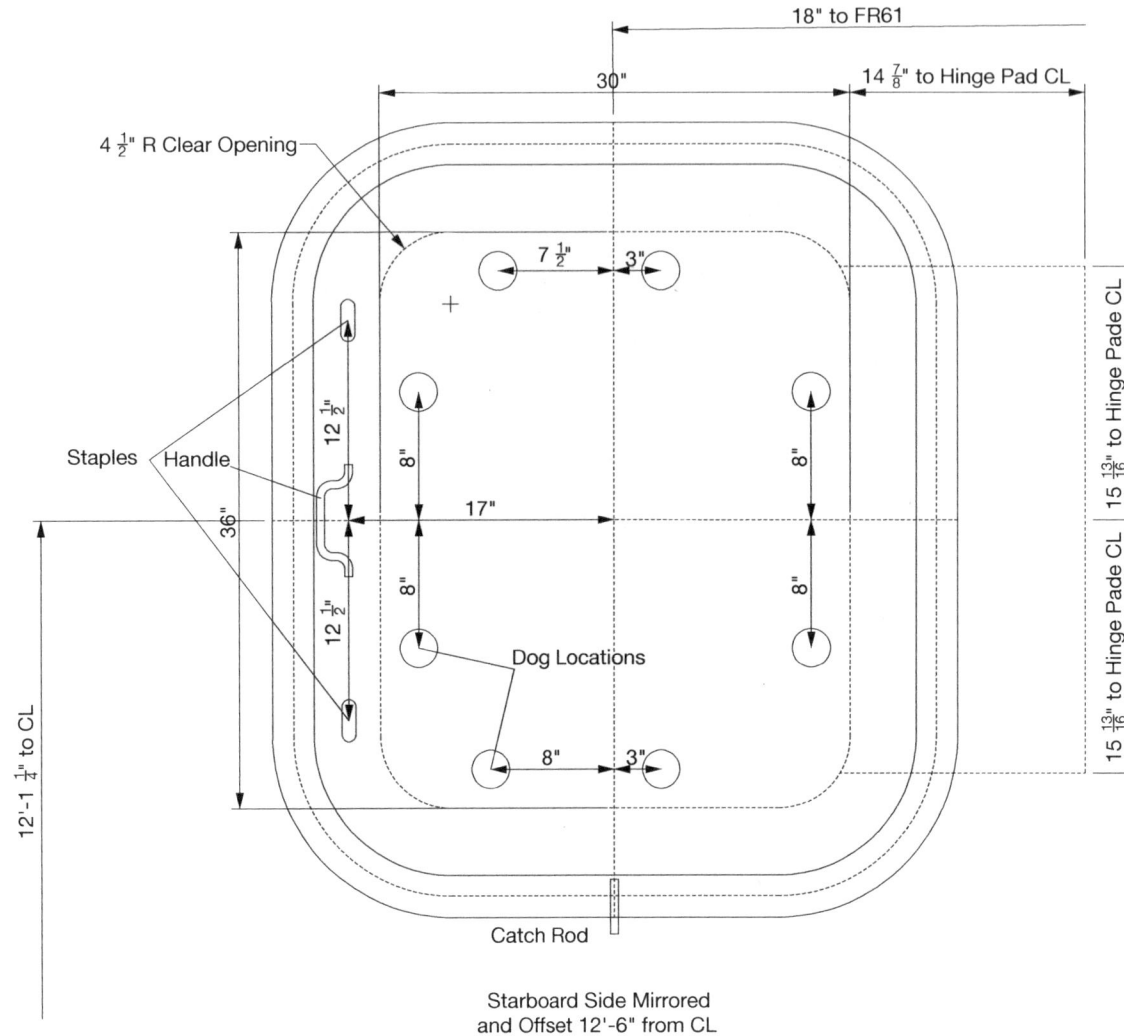

18" to FR61

30"

14 $\frac{7}{8}$" to Hinge Pad CL

4 $\frac{1}{2}$" R Clear Opening

7 $\frac{1}{2}$"

3"

Staples   Handle

12 $\frac{1}{2}$"

12 $\frac{1}{2}$"

36"

8"

8"

8"

8"

17"

8"

3"

Dog Locations

15 $\frac{13}{16}$" to Hinge Pade CL

15 $\frac{13}{16}$" to Hinge Pade CL

15 $\frac{13}{16}$" to Hinge Pade CL

12'-1 $\frac{1}{4}$" to CL

Catch Rod

Starboard Side Mirrored
and Offset 12'-6" from CL

Figure 10.5: Second Deck 30x36 FR61 Hatch (1:12). The port side hatch is shown. The starboard hatch is a mirror. These hatches access watertight trunks along Turret 1. Hatches above and below create an ammunition loading path running from main deck to second platform. These covers were made from plugs cut from 36x42 third deck armor hatch openings.

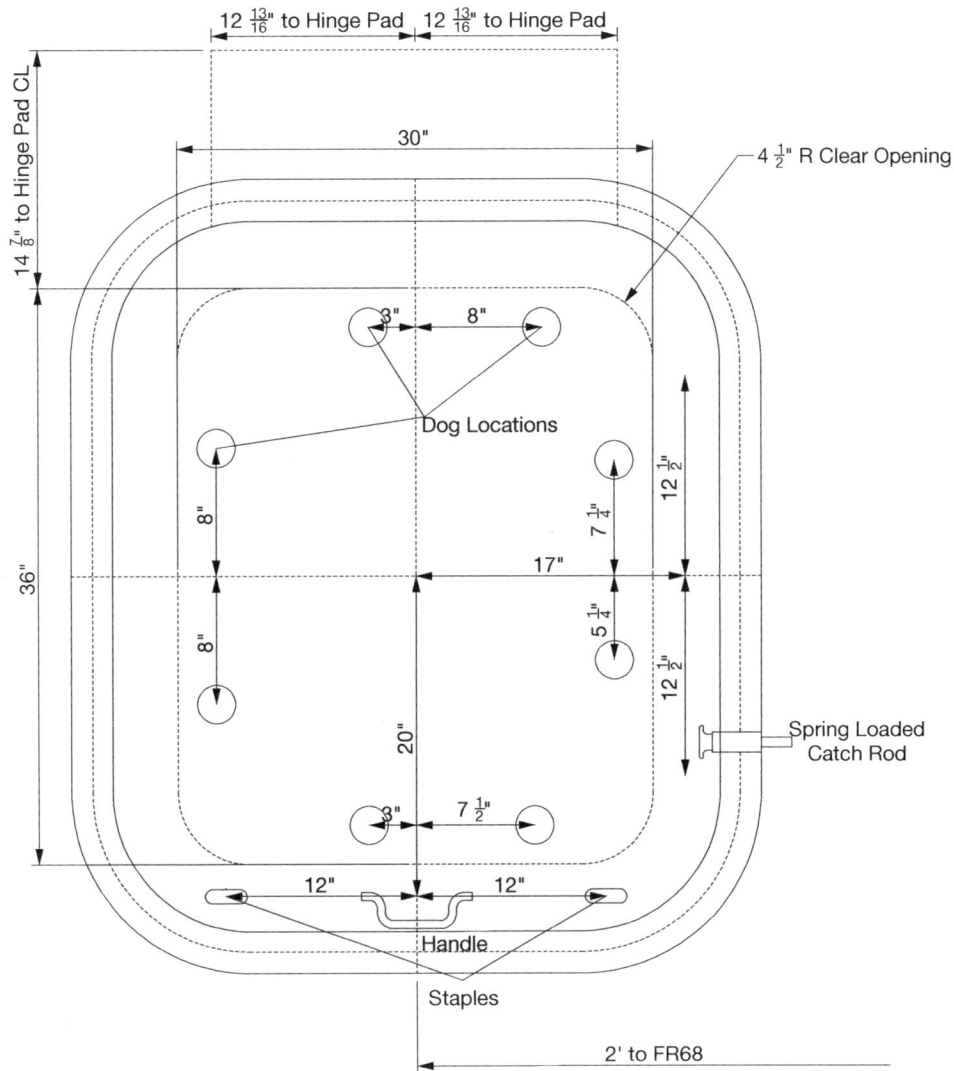

Figure 10.6: Second Deck 30x36 FR68 Starboard Hatch Cover (1:12). This hatch opens into a watertight trunk forward of Turret 2. This is part of a series of ammunition loading hatches that extends from main deck to second platform.

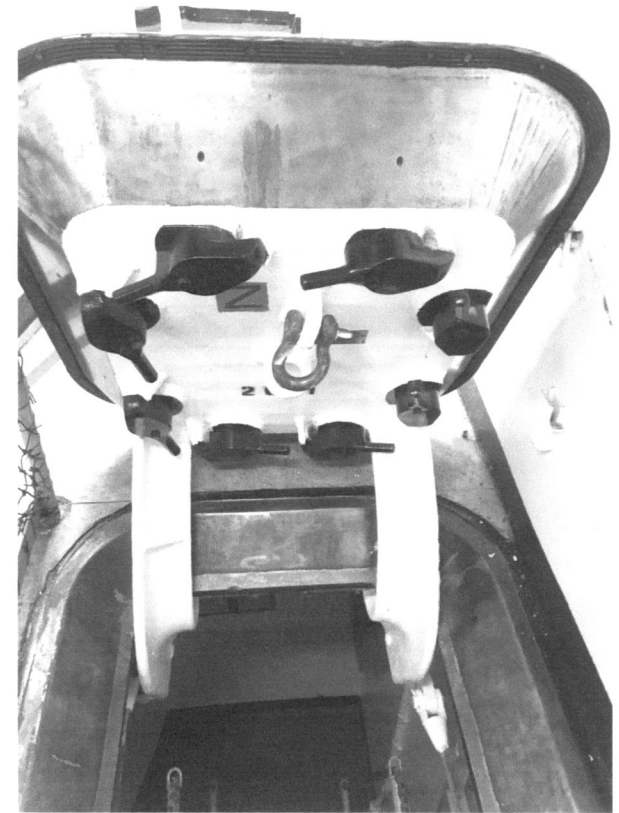

Photograph 10.3: FR68 Hatch Cover on USS New Jersey. The lubricating ports above the two closest dogs are visible. The hatch cover is being held open by a spring loaded catch rod extending into a fixtured attached to the bulkhead at the right. The eyes at the center and the rings at the base of the dogs are not shown on the plans. The stop wedges are much thicker than shown on the plans to compensate.

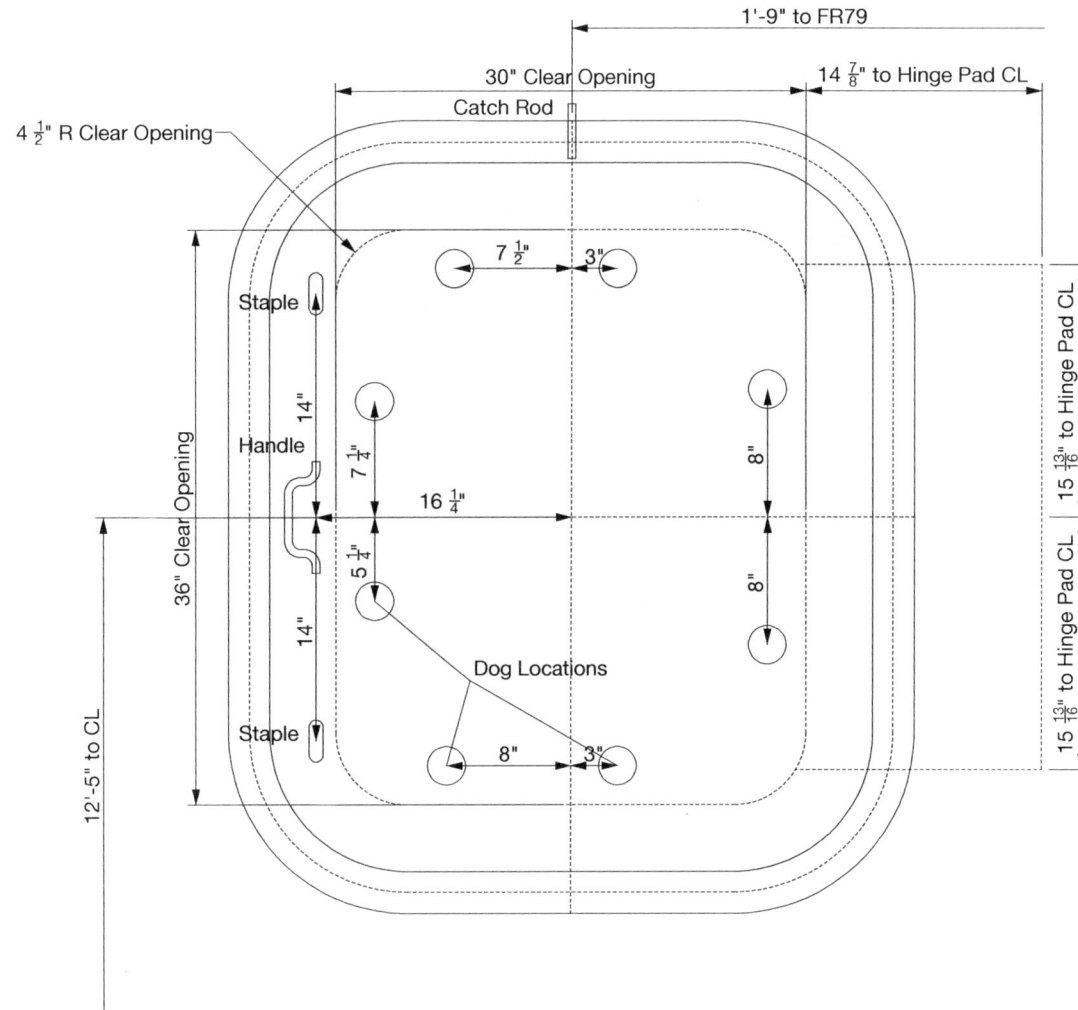

Figure 10.7: Second Deck 30x36 FR79 Port Hatch Cover (1:12). This hatch is part of a series of ammunition loading openings that runs from main deck (within the deckhouse) to second platform.

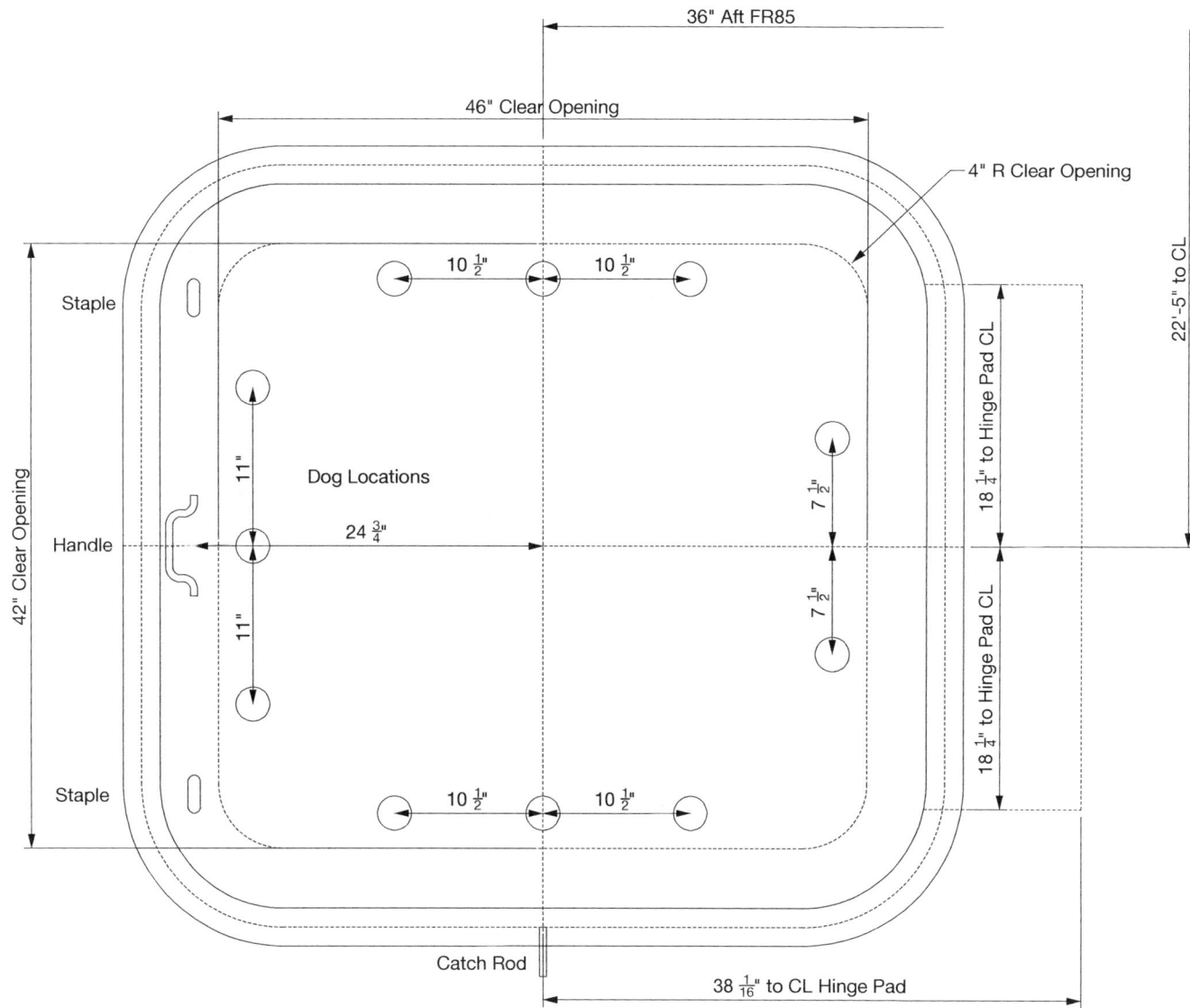

Figure 10.8: Second Deck 46x32 FR85 Starboard Hatch Cover (1:12). The larger hatch at this location provided the main access point from second deck to the forward end of the citadel.

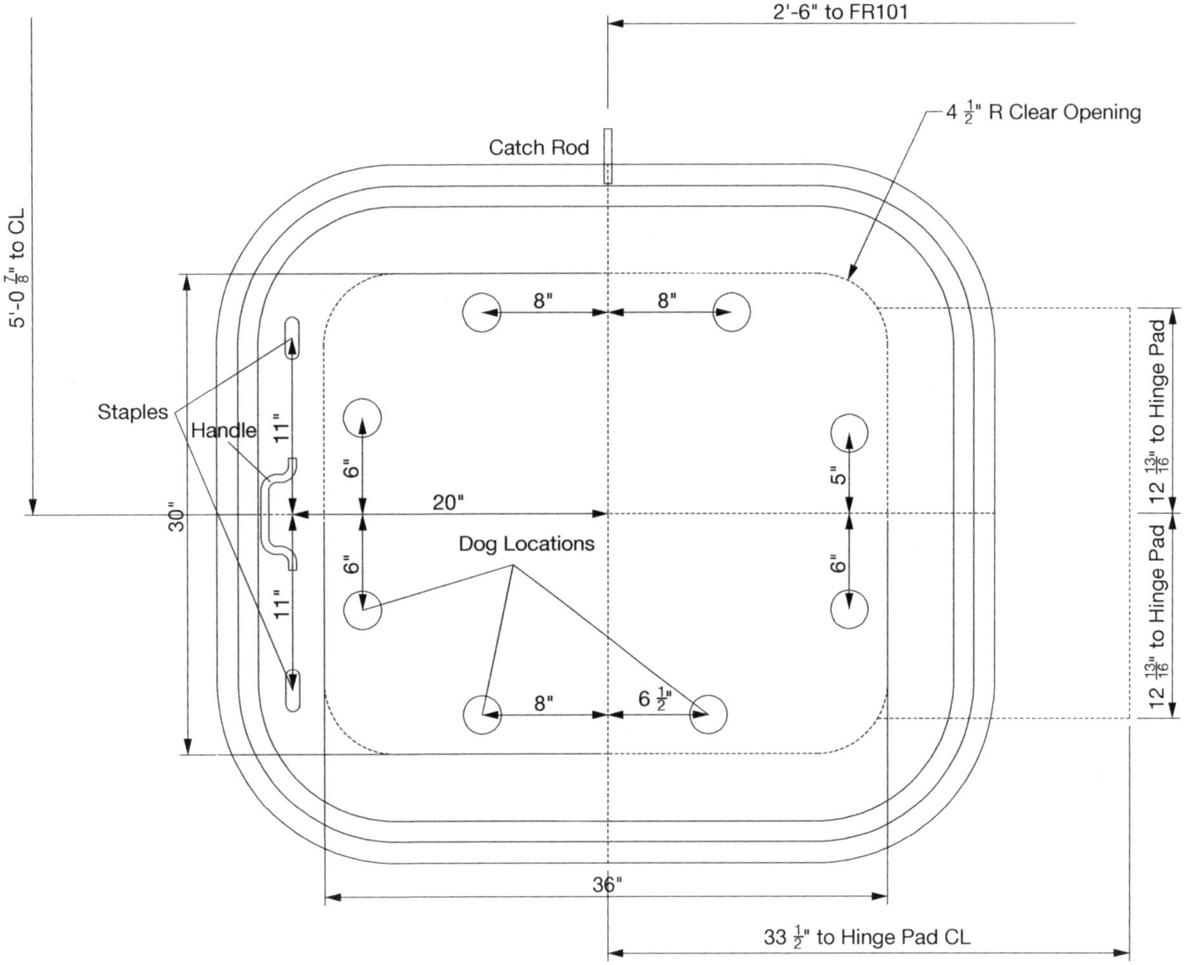

Figure 10.9: Second Deck 36x30 FR101 Starboard Hatch Cover (1:12). Each engine room compartment has a hatch leading from second deck to Broadway. This hatch is for engine room 1.

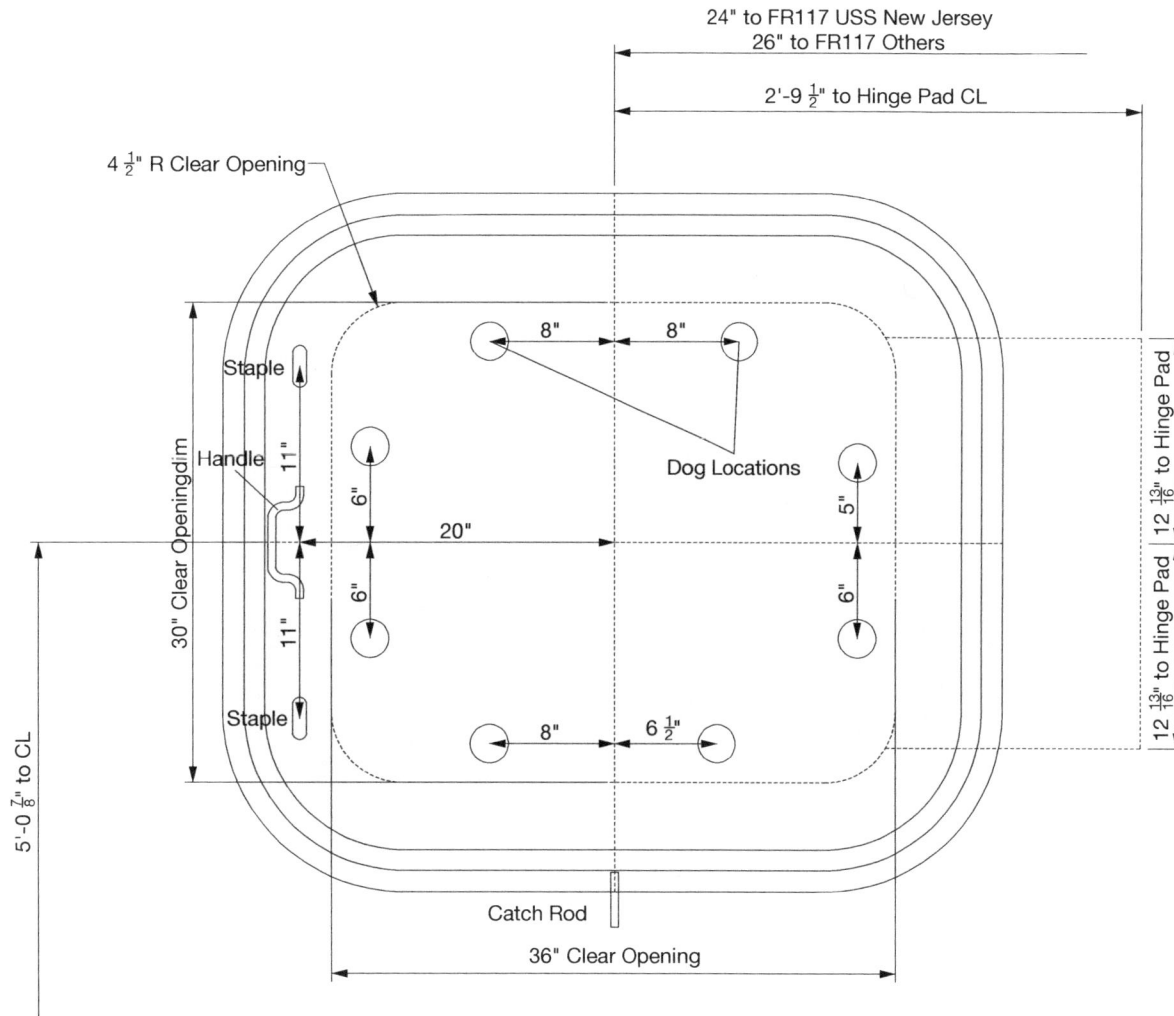

Figure 10.10: Section Deck 36x30 FR117 Port Hatch Cover (1:12). This hatch opens into the watertight trunk accessing engine room 2.

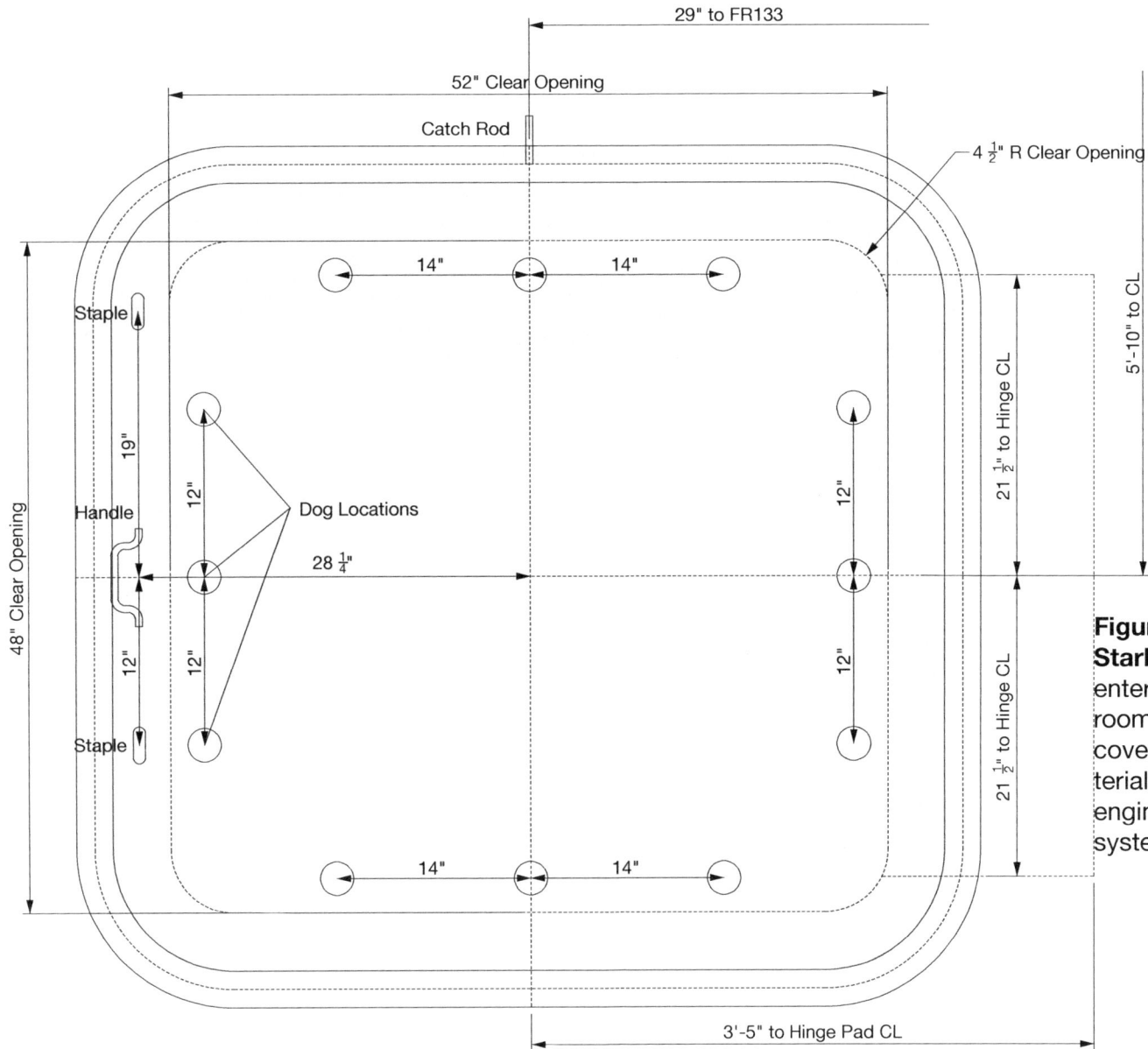

Figure 10.11: Second Deck 52x48 FR133 Starboard Hatch Cover (1:12). This hatch enters the watertight trunk accessing engine room 3. This is the largest armored hatch cover because it is was intended to move materials from the second deck workshop to the engineering spaces using the overhead trolley system.

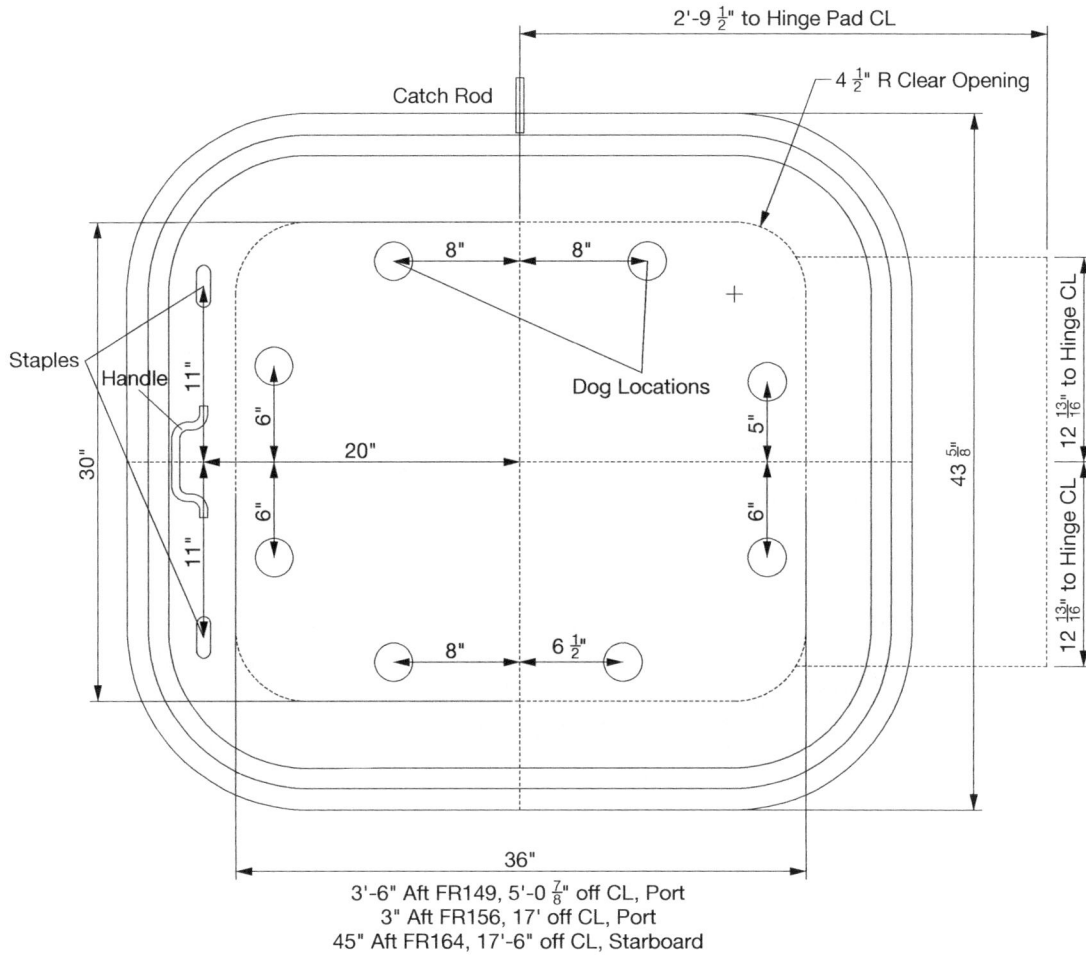

Figure 10.12: Second Deck FR149 Port, FR156 Port, and FR164 Starboard Hatch Covers (1:12). These three hatch covers are identical. However, the arrangement of their mounting equipement is different. The hatch at FR149 leads to the watertight trunk accessing engine room 4. The hatches at FR156 and FR164 are arranged as part of two series of hatches for ammunition loading that extend from main deck to Second Platform.

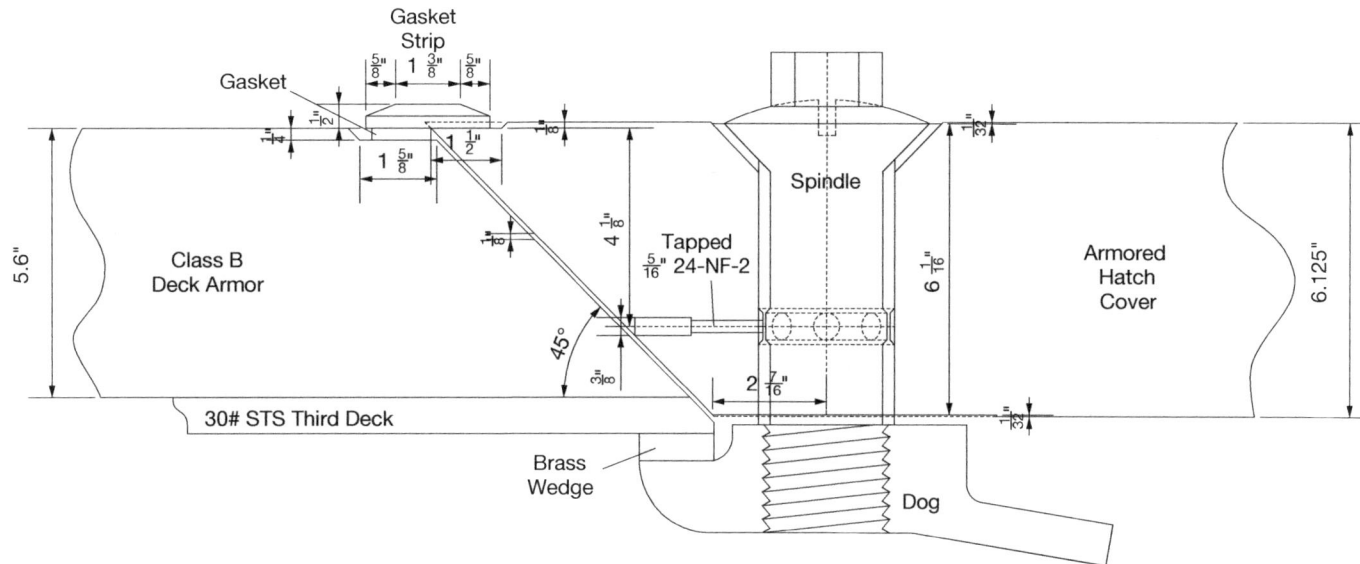

Figure 10.13: Cross Section of Third Deck Armored Hatch Cover (1:4). All third deck armored hatches have this cross section. It is very similar to those of the second deck except that the deck armor and hatch cover are thicker.

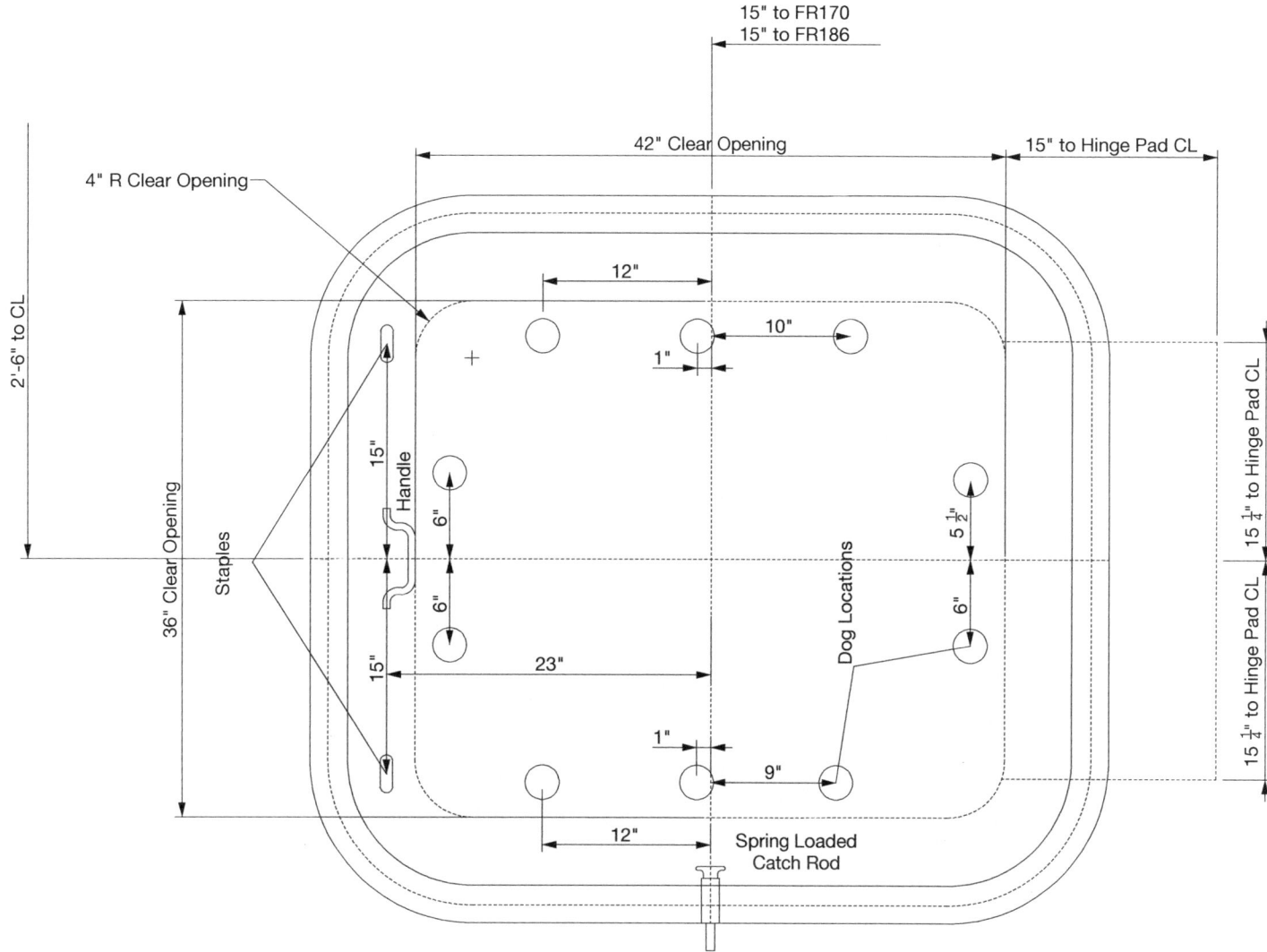

Figure 10.14: Third Deck 36x42 Hatch Cover FR169 and FR185 Starboard (1:12). These hatch access food storerooms. The hatch covers are identical but the external mounting fixtures are arranged slightly differently.

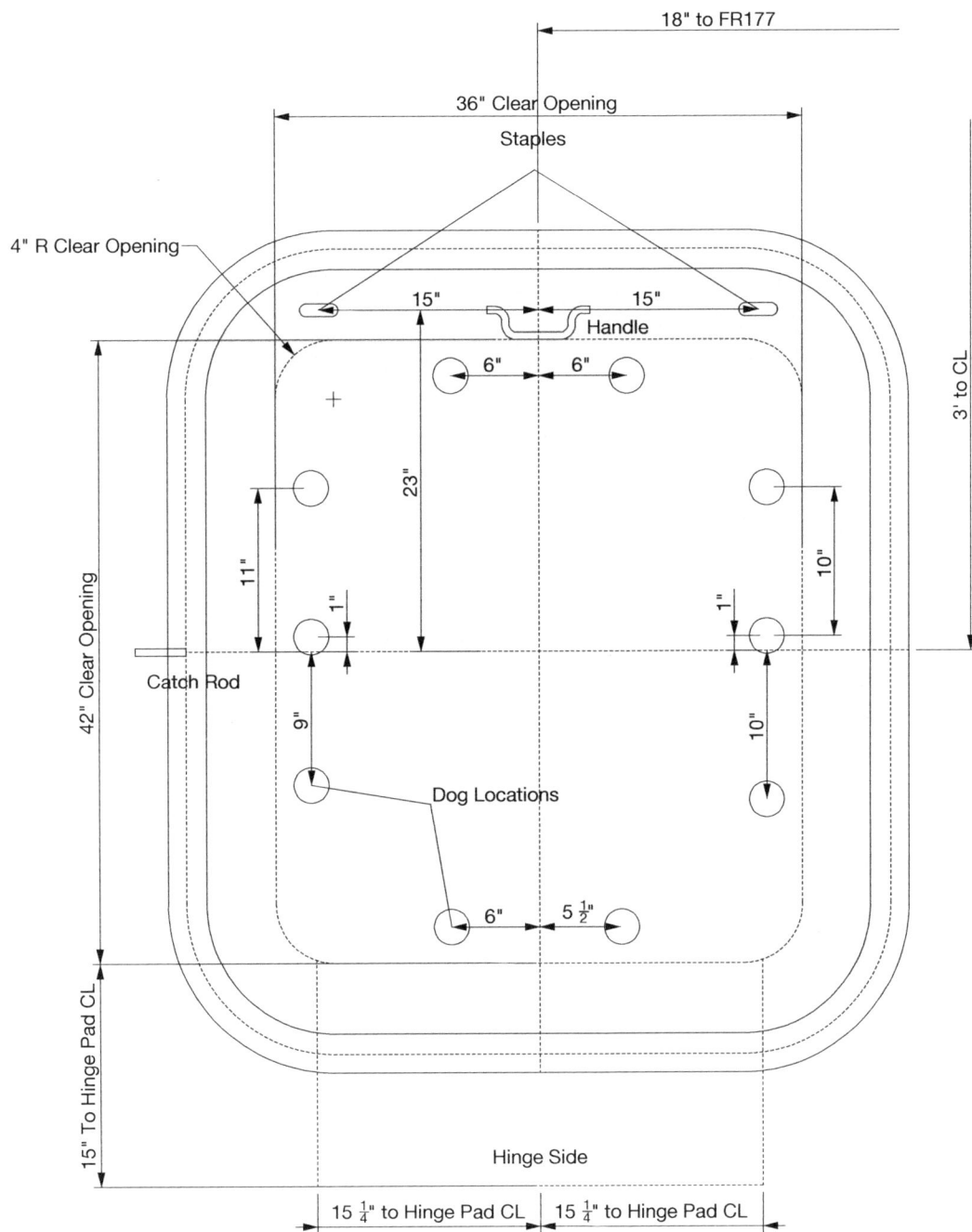

Figure 10.15: Third Deck 36x42 Hatch Cover FR177 Starboard (1:12). This hatch accesses food storerooms.

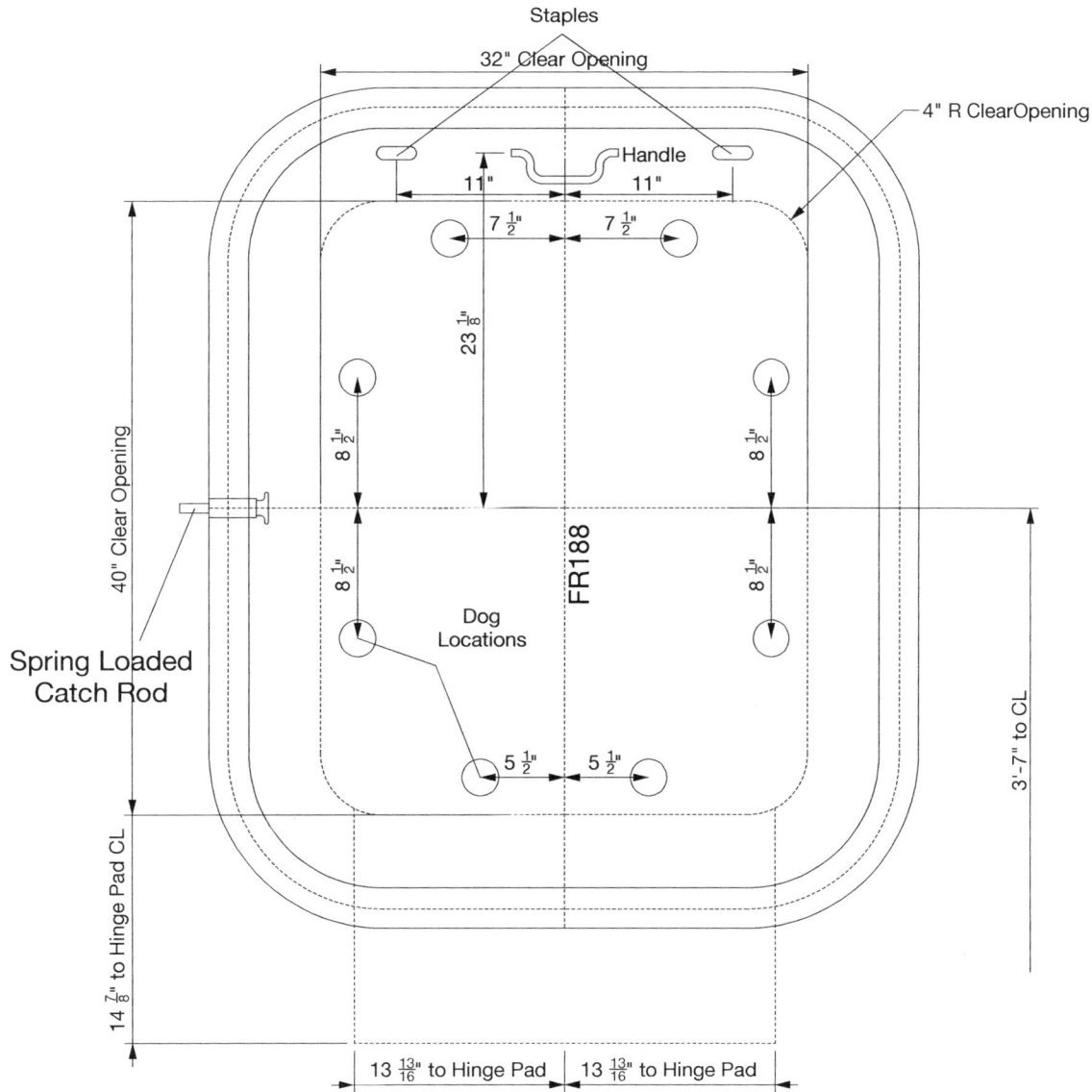

Figure 10.16: Third Deck 32x40 Hatch Cover FR188 Port (1:12). This hatch is located in the watertight trunk that gives direct access from Second Deck to the steering gear at First Platform. It is located just forward of where the deck armor increases from 5.6" to 6.2" There are no hatch openings in the 6.2" deck armor.

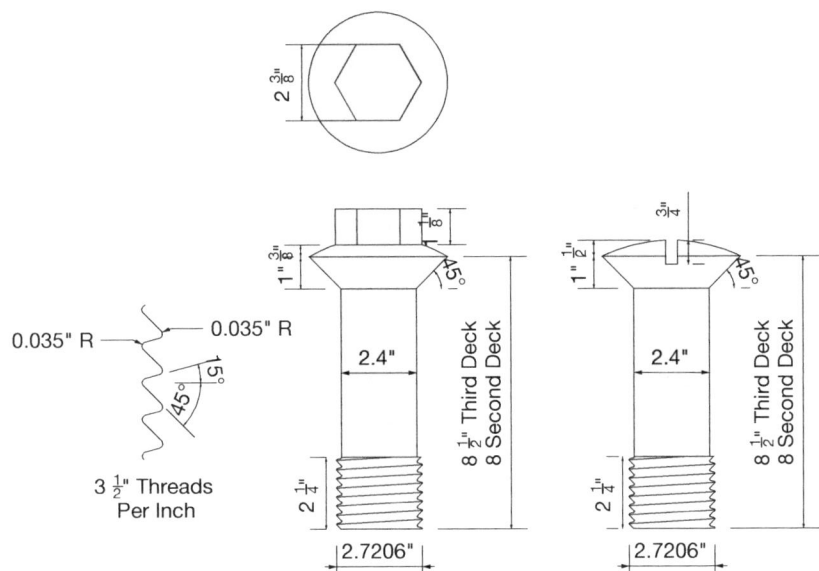

Figure 10.17: Spindles for the Dogs (1:6). At the left is a 4x enlargment of the thread. At the center is the hex head spindle designed for USS Missouri and USS Wisconsin (and found on USS New Jersey). At the right is the blade head spindle found on USS Iowa. It is not known whether the spindles were replaced on USS New Jersey or whether it was build differently from the plans. The second deck and third deck spindles differ only in length.

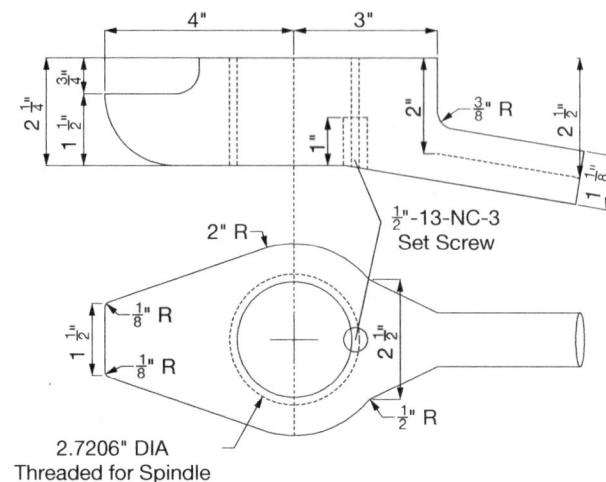

Figure 10.18: Dogs for Armored Hatch Covers (1:4). The dogs were screwed on the end of the spindles. A hole for a set screw was driled and tapped to keep the dog in place.

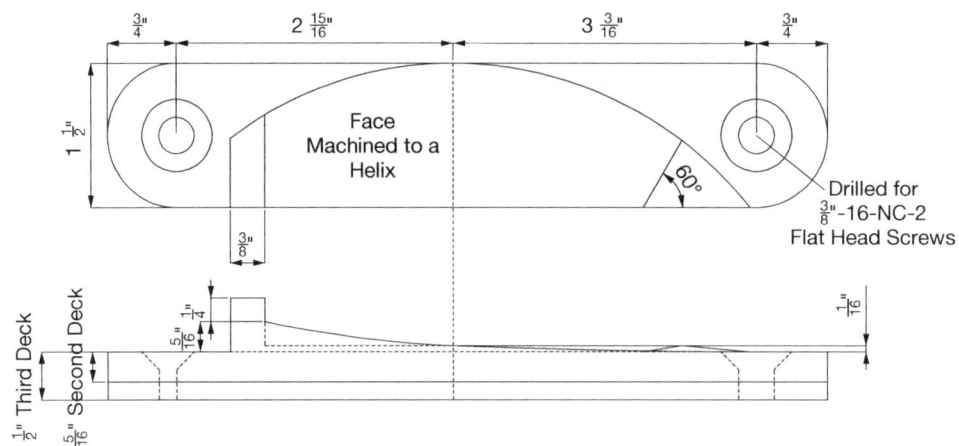

Figure 10.19: Wedge (1:2). Dog wedges are made of brass. The wedge face was machined to a helix with the wedge face parallel to the base at the reference line,

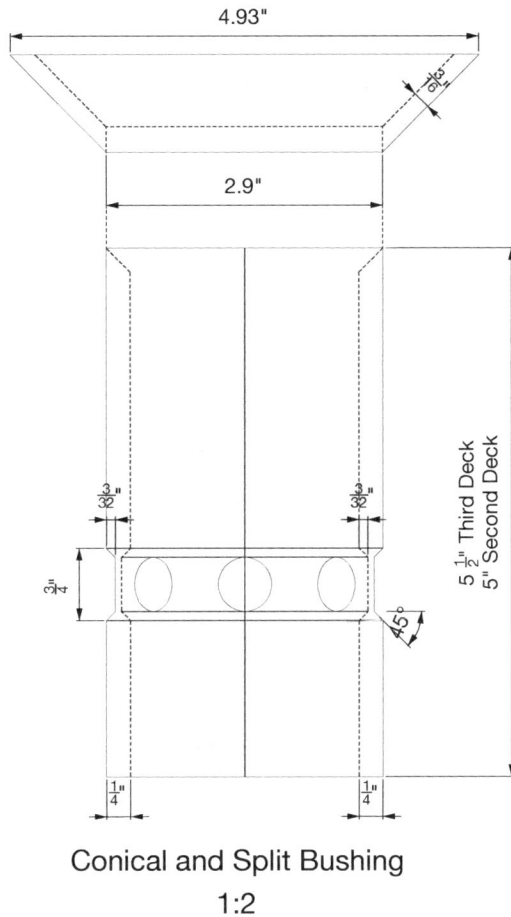

Conical and Split Bushing

1:2

Figure 10.21: Conical and Split Bushing (1:2). These bushings provided a smooth surface for the splindle to rotate. The split bushing first inside the conical bushing. The openings in the split bushing allow lubrication to reach the spindle.

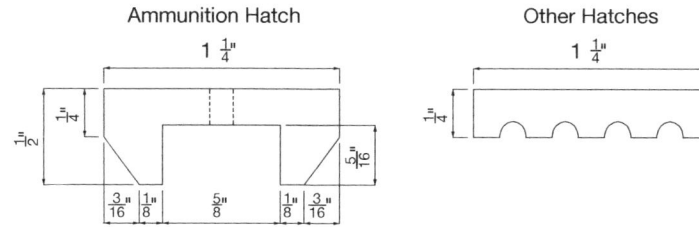

Ammunition Hatch   Other Hatches

Figure 10.20: Gaskets (1:1). The gasket for the circular ammunition hatch cover is at the left and the gasket for other armored hatch covers is at the right

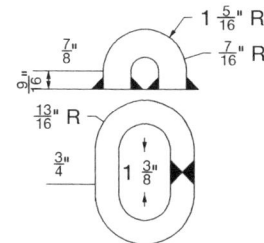

Figure 10.22: Staple and Link (1:6). Hatch covers had staples welded over a link to create an attachment point for lifting the hatch in case the counter-balancing springs failed.

Chapter 11: Plate Lifting and Bolts

This chapter consolidates the various devices used to crane plates into position and the various bolts used to attach plates because there is much duplication.

Most armor lacked natural attachment points to be lifted by a crane. Such attachments were created by drilling and tapping holes for screwing lifting eyes or lifting pads to an armor plate or group of armor plates. The screw holes were plugged and grinded flush once the plate was in position. Many of the lifting pads were reused from the earlier *North Carolina-* and *South Dakota-*classes.

This chapter also includes details of how armor was attached to backing bulkheads. Diagrams for the adaptors and bolts are here.

Finally, this chapter includes diagrams of various bolts used to join armor plates into assemblies.

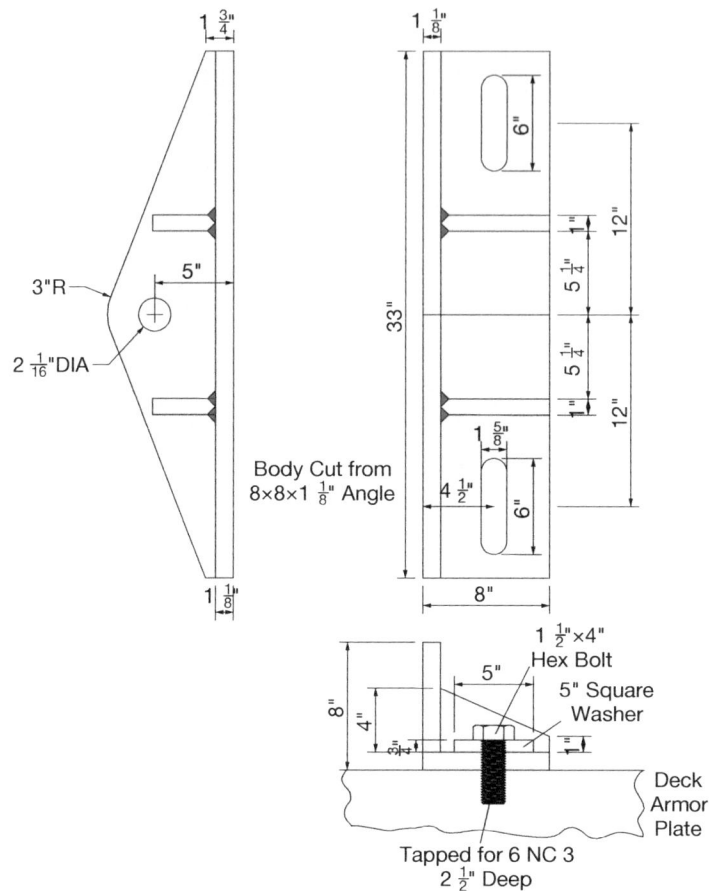

Figure 11.1: Clips for Moving Deck Armor Plates (1:12). This type of clip was used to create attachment points for craning deck armor plates.

Figure 11.2: Typical Position of Clips on a Deck Armor Plate (1:72). After the plate was in position, the clips were removed. For plates using quilting pins, the bolt holes were reamed for quilting pins. For plates without quilting pins, the holes were plugged with $1\frac{1}{2}$" 6 NC 3 bolts.

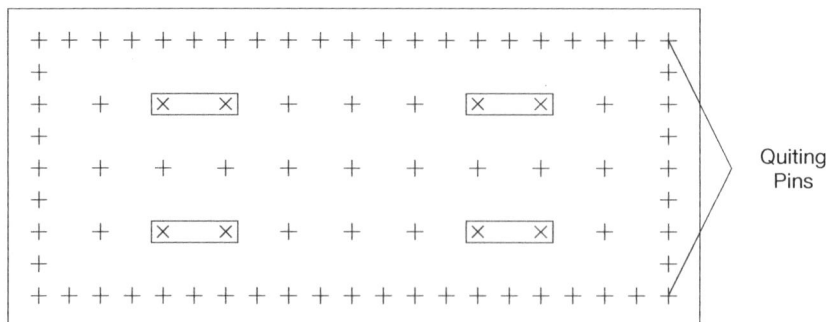

Plan 11.1: (1302BV). The Type "E" eye bolt was used for the upper conning tower tube.

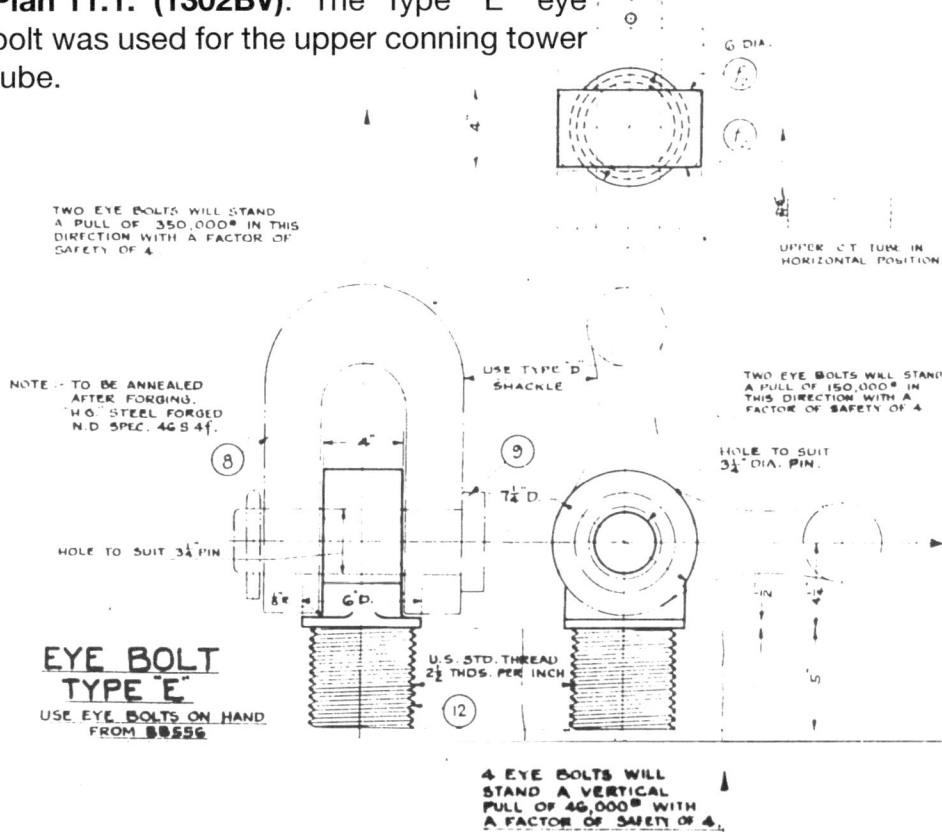

TWO EYE BOLTS WILL STAND A PULL OF 350,000# IN THIS DIRECTION WITH A FACTOR OF SAFETY OF 4.

UPPER CT TUBE IN HORIZONTAL POSITION

NOTE:- TO BE ANNEALED AFTER FORGING. H.S. STEEL FORGED N.D. SPEC. 46 S 4f.

USE TYPE "D" SHACKLE

TWO EYE BOLTS WILL STAND A PULL OF 150,000# IN THIS DIRECTION WITH A FACTOR OF SAFETY OF 4

HOLE TO SUIT 3¼" DIA. PIN.

HOLE TO SUIT 3¼" PIN

U.S. STD. THREAD 2½ THDS. PER INCH

EYE BOLT TYPE "E"
USE EYE BOLTS ON HAND FROM BB556

Plan 11.2: (1302BV). The Type "R" eye bolt was used for conning tower doors.

DIE 1098

FOR INSTALLING DOORS EYE-BOLTS TO BE PLACED IN POSITION SHOWN.

CT. SIDE PLATE

PIN FOR EYE BOLT ON INNER SURFACE

U.S. STD THREAD 6 THDS. PER INCH NC-3 FIT.

METHOD OF INSTALLING DOOR.

EYE BOLT TYPE "R"

USE EYE BOLTS FOR INSTALLING CONNING TOWER DOORS.

ENLARGED SECT. OF THREAD FOR 3"⌀ DIA. BOLTS

TYPE "D" SHACKLE

HOLE TO SUIT 3½" DIA. PIN

TAPERED LINER

EYE BOLT TYPE "P"

4 EYE BOLTS WILL STAND A VERTICAL PULL OF 46,000# WITH A FACTOR OF SAFETY OF 4.

Plan 11.3: The Type "G" eye bolt was used for the top and bottom plates of the conning tower.

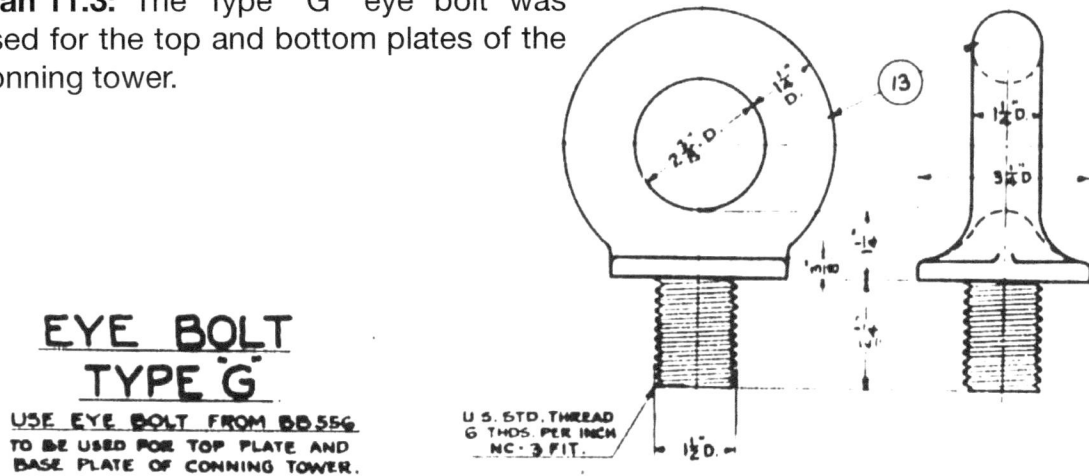

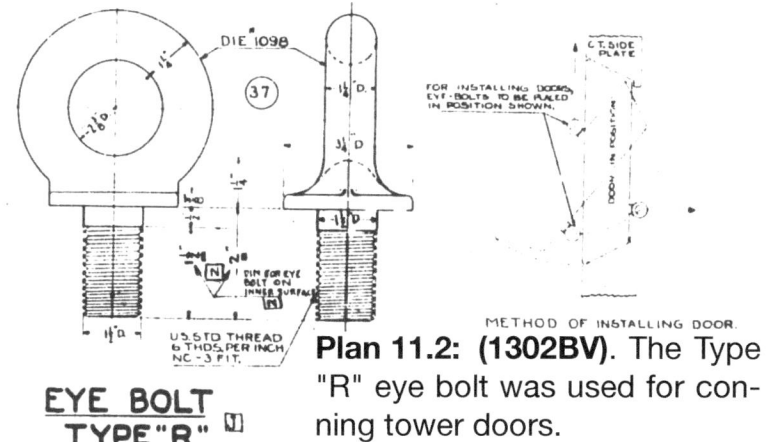

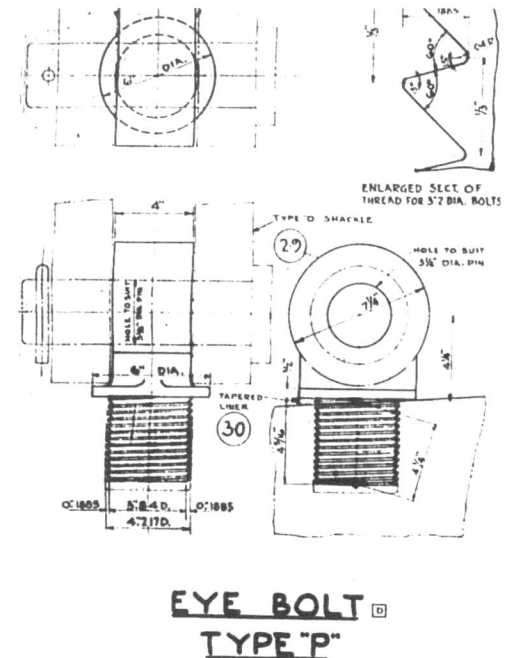

EYE BOLT TYPE "G"

USE EYE BOLT FROM BB556 TO BE USED FOR TOP PLATE AND BASE PLATE OF CONNING TOWER.

U.S. STD. THREAD 6 THDS. PER INCH NC-3 FIT.

Plan 11.4: (1302BV). The Type "P" eye bolt was used for the FR50 transverse plates on USS Missouri, USS Wisconsin, USS Illinois, and USS Kentucky.

TYPE "D" SHACKLE & PIN
FOR TYPE "E" & "P" EYE BOLT & TYPE "Q" PAD

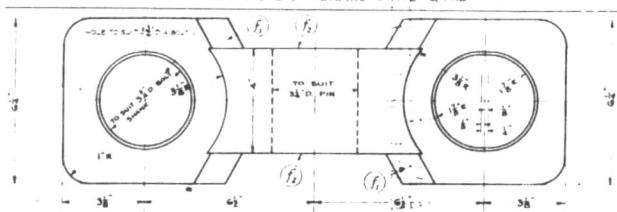

TYPE "J" SHACKLE

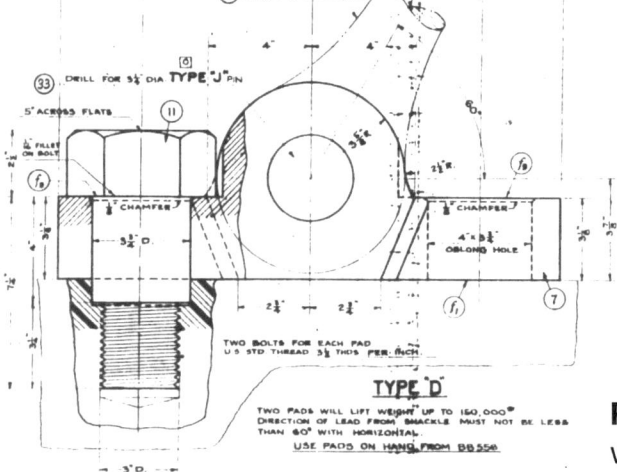

TYPE "D"

TWO PADS WILL LIFT WEIGHT UP TO 150,000#
DIRECTION OF LEAD FROM SHACKLE MUST NOT BE LESS
THAN 60° WITH HORIZONTAL.
USE PADS ON HAND FROM BB556

Plan 11.6: (1302BV). The Type "C" pad was used for the lower plate of the FR50 transverse armor (B-50-6)

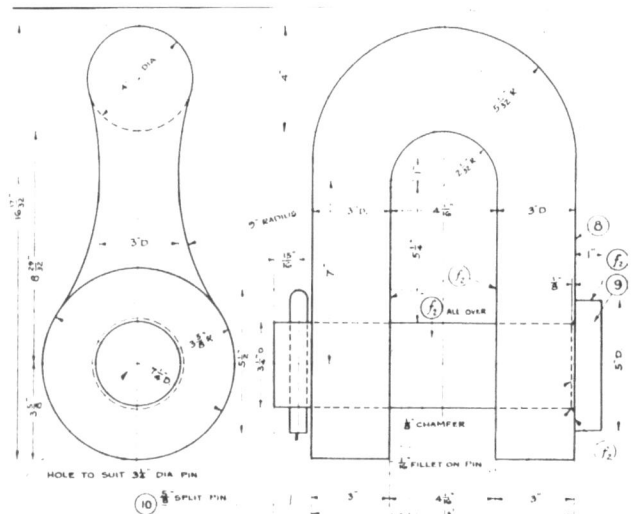

TYPE "C"

TWO PADS WILL LIFT WEIGHT UP TO 88,000#
DIRECTION OF LEAD TO SHACKLE MUST NOT BE LESS
THAN 60° WITH HORIZONTAL.
USE PADS ON HAND FROM BB556

Plan 11.5: (1302BV). The Type "D" pad was used for the upper course of the FR50 transverse armor for USS Iowa and USS New Jersey and the FR166 transverse armor.

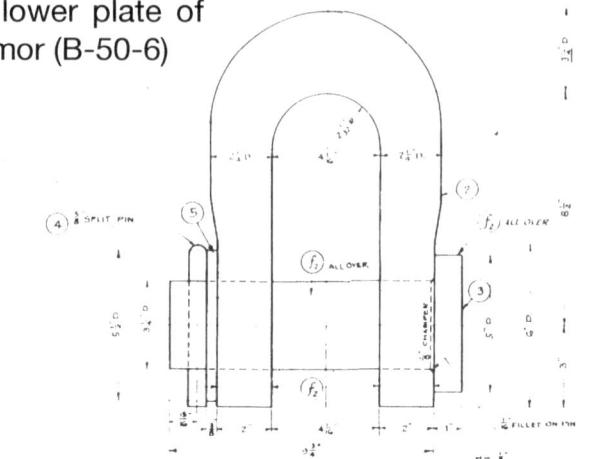

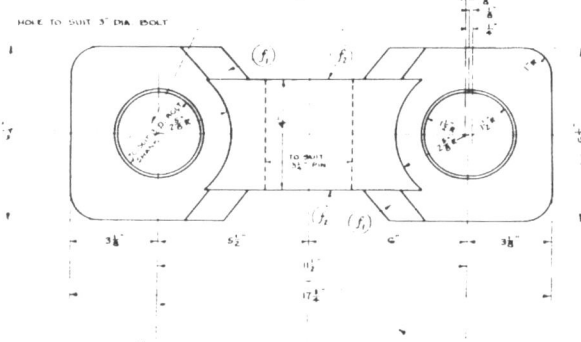

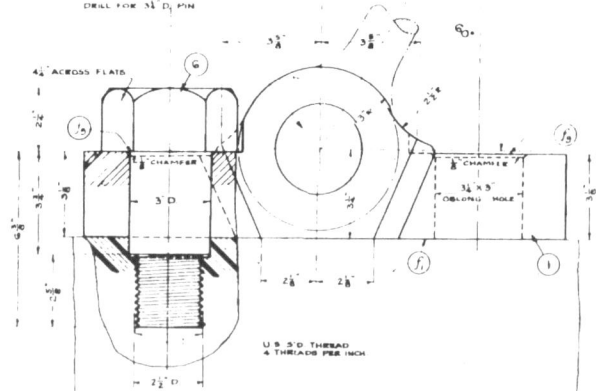

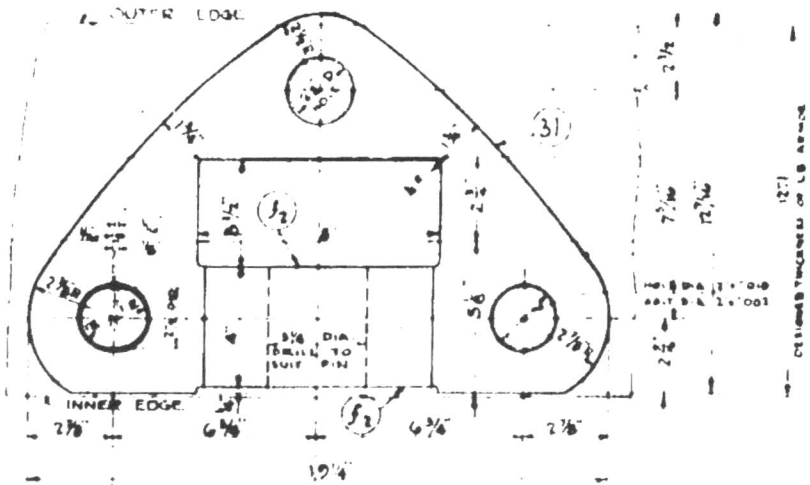

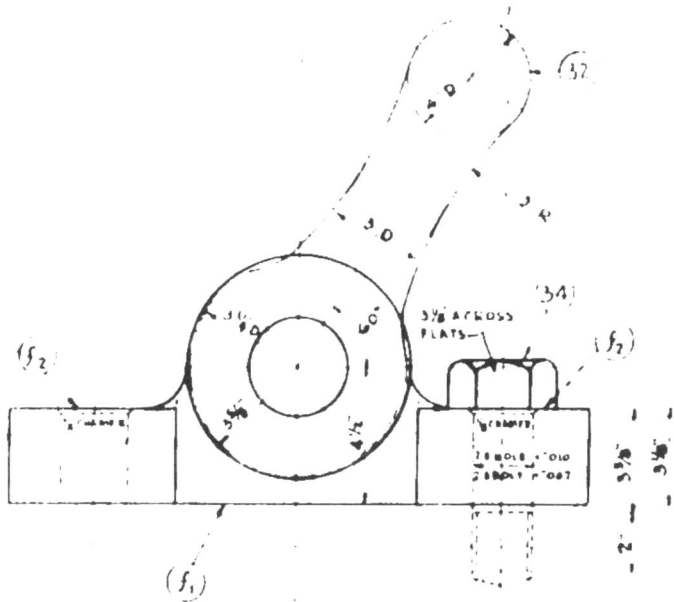

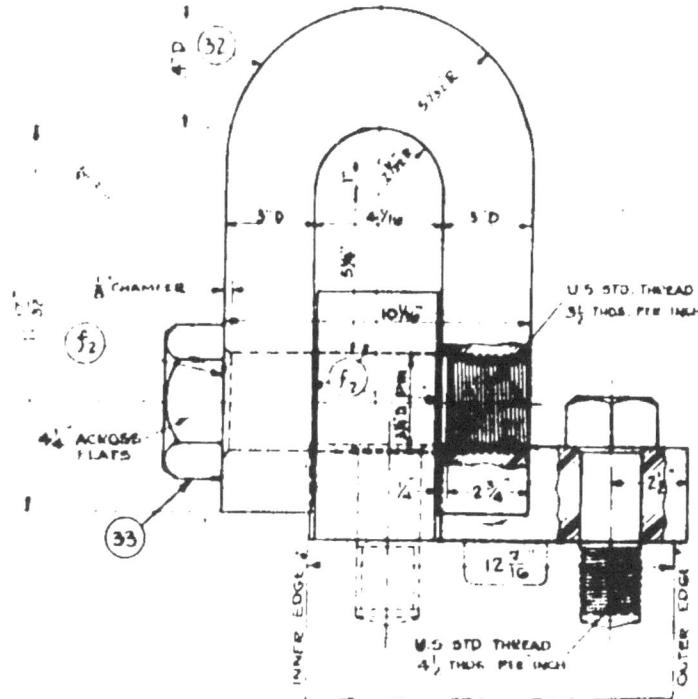

Plan 11.7: (1302BV). The Type "J" pad was used for lower belt plates.

TYPE "J"

TO BE USED FOR LOWER SIDE ARMOR PLATES

Plan 11.8: (1302BV). The Type "K" pad was used for the belt between FR166 and FR203 and the transverse armor at FR203.

Plan 11.9: (1302BV). The Type "M" pad was used for the side plates. When a Type "D" pad shackle was used it was called a Type "Q" pad used for the lower conning tower tube.

NC-3
4 $\frac{1}{2}$ Theads
Per Inch

Plug
Steering Gear Belt
FR166–F189
Type K Pad

Plugs Grinded Flush to Armor
After Installation

U.S. STD
4 Threads
Per Inch

Conning Tower Sides
Type M Pad

US STD
6 Threads/Inch

Conning Tower Top
and Bottom Plate
Type G Lifting Eye

2 $\frac{31}{32}$" DIA

U.S. STD
4 Threads
Per Inch

Plate B-50-6
Type C Pad

0.1443"

0.024"R

0.222"

0.222"

0.222"

30°

30°

0.024"R

Enlarged Thread
For Turret Roof

Turret Roof
Pads Reused from USS North Carolina

Plug
Belt FR166–FR189

NC-3
4 $\frac{1}{2}$ Theads
Per Inch

Common
Bolt Hole
Type K Pad

Plug
FR203 Transverse

U.S. STD
3 $\frac{1}{2}$ Threads
Per Inch

Upper Belt, FR50 (Iowa, New Jersey)
and FR166 Transverse
Type D Pad

Plug Weighs 9 lbs.

U.S. STD 2 $\frac{1}{2}$
Threads/Inch

2" DIA, 4 $\frac{1}{2}$ Threads Per Inch
STD Thread NC-3
FR203 Transverse Plate

Holes Drilled Normal to the
Face for Belt Plates and
Vertically for Transverse Plate

2" DIA, 4 $\frac{1}{2}$ Threads Per Inch
STD Thread NC-3
FR189–FR203 Belt

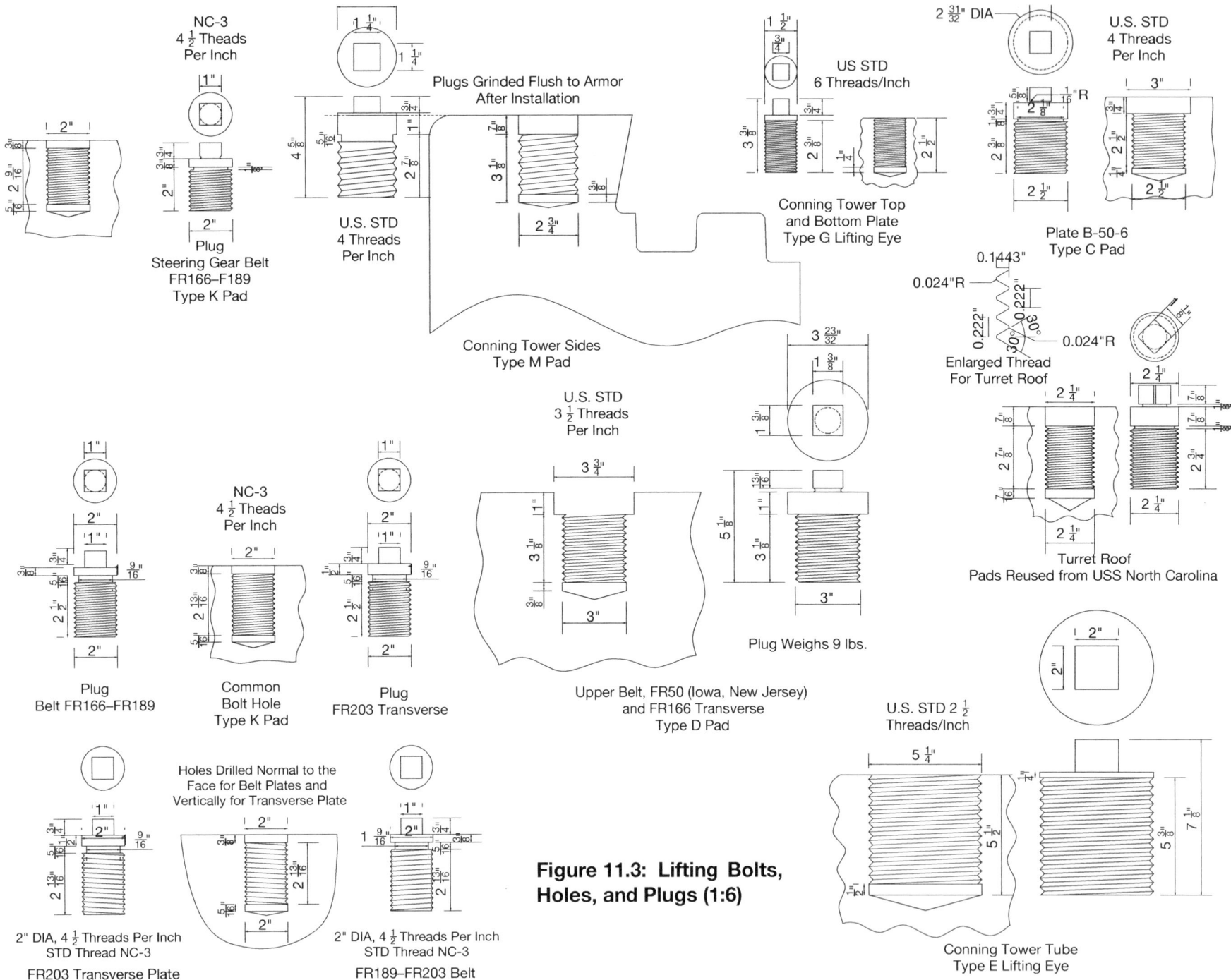

**Figure 11.3: Lifting Bolts,
Holes, and Plugs (1:6)**

Conning Tower Tube
Type E Lifting Eye

Figure 11.4: Brass Adapter Plugs (1:2). Three sizes of adaptor plugs were used for installing armor. The adaptor size is the diameter of the bolt used to secure the armor at the opening in the backing bulkhead.

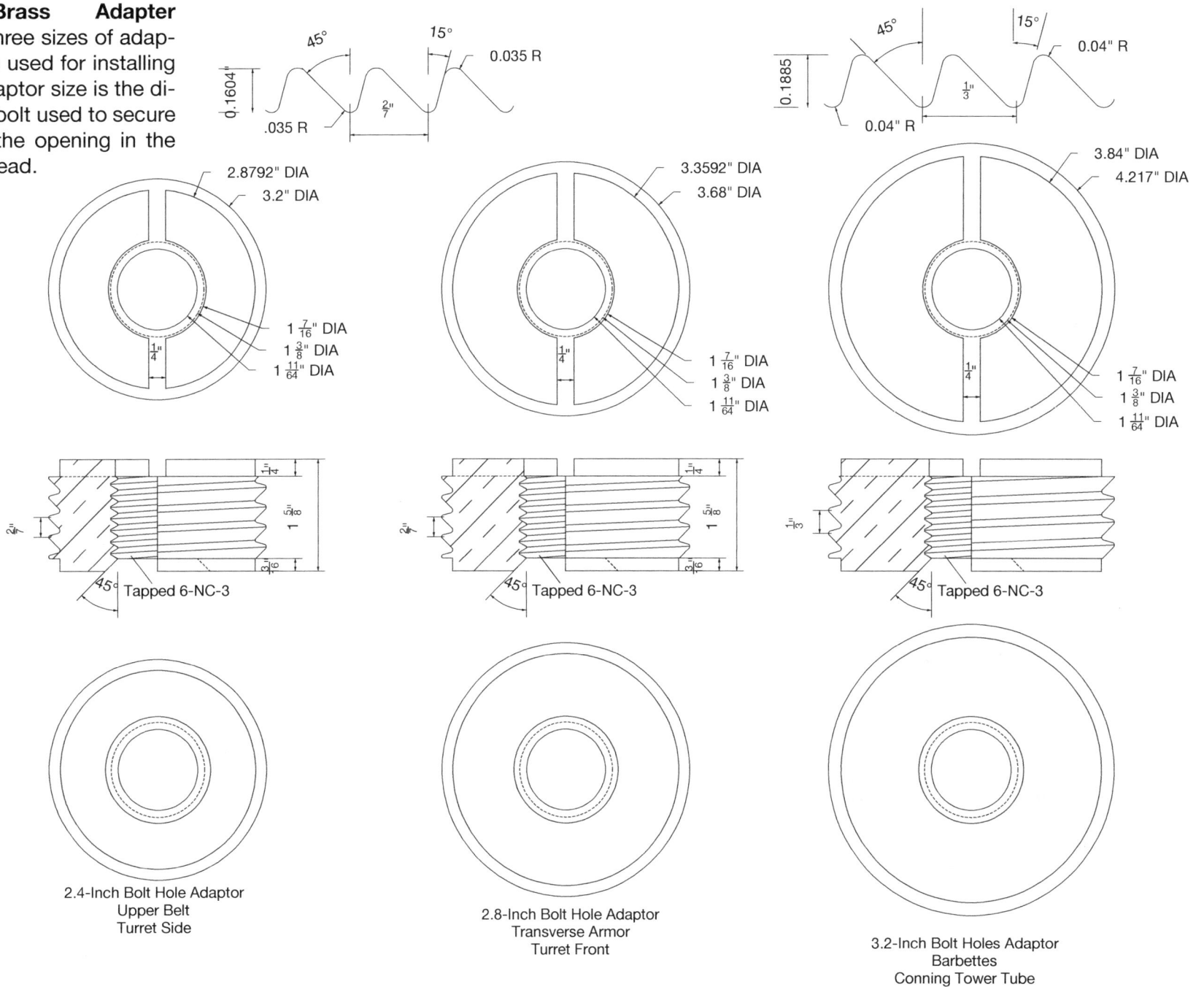

2.4-Inch Bolt Hole Adaptor
Upper Belt
Turret Side

2.8-Inch Bolt Hole Adaptor
Transverse Armor
Turret Front

3.2-Inch Bolt Holes Adaptor
Barbettes
Conning Tower Tube

Draw bolts were placed in every other bolt hole during upper belt positioning.

Draw Bolt With 2.4-Inch Adaptor Plug Upper Belt Only

Threaded 6-NC-3

Washer

Backing BHD

2.4" Adaptor Plug

Upper Belt

Marking Plug for 2.4-Inch Adaptor Plugs

Marking Plug for 2.8- and 3.2-Inch Adaptor Plugs

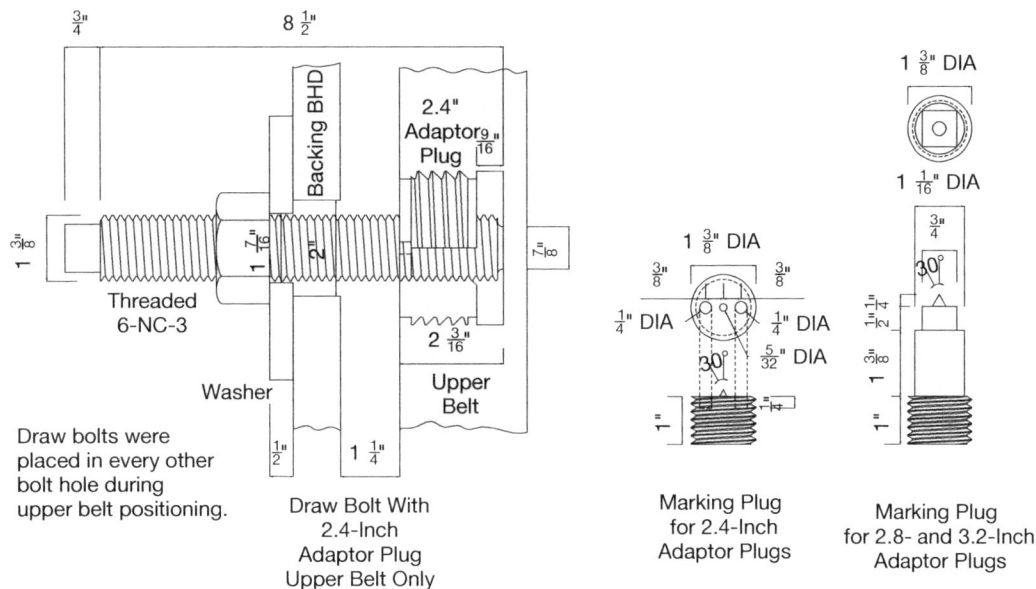

Figure 11.5: Use of Adaptor Plugs (1:4). Various devices could be screwed into adaptor plugs. The draw bolt was used to pull an armor plate toward its backing plate. Marking plugs were used to create templates for drilling holes in backing bulkheads.

Wire Rope

Threaded 5 NC 3

Wire Rope Socket

Wire Rope Socket

Adaptor Plug for Wire Rope Socket and 2.4-Inch Bolt Hole

Figure 11.6: Wire Rope Socket and Adaptor Plug (1:6). Wire ropes could be screwed into these adaptor plugs. Four wire ropes were attached at the top and bottom row of each upper belt plate. The ropes were passed through the corresponding holes in the backing plate and were used to pull the plate towards the backing plate.

Four Adjusting Bolts were used to position each armor plate. These were located 6"–9" inches below the upper corner bolt holes in the plate and above the lower corner bolt holes.

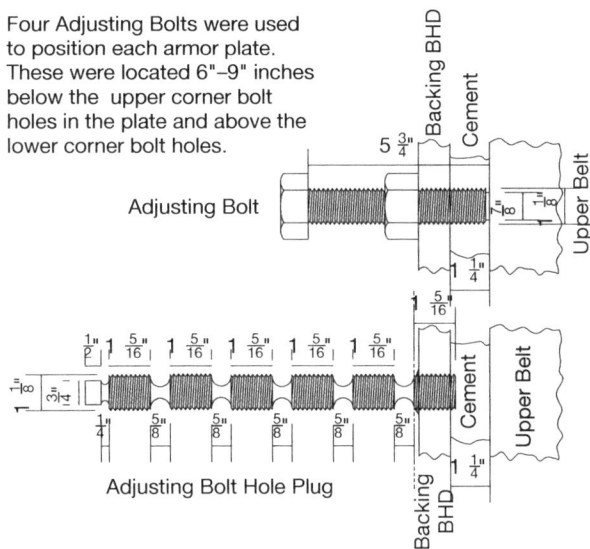

Adjusting Bolt

Backing BHD

Cement

Upper Belt

Adjusting Bolt Hole Plug

Backing BHD

Cement

Upper Belt

Figure 11.7: Adjusting Bolts and Plugs (1:6). Adjusting bolts pushed the armor away from the backing plate. After the plate was in position and cement filled the gap, the adjusting bolt was removed and a plug was screwed into the hole and grinded flush. The plugs were machined from a long screw so that each screw produced multiple plugs.

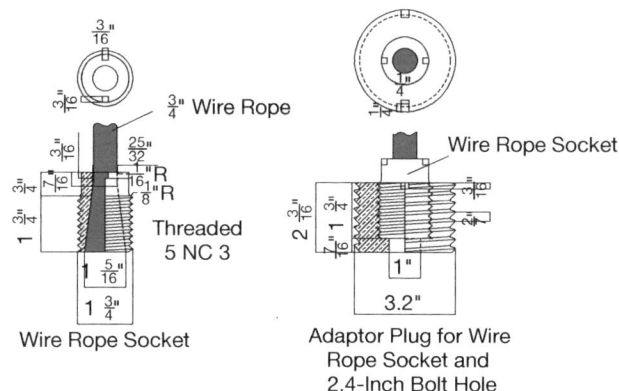

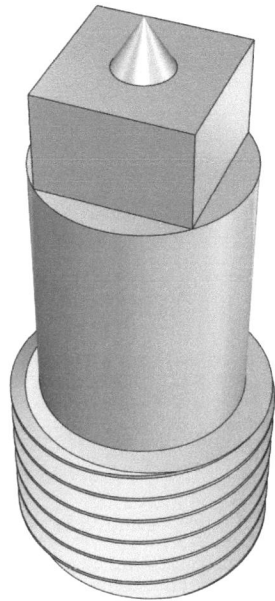

Figure 11.8: Marking Plug.

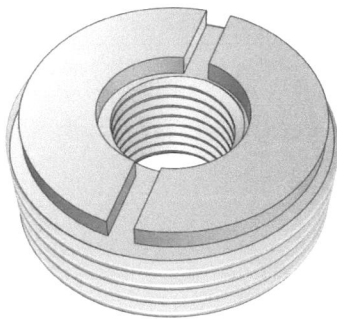

Figure 11.10: Rending of a 2.4-inch adaptor plug. The other sized plugs look similar.

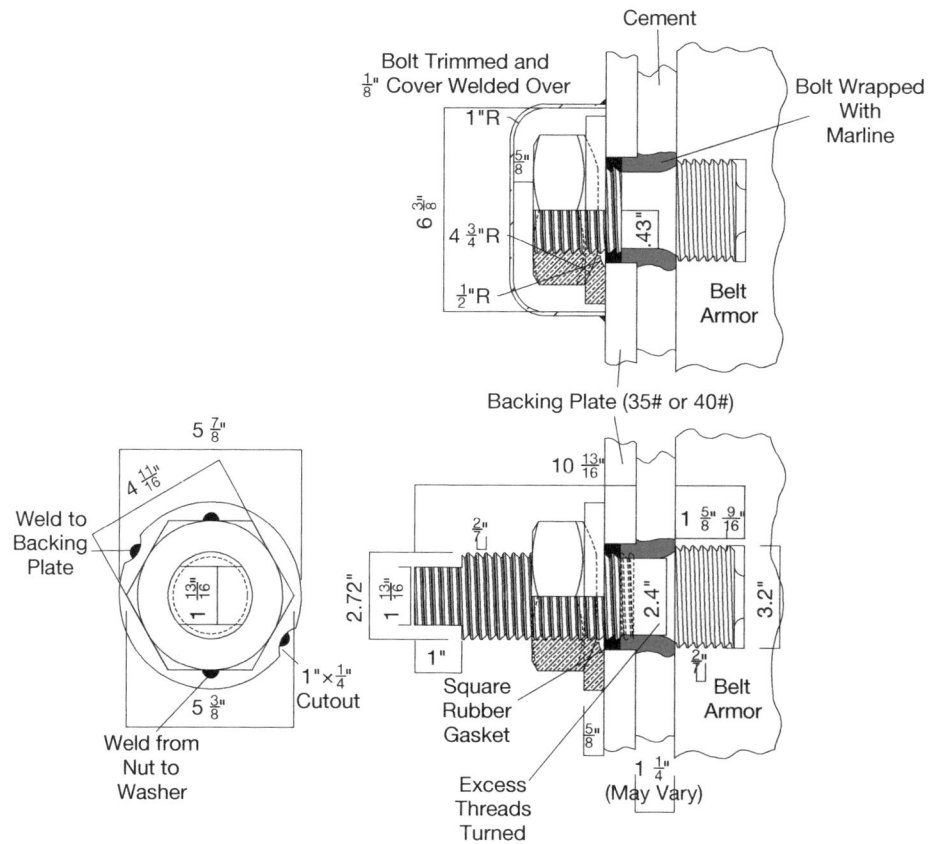

Figure 11.9: 2.4-Inch Bolt Connecting Belt Armor to Backing Plate (1:6).
This figure shows the final connection of the belt armor to the backing plate. The opening in the backing plate was drilled out from 2-inches to 3.2-inches. The 2.4-inch adaptor plug could be removed through the larger opening. The 2.4-inch bolts were machined to remove the threads where they would contact cement. The bolt was screwed in place with a nut and washer. After cement was poured into the gap, final adjustments could be made. The excess bolt was trimmed. The nut and washer were welded together and to the backing plate. Finally a cover was welded over the bolt.

Figure 11.11: 2.4-Inch Bolt Connecting Turret Armor to Backing Plate (1:6)

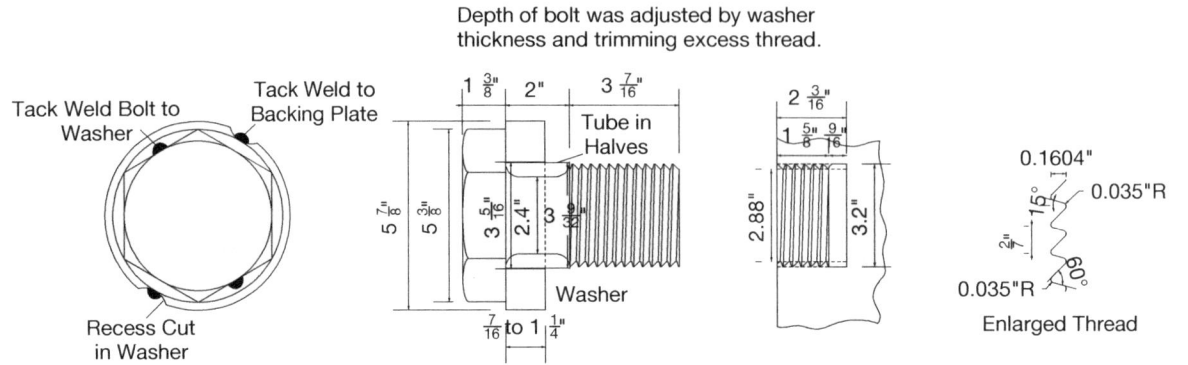

Depth of bolt was adjusted by washer thickness and trimming excess thread.

438 Per Ship

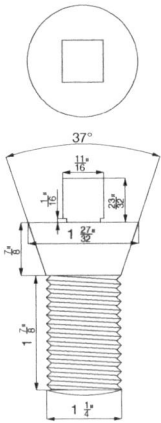

H Tap
7-NC-3

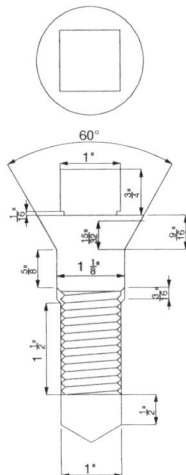

1 1/8" SPEC Tap
8-NC-3

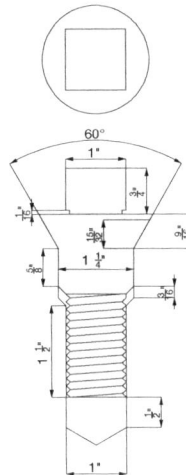

1 1/4" SPEC Tap
8-NC-3

Taps for Steering Gear Armor
1:2

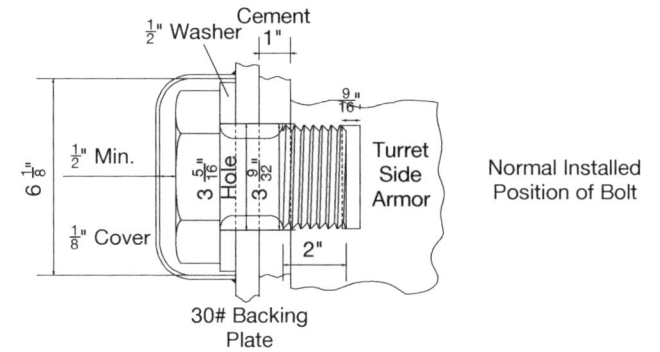

Normal Installed
Position of Bolt

Figure 11.12: Tap Bolts for Steering Gear Armor (1:2)

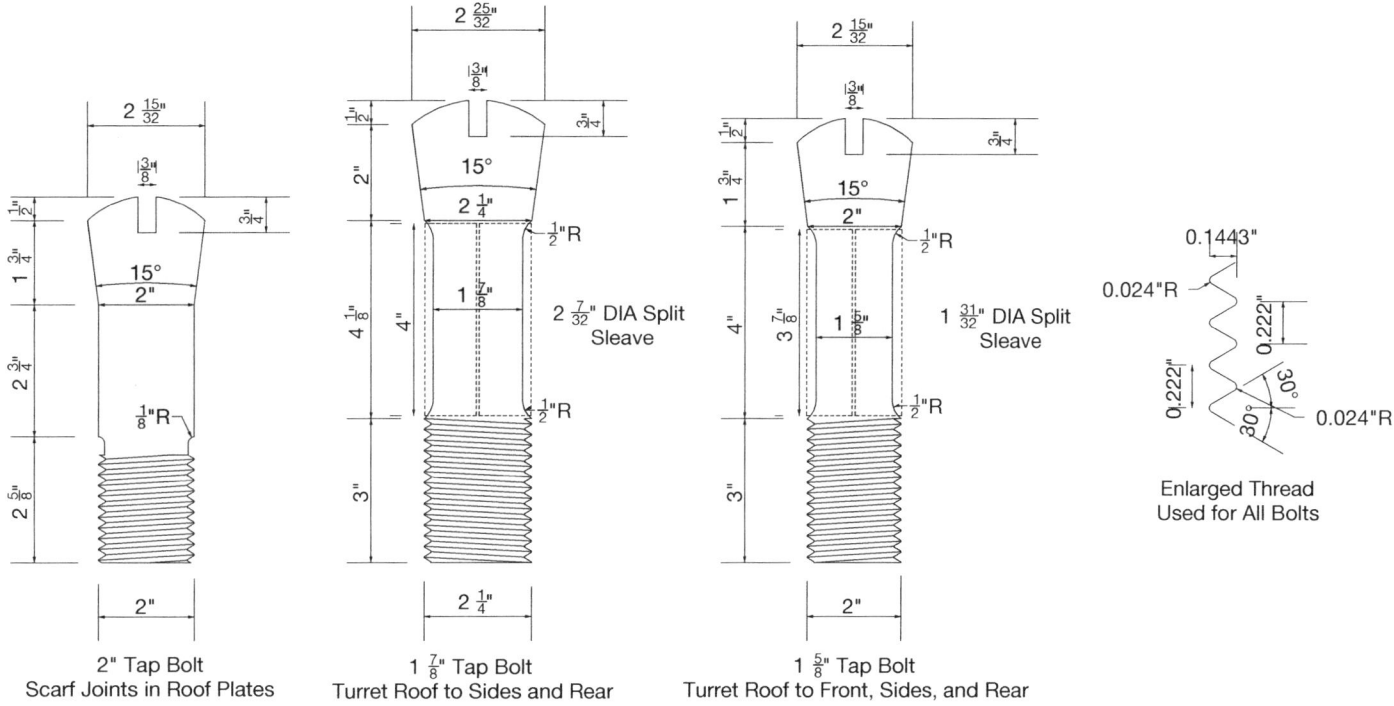

Figure 11.13: Turret Roof Bolts (1:4)

2" Tap Bolt
Scarf Joints in Roof Plates

1 7/8" Tap Bolt
Turret Roof to Sides and Rear

1 5/8" Tap Bolt
Turret Roof to Front, Sides, and Rear

Enlarged Thread
Used for All Bolts

Figure 11.14: 2-Inch Tap Bolts (1:4). This 2-inch tap bolt was used to connect third deck to the belt armor between FR166 and FR203.

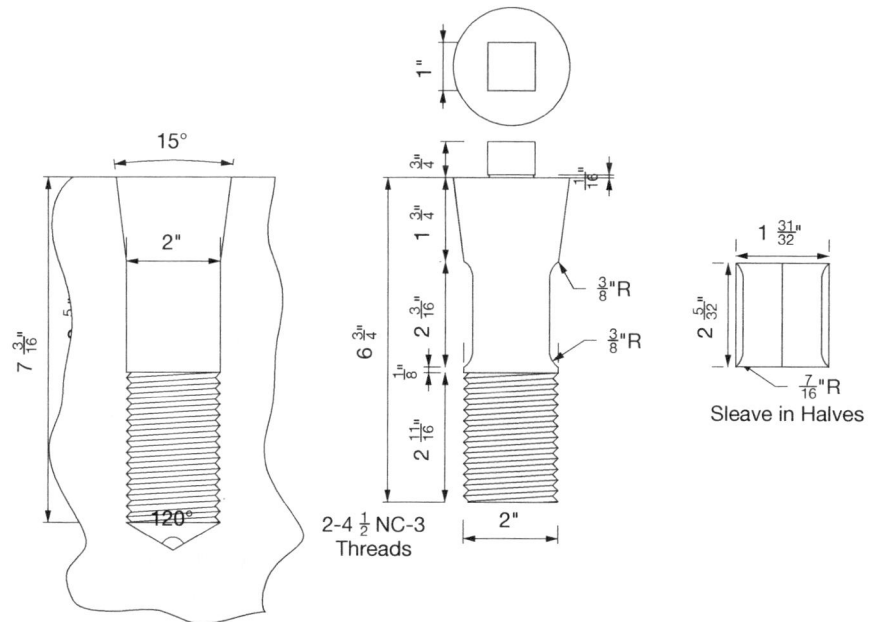

2-4 1/2 NC-3
Threads

Sleeve in Halves

Figure 11.15: 2.8-Inch Bolts (1:6)

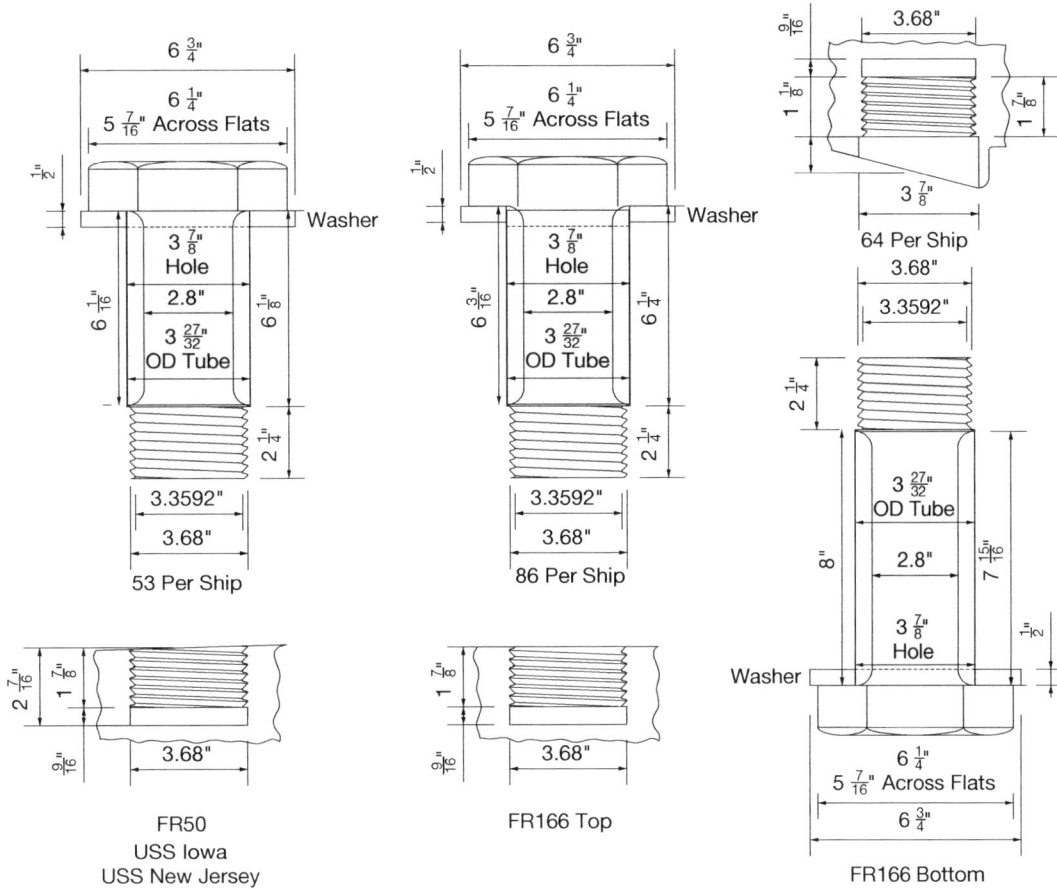

6 $\frac{3}{4}$"
6 $\frac{1}{4}$"
5 $\frac{7}{16}$" Across Flats
$\frac{1}{2}$"
Washer
3 $\frac{7}{8}$" Hole
2.8"
3 $\frac{27}{32}$" OD Tube
6 $\frac{1}{16}$"
6 $\frac{1}{8}$"
3.3592"
3.68"
2 $\frac{1}{4}$"
53 Per Ship

6 $\frac{3}{4}$"
6 $\frac{1}{4}$"
5 $\frac{7}{16}$" Across Flats
$\frac{1}{2}$"
Washer
3 $\frac{7}{8}$" Hole
2.8"
3 $\frac{27}{32}$" OD Tube
6 $\frac{3}{16}$"
6 $\frac{1}{4}$"
3.3592"
3.68"
2 $\frac{1}{4}$"
86 Per Ship

$\frac{9}{16}$"
3.68"
1 $\frac{1}{8}$"
1 $\frac{7}{8}$"
3 $\frac{7}{8}$"
64 Per Ship

3.68"
3.3592"
3 $\frac{27}{32}$" OD Tube
2.8"
3 $\frac{7}{8}$" Hole
Washer
2 $\frac{1}{4}$"
8"
7 $\frac{15}{16}$"
$\frac{1}{2}$"
6 $\frac{1}{4}$"
5 $\frac{7}{16}$" Across Flats
6 $\frac{3}{4}$"
FR166 Bottom

2 $\frac{7}{16}$"
1 $\frac{7}{8}$"
$\frac{9}{16}$"
3.68"
FR50
USS Iowa
USS New Jersey

1 $\frac{7}{8}$"
$\frac{9}{16}$"
3.68"
FR166 Top

Bolts are welded to washers which are welded to the plating.

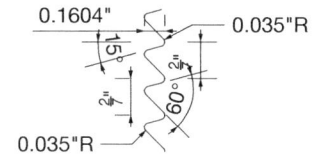

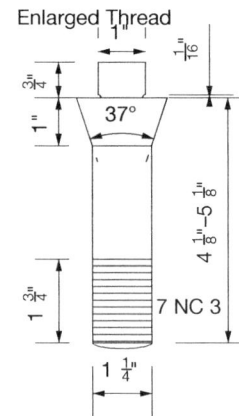

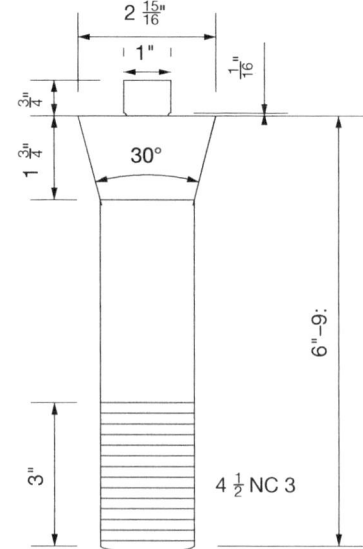

0.1604"
0.035"R
15°
60°
$\frac{1}{2}$"
$\frac{1}{2}$"
0.035"R
Enlarged Thread

Figure 11.16: Transverse Tap Bolts (1:4). The tap bolts for the FR50 transverse armor were made in various lengths to account for armor being thinner towards the bottom of the ship

1"
$\frac{1}{16}$"
$\frac{3}{4}$"
1"
37°
1 $\frac{3}{4}$"
4 $\frac{1}{8}$"–5 $\frac{1}{8}$"
7 NC 3
1 $\frac{1}{4}$"

2 $\frac{15}{16}$"
1"
$\frac{1}{16}$"
$\frac{3}{4}$"
1 $\frac{3}{4}$"
30°
6"–9:
3"
4 $\frac{1}{2}$ NC 3

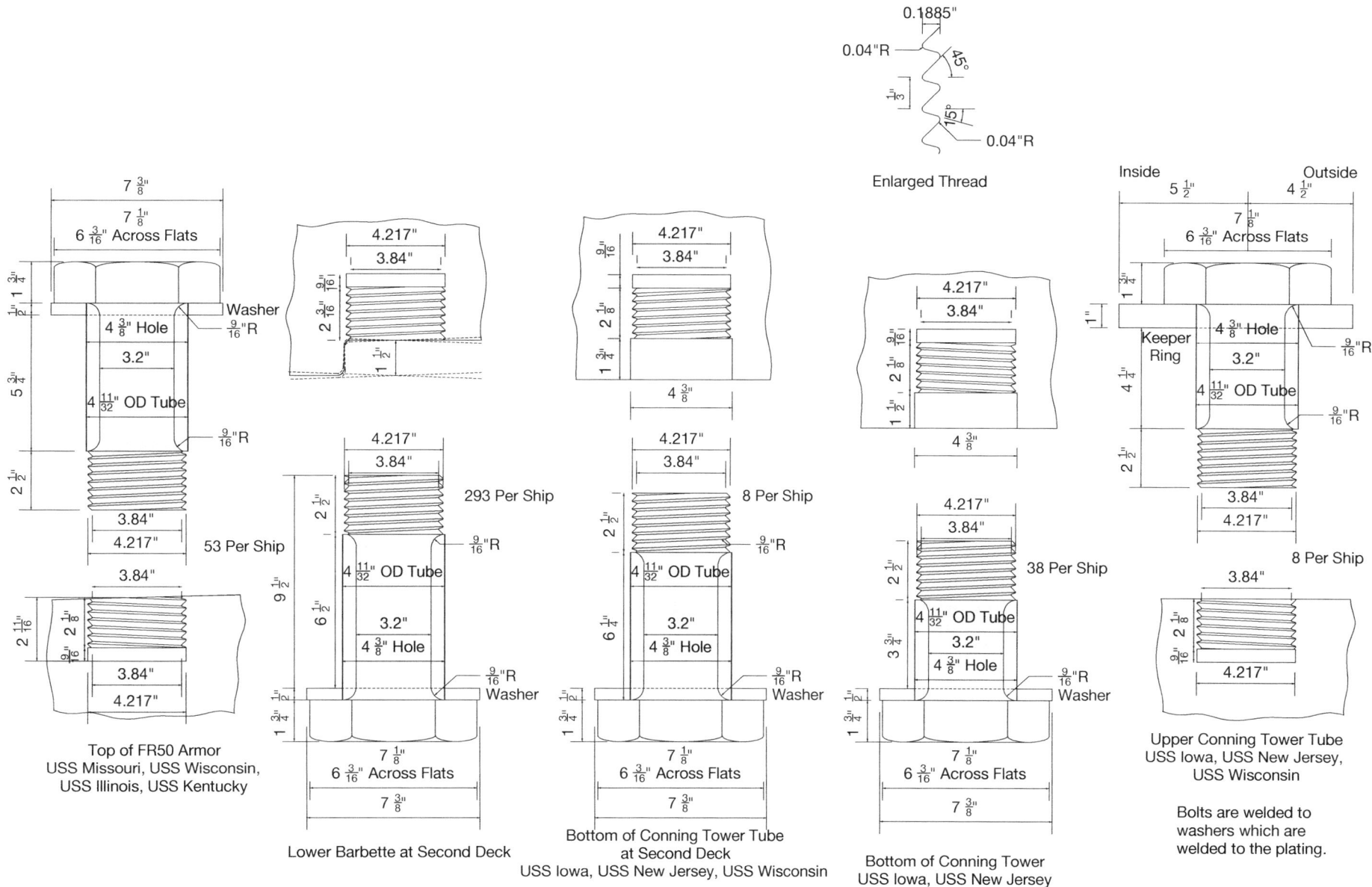

Figure 11.17: 3.2-Inch Bolts (1:6)

Abbreviations

ABL	Above Baseline
ABT	About
ABV	Above
ADDEN	Addendum
BHD	Bulkhead
BL	Baseline
C	Center
C/CK	Countersunk Head and Point
CHT	Collection and Holding Tank
CL	Centerline
CTC	Center to Center
CTS	Centers
D	Diameter
DBLR	Doubler
DEDEM	Dedendum
DIA	Diameter
DIAS	Diameters
DP	Deep
FB	Flat Bar
FLG	Flange
FR	Frame
FWD	Forward
HB	Halfbreadth
Horiz.	Horizontal
HT	Height
HTS	High Tensile Steel
INB'D	Inboard
KN	Knuckle
MIN	Minimum
ML	Molded Line
MS	Medium Steel

NC	National Coarse
NF	National Fine
OD	Outside Diameter
OUTB'D	Outboard
P	Port
PLT	Plate
P/S	Port and Starboard
R	Radius
RAD	Radius
RF	Rangefinder
RM	Room
S	Starboard
STD	Standard
STS	Special Treatment Steel

Bibliography

- P.R. Alger, Development of Ordinance and Armor in the Immediate Past and Future, Proceedings of the United States Naval Institute, Vol. 22, p. 793 (1886)
- Cleland Davis, Modern Armor; Its Influence on the Development of Ordnance, Proceedings of the United States Naval Institute, Vol. 27, p. 556 (1901
- William H. Garzke, Jr. & Robert O. Dulin, Jr. United States Battleships, 1935–1992, Naval Institute Press, 1995
- John W. Gulik, Armor and Ships, Journal of the United States Artillery, Vol. 38, No. 3, p. 265 (1912).
- Memoir of Hayward Augustus Harvey by His Sons, 1900
- Paul J. Hazell, Armour: Materials, Theory, and Design. CRC Press, 2015
- Hand Lengerer & Lars Ahlberg, The Yamato Class and Subsequent Planning, Nimble Books, 2014
- John Roberts, The Battleship Dreadnought, Naval Institute Press, 1992
- Robert Sumrall, Iowa Class Battleships, Naval Institute Press, 1989
- Naval Progress, July 1900, U.S. Government Printing Office
- 16-Inch Range Table, Ordnance Pamphlet No. 770, U.S. Navy, 1941
- Naval Ordinance, U.S. Naval Institute, 1939
- Principles of Armor Protection, Watertown Arsenal Laboratory, June 28, 1944